The Mycota

A Comprehensive Treatise on Fungi as Experimental Systems for Basic and Applied Research

Volume 6

The fungi represent a heterogenous assemblage of eukaryotic microorganisms and have become favored organisms for research at the cellular and molecular level. Such research involvement has been stimulated by interest in the biotechnological application of fungi in processes related to industry, agriculture and ecology. Considering both yeasts and mycelial fungi, THE MYCOTA highlights developments in both basic and applied research and presents an overview of fungal systematics and cell structure. Foremost authorities in research on mycology have been assembled to edit and contribute to the volumes.

All volumes/chapters are systematically reviewed by the series editor(s) and respective volume editor(s).

Axel A. Brakhage • Olaf Kniemeyer •
Peter F. Zipfel
Editors

Human and Animal Relationships

Third Edition

Springer

Editors
Axel A. Brakhage
Department of Molecular and Applied Microbiology
Leibniz Institute for Natural Product Research and Infection Biology (Leibniz-HKI)
Jena, Germany

Olaf Kniemeyer
Department of Molecular and Applied Microbiology
Leibniz Institute for Natural Product Research and Infection Biology (Leibniz-HKI)
Jena, Germany

Peter F. Zipfel
Department of Infection Biology
Leibniz Institute for Natural Product Research and Infection Biology (Leibniz-HKI)
Jena, Germany

ISSN 2945-8048 ISSN 2945-8056 (electronic)
The Mycota
ISBN 978-3-031-64852-6 ISBN 978-3-031-64853-3 (eBook)
https://doi.org/10.1007/978-3-031-64853-3

This Springer imprint is published by the registered company Springer Nature Switzerland AG
The registered company address is: Gewerbestrasse 11, 6330 Cham, Switzerland

Series Preface

The Mycota is an encyclopaedic book series published by Springer as a comprehensive treatise spanning the fungal kingdom through a focus on fungi as experimental systems for basic and applied research. Articles in this series present introductory information in a comprehensive manner to broaden our general knowledge, along with state-of-the-art information on selected topics in fungal biology. Thus, these articles serve as a timeless reference source for historically relevant discoveries in fungal biology to drive further advances in the field.

The Mycota was founded in 1994 by Karl Esser and Paul Lemke. Since then 15 volumes have been published, with several now in their third edition. The steadily growing interest in *The Mycota*, reflected in particular by the high number of e-book downloads, encouraged Springer to continue this publication with a new international Series Editorial Board beginning in 2022.

Historically, the study of fungi originated as a sub-discipline of botany and was a largely descriptive discipline until the early nineteenth century. Subsequent experimental research has propelled the field of mycology, and achievements in the genetics and molecular biology of fungi have benefited studies in the related fields of fungal biochemistry and biophysics, plant pathology, medical mycology, and systematics. Ground-breaking research has been carried out using fungi as recognized by receipt of nine Nobel Prizes, including the revolutionary 1945 Nobel Prize in Physiology or Medicine for the discovery of penicillin and the foundational 2006 Nobel Prize in Chemistry for establishing the molecular basis of eukaryotic transcription. All such studies expand our knowledge of fungi and of fungal processes and improve our ability to understand, utilize, and control fungi for the benefit of humankind.

Fungi have invaded every conceivable ecological niche. Saprobic forms abound, especially in the decay of organic debris. Pathogenic forms exist with both plant and animal hosts. Fungi even grow on other fungi. They are found in aquatic as well as soil environments, and their spores have been found in the air of extreme environments. Fungi can be variously associated with plants as symbionts in the form of lichens, mycorrhizae, and endophytes and also occur as overt pathogens. Association with animal systems varies; examples include the predaceous fungi that trap nematodes, the micro-fungi that grow in the anaerobic environment of the rumen, and medically important pathogens afflicting humans. Currently, taxonomists have identified about 150,000 species, but it is thought that over 90% of fungal species remain undescribed, with conservative estimates ranging from 2 to 5 million fungal species on earth.

From this perspective, there are new topics in fungal biology to explore, which will expand the current volumes and take them in new directions. Further, there will be new volumes in areas (e.g., Cryptomycota) that were not covered before.

Our understanding of the evolution of fungi is still incomplete and mainly based on species that can be grown in culture. But recent environmental DNA analyses have revealed a highly diverse form of eukaryotic life that branches with the Fungi, and thus the resulting and highly diverse clades were named the cryptomycota. The discovery of novel intermediate forms will redefine the fungal tree of life. Other topics will consider the adaptation of fungi to climate changes, the occurrence of fungal pathogens in the environment, and the dispersal of fungi during global pandemics.

Fungal constituents of the microbiome have received much less attention thus far, yet recent findings from clinical and animal studies clearly establish fungi as a significant component of the oral, gastrointestinal, pulmonary, and skin microbiomes. Finally, the relevance of fungal biology to society is reflected by the increasing number of fungal related human diseases.

In the history of pharmacology, fungi have always been sources of useful molecules for humans, but the diversity of molecules that can be obtained from fungi is still largely under-explored. However, efforts in fungal genomics provide resources for data-driven genome mining and large-scale comparisons to explore the molecular repertoires produced by fungi. The result will be new compounds with applications in the pharmaceutical and agricultural science industries.

Fungi can also help the entire planet and may for example be relevant in specific sectors, such as that linked to pollution from plastic. Fungi produce a wide range of enzymes that have the potential to break down the chemical bonds of plastic polymers, and in this context the potential role of marine fungi in plastic degradation may be of major relevance. Finally, new biomaterials from fungal species may open the door to alternatives to fossil-based materials and thus reduce environmental pollution.

For consistency throughout this series of volumes, the names adopted for major groups of fungi should be followed according to the following paper, which gives an overview of all the orders in the fungal kingdom: https://pubmed.ncbi.nlm.nih.gov/32660385/

We are grateful to Springer for continuing *The Mycota* series and are especially thankful to all of the Volume Editors in selecting topics and assembling experts from diverse fields of fungal biology.

Sydney, Australia — Dee Carter
Delhi, India — Anuradha Chowdhary
Durham, NC, USA — Joseph Heitman
Bochum, Germany — Ulrich Kück
July 15, 2022

Volume Preface to the Third Edition

Fungi have a significant impact on the life of humans and animals, both directly and indirectly. They play an important role as saprophytes in the soil carbon cycle, and most plants form a symbiotic relationship with fungi. On the other side, fungi are one of the major causal agents of plant diseases. On the positive side, fungi are important cell factories whose products range from organic acids, chemicals, antibiotics, proteins and enzymes, vitamins, and polyunsaturated fatty acids, to even meat alternatives and composite material, not to forget fermented food and alcoholic beverages. In a recent study, a first attempt was made to quantify the global monetary value of fungi, which is estimated to be around 54.57 trillion $ (*Fungal Diversity* **121**, 95–137, 2023).

Of course, there are also direct interactions between fungi and humans or animals. Fungi are common members of the microbiome on mucosal surfaces and the skin but can also cause diseases ranging from life-threatening, systemic infections to allergic reactions. Recent estimates suggest an annual incidence of 6.5 million invasive fungal infections worldwide, which often lead to death due to their difficult diagnosis and the limited therapeutic options available in the clinics. A particular concern is the growing number of fungal strains resistant to common antifungal drugs.

In the year 2022, the WHO reported on the relevance and importance of fungal infections for health and antifungal resistance to existing drugs. Fungal pathogens are included in the priority list of the WHO showing an unmet need for extended research and development. Therefore, more research is necessary to extend our knowledge about the molecular mechanisms of host–fungal pathogen interactions, to improve the diagnosis of fungal diseases, and overall to expand the antifungal armamentarium. The intention of Volume VI of the series Mycota is to exemplify recent aspects of the interactions of fungi with their human or animal host. The first part summarizes our current understanding of important pathogenic fungi such as *Candida species*, *Malassezia* yeasts, *Aspergillus fumigatus*, and fungi of the order Mucorales. The second part addresses the characterization of the host response toward pathogenic fungi, whereby the focus is placed on RNA as mediator of host–pathogen interaction, the human gut mycobiome, the role of the innate immune system to defeat infections, pattern recognition receptors involved in fungal infections, and a summary of established infection models for studying host–fungal pathogen interactions. The third part of this volume provides insights into the impact transcriptomics and proteomics technologies have on the research of human-pathogenic fungi.

We hope that the chapters are of interest for the readers, as they provide state-of-the-art knowledge of pathogenic fungi and describe new developments, concepts, and ideas. We are in particular grateful to all authors for the excellent contributions to this Mycota volume.

Jena, Germany — Axel A. Brakhage
Jena, Germany — Olaf Kniemeyer
Jena, Germany — Peter F. Zipfel
April 2024

Contents

Part III Techniques

Editors and Contributors

About the Series Editors

Dee Carter is Professor of Microbiology at the University of Sydney. She has a degree in microbiology and biochemistry from Otago University, New Zealand, and a PhD from Imperial College, London, UK. After postdoctoral fellowships in Montpellier, France, and UC Berkeley, California, USA, she moved to Australia in 1995 to take up a lectureship at the University of Sydney. Dee's research has encompassed 1) the population genetics and ecology of medically important fungi, including yeast and mold pathogens; 2) responses of fungi to host and antifungal stress, including the production of variant morphological forms; and 3) the transcriptome and proteome response to antifungal therapy, particularly during synergistic interactions between antifungals and natural products. As well as research, Dee teaches microbiology including mycology, epidemiology, and molecular biology at Sydney University.

Anuradha Chowdhary Anuradha Chowdhary is Professor of Medical Mycology at the Vallabhbhai Patel Chest Institute, University of Delhi, India. She received her medical degree in 1992 and MD (Microbiology) degree in 1996 from the University of Delhi, India, and PhD, from the Faculty of Medical Sciences, Radboud University Medical Center, Netherlands. Dr Chowdhary's research interests include molecular ecology and population genetics of pathogenic fungi, antifungal drug resistance mechanisms, especially *Aspergillus* and *Candida* spp., and the epidemiology of systemic mycoses. She is currently working on molecular epidemiology of *Candida auris*, terbinafine-resistant dermatophytes, and azole-resistant *Aspergillus fumigatus*.

Joseph Heitman Joseph Heitman is James B. Duke Professor and Chair, Department of Molecular Genetics and Microbiology, Duke University, Durham, NC, USA. His research studies model and pathogenic fungi addressing fundamental questions of scientific and medical importance. Pioneering studies with Baker's yeast revealed how immunosuppressive natural products interdict signaling cascades via FKBP12-drug complexes and discovered TOR as a globally conserved nutrient sensor targeted by the immunosuppressive, antiproliferative drug rapamycin. His research discovered unisexual reproduction of pathogenic microbes, with implications for pathogen emergence, how sex generates diversity, and how sex evolved. Dr Heitman's lab has further developed genetic and genomic approaches elucidating molecular principles of fungal virulence, identifying therapeutic targets, and illustrating convergent evolution of fungal mating-type loci with mammalian, insect, and plant sex chromosomes, defined the calcium-activated protein phosphatase calcineurin as a globally conserved fungal virulence factor, and elucidated functions of RNAi in microbial pathogen genome integrity, hypervirulent outbreak lineages, and drug resistance via epimutation.

Ulrich Kück is Professor of General and Molecular Botany at the Ruhr University in Bochum, Germany. He has graduated in Biology and Chemistry and has a long-standing experience in the molecular biology of fungi, algae, and plants. His research with fungi has focused on two general aspects: first, genetic engineering of the secondary metabolism of biotechnically relevant filamentous fungi, including functional genomics and proteomics, and second, molecular genetic analysis of cellular growth in filamentous fungus, with the focus on sexual development. In particular, he is interested in the function of mating-type loci, and the involvement of conserved signaling complexes, such as the striatin-interacting phosphatases and kinases (STRIPAK) complex, in the control of cellular and developmental processes.

About the Volume Editors

Axel A. Brakhage earned a Master's degree in Biology with the major subject Microbiology at the University of Münster. During his PhD, he spent a year at the Institut de Biologie Physico-Chimique in Paris. After obtaining a PhD on the regulation of the phenylalanyl-tRNA synthetase in *Bacillus subtilis*, he joined the Biotechnology Department of BASF as a staff scientist. He then started work on the molecular biology of filamentous fungi as a postdoctoral student at the University of Sheffield (UK). After two years, he became Assistant Professor at the Institute of Genetics and Microbiology at the University of Munich where he earned his habilitation in Microbiology. In 1998, he became Associate Professor at the Darmstadt University of Technology. During 2001–2004, he was chair and Full Professor of Microbiology at the University of Hanover. From 2004, he has held a chair of Microbiology and Molecular Biology at the Friedrich Schiller University of Jena. In 2005, he became head of the Department of Molecular and Applied Microbiology at the Leibniz Institute of Natural Product Research and Infection Biology—Hans-Knoell-Institute (Leibniz-HKI)—in Jena and, simultaneously, he serves as the Director of the Institute. He is a member of several scientific boards and a Vice President of the Deutsche Forschungsgemeinschaft. In addition, he is Speaker of the DFG Collaborative Research Centre/Transregio 124 "FungiNet," was founding speaker of the cluster of excellence "Balance of the Microverse," and serves on the editorial boards of several journals. He has received the Seeliger-Award for Bacteriology and Mycology (2006) and the Main Research Award of the German Society for Hygiene and Microbiology (DGHM). In addition, he is elected member of the German Academy of Sciences Leopoldina and of both the American and European Academy of Microbiology. His research addresses two main areas: (1) pathobiology of the most important airborne fungal pathogen *Aspergillus fumigatus* and (2) molecular regulation of the biosynthesis of fungal secondary metabolites with an emphasis on microbial communication and shaping of microbiomes.

Olaf Kniemeyer studied biology with the major subject Microbiology at the University of Bremen. He completed his Master's thesis (Diplom) about anaerobic cholesterol degradation in 1998.He performed his dissertation and a short postdoctoral time under the supervision of Friedrich Widdel at the Max Planck Institute for Terrestrial Microbiology in Bremen in the field of anaerobic hydrocarbon degradation. After a short stay as visiting scientist at the University of Sheffield (UK) in the lab of Geoffrey Turner, he joined the laboratory of Axel A. Brakhage at the Leibniz University of Hanover in 2003. Since 2005, he is scientific co-worker at the Leibniz Institute for Natural Product Research and Infection Biology—Hans-Knoell-Institute (HKI) in Jena and since 2009 Deputy Head of the Department of Molecular and Applied Microbiology of Axel Brakhage. His work focuses on the application of proteomics methods to study the human-pathogenic fungus *Aspergillus fumigatus* and other Aspergilli. Olaf Kniemeyer serves on the editorial board of several journals and has received funding for his research from funding agencies such as DFG, GIF, and BMBF.

Peter F. Zipfel studied Biology and Biochemistry with specialization in Cellular Biology. In 1984, he graduated from the University in Bremen. He performed his postdoctoral studies at the Laboratory of Immunoregulation at the NIAID, National Institutes of Health in Bethesda (MD, USA). Then he moved to the Bernhard Nocht Institute for Tropical Medicine in Hamburg where he headed a junior research group. He received his habilitation and *venia legendi* from the University of Hamburg in 1993 in the areas of immunology and genetics. In 2000, he was appointed University Professor for Infection Biology at the Friedrich Schiller University in Jena and at the same time chairman of the Department of Infection Biology at the Leibniz Institute for Natural Product Research and Infection Biology—Hans-Knoell-Institute (HKI)—in Jena. He is the author of over 400 research papers, book chapters, and review articles; holds several patents; and has edited a book on the role of the complement system in human renal diseases. He has received the Research Prize for Basic Science from the State of Thueringen (2004), the "Heinz-

Spitzbart Preis" of the European Society for Infectious Diseases in Obstetrics and Gynaecology (ESIDOG; 2007), the EFIS Lecture Award of the European Federation of Immunological Societies (2008), and the Main Research Award of the German Society of Hygiene and Microbiology (DGHM) in 2009. In addition, he received the Galenus von Pergamon Award in Basic Sciences (2014) and the Medal in Gold for lifetime research achievements from the European Complement Network, ECN (2023). His scientific interests include (1) understanding the role of the complement system as part of the innate immune system in health and disease and (2) how human pathogenic fungi such as *Candida albicans* and *Aspergillus fumigatus* and numerous other pathogens utilize host complement components for immune evasion.

Contributors

Jaime David Acosta-España Jena Microbial Resource Collection, Leibniz Institute for Natural Product Research and Infection Biology – Hans Knöll Institute, Jena, Germany

Institute of Microbiology, Friedrich Schiller University Jena, Jena, Germany

Faculty of Medicine, Pontifical Catholic University of Ecuador, Quito, Ecuador

Faculty of Health Sciences, Universidad Particular Internacional SEK, Quito, Ecuador

Alexander Bruch Leibniz Institute for Natural Product Research and Infection Biology–Hans Knöll Institute (Leibniz-HKI), Jena, Germany

Rida Ali Institute of Microbiology, Friedrich Schiller University Jena, Jena, Germany

Elaine Bignell Medical Research Council Centre for Medical Mycology, University of Exeter, Exeter, UK

Matthew G. Blango Leibniz Institute for Natural Product Research and Infection Biology–Hans Knöll Institute (Leibniz-HKI), Jena, Germany

M. Rosa Bragulat Department of Animal Health and Anatomy, Veterinary Mycology Group, Universitat Autònoma de Barcelona, Bellaterra, Catalonia, Spain

Axel A. Brakhage Department of Molecular and Applied Microbiology, Leibniz Institute for Natural Product Research and Infection Biology – Hans Knöll Institute (Leibniz-HKI), Jena, Germany

Department of Microbiology and Molecular Biology, Institute of Microbiology, Friedrich Schiller University, Jena, Germany

Gordon D. Brown University of Exeter, Exeter, UK

F. Javier Cabañes Facultat de Veterinària, Departament de Sanitat i d'Anatomia Animals, Grup de Micologia Veterinària, Universitat Autònoma de Barcelona, Bellaterra (Barcelona), Spain

Gemma Castellá Department of Animal Health and Anatomy, Veterinary Mycology Group, Universitat Autònoma de Barcelona, Bellaterra, Catalonia, Spain

Jiarui Chen Department of Microbiome Dynamics, Leibniz Institute for Natural Product Research and Infection Biology – Hans Knöll Institute, Jena, Germany

State Key Laboratory of Pharmaceutical Biotechnology, The University of Hong Kong, Hong Kong, SAR China

Department of Medicine, The University of Hong Kong, Hong Kong, SAR ChinaGuangdong Provincial Key Laboratory of Food, Nutrition and Health, and Department of Nutrition, School of Public Health, Sun Yat-sen University, Guangzhou, China

Leyna Díaz Department of Animal Health and Anatomy, Veterinary Mycology Group, Universitat Autònoma de Barcelona, Bellaterra, Catalonia, Spain

Mark S. Gresnigt Junior Research Group Adaptive Pathogenicity Strategies, Leibniz Institute for Natural Product Research and Infection Biology – Hans Knöll Institute (HKI), Jena, Germany

Bernhard Hube Department of Microbial Pathogenicity Mechanisms, Leibniz Institute for Natural Product Research and Infection Biology – Hans Knöll Institute (HKI), Jena, Germany

Institute of Microbiology, Friedrich-Schiller University, Jena, Germany

Ilse D. Jacobsen Research Group Microbial Immunology, Leibniz Institute for Natural Product Research and Infection Biology – Hans Knoell Institute, Jena, Germany

Institute of Microbiology, Friedrich Schiller University Jena, Jena, Germany

Lei-Jie Jia Department of Molecular and Applied Microbiology, Leibniz Institute for Natural Product Research and Infection Biology – Hans Knöll Institute (Leibniz-HKI), Jena, Germany

Olaf Kniemeyer Department of Molecular and Applied Microbiology, Leibniz Institute for Natural Product Research and Infection Biology – Hans Knöll Institute, Jena, Germany

Thomas Krüger Department of Molecular and Applied Microbiology, Leibniz Institute for Natural Product Research and Infection Biology – Hans Knöll Institute, Jena, Germany

Dolly Montaño Jena Microbial Resource Collection, Leibniz Institute for Natural Product Research and Infection Biology – Hans Knöll Institute, Jena, Germany

Institute of Microbiology, Friedrich Schiller University Jena, Jena, Germany

Jana-Sophie Niegisch Jena Microbial Resource Collection, Leibniz Institute for Natural Product Research and Infection Biology – Hans Knöll Institute, Jena, Germany

Institute of Microbiology, Friedrich Schiller University Jena, Jena, Germany

Gianni Panagiotou Department of Microbiome Dynamics, Leibniz Institute for Natural Product Research and Infection Biology – Hans Knöll Institute, Jena, Germany

Department of Medicine, The University of Hong Kong, Hong Kong, SAR China

Department of Pharmacology and Pharmacy, The University of Hong Kong, Hong Kong, SAR China

Cluster of Excellence Balance of the Microverse, Friedrich Schiller University, Jena, Germany

Faculty of Biological Sciences, Friedrich Schiller University, Jena, Germany

Marina Pekmezovic Junior Research Group Adaptive Pathogenicity Strategies, Leibniz Institute for Natural Product Research and Infection Biology – Hans Knöll Institute (HKI), Jena, Germany

Emily Sey University of Exeter, Exeter, UK

Christine Skerka Department of Infection Biology, Leibniz Institute for Natural Product Research and Infection Biology, Hans-Knöll-Institute, Jena, Germany

Kerstin Voigt Jena Microbial Resource Collection, Leibniz Institute for Natural Product Research and Infection Biology – Hans Knöll Institute, Jena, Germany

Institute of Microbiology, Friedrich Schiller University Jena, Jena, Germany

Yao Wang State Key Laboratory of Pharmaceutical Biotechnology, The University of Hong Kong, Hong Kong, SAR China

Department of Medicine, The University of Hong Kong, Hong Kong, SAR China

Janet A. Willment University of Exeter, Exeter, UK

Aimin Xu State Key Laboratory of Pharmaceutical Biotechnology, The University of Hong Kong, Hong Kong, SAR China

Department of Medicine, The University of Hong Kong, Hong Kong, SAR China

Department of Pharmacology and Pharmacy, The University of Hong Kong, Hong Kong, SAR China

Phoebe Chin Xiu Yin Jena Microbial Resource Collection, Leibniz Institute for Natural Product Research and Infection Biology – Hans Knöll Institute, Jena, Germany

Institute of Microbiology, Friedrich Schiller University Jena, Jena, Germany

Peter F. Zipfel Department of Infection Biology, Leibniz Institute for Natural Product Research and Infection Biology, Hans-Knöll-Institute, Jena, Germany

Institute of Microbiology, Friedrich Schiller University, Jena, Germany

Part I

Pathogens

Trinity of Environment, Animals, and Humans: A Résumé in the Case of the Fungal Order Mucorales

1

Jaime David Acosta-España, Rida Ali, Dolly Montaño, Phoebe Chin Xiu Yin, Jana-Sophie Niegisch, and Kerstin Voigt

Abstract

The Mucorales are fungal microorganisms found in the environment, mostly in soil and organic decaying material. Humans have used these fungi for industrial production purposes, such as the fermentation of food or beverages. Nonetheless, these fungi can become opportunistic pathogens when encountering individuals with immunodeficiency. Environmental Mucorales can infect humans and animals through various pathways; once they have invaded the host, Mucorales display a set of virulence factors that enhance further dissemination from the infected tissue. In that regard, the host immune response plays an essential role in the resolution of the infection, especially for individuals with a higher risk of infection due to pre-existing medical conditions, such as diabetes, hematological malignances, and organ transplantation. Current diagnosis of mucormycosis is based on the identification of the fungi in infected tissues, and actual treatments include administration of liposomal amphotericin B and itraconazole. However, as many other pathogenic fungi, Mucorales present resistance to antifungal agents. Owing to the worldwide relevance of Mucorales, we will resume in this chapter the trinity of environment, animals, and humans, along with beneficial and deleterious factors affecting human life.

Keywords

Mucormycosis · Mucorales · Zygomycosis · Animal · Human · Virulence factor · Ecology

1.1 Introduction

Mucormycosis is an infection caused by the fungal order Mucorales. The term "zygomycosis" is no longer used after molecular phylogenetic

J. D. Acosta-España
Jena Microbial Resource Collection, Leibniz Institute for Natural Product Research and Infection Biology – Hans Knöll Institute, Jena, Germany

Institute of Microbiology, Friedrich Schiller University Jena, Jena, Germany

Faculty of Medicine, Pontifical Catholic University of Ecuador, Quito, Ecuador

Faculty of Health Sciences, Universidad Particular Internacional SEK, Quito, Ecuador

R. Ali
Institute of Microbiology, Friedrich Schiller University Jena, Jena, Germany

D. Montaño · P. C. X. Yin · J.-S. Niegisch · K. Voigt (✉)
Jena Microbial Resource Collection, Leibniz Institute for Natural Product Research and Infection Biology – Hans Knöll Institute, Jena, Germany

Institute of Microbiology, Friedrich Schiller University Jena, Jena, Germany
e-mail: Kerstin.Voigt@leibniz-hki.de

A. A. Brakhage et al. (eds.), *Human and Animal Relationships*, The Mycota 6,
https://doi.org/10.1007/978-3-031-64853-3_1

analysis indicated differences between the fungal lineages of Zygomycetes as paraphyletic class of the fungal kingdom (Hibbett et al. 2007; Spatafora et al. 2016). Mucorales fungi were reclassified in the phylum Mucoromycota (Index Fungorum 2022), which include eleven different genera that can infect humans. These fungi are found throughout the environment, especially in soil and air (Richardson 2009). In recent years, mucormycosis has gained attention due to its high mortality rate and limited antifungal treatment (Lax et al. 2020). However, Mucorales have also contributed to the exploration of industrial applications, such as fermentation in food and beverages, and the production of enzymes and secondary metabolites (Karimi and Zamani 2013). Besides humans, several animals have been infected worldwide, given our proximity to animals in different contexts, such as domestic, cattle raising, and cultural. Itis therefore important to understand the interplay among man, animal, and environment (Fig. 1.1). In humans, studies have shown that the epidemiology of mucormycosis differs between developed and developing countries. However, in developed countries, the occurrence of mucormycosis is still relatively rare, whereas in developing countries occasional occurrence of mucormycosis has been reported in people with risk factors such as uncontrolled diabetes, oncologic disease, organ transplantation, or trauma, particularly with high burden in India (Petrikkos et al. 2012). Especially with the ongoing global pandemic due to SARS-CoV-2, several countries are facing an increase in COVID-19-associated mucormycosis (CAM) (Hoenigl et al. 2022).

The pathogenesis of mucormycosis is favored by several virulence factors. The spore coat protein CotH is one of the virulence factors that acts as a fungus-specific ligand for the host receptor

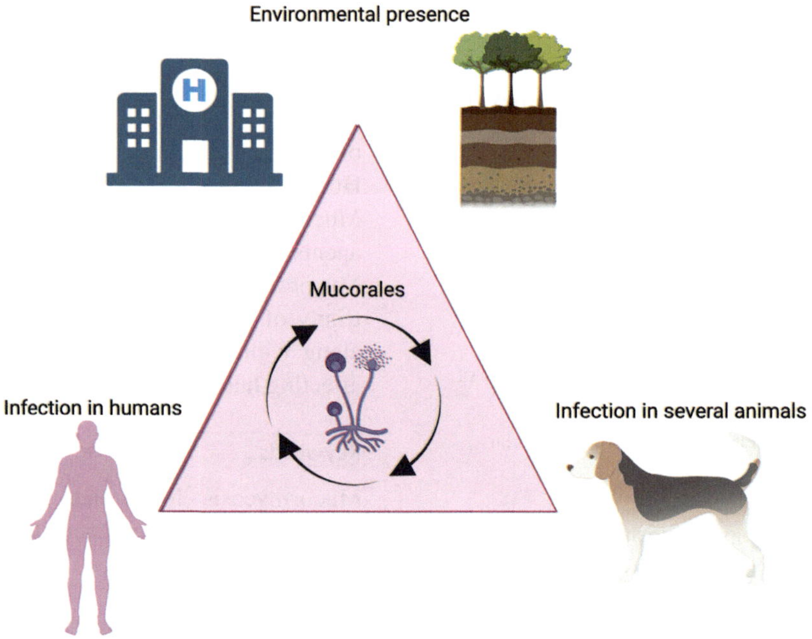

Fig. 1.1 Trinity of environment, animals, and humans: fungal order Mucorales. Mucorales have a wide distribution in the environment. This results in animals and humans being exposed to these fungi. They can cause infections through various pathways. In humans, individuals with certain underlying diseases, such as diabetes, have higher risk of infection. In animals, traumatic inoculation of spores with organic material carrying the microorganism results in infection of the extremities and orofacial region. Mucormycosis has been documented in both humans and animals, often associated with underlying comorbidities compromising the immune system and instances of traumatic inoculation. Created with BioRender.com

such as glucose-regulated protein 78 (GRP78) (Alqarihi et al. 2020). Iron also plays an important role in fungal survival. In Mucorales, the mechanism of high-affinity iron uptake has been intensively studied. The iron uptake mechanism enables the survival of Mucorales in the host. This mechanism is composed of three proteins: an iron reductase (Fre), a ferroxidase (Fet3), and an iron permease (Ftr1) (Stanford and Voigt 2020). One of the most important preventive measures and clinical approaches to treating mucormycosis is to replace deferoxamine with other iron chelators such as deferasirox (Lax et al. 2020). Some other virulence factors are *Rhizopus* alkaline protease enzyme (Arp), ADP-ribosylation factor (Arf), dihydrolipoyl dehydrogenase, and calcineurin (CaN) (Hassan and Voigt 2019).

After candidiasis and aspergillosis, mucormycosis is the third most common angioinvasive fungal infection in Europe. The majority of mucormycosis is caused by *Rhizopus* spp., *Mucor* spp., and *Lichtheimia* spp., which together account for approximately 70% of infections (Torres-Narbona et al. 2007). Patients with impaired cellular immunity are at increased risk of developing mucormycosis, and common predisposing risk factors include organ transplantation, chemotherapy, diabetes mellitus with ketoacidosis (DKA), hematologic disorders, and elevated serum iron levels (Skiada et al. 2011; Ghuman and Voelz 2017). Susceptibility to mucormycosis is increased when these individuals are treated with immunosuppressive therapies that impair the innate immune system (Ghuman and Voelz 2017), which is the body's first line of defense against germs and foreign substances that enter the body through the skin, respiratory tract, and gastrointestinal tract (Caffrey and Obar 2016). Whereas the adaptive immune response creates immunological memory to respond to a previously encountered antigen (Kumar et al. 2018). Therefore, it is critical to understand the interaction of the host immune response and Mucorales to elucidate novel and faster diagnosis and antifungal treatments.

Clinical presentations of invasive mucormycosis are normally found in six forms: (1) rhino-orbital-cerebral mucormycosis (ROCM), (2) pulmonary mucormycosis, (3) cutaneous mucormycosis, (4) gastrointestinal mucormycosis, (5) disseminated mucormycosis, (6) rare forms of mucormycosis such as endocarditis, osteomyelitis, peritonitis, and pyelonephritis (Petrikkos et al. 2012; Skiada et al. 2018). Nonspecific symptoms and lack of diagnostic tools delay diagnosis of Mucorales infections. Current diagnostic methods include analysis of biopsies of infected tissue using culture media, histopathology, and molecular biology. Imaging studies such as computed tomography and magnetic resonance imaging are useful to assess the involvement of various organs over infection. Conventional treatment for mucormycosis includes surgical debridement, antifungal treatment with amphotericin B, and long-chain azoles (Cornely et al. 2019). Therefore, future advances in targeted immunotherapy, reversal of tissue hypoxia, host immunologic, and metabolomic analysis may improve the treatment of this dreadful disease (Skiada et al. 2018; Cornely et al. 2019).

1.2 History, Etiological Agents, and Taxonomy: Mucorales

1.2.1 History

The order of Mucorales has accompanied mankind since time immemorial, given the fermentation of food. This historical review begins with the recollection that *Sufu* or *Furu* is a fermented soybean originally from China. This product was associated with fermentation processes involving *Rhizopus* spp., *Actinomucor* spp., and *Mucor* spp. The earliest written records come from China, from the period between 220 and 265 CE, from the Wei Dynasty. These fungi served for this fermentation process and were described in several books. It is known that the popularity of these products increased between 1368 and 1644 at the time of the Ming Dynasty. Since then, commercial and domestic manufacturing processes by fermentation with Mucorales are used in

China and worldwide for various purposes (Han et al. 2001).

The first description of a fungal microorganism with the appearance of *Mucor* sp. dates from 1665 by Robert Hooke. In *Micrographia*, he illustrated his observation of a colony that had the appearance of a "hairy mold" (Gest 2004). These fungi are cosmopolitan and closely associated with environments where humans and other animals can come into contact and become infected with mucormycosis. Chronologically, the first case referred to as mucormycosis in today's scientific literature is the 1885 publication by Dr. Arnolf Paltauf entitled "*Mycosis Mucorina*". Interestingly, Paltauf already mentioned the incidental findings of fungi in humans and pointed out that these should be distinguished from cases where these microorganisms are the main cause of the disease. He also cited the report of Lichtheim (1884) on the infection of rabbits with *M. corymbifer* (now *Lichtheimia corymbifera*) and compared his findings with those of a 52-year-old patient whose autopsy revealed unseptate hyphae with sporangiospores in infected brain, abdominal, and lung tissue (Paltauf 1885).

Scientific literature has increasingly reported cases of mucormycosis in humans and animals since the 1950s. In a 1955 study conducted by Ainsworth and Austwick on farms in the United Kingdom, 1200 isolates were analyzed, and *Lichtheimia ramosa* (formerly *Absidia ramosa*) was found in 8.6% of animal cases (Ainsworth and Austwick 1955). Over time, the prevalence of the disease has increased. Recently, mucormycosis became relevant in the context of patients with coronavirus disease-19 (COVID-19). In 2021 and early 2022, several cases were reported worldwide, with a clear focus in India (Hoenigl et al. 2022). For example, a study conducted by Arunaloke Chakrabarti focused on estimating the burden of fungal infections in India through computational modeling, with mucormycosis as a specific case study. The cumulative burden was estimated to fall within the range of 137,807 to 208,177 cases, with a mean of 171,504. The average attributable mortality was calculated to be 65,500 deaths per year, accounting for 38.2% (95% CI: 57,989 to 73,402) of the total (Chakrabarti et al. 2023).

1.2.2 Etiological Agents and Taxonomy

The Mucorales have undergone significant taxonomic changes. This explains why the term zygomycosis is no longer used but has been replaced by the term mucormycosis. The Entomophthorales were separated, and infections with these pathogens should now be referred to as entomophthoromycosis. Previously, zygospore-forming fungi were assigned to the phylum Zygomycota because of structural differences. The Mucorales produce these zygospores, which are thick-walled, ornate, pigmented zygotes. However, due to the lack of support for molecular phylogenies, the phylum Zygomycota has been abandoned (Walther et al. 2019).

In 2016, Spatafora and his collaborators examined 25 taxa for zygomycetes. This resulted in two clades (equivalent to fungal phyla: Mucoromycota and Zoopagomycota) based on their phylogenetic analyses of 192 proteins (Fig. 1.2). Mucoromycota mainly attacks plants and consists of three subphyla (Glomeromycotina, Mortierellomycotina, and Mucoromycotina). The Mucoromycotina comprise three orders (Mucorales, Umbelopsidales, and Endogonales). The Umbelopsidales include the endophytes and saprobes of forest soils. The Endogonales are saprobic or form ectomycorrhizae. The Mucorales include many decomposers of dead plant material and plant parasites (Spatafora et al. 2017).

Although these microorganisms are abundant destruents playing a major role in the decomposition of organic matter in the environment, they can also become opportunists infecting animals and humans if their immune system is unbalanced. The most common genera are *Rhizopus*, *Mucor*, and *Lichtheimia* (see Table 1.1). However, the frequency may vary in different regions due to local epidemiological conditions. About 45 different species have been associated with clinically relevant infections. Of these, *Rhizopus arrhizus*, *Rhizopus microsporus*, *Mucor irregula-*

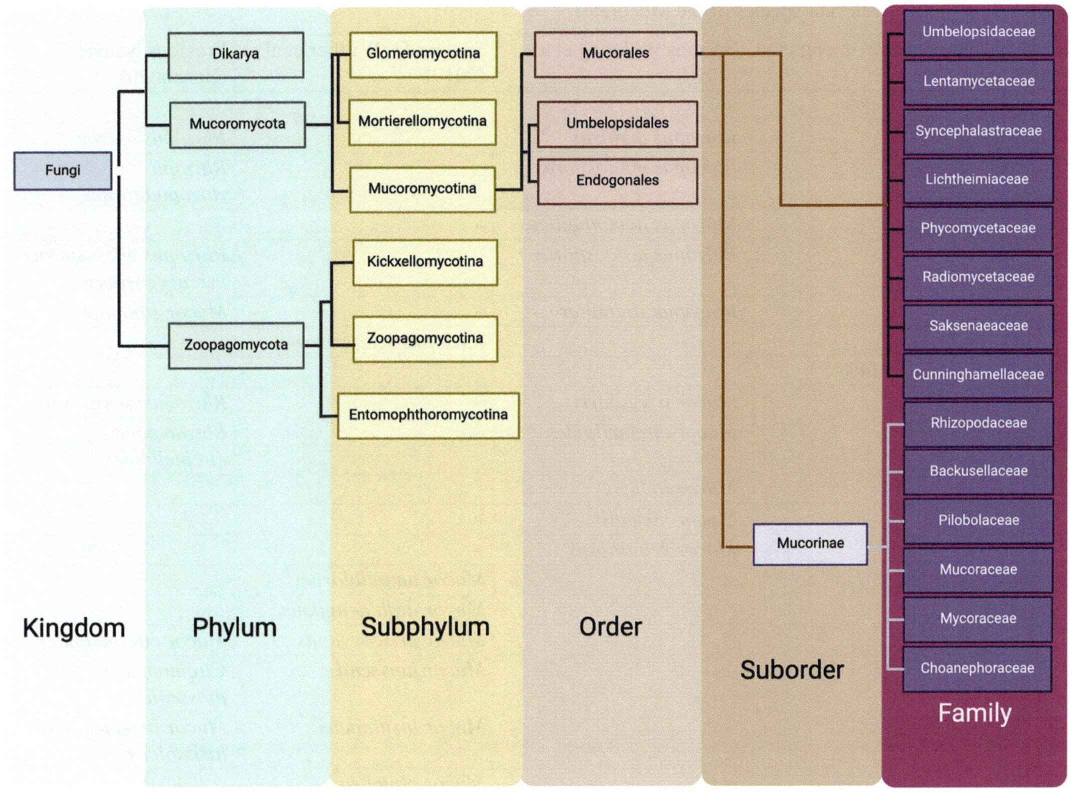

Fig. 1.2 Taxonomic update of the Mucorales based on Voigt (2012) and Hoffmann et al. (2013) (family level), Hibbett et al. (2007) (subphyla level), and Spatafora et al. (2017) (phylum-level classification). Created with BioRender.com

ris, *Mucor circinelloides*, *Lichtheimia corymbifera*, *Cunninghamella bertholletiae*, *Saksenaea vasiformis*, and *Rhizomucor pusillus* stand out because of their frequency (Jeong et al. 2019). Table 1.1 provides a detailed overview of the causative agents of mucormycosis.

1.3 Environmental Presence of Mucorales (Geographic Distribution and Ecological Conditions)

Mucorales are fast-growing, thermotolerant molds that spread rapidly. They are also ubiquitous, which is explained by their ability to be oligotrophic to mesotrophic. They are widely distributed in organic substrates such as bread, vegetable matter, crop residues, soil, animal excrement, compost piles, or even typically decomposers of fruits as post-harvest pathogens. The optimal conditions for growth and sporulation are a temperature of 27 °C and high humidity. The fact that these fungi compete successfully in the environment is partly due to their chitinolytic enzymes (Richardson 2009).

The sporangiospore is the dissemination unit of the Mucorales which is mitotically developed (mitospore). These molds are capable of aerial growth in a variable range of humidity in the environment. At high humidity, water passes through the columella and forms a “sporangia droplet”. This is larger than the sporangium and is dispersed by spraying. Other species in dry environments produce dry, powdery spores. The sporangiospores break open, and the dry sporangiospores are released to be dispersed in low wind speeds. The sporangiospores range from 3 to 11 micrometers in diameter and move easily in the environment. In addition, Mucorales spores

Table 1.1 Clinically relevant species of the Mucorales

Prevalence based on W. Jeong et al. (2019)		Species W. Jeong et al. (2019)	Species G. Walther et al. (2019)	Previous Names (Basionym)
Rhizopus	48%			
		Rhizopus arrhizus	–	*Rhizopus oryzae*
		Rhizopus microsporus	–	*Rhizopus rhizopodiformis*
		Rhizopus homothallicus	–	
		Rhizopus azygosporus	–	*Rhizopus microsporus* var. *azygosporus*
		Rhizopus stolonifer	–	*Mucor stolonifer*
		–	*Rhizopus schipperae*	
Mucor	14%			
		Mucor irregularis	–	*Rhizomucor variabilis*
		Mucor circinelloides	–	*Rhizomucor circinelloides*
		Mucor indicus	–	
		Mucor hiemalis	–	
		Mucor velutinosus	–	
		–	*Mucor amphibiorum*	
		–	*Mucor ardhlaengiktus*	
		–	*Mucor griseocyanus*	*Mucor circinelloides*
		–	*Mucor janssenii*	*Circinomucor janssenii*
		–	*Mucor lusitanicus*	*Mucor racemosus* var. *lusitanicus*
		–	*Mucor plumbeus*	
		–	*Mucor racemosus*	*Rhizomucor racemosus*
		–	*Mucor ramosissimus*	
		–	*Mucor variicolumellatus*	
Lichtheimia	13%			
		Lichtheimia corymbifera	–	*Mucor corymbifer, Absidia corymbifera, Mycocladus corymbifer*
		Lichtheimia ramosa	–	*Absidia ramosa, Rhizopus ramosus, Mucor ramosus*
		–	*Lichtheimia ornata*	
Cunninghamella	7%			
		Cunninghamella bertholletiae	–	
		–	*Cunninghamella blakesleeana*	
		–	*Cunninghamella echinulata*	
		–	*Cunninghamella elegans*	
Apophysomyces	8%			
		Apophysomyces elegans	–	
		Apophysomyces trapeziformis	–	

(continued)

Table 1.1 (continued)

Prevalence based on W. Jeong et al. (2019)		Species W. Jeong et al. (2019)	Species G. Walther et al. (2019)	Previous Names (Basionym)
		Apophysomyces variabilis	–	
		–	*Apophysomyces mexicanus*	
		–	*Apophysomyces ossiformis*	
Rhizomucor	6%			
		Rhizomucor pusillus	–	*Mucor pusillus*
		Rhizomucor variabilis	–	
		–	*Rhizomucor miehei*	
Saksenaea complex	3%			
		Saksenaea vasiformis	–	
		Saksenaea erythrospora	–	
		–	*Saksenaea loutrophoriformis*	
		–	*Saksenaea trapezispora*	
		–	*Saksenaea vasiformis*	
Others	1%			
		Syncephalastrum racemosum		
		Thamnostylum lucknowense		
		Cokeromyces spp.		
		Actinomucor elegans		*Rhizopus elegans*
		Cokeromyces recurvatus		

Information from Walther et al. (2019) and Jeong et al. (2019) has been adapted. Species that were repeated between studies were not included

can also be transmitted by insects (Ingold and Zoberi 1963).

One study examined 12,026 fungal air samples in 1717 buildings in the United States. Only *Mucor* spp., *Cunninghamella* spp., and *Rhizopus* spp. were found in low numbers (Shelton et al. 2002). It used to be assumed that most fungal species spend part of their lives in the soil. The Mucorales are abundant in accumulated bird droppings in an aviary (Ribes et al. 2000). *R. oryzae* has been isolated from the stored grain, onions, various nuts, and seed potatoes (Ribes et al. 2000). They are occasionally found in buildings damaged by water, as evidenced by air sampling and analysis of deposited dust by quantitative PCR. Outbreaks of rhinocerebral or pulmonary mucormycosis have been associated with inhalation of sporangiospores in dust caused by contaminated excavations, construction, or air conditioning filters (Connolly et al. 1999).

Although most Mucorales infections are acquired in the community. Nosocomial infections are possible. Sporadic cases and pseudo-outbreaks have been associated with contaminated dressings, needles, and tongue depressors used to make splints for arterial cannula and injection sites in premature infants (Ribes et al. 2000). Wooden materials such as tongue depressors have been repeatedly found to be a source of *Rhizopus inoculum* in hospitals. However, *R. arrhizus* (formerly *R. oryzae*) and *L. corymbifera* (formerly *Absidia corymbifera*) were found in potted topsoil in a hematology department where a leukemia patient developed invasive mucormycosis (Ribes et al. 2000).

The SARS-CoV-2 pandemic had multidimensional effects. And the coronavirus disease of

2019 (COVID-19) was linked to a Mucorales superinfection reported in early 2021. This association has been termed COVID-19-associated-mucormycosis (CAM). This superinfection is associated with multiple risk factors. A multi-center study in India showed that Mucorales grew in 21.7% of indoor air samples and 53.8% of outdoor air samples (Biswal et al. 2022). This may explain why India was one of the countries with the most reported cases of CAM. Despite this, the prevalence of mucormycosis was always higher in India. Another hypothesis is that they may cause a high mucormycosis burden due to their socio-cultural activities such as cow dung burning and other agricultural practices. They may lead to high environmental contamination with Mucorales, given that biomass burning can spread viable fungal spores (Skaria et al. 2022). This could be supported by a sample from a manure investigation in Brazil, where 24 Mucorales taxa were isolated, distributed among the genera *Absidia*, *Circinella*, *Cunninghamella*, *Lichtheimia*, *Mucor*, *Pilobolus*, *Rhizopus*, and *Syncephalastrum* (de Souza et al. 2017).

1.4 Industrial Applications

Fungi are abundant in nature and their biotechnological potential has been used for many purposes. An example known to all is *Saccharomyces cerevisiae*. This particular yeast is utilized globally in the fermentation process of various alcoholic beverages, including wine, beer, and cider. It is also used in the production of distilled spirits like rum, vodka, whiskey, brandy, and sake (Parapouli et al. 2020). On the other hand, less is known about the Mucorales and their biotechnological potential. The *Mucor* genus stands out for its relevance, and studies have focused on the indicus species. *M. indicus* is a dimorphic organism that is generally non-pathogenic.

M. indicus can grow on a variety of substrates, including lignocellulosic hydrolysates, which are a mixture of hexoses, pentoses, and various strong fermentation inhibitors. In addition, various possibilities as a fish feed source up to the production of polyunsaturated fatty acids have been proposed (Lennartsson et al. 2011; Satari et al. 2016). Since 1838 microorganism has been associated with the production of ethanol from pure glucose since 1838 (Takahashi and Asai 1988), and it has drawn attention to its fermentation capacity in the presence of substances that hinder it, such as lignocellulosic hydrolysates (Karimi et al. 2005).

The genus *Rhizopus* is another member of the Mucorales with biotechnological potential. Several species have been used to produce enzymes such as glucoamylases, cellulases, tannases, organic acids such as lactic acid, and fumaric acid. They are also used in the production of traditional foods such as tempeh, peka, ragi, and loog-pang. *Rhizopus arrhizus* is used for the development of food (koji, tempeh), antioxidants, organic acids, bioethanol, enzyme production (pyruvate decarboxylase, lipases, etc.) and genetic engineering. Currently, the main industrial use is enzyme production and food fermentation (Londoño-Hernández et al. 2017).

Cortisone was usually synthesized from deoxycholic acid in a 31-step chemical process. This production is characterized by low mass yield (0.16%) and high cost ($200/g, 1949). The integration of a biotransformation step with *Aspergillus niger* ATCC 9142 and *Rhizopus arrhizus* ATCC 11145 significantly reduced the number of biochemical steps required (11 steps) and the production cost ($1/g, 1979) of the industrial process (Carballeira et al. 2009). Another representative is *Mucor circinelloides*, which naturally accumulates lipids. Polyunsaturated fatty acids are important for the maintenance of human health. *M. circinelloides* is a model oleaginous fungus. It is considered as an exemplary organism to produce fatty acids (linoleic acid, oleic acid, etc.) and other essential bioactive compounds (Fazili et al. 2022).

1.5 Risk Factors for Mucormycosis

Mucorales are capable to switch from environmental commensal to pathogenic and from saprotrophic to parasitic lifestyles, respectively.

Several animals, including humans, can be infected (Tables 1.3 and 1.4). Patients with certain predisposing factors are more susceptible to infection with Mucorales. These characteristics, with predominance in patients with mucormycosis, are called risk factors. Most cases occur in people with reduced immunity. Specifically, when they lack phagocytes or have altered phagocytosis function (reviewed by Acosta-España and Voigt 2022).

Patients with diabetes are one of the main risk factors for mucormycosis. Diabetes is an endocrine disease caused by insulin deficiency or resistance leading to hyperglycemia (American Diabetes Association 2022). Hyperglycemia in diabetes leads to immune response dysfunction. For this reason, the spread of invading pathogens is not controlled in diabetic patients. Mechanisms explaining this failure include suppression of cytokine production, defects in phagocytosis, immune cell dysfunction, and inability to kill microbes (Berbudi et al. 2020).

However, it is also possible to observe mucormycosis incidences in people without a weakened immune system. Especially in patients who have previously taken antifungal drugs (voriconazole and fluconazole), and in patients with different types of traumas. Thus, environmental Mucorales can invade by different routes, and bypass the first line of defense, namely the barriers (Lelievre et al. 2014). These and other risk factors are discussed in detail below.

1.5.1 Patient Conditions and Genetic Factors

In a study of 465 patients with mucormycosis, cases were observed between the ages of 35 and 58 years. The average age was 48 years. And the disease occurred predominantly in men. 37.5% of the population had comorbidities. The most common clinical presentation is rhino-orbital, which occurred in 67% of the patients studied. However, not all patients with the above-mentioned risk factors developed the disease (Patel et al. 2020), which implies that multiple factors interact in the development of this infection.

Another study suggests that alterations in Toll-Like receptors (TLRs) may play a role as a risk factor. These cell-surface receptors are important in pathogen recognition by many cell types. Activation of different TLRs controls the type of immune response to different microorganisms (Acosta-España and Voigt 2022). The interplay of single nucleotide polymorphisms (SNPs) in TLR's has been shown to increase susceptibility to invasive candidiasis and invasive aspergillosis (Pana et al. 2014), but there is currently non-related information about Mucorales, which opens a fundamental question about the relation between SNPs in TLRs and susceptibility of certain patients to mucormycosis.

1.5.2 Comorbidities

Patel et al. showed that 37.6% of patients with mucormycosis had at least one comorbidity. The most important comorbidity was chronic kidney disease. Nevertheless, various diseases can lead to renal damage. In fact, 13% of patients had two risk factors and only 6% had one risk factor. Analysis of individual risk factors showed that diabetes was present in 73% of patients. Most cases occurred in a decompensated diabetic state. It has been described previously that immunity is impaired in diabetic patients by several mechanisms (Banerjee et al. 2021).

1.5.2.1 Diabetes and Ketoacidosis

There is evidence that hyperglycemia and acidosis may impair the ability of phagocytes to move toward spores. These medical conditions decrease the clearance of Mucorales spores *via* oxidative and nonoxidative mechanisms (Chinn and Diamond 1982). In addition, serum from diabetic mice promoted spore germination and impaired macrophage binding (Waldorf et al. 1984). Beginning in 1973, diabetic ketoacidosis was highlighted as a factor in the development of mucormycosis (Beigelman and Warner 1973). Interestingly, in mice with diabetic ketoacidosis, glucose-regulated protein (GRP-78) expression is increased in the sinuses, lungs, and brain (Liu et al. 2010). GRP-78 mediates invasion and damage of human endothelial cells by *R. arrhizus*, and iron is a basic cofactor in microorganisms. In *R. arrhizus*, iron uptake occurs through a high-affinity iron permease mediated by the gene "FTR1". Expression of the "FTR1" gene is expressed in the mouse model of diabetic ketoacidosis (Ibrahim et al. 2010).

Decreased pH in serum, as occurs in ketoacidosis, enables fungal growth by impairing the iron binding capacity of transferrin. Therefore, iron metabolism, which is known to be impaired in diabetes, plays a central role in the virulence of Mucorales and *R. arrhizus*. On the other hand, platelets play a role in innate immunity to Mucorales. Damage to hyphae by platelets reduces their growth. However, platelet function is impaired in diabetics due to structural and functional changes (Banerjee et al. 2021).

1.5.2.2 Organ and Stem Cell Transplant Recipients

Patients after solid organ transplantation are susceptible to mucormycosis. Patients in whom the infection occurs are associated with renal failure, diabetes mellitus, and previous use of antifungal agents (voriconazole and/or caspofungin). The age of the patients ranged from 47 to 59 years. Most of these patients were men. The most affected transplanted organ was the kidney, and half of the patients had base renal failure. Other organs affected were the liver, lung, heart, pancreas, and small intestine. The disseminated disease was found in 26% of the cases studied by Singh et al. The disseminated infections mainly affected the liver (Singh et al. 2009).

Risk factors for mucormycosis in hematopoietic stem cell transplant recipients have been shown to include severe graft-versus-host disease, high-dose steroids, previous cytomegalovirus or respiratory viral disease, and older age. The most common sites of infection were the lungs and rhinocerebral. The mean age was 37 years, with male patients in the majority. The major underlying diseases were acute myeloid leukemia, acute lymphoblastic leukemia, and chronic myeloid leukemia. Fifty-seven percent of patients received antifungal prophylaxis. The main drugs used were voriconazole and fluconazole (Rehman et al. 2022).

1.5.2.3 Other Factors

In a smaller percentage, use of immunosuppressants such as corticosteroid, trauma, burns, HIV infection, intravenous drug use, low birth weight infants, malnutrition, chronic alcoholism, liver disease, chemotherapy and calcineurin inhibitor use, environmental disasters, antifungal prophylaxis, iron overload, and deferoxamine therapy was associated with mucormycosis (Prakash and Chakrabarti 2019; Patel et al. 2020).

1.5.3 COVID-19-Associated Mucormycosis (CAM)

CAM has been reported in several countries around the world. It was most observed in India and the USA. The reported age range was from 10 to 86 years. The average age was 55 years, with males outnumbering females. Concomitant diseases included diabetes, hematologic malignancy, lymphopenia, hypertension, chronic kidney disease, and severe obesity. According to studies, the proportion of moderate and severe cases was equal in COVID-19. In these patients, corticosteroids were used in 79% of the cases studied. For this reason, special attention must be paid to patients with COVID-19, diabetes, poorly controlled hyperglycemia, and the use of corticosteroids (Hoenigl et al. 2022).

However, environmental factors may also have played a role. This is because many countries with a high prevalence of diabetes and COVID-19 did not have the same CAM burden. Other authors have highlighted cytokine storm (interleukin 6), elevated ferritin, and endotheliitis as factors in the development of Mucorales infection (Aranjani et al. 2021; Pal et al. 2021). In COVID-19 infection, there is an alteration in iron metabolism. Diabetic ketoacidosis can induce ferritin synthesis and increase intracellular iron concentrations. Together, these factors place patients with COVID-19 at increased risk of developing the infection.

1.6 Mucormycosis in Humans

1.6.1 Epidemiology

It is difficult to determine the true epidemiologic burden of mucormycosis because it is a rare infection, without mandatory reports. Since 1980, cases have gradually increased worldwide (Roden et al. 2005). A study that analyzed 851 reported cases showed that 34% were from Europe, 31% from Asia, 28% from the Americas, 3% from Africa, and 3% from Australia and New Zealand. The countries from which increased cases are reported are China and India. This could be related to environmental conditions, population size, presence of comorbidities such as diabetes, and other comorbidities. The risk and prevalence of mucormycosis are 70 times higher in India than in global data (Prakash and Chakrabarti 2021).

A surveillance study conducted between 1992 and 1993 in San Francisco, California, USA, found that the annual incidence of mucormycosis was 1.7 cases per million population (Rees et al. 1998). Another study in a more general population in Spain found a lower incidence of 0.43 cases/one million population (Torres-Narbona et al. 2007). Meanwhile, analysis of hospital records in France showed an increase in incidence from 0.7/million in 1997 to 1.2/million in 2006 (Bitar et al. 2009). Of course, the incidence in different countries may vary depending on several factors.

1.6.2 Virulence Factors and Pathophysiology

Humans can become infected with mucormycosis by inhaling the sporangiospores (spores) of the Mucorales. *Via* cutaneous route through injury or trauma with skin-opened wounds. Through the gastrointestinal tract by consumption of contaminated food. After entry through any of these routes Mucorales spores encounter the epithelium. Regarding the interaction of host cells and Mucorales, the spore coat protein homologs (CotH1, CotH2, CotH3) are fungal ligands that interact with the receptor glucose-regulated protein 78 (GRP78) in the host. Indeed, elevated glucose and iron concentrations, as occurs in hyperglycemia, diabetic ketoacidosis, or other forms of acidosis, increase the expression of GRP78. This leads to fungal invasion and endothelial cell damage (Gebremariam et al. 2014). Experiments with *R. arrhizus* indicated that CotH2 and CotH3 are required for its virulence *in vivo,* given that CotH3 and CotH2 encode endothelial cell invasins. In contrast, *R. arrhizus* strains with disrupted or blocked CotH3 and CotH2 function did not show adherence. Notably, CotH3 showed the highest affinity for GRP78 (Gebremariam et al. 2014). *Rhizopus* spp., *Mucor* spp., and *Lichtheimia* spp. contain 3 to 7 copies of CotH. In contrast, genera that are less abundant in mucormycosis (*Apophysomyces* spp., *Cunninghamella* spp., *Saksenaea* spp., *Syncephalastrum* spp.) have only 1 to 2 copies of CotH (Chibucos et al. 2016). On the other hand, basement membranes are composed of extracellular protein matrices. These are mainly composed of laminin and collagen type IV.

Exposure of these extracellular matrix proteins may promote direct interaction with fungal spores. In this context, diabetes, chemotherapies, and other diseases can facilitate epithelial damage, and exposure to the fungus, as it has been shown with *Rhizopus* spores and its adhesion to collagen type 4 and laminin (Bouchara et al. 1996). Another important factor are the platelet-derived growth factor receptors (PDGFR), which are serum proteins that stimulate cell migration, and play an important role in angiogenesis, a hallmark of cancer and atherosclerosis (Ross et al. 1974; Demoulin and Essaghir 2014). Inhibition of PDGFR in an infection model with *R. delamar* significantly reduced cell injury to endothelial cells (Chibucos et al. 2016) suggesting that *Rhizopus* utilizes the PDGF signaling pathway to inflict epithelial damage and angiogenesis, enabling its hematogenous spread. Once in the tissue, these dormant spores can swell and germinate to produce rapidly growing hyphae. Interestingly, various studies indicated that germination and filamentous growth into the host cause angioinvasion, thrombosis of vessels, and necrosis of the affected tissue (Patil et al. 2016).

One of the most remarkable virulence factors is the ability of the Mucorales to absorb iron. The importance of iron acquisition by pathogenic fungi is well known (Howard 1999; Bairwa et al. 2017). In humans, iron is bound to transferrin, ferritin, and lactoferrin. This sequestration prevents the toxic effects of free iron and is a defense mechanism against infection. For example, *R. arrhizus* has been shown to significantly increase its growth in serum samples when exogenous iron is added (Artis et al. 1982). Fungi can take up iron from the host using high-affinity iron permeases or low molecular weight iron chelators (siderophores). One study showed that *R. arrhizus* and *L. corymbifera* express genes for the high-affinity iron permease (FTR1), which enables iron uptake (Ibrahim et al. 2010; Stanford et al. 2021). High-affinity iron permeases are present in fungi as part of a reductive system that includes redundant surface enzymes (Fig. 1.3a). These convert iron from ferric to its ferrous form, which is a better soluble compound. The reduced iron produced by the surface reductase is then captured by a protein complex consisting of a multi-copper oxidase and an iron permease, which is crucial for mucormycosis. In accordance with this, blockade, or inhibition of FTR1 reduces the virulence of *R. arrhizus* in mouse models (Ibrahim et al. 2010). On the other hand, iron chelators such as deferoxamine used in dialysis patients increase susceptibility to mucormycosis, since bacterial siderophore deferoxamine predisposes patients to *Rhizopus* infection by acting as a xenosiderophore (Fig. 1.3c) and facilitating the acquisition of iron by the fungus. In an experiment on infected guinea pigs, the survival rate of the *Rhizopus*-infected group deteriorated. In contrast, deferoxamine did not worsen the survival rate of the *Candida*-infected group of guinea pigs (Boelaert et al. 1991, 1993; De Locht et al. 1994). The uptake of radiolabeled iron from deferoxamine in serum was compared *in vitro* in different fugal pathogens. *Rhizopus* was able to absorb 8 times more iron than *Aspergillus fumigatus* and 40 times more than *C. albicans* (Boelaert et al. 1991).

Rhizopus secretes a siderophore belonging to the polycarboxylate family rhizoferrin. It supplies iron through a receptor-mediated, energy-dependent process. The presence of rhizoferrin has been confirmed for *Rhizopus delemar*, *Mucor circinelloides*, *Lichtheimia* (formerly *Absidia*) *corymbifera*, *Syncephalastrum racemosum*, *Mucor hiemalis*, *Rhizomucor pusillus*, and *Cunninghamella echinulata* (Carroll et al. 2017). However, it is currently unknown whether rhizoferrin transports iron via extracellular iron release or whether the siderophore is internalized before iron is released into the cytoplasm. It also appears to play a secondary role in iron uptake, being inefficient in extracting iron from serum. This suggests that the fungus has adapted to the use of xenosiderophores such as deferoxamine, which are more efficient at extracting iron from the host (Thieken and Winkelmann 1992a, b). Finally, *R. arrhizus* could enable the extraction of iron from host hemoglobin via heme oxygenase, which may explain its angioinvasive nature (Fig. 1.3b) (Ma et al. 2009).

A biological process called RNA interference (RNAi) involves double-stranded RNA molecules, which repress gene expression by sequence-specific means through translation or transcription. RNAi was previously known as co-suppression, posttranslational gene silencing (PTGS), or silencing. Thus, RNAi can silence cellular mechanisms. It has been shown, for instance, that *M. circinelloides* can become resistant to antifungal drugs through an RNAi-based mechanism. There are three RNAi pathways established in this pathogen: small interfering RNA (siRNA)-exosomal shuttle RNA (esRNA), rdrp-dependent, dicer-independent pathways, and epimutation. The siRNAs-esRNAs and the epimutational pathways are canonical in the sense that they require the participation of at least one Dicer and one Argonaute protein to produce sRNAs (Calo et al. 2014). As a defense mechanism against invaders, siRNAs are produced *via*

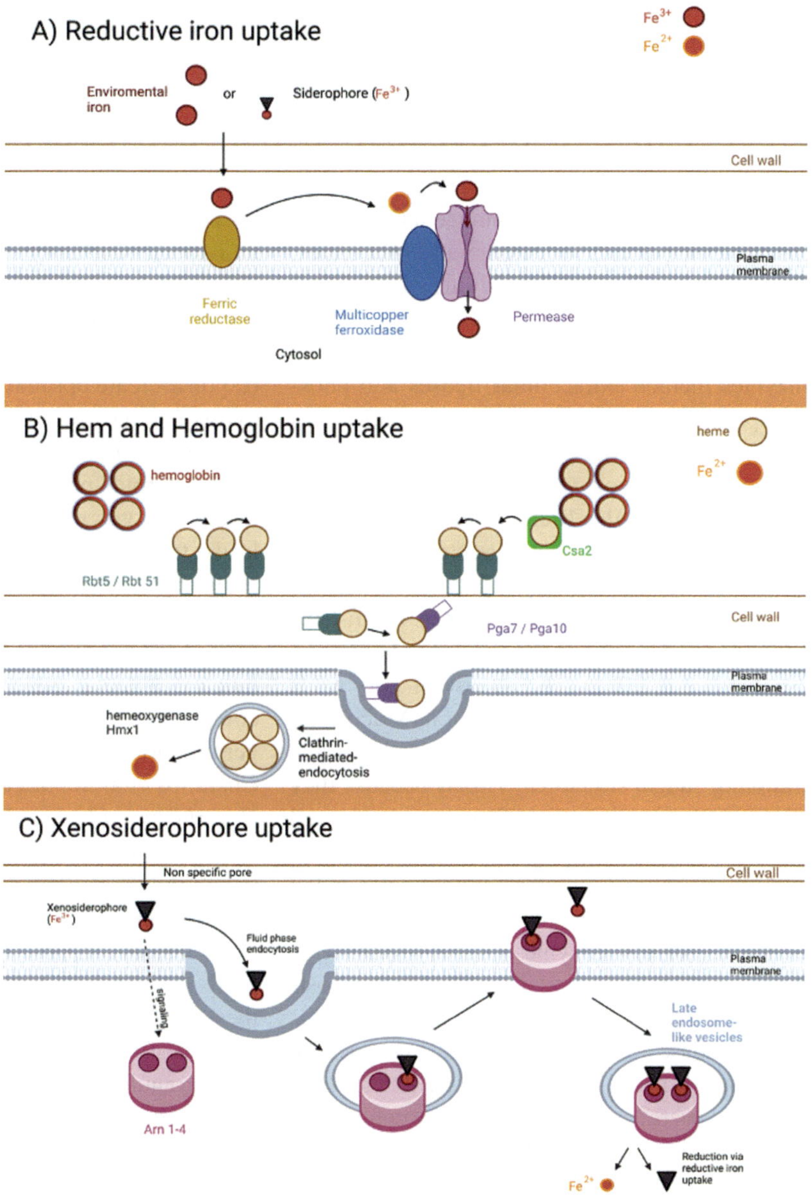

Fig. 1.3 Iron uptake in Mucorales. (**a**) Reductive pathway through iron reductases and permeases. (**b**) Assimilation of heme and hemoglobin. (**c**) Use of Xenosiderophore (deferoxamine) for the acquisition of iron. Created with BioRender.com

the siRNAs-esRNAs pathway. On the other hand, esRNAs are produced to regulate physiological processes such as growth, sporulation, and autolysis (de Haro et al. 2009; Nicolas et al. 2010; Cervantes et al. 2013). Most proteins involved in the epimutation pathway are also present in the canonical pathway, which explains the fungus' plasticity under stressful conditions, such as antifungal drugs. In these circumstances, epimutants produce sRNA against the target gene and suppress its expression temporarily through mRNA degradation. After passages in a free-drug environment, epimutants re-expressed the target gene and became sensitized to the antifungal drugs (Calo et al. 2014).

Polygalacturonase, amylase, protease, lipase, and laccase have been studied in an industrial setting (Alves et al. 2002; Geethanjali et al. 2020). Many of these enzymes have not been studied in Mucorales from an infection point of view.

Indeed, pathogenic fungal species can produce proteases during the infection process (Monod et al. 2002). Protein denaturation and amino acid acquisition is essential for fungi. And these proteases might play a role in tissue invasion in patients with mucormycosis. These proteases might be involved in tissue damage and necrosis observed in patients infected with Mucorales. The secretion of lipolytic enzymes is a factor contributing to the virulence of *Candida albicans* and *Cryptococcus neoformans*. Lipolytic enzymes include extracellular lipases and phospholipases. *Candida albicans*, for example, uses these types of enzymes for morphological modification, colonization, cytotoxicity, and host invasion (Park et al. 2013; Richardson et al. 2022). In Mucorales, the use of lipases contributing to cell membrane damage as virulence factors has not yet been investigated, but rather a Ricin-Like Toxin secreted by Mucorales known as Mucoricin (Soliman et al. 2021). The release of fatty acids could contribute to tissue invasion in Mucorales. Extraction of fatty acids also promotes the proliferation of fungal structures. This is because they are important for the formation of cell membranes.

As a reference, *C. neoformans* is known in the production of laccase as a virulence factor. Deficiency of this enzyme reduces virulence in mouse and rabbit models of cryptococcosis (Polacheck et al. 1982). In *C. neoformans*, this phenoloxidase supports melanin production. Melanin synthesis has been linked to the virulence of several pathogens. Melanin contributes to microbial virulence by reducing the susceptibility of the pathogen to host antimicrobial mechanisms. Thus, this phenoloxidase hampers the antifungal immune response (Nosanchuk and Casadevall 2006). Therefore, it would be interesting to analyze these mechanisms in mucormycosis infection models. A summary of the virulence factors can be seen in Table 1.2.

1.6.3 Immunology

Mucormycosis is usually associated with immunocompromised individuals. However, patients infected with human immunodeficiency virus (HIV) in the context of acquired immunodeficiency syndrome (AIDS) are not particularly vulnerable to mucormycosis (Antinori et al. 2009). Hence, susceptibility to infection by Mucorales relies on the innate immunity as the decisive factor and on previous immune conditions of each individual. Certainly, inhalation of sporangiospores does not result in mucormycosis in immunocompetent animals (Waldorf et al. 1984), pre-disposing factors and an unbalanced immune response create optimal conditions for Mucorales to disseminate within the host. For instance, animals whose immune systems are suppressed by corticosteroids present pulmonary infection with hematogenous dissemination, which normally leads to death. Furthermore, *in vitro* studies showed that bronchoalveolar macrophages from immunocompetent mice can inhibit germination of *R. arrhizus* spores. This is in contrast to pulmonary alveolar macrophages from immunocompromised mice, which did not prevent spore germination (Waldorf et al. 1984). Moreover, Chamilos et al. reported that neutrophils damage *R. arrhizus hyphae*, and elicited a strong response through selective upregulation of TLR2 mRNA (Chamilos et al. 2008). In addition, efficiency of the neutrophils response to hyphae of *A. fumigatus* is faster compared to R. *arrhizus*. This could partly explain the relatively high pathogenicity of these fungi. One of the differences is that the production of reduced superoxide anions by neutrophil granulocytes was lower in *R. arrhizus* compared to *A. fumigatus* (Liles et al. 1997; Chamilos et al. 2008). This suggests that the activity of these immune cells is altered in patients who develop mucormycosis and have cofactors such as ketoacidosis, diabetes mellitus, and/or corticosteroid use (Andrianaki et al. 2018).

1.6.3.1 The Innate Immune Response to Mucorales

Spores of pathogenic Mucorales typically cross the physical barriers such as skin *via* trauma (Wilson et al. 2019), inhalation or ingestion (Spellberg 2012). The intact barriers such as skin, respiratory, and intestinal mucosa are the first

Table 1.2 Virulence factors in Mucorales

Name	Abbreviation	Mechanism	References
Proven virulence			
High-affinity iron permease	**FTR1**	**Iron uptake and transport in Mucorales fungi especially during the lack of iron in the environment**	**Stanford and Voigt** (2020)
Spore coat protein	**CotH**	**Invasins in the pathogenesis of mucormycosis**	**Gebremariam et al.** (2014)
Eumelanin		**Inhibition of phagocytosis and intracellular clearance by macrophages**	**Soliman et al.** (2020)
Alkaline *Rhizopus* protease enzyme	**Arp**	**Enhancing the coagulation process in patients suffering from mucormycosis**	**Spreer et al.** (2006)
ADP-ribosylation factor	**Arf**	**Fungal dimorphism and virulence in *M. circinelloides***	**Patiño-Medina et al.** (2018)
Dihydrolipoyl dehydrogenase	–	**Most abundant antigen in the serum of patients suffered from FLD compared to healthy donors**	**Hassan and Voigt** (2019)
Calcineurin	**CaN**	**Tangible role in the transition from yeast to hyphae in *M. circinelloides***	**Lee et al.** (2013)
Serine and aspartate proteases	**SAPs**	**Most common candidates of the secreted proteases in the genome of *L. corymbifera*in *Rhizopus* act as secrete lytic enzymes**	**Hassan and Voigt** (2019)
Heme oxygenase		**Obtain iron from host hemoglobin and might explain the angioinvasive nature**	**Stanford and Voigt** (2020)
Rhizoferrin		**Obtain iron through a receptor-mediated, energy-dependent process**	**Thieken and Winkelmann** (1992a, b)
Rhizoxin		**An antimitotic macrocyclic polyketide metabolite**	**Ibrahim et al.** (2012)
RNA interference pathways	**RNAi**	**A way of negatively regulating gene expression through small non-coding RNAs or short RNAs (sRNAs)**	**Calo et al. (2014)**
Unproven mechanism of virulence			
Polygalacturonase		*Enzyme that hydrolyzes the alpha-1,4 glycosidic bonds between galacturonic acid residues*	–
Amylase		*Enzyme that catalyzes the hydrolysis of starch into simple sugars*	–
Protease		*Family of enzymes that break down protein*	–
Lipase		*Family of enzymes that catalyze the hydrolysis of fats.*	–
Laccase		*Multi-copper oxidase that catalyzes the oxidation of one electron of a wide range of phenolic compounds*	–

This table includes a division between proven virulence factors (bold) and other factors (italic) that might also influence virulence based on studies of other fungal pathogens

host defenses against mucoralean sporangiospores (Roilides et al. 2014). This physical barrier prevents the microbe from entering the host system. In addition, several complement proteins help effector cells to recognize the pathogen (Ghuman and Voelz 2017). The spores of the Mucorales directly interact with innate immune cells such as macrophages, neutrophils, and dendritic cells (DCs) (Ibrahim et al. 2012). These immune effectors prevent pathogenic propagation and activate the adaptive immune response (Ghuman and Voelz 2017). The infection is established during fungal germination due to a failure of the innate immune response against the fungus (Ibrahim et al. 2012), either in the recognition, clearance, or activation of the inflammatory response *via* production of pro-inflammatory cytokines (reviewed by Montaño and Voigt 2020).

Epithelial Cells

Epithelial cells are the first line of the external defense against various fungal infections (Hassan and Voigt 2019). Therefore, skin lesions in immunocompromised individuals represent a risk factor to develop Mucormycosis (Bouchara et al. 1996). Epithelial damage extending to the basement membrane and extracellular matrix proteins has also been observed in patients at high risk for invasive mucormycosis (Bouchara et al. 1996). Studies with *Rhizopus arrhizus* (formerly oryzae) suggest that spore adhesion to basement membrane proteins following epithelial damage is responsible for active spore germination and host infection (Liu et al. 2010). Lytic enzymes encoded by *R. arrhizus*-specific genes were observed in individuals with mucormycosis, which suggests infection with Mucorales *via* tissue invasion (Spreer et al. 2006). Analyses of host transcriptional response due to Mucorales infection suggest activation of signaling pathways such as interleukins (IL), tumor necrosis factor (TNF), and nuclear factor kappa B (NF-κB) (Chibucos et al. 2016). Furthermore, Belic et al. reported increased production of IL-6 and IL-8 by epithelial cells, and reduction of the chemokines CCL2 and CCL5 in response to *R. arrhizus*, *C. bertholleitiae*, and *R. pusillus*. Moreover, inhibition of the platelet-derived growth factor receptor B reduces epithelial damage caused by Mucorales (Chibucos et al. 2016), while epidermal growth factor receptor (EGFR) inhibition in epithelial cells reduces invasion during *Rhizopus delemar* infection (Nicolás et al. 2020).

Macrophages

Macrophages are a crucial component of the early immune response and fungal clearance and are actively recruited to the site of infection (Nicolás et al. 2020). In addition, macrophages interact, phagocytose, degrade, and present antigens for specific adaptive immune responses (Ghuman and Voelz 2017).

Spores of Mucoralean fungi can be found in three different stages: dormant, resting, and swollen, each of them differently recognized by macrophages (Fig. 1.4) (Thanh et al. 2005). For instance, studies conducted on murine macrophages reported that resting and swollen spores of *Lichtheimia corymbifera* are less phagocytosed compared to opsonized spores, while alveolar macrophages fail to eliminate *Rhizopus* resting spores (Kraibooj et al. 2014). Furthermore, Levitz et al. (1986) demonstrated that the macrophage response against Mucorales is higher in the case of swollen spores compared to the resting spores. Inhibition of fungal clearance may be related to an impaired phagosome biogenesis pathway (LAP), which later promote fungal resistance within macrophages. In fact, Andrianaki et al. reported that *Rhizopus* survives within alveolar macrophages *via* melanin-induced phagosome maturation arrest, while its growth is inhibited under intracellular iron starvation (Andrianaki et al. 2018). Nonetheless, in an adult zebrafish model Lopez et al. described that macrophages are able to response to *M. circinelloides* by increasing mRNA levels of IL-1β, TNF-α, and IL-22 (López-Fernández et al. 2018; López-Muñoz et al. 2018).

Neutrophils

Polymorphonuclear leukocytes (PMNs) or neutrophils are the most abundant leukocytes in the blood, are actively recruited to the site of infection and participate in intracellular and extracellular fungal clearance *via* phagocytosis, oxidative bursts, producing neutrophil cationic peptides (NP), and extracellular traps (NETs), respectively (Kobayashi and DeLeo 2009; Urban and Nett 2019). In addition, neutrophils mediate inflammation against *R. arrhizus*, *R. microsporus*, and *L. corymbifera* by producing IL-1β, TNF-α, (IFN)-γ, and IL-8 (Gil-Lamaignere et al. 2005; Chamilos et al. 2008). A previous study reported that neutrophil recovery after chemotherapy in patients with hematological malignancies significantly correlated with the resolution of the infection (Pagano et al. 1997). Furthermore, underlying health conditions such as neutropenia are com-

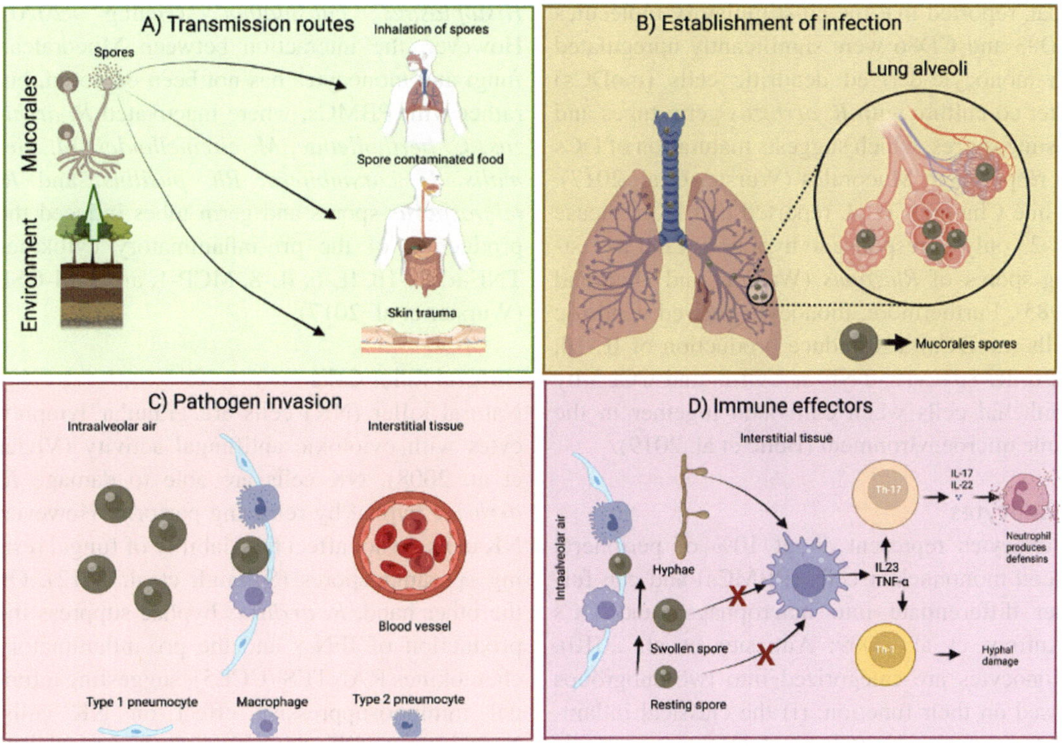

Fig. 1.4 Activity of immune effectors during interaction with Mucorales spores. The performance of immune effectors depends on the developmental stage of the fungus. Environmental Mucorales spores may enter the host after inhalation of the fungus, intake of contaminated food or a severe skin trauma (panel **a**), once in the airway or other site of infection, Mucorales rapidly invade further tissues and organs, such as the lung (panel **b**), where immune cells recognize the pathogen (panel **c**) and execute a response according to the fungal invading type, hyphae, or spores (panel **d**). Image created in BioRender.com

monly known as a risk factor for mucormycosis (Walsh and Gamaletsou 2013). Studies conducted on immunocompetent mice infected with *R. oryzae* demonstrate a significant influx of neutrophils in the lungs. However, neutrophils had a lower phagocytic rate of *R. arrhizus* compared to *A. fumigatus* (Andrianaki et al. 2018).

Studies on *Rhizopus arrhizus* indicate that neutrophils damage both fungal spores and hyphae by releasing reactive oxygen metabolites, perforins, and cationic peptides (Waldorf et al. 1984) such as NP-1 and NP-2, which possess redundant activities against resting and swollen spores (Waldorf et al. 1984). Furthermore, each developmental stage of Mucorales influences neutrophils' activity, swollen spores of *Rhizopus arrhizus* increase neutrophils chemotaxis compared to resting spores, while hyphal stage limits it (Waldorf and Diamond 1985).

Dendritic Cells

Dendritic cells (DCs) play an instrumental role in the host response by acting as an intermediate between innate and adaptive immunity (Lambrecht et al. 2001). DCS phagocyte pathogens secrete pro-inflammatory cytokines and present antigens to naive T cells and B lymphocytes, which results in immunological memory against a specific pathogen (Lambrecht et al. 2001). In the case of mucoralean fungi, Wuster et al. reported that the co-stimulatory molecules CD83 and CD86 were significantly upregulated on monocyte-derived dendritic cells (moDCs) after co-culture with *R. arrhizus* germ tubes and resting spores, which suggests maturation of DCs in response to Mucorales (Wurster et al. 2017). While Chamilos et al. reported that DCs release IL-23 only in response to hyphae but not to resting spores of *Rhizopus* (Waldorf and Diamond 1985). Furthermore, monocyte-derived dendritic cells (moDCs) also induce production of IL-1β, IL-6, IL-8, IL-12, TNF-α, CCL2, and CCL5 by epithelial cells when cultivated together in the same microenvironment (Belic et al. 2019).

Monocytes

Monocytes represent about 10% of peripheral blood mononuclear cells (PBMCs) and can further differentiate into macrophages and DCs (Auffray et al. 2009; Autissier et al. 2010). Monocytes are categorized into two subgroups based on their function: (i) the classical inflammatory monocytes (CD14+ CD16-) and (ii) the non-classical patrolling monocytes (CD14+ CD16+) (Yona et al. 2013). Studies show that the non-classical monocytes are less abundant in the human population. However, the population increases as a response to chronic inflammation stimulus (Yona et al. 2013).

Fungal cells can be eliminated by monocytes and their derivatives immune cells, macrophages and DCs, mostly through phagolysosome formation, nutrient sequestration, and generation of free radicals such as reactive oxygen species (ROS), in particular by the NADPH oxidase (NOX) that generates ROS (Uribe-Quero and Rosales 2017). Monocytes patrol the body and are engaged in clearing dead cells. During an infection, the classical monocytes are mobilized from bone marrow as a response to chemokines, and upon entering the affected tissue, they differentiate into macrophages and monocyte-derived dendritic cells (Heung 2020). Monocyte-deficient mice have shown increased susceptibility toward infections with fungi like *Aspergillus fumigatus*, *Candida albicans*, and *Histoplasma capsulatum* (Heung 2020). However, the interaction between Mucoralean fungi and monocytes has not been explored, but rather with PBMCs, where inactivated *R. arrhizus*, *C. bertholletiae*, *M. circinelloides*, *M. hiemalis*, *L. corymbifera*, *Rh. pusillus*, and *R. microsporus* spores and germ tubes induced the production of the pro-inflammatory cytokines TNF-α, IL-1β, IL-6, IL-8, MCP-1, and GM-CSF (Wurster et al. 2017).

Natural Killer Cells

Natural killer (NK) cells are granular lymphocytes with cytotoxic antifungal activity (Vivier et al. 2008). NK cells are able to damage *R. arrhizus* hyphal by releasing perforin. However, NK cells do not affect the viability of fungal resting sporangiospores (Schmidt et al. 2012). On the other hand, *R. arrhizus* hyphae suppress the production of IFN-γ and the pro-inflammatory chemokines RANTES (CCL5), suggesting a fungal immunosuppressive effect on NK cells. Interestingly, different mucoralean clinical isolates produced IFN-γ by NK cells. Therefore, finding NK receptors involved in the recognition of Mucorales may represent a potential therapeutic target, since different Mucorales species induce similar and effective antifungal responses by NK cells (Schmidt et al. 2016).

1.6.3.2 The Adaptive Immune Response to Mucorales

The adaptive immune system plays a subordinated role in Mucorales-induced invasive fungal infections (Ghuman and Voelz 2017). Initially, it was believed that T cells did not significantly impact the outcome of the immune response as their deficiency was not seen to be a risk factor for Mucorales-induced infections (Ibrahim and Kontoyiannis 2013). However, later studies show that T cells play a role in the immune response against Mucorales (Chamilos et al. 2010; Potenza et al. 2016; Page et al. 2018). For instance, *R. arrhizus* promoted T_h-17 response as a result of *in vitro* stimulation by DCs-mediated IL-23 (Chamilos et al. 2010). Moreover, T cells of patients with hematological malignancies produced IL-10 at the beginning of the infection, followed by IFN-γ and IL-4 at later stages (Potenza et al. 2011). Similar to the innate immune response, T cells only respond toward hyphae (Schmidt et al. 2012), and were lost soon after the resolution of the infection (Potenza et al. 2011). As revealed by Castillo et al., T cells-mediated therapy may be promising in the field of immunotherapy and diagnostics. For example, T cells with a combination of IL-2/IL-7 induced expansion of *R. arrhizus*-specific T cells along with production of IL-13, IL-5, TNFα, and IL-10 (Castillo et al. 2018). Therefore, the authors suggested the expansion of rare Mucorales-specific T cells from healthy donors for potential immunotherapy.

1.6.4 Clinical Presentations

Invasive mucormycosis is classified into six clinical forms depending on the site of infection and clinical presentation (Table 1.3). Any Mucorales species can cause infection at these sites. In the susceptible population, the rhino-orbital-cerebral form is the most observed infection, accounting for 34% of cases. This is followed by the cutaneous form with 22%. And finally, the pulmonary form with 20% of cases (Jeong et al. 2019). Dissemination is possible in 23% of cases, and mortality in these patients can be as high as 96%. Mortality from gastrointestinal infections is 85%. Pulmonary infection ranks third, with a mortality of 76% of infected individuals. Cutaneous and intestinal infections of mucormycosis are more likely in the pediatric population.

On the other hand, patients with CAM have been reported in at least 18 countries, and 95% of patients had concomitant disease. The rhino-orbital-cerebral form was the most common in 80% of patients who became infected, with evidence of central nervous system involvement in only one-third of these patients, and the pulmonary form accounted for 20% of the population studied (Hoenigl et al. 2022). Most patients with hematologic malignancies developed pulmonary Mucorales infection. The diagnosis was made between 0 and 90 days, time in which these patients were hospitalized with a diagnosis of rhino-orbital-cerebral mucormycosis. Early diagnosis plays an essential role in the survival of these patients.

1.6.4.1 Rhino-Orbital-Cerebral Mucormycosis (ROCM)

Sinus infections are caused by inhalation of fungal sporangiospores. As a result, the infection can spread rapidly to surrounding tissues. Once germinated, the fungi may spread downward and infect the palate, invade the sphenoid sinus from behind, invade the cavernous sinus laterally and infect the orbits, or invade the brain from above once they have begun to grow. Fungi invade the cranium through the orbital apex or cribriform plate of the ethmoid bone, killing the host. Hematogenous dissemination may occur with or without mycotic aneurysms resulting from cerebral vascular invasion (Orguc et al. 2005; Hosseini and Borghei 2005).

The initial clinical presentation of ROCM is nonspecific and focused on the head area. Initially, patients often report unilateral retroocular pain and lethargy. This may be accompanied by fever, facial pain and numbness, nasal congestion, rhinorrhea, nasal hypoesthesia, and epistaxis. At the ocular level, amaurosis, diplopia, and blurred vision may also occur. If seizures,

Table 1.3 Etiologic agents of lagenidiosis, pythiosis, entomophthoromycosis, and mucormycosis

Disease	Phylum	Microorganisms	Similar characteristics
Lagenidiosis	Oomycota	*Lagenidium* sp.	The microorganism produces lesions characterized by pyogranulomatous and eosinophilic inflammation and hyphae with sparse septum and broad branching patterns
Pythiosis	Oomycota	*Pythium insidiosum*	
Entomophthoromycosis	Zoopagomycota	Entomophthorales	
Mucormycosis	Mucoromycota	Mucorales	

dizziness, altered mental status, and/or gait disturbances occur, central nervous system (CNS) disease should be suspected (Ribes et al. 2000).

Physical examination usually reveals nasal and orbital cellulitis. It is associated with edema of the dorsum of the nose and cheeks. In advanced stages, necrosis due to angioinvasion is observed. This results in progressive black, visible detachment of the palatal or nasal mucosa. It is important to perform an inspection of the oral cavity, where ulceration of the palate can be detected. Proptosis is the most common ocular sign, followed by ophthalmoplegia and loss of visual acuity. Patients may also present with conjunctival chemosis, nygtasms, and a fixed pupil. Patients with CNS infection may have an altered state of consciousness. Patients with central nervous system infections may exhibit changes in consciousness and paralysis caused by damage to cranial nerves II through VII (Ribes et al. 2000).

1.6.4.2 Pulmonary Mucormycosis

Once the fungal sporangiospores reach the lower respiratory tract, it can cause endotracheal and/or bronchoalveolar infection. In case of endotracheal mucormycosis, airway obstruction with subsequent pulmonary collapse may occur. In addition, hilar angioinvasion is possible, leading to dissemination to the mediastinum, pericardium, and chest wall. Again, the clinic is nonspecific, so it can be confused with other etiologies. For example, bacterial infection or pulmonary aspergillosis is often suspected. Patients with pulmonary mucormycosis have fever greater than 38 °C (Connor et al. 1979; Passamonte and Dix 1985).

Less commonly, there is dyspnea, hemoptysis, and pleuritic pain. One of the problems is that the symptoms are not always noticeable. This may be due to immunosuppression, which is common in these patients. Neutropenic cancer patients undergoing induction chemotherapy and patients with graft-versus-host disease due to hematopoietic stem cell transplantation are most affected. In pulmonary mucormycosis, symptoms are usually mild and diagnostic imaging shows infiltrates suggestive of infection. Wang et al. describe a 15-year-old male patient who was diagnosed with T-cell acute lymphoblastic leukemia (Wang et al. 2016).

Therefore, he received induction chemotherapy with idamycin, vindesine, and dexamethasone in therapeutic doses. On the fifteenth day of hospitalization, the patient had a fever of 39 °C. A lung computed tomography scan was performed. The right upper lobe showed an infiltrate that expanded into a cavity. However, laboratory tests for fungi, mycobacteria, and bacteria were negative. He received therapy with meropenem, teicoplain, amphotericin B, and caspofungin in therapeutic doses. He received therapeutic doses of meropenem, teicoplanin, amphotericin B, and caspofungin due to the worsening radiological progression of the pulmonary infiltrate.

On the 14th day of hospitalization, a lobectomy was performed at the affected site. During surgery, a 5 × 5 × 5 cm mass was noted arising from the posterior segment of the right upper lobe. Histopathological analysis of the mass exhibited numerous wide, non-septate, hyphae with right-angled branching. Regrettably, the patient succumbed to septic shock, underscoring the critical significance of early diagnosis, meticulous monitoring of vulnerable groups, and the necessity for suitable surgical intervention due to the limited effectiveness of antifungal agents in the affected tissue.

1.6.4.3 Cutaneous Mucormycosis

Infection with sporangiospores is associated with skin trauma. Cutaneous manifestation of a disseminated form of mucormycosis is less likely. It is classified as superficial if it affects only the skin or subcutaneous cellular tissue. Deep dissemination, on the other hand, is present when tendons, muscles, and bones are affected. And a cutaneous focus can also spread and infect distant tissues. The course of the disease is slowly progressive but inconspicuous. Initially, there is inflammation of the affected tissue, which later develops into a necrotic scab. The tissue surrounding the necrotic eschar is indurated and erythematous (Spellberg et al. 2005; Roden et al. 2005).

Cases have been reported in which the symptoms resemble those of tinea corporis. This clinical picture is rare and is characterized by target-like plaques with outer erythematous margins and ecchymotic or blackened necrotic centers (Paes De Oliveira-Neto et al. 2006). In open lesions, a cotton-wool appearance reminiscent of bread mold may be observed (Kordy et al. 2004). Differential diagnosis with pyoderma gangrenosum, synergistic bacterial gangrene, or other infections caused by bacteria or fungi is also important. If the infection is not controlled, it may invade deeper tissues and cause gangrene and hematogenous spread (Rubin and Grossman 2004).

1.6.4.4 Gastrointestinal Mucormycosis

Mucorales spores can be consumed via fermented milk, dried bread products, fermented porridges, and alcoholic beverages made from corn. Herbal and homeopathic remedies that have been contaminated with fungi. Anecdotally, a hematology-oncology clinic reported cases of gastrointestinal mucormycosis presumably transmitted orally by tongue depressors contaminated with sporangiospores (Mitchell et al. 1996; Roden et al. 2005). Only ¼ of cases are diagnosed antemortem. Preterm infants, malnourished children, and people with malignant hematological diseases, diabetes mellitus, or a history of corticosteroid use are particularly affected (Michalak et al. 1980; Mitchell et al. 1996; Garg et al. 2008).

The most affected organs include the stomach, colon, and ileum. Masses in the cecum, ileum, or appendix are usually the result of infection. Massive bleeding from the upper gastrointestinal tract may be associated with gastric perforation. In premature infants, it occurs as necrotizing enterocolitis. Neutropenic patients may also present with typhlitis, hematochezia, and neutropenic fever. The fungus can invade the intestinal walls and blood vessels, leading to intestinal perforation, peritonitis, sepsis, and massive gastrointestinal hemorrhage, which is the most common cause of death (Michalak et al. 1980; Cherney et al. 1999; Garg et al. 2008).

1.6.4.5 Disseminated Mucormycosis

The development of disseminated mucormycosis (DM) is rare but typically occurs in patients with severely weakened immune systems, after burns, in premature infants, or in patients who have received deferoxamine. In the series of 929 cases described by Roden et al., it was found that disseminated disease occurred in 40–50% of patients with cerebral or pulmonary mucormycosis. A mortality rate of 96% was noted in patients with disseminated mucormycosis (Roden et al. 2005). The brain is the most common site of spread, but we may also find metastatic lesions in the liver, spleen, heart, and in other organs. DM varies widely in symptomatic and evolutionary characteristics, depending on the host as well as the location and degree of vascular invasion and tissue infarction.

1.6.4.6 Uncommon Forms of Mucormycosis

Mucormycosis can develop in any organ or part of the body in patients with risk factors that allow invasion of sporangiospores. A variety of less common or unusual forms of mucormycosis can occur, including endocarditis, osteomyelitis, peritonitis, and pyelonephritis. The infection with Mucorales is a rare cause of endocarditis of prosthetic or native valves and can cause aortic thrombosis. The most common risk factor is intravenous drug use (Virmani et al. 1982). Osteomyelitis usually occurs after traumatic inoculation or surgery (e.g., tibial pin insertion, anterior cruciate ligament repair). Some case reports have described osteomyelitis of the femur, humerus, scapula, calcaneus, tibia, cuboid, metacarpals, sternum, and phalanges. Hematogenous osteomyelitis is extremely rare (Echols et al. 1979; Aymoré et al. 2005). In patients undergoing continuous ambulatory peritoneal dialysis, infection of the peritoneal cavity by Mucorales is extremely infrequent (Ram et al. 2007).

1.6.5 Diagnostic Methods

Diagnosis of mucormycosis requires analysis of clinical criteria, risk factors, histopathologic findings, cultures, and imaging studies. When symptoms, clinical signs, and risk factors are supported by findings on diagnostic nasal endoscopy, MRI (magnetic resonance imaging), or contrast-enhanced computed tomography, the patient is considered as a probable case. Clinic-radiologic features, as well as microbiologic confirmation by direct microscopy, culture, histopathology with special stains, or molecular diagnosis, are essential for classification of a patient with proven mucormycosis. Therefore, biopsy or fine needle aspiration of infected tissue should be performed for direct microscopic analysis, culture, histopathology, and molecular studies (Cornely et al. 2019).

Fungal culture has variable sensitivity for diagnosis. It is always preferable to send a biopsy fragment of infected tissue. Unfortunately, Mucorales' culture from clinical specimens is low. In addition, fungal growth can be slow. A disadvantage of biopsy is that invasive endoscopic procedures may be required to obtain tissue, especially for infections of the lung and gastrointestinal tract, etc. Histopathologic examination of tissues from patients with mucormycosis shows tissue damage and invasion of blood vessels, as described in the section on virulence factors. It is recommended to use special techniques to detect fungal pathogens such as periodic acid-Schiff (PAS), Grocott's methenamine silver stain (GMS), etc. The presence of hyphae with the following characteristics may be observed:

1. Broad, small quantity.
2. Coenocytic, irregular branching of hyphae.
3. Large branching angle, approximately rectangular.

Neutropenia in white blood cells may be detected with additional laboratory tests. An elevated urinary ketones level and low blood pH are the characteristics of diabetic ketoacidosis. The diagnosis may be made using imaging studies, although computed tomography (CT) on the sinuses is less sensitive to soft tissue invasion than magnetic resonance imaging (MRI). Chest CT examination with intravenous contrast is a sensitive test for detecting abnormalities in pulmonary mucormycosis. Nonetheless, CT of the brain cannot distinguish between abscesses and early infarcts (Therakathu et al. 2018).

Laboratory and imaging studies are useful, but they are not infallible since other pathologies can show similar results. Treatment and prognosis can be improved by early diagnosis. There are few clinically approved options for biomarkers and little research has been conducted on them. Therefore, early diagnosis and monitoring of mucormycosis by detection of circulating DNA in serum has been considered a possibility. Detection of circulating DNA in a patient with mucormycosis was provided in a cohort study of 44 patients, demonstrating the diagnostic utility and accurate quantification of fungal DNA load for treatment monitoring (Millon et al. 2016).

1.6.6 Basics of Treatment

Treatment must follow a multidisciplinary approach. Therefore, administration of antifungal agents, surgical debridement of infected tissue, and control of modifiable risk factors should be combined. Homeostasis is important in patients with mucormycosis. Therefore, metabolic acidosis and hyperglycemia must be controlled, deferoxamine administration interrupted or replaced, immunosuppressive therapies modulated, and, when possible, neutropenia controlled. Antifungal resistance in Mucorales should be considered in the choice of pharmacological treatment. Many fungal diseases are treated with short-tailed azoles as the initial treatment of the infection. This is not the case with mucormycosis. Drugs such as fluconazole and voriconazole are not helpful (Caramalho et al. 2017).

The agent of choice is liposomal amphotericin B. Long-tailed triazoles such as posoconazole and isavuconazole can be used for de-escalation in patients who have responded to initial treatment with amphotericin (Cornely et al. 2019). Another indication for the use of posoconazole and isavuconazole is salvage therapy in patients who have not responded to amphotericin at therapeutic doses (Cornely et al. 2019). Surgical removal of necrotic tissue is critical in all patients with mucormycosis because the hyphae of the Mucorales may remain viable in this tissue due to the low or lack of penetrance of the antifungal agent in this tissue, it is considered a focal point for maintenance of infection. This aggressive debridement of infected tissue has been associated with higher survival in patients with the pulmonary form and ROCM (Sun et al. 2009, 2010).

1.7 Mucormycosis in Animals

Mucormycosis can also occur in animals. However, due to taxonomic changes, it is important to pay attention to the differential diagnosis of entomophthoromycosis, pythiosis, and lagenidiosis. Because of the clinical and histopathological similarities, the term phycomycosis is used. Especially since confirmation by culture or molecular methods is often not possible. Table 1.3 shows a compilation of diseases and associated microorganisms. In each case, a specific diagnosis helps in the targeted treatment and prognosis of the disease. Therefore, microbiological confirmation by culture and molecular biology is important. With these techniques, the pathogen can be identified.

Mucormycosis is usually angioinvasive and leads to dissemination. Therefore, mortality in animals is high. Most case reports are postmortem by histopathology. After searching academic search engines and databases, at least 28 animal species associated with natural mucormycosis were identified (Table 1.4). In the scientific literature, the terms mucormycosis and zygomycosis are confused with each other, and many animals reported as having mucormycosis by the above extrapolation are actually cases of entomophthoromycosis.

In other animals, infection with Mucorales is like that in humans. Trauma is an important route of transmission. In addition, airborne and gastrointestinal transmission have also been described. Regarding virulence and immunity factors, much of the research is based on animal models (Jacobsen 2019). Therefore, these factors are extrapolated for all animals. Only a few animals are treated in a timely manner. Spread to other organs is possible. The Huggins et al. (2020) study describes marine mammals in which mucormycosis has been detected in the brain, lungs, spleen, pancreas, kidneys, muscles, lymph nodes, and skin (Huggins et al. 2020). Treatment is available for some species. A good response to posoconazole has been described in infected animals (Wray et al. 2008).

1.8 Future Perspectives

Mucormycosis is a life-threatening fungal infection with often fatal outcome. The consequences of the disease in human populations and animal life stocks are far from being fully understood. The occurrence of the disease is directly linked to immunosuppression, so it can be used as indicator for the immune status of a population. The type and the clinical manifestation of mucormycosis may serve sentinel functions for the management of the underlying primary diseases. During the COVID-19 pandemic large outbreaks of mucormycosis were observed in direct connection with SARS-CoV2- infections, and their underlying complications regarding treatment and maintenance of the basic health conditions. If the primary diseases are controlled, incidence of mucormycosis will decrease in the future. Furthermore, novel and inexpensive diagnoses could decrease mortality rates and promote faster medical intervention and treatments. In that regard, it is essential to understand the interaction between the host immune response and

Table 1.4 Mucormycosis in animals

Common name	Scientific name	Pathogen	Outcome	Treatment	References
Cat	*Felis catus*	*Mucor* species	Survived	Posoconazale	Wray et al. (2008)
Dog	*Canis lupus familiaris*	*Saksenaea vasiformis*	Not survived	Griseofulvin, itraconazole, and surgical debridemen	Reynaldi et al. (2017)
Bat	*Leptonycteris yerbabuenae*	*Apophysomyces elegans*	Not survived	Post-mortem report	López-Romero et al. (2021)
Bull	*Bos taurus*	*Mucor pusillus*	–	–	Vítovec et al. (1975)
Cynomolgus monkey	*Macaca fascicularis*	–	Not survived	Post-mortem report	Torres-Urbano et al. (2010)
Pig	*Sus scrofa domesticus*	–	Not survived	Post-mortem report	Abbott et al. (1976)
Roe deer	*Capreolus capreolus*	*Rhizopus microsporus*	–	–	Peter et al. (2014)
Bottlenose dolphin	*Tursiops truncatus*	*Rhizopus microsporus*	Not survived	Antibiotics and corticosteroids	Chang et al. (2021)
Asian elephant	*Elephas maximus*	Apophysomyces elegans	–	–	Thomas et al. (2008)
American bison calf	*Bison bison*	*Lichtheimia corymbifera*	Not survived	–	Espinosa de los Monteros et al. (1999)
Goat	*Capra aegagrus hircus*	–	–	–	Jensen et al. (1996)
Guinea-pig	*Cavia porcellus*	*Lichtheimia corymbifera*	–	–	Rigby (1976)
Horse	*Equus caballus*	*Lichtheimia corymbifera*	Not survived	–	Guillot et al. (2000)
Platypus	*Ornithorhynchus anatinus*	*Mucor amphibiorum*	Not survived	–	Obendorf et al. (1993)
Ringed Seal	*Pusa hispida*	*Rhizomucor pusillus*	not survived	Antibiotics and corticosteroids	Nakagun et. al. (2018)
Gray squirrel	*Sciurus carolinensis*	–	Not survived	–	Novilla et al. (1981)
Killer whale	*Orcinus orca*	*Mucor* spp.	Not survived	–	Abdo et al. (2012)
Blue tit	*Cyanistes caeruleus*	Mucor racemosus	Not survived	–	Schmitz et al. (2018)
Blackbird	*Turdus merula*	Mucor racemosus	Not survived	–	Stenkat et al. (2013)
Atlantic canary	*Serinus canarius*	Mucor ramosissimus	Not survived	–	Quesada et al. (2007)
Humboldt penguin	*Spheniscus humboldtii*	Mucor spp.	Not survived	–	Bigland et al. (1961)
Duck	*Anatidae*	Mucor spp.	Not survived	–	McCasey et al. (1984)
Ostriches	*Struthio camelus*	Mucor spp.	Not survived	–	Gulbahar et al. 2000
Lovebirds	*Agapornis roseicollis*	*Lichtheimia corymbifera*	Not survived	–	Carrasco et al. (1993)

Table 1.4 (continued)

Common name	Scientific name	Pathogen	Outcome	Treatment	References
African gray parrot	*Psittacus erithacus*	–	Not survived	–	Dawson et al. (1976)
Budgerigar	*Melopsittacus undulatus*	–	Not survived	–	Barros et al. (1972)
Chickens	*Gallus gallus domesticus*	–	Not survived	–	Steinlage et al. (2003)
Cane toads	*Rhinella marina*	*Mucor amphibiorum*	Not survived	–	Speare et al. (1997)

A search was made in academic databases with the terms "animal" mucormycosis or zycomycosis. Cases of lagenidiosis, pythiosis, and entomophthoromycosis with erroneous reporting were excluded. The term "animal" was individually replaced with all the terms in the "Common name" column and the other two terms described were added. A representative animal with confirmed detection for mucormycosis was included in this table.

Mucorales. Studies on the immunological field could help to elucidate earlier signs of the infection rather than late tissue necrosis, finding new strategies to hamper initial host-pathogen interactions, and treatments to enhance fungal clearance. Altogether, battling mucormycosis not only depends on preventive measures, but also on the development of new diagnosis, treatment, and educational programs in the medical and scientific fields. This would help to create more awareness of such a rare fungal infection.

References

Abbott SE, van den Ende J, Hickman R, Terblanche J (1976) Gastric mucormycosis and moniliasis in an unimmunosuppressed pig following renal transplantation. J S Afr Vet Assoc 47:47–48

Abdo W, Kawachi T, Sakai H et al (2012) Disseminated mycosis in a killer whale (Orcinus orca). J Vet Diagn Invest 24:211–218. https://doi.org/10.1177/1040638711416969/ASSET/IMAGES/LARGE/10.1177_1040638711416969-FIG6.JPEG

Acosta-España JD, Voigt K (2022) Mini review: risk assessment, clinical manifestation, prediction, and prognosis of mucormycosis: implications for pathogen- and human-derived biomarkers. Front Microbiol 13:1–12. https://doi.org/10.3389/FMICB.2022.895989

Ainsworth GC, Austwick PKC (1955) A survey of animal mycoses in Britain: mycological aspects. Trans Br Mycol Soc 38:369–386. https://doi.org/10.1016/S0007-1536(55)80040-X

Alqarihi A, Gebremariam T, Gu Y et al (2020) GRP78 and integrins play different roles in host cell invasion during mucormycosis. MBio 11. https://doi.org/10.1128/MBIO.01087-20

Alves MH, Campos-Takaki GM, Figueiredo Porto AL, Milanez AI (2002) Screening of Mucor spp. for the production of amylase, lipase, polygalacturonase and protease. Braz J Microbiol 33:325–330. https://doi.org/10.1590/S1517-83822002000400009

American Diabetes Association (2022) Introduction: standards of medical care in diabetes 2022. Diabetes Care 45. https://doi.org/10.2337/dc22-SINT

Andrianaki AM, Kyrmizi I, Thanopoulou K et al (2018) Iron restriction inside macrophages regulates pulmonary host defense against Rhizopus species. Nat Commun 9. https://doi.org/10.1038/S41467-018-05820-2

Antinori S, Nebuloni M, Magni C et al (2009) Trends in the postmortem diagnosis of opportunistic invasive fungal infections in patients with AIDSA retrospective study of 1,630 autopsies performed between 1984 and 2002. Am J Clin Pathol 132:221–227. https://doi.org/10.1309/AJCPRAAE8LZ7DTNE

Aranjani JM, Manuel A, Razack HIA, Mathew ST (2021) COVID-19–associated mucormycosis: Evidence-based critical review of an emerging infection burden during the pandemic's second wave in India. PLoS Negl Trop Dis 15:e0009921. https://doi.org/10.1371/JOURNAL.PNTD.0009921

Artis WM, Fountain JA, Delcher HK, Jones HE (1982) A mechanism of susceptibility to mucormycosis in diabetic ketoacidosis transferrin and iron availability. Diabetes 31:1109–1114. https://doi.org/10.2337/DIACARE.31.12.1109

Auffray C, Sieweke MH, Geissmann F (2009) Blood monocytes: development, heterogeneity, and relationship with dendritic cells. Annu Rev Immunol 27:669–692. https://doi.org/10.1146/ANNUREV.IMMUNOL.021908.132557

Autissier P, Soulas C, Burdo TH, Williams KC (2010) Evaluation of a 12-color flow cytometry panel to study lymphocyte, monocyte, and dendritic cell subsets in humans. Cytometry A 77A:410–419. https://doi.org/10.1002/CYTO.A.20859

Aymoré IL, Meohas W, Brito De Almeida AL, Proebstner D (2005) Case report: periosteal Ewing's sarcoma - case report and literature review. Clin Orthop Relat

Res 434:265–272. https://doi.org/10.1097/01.BLO.0000151439.18746.F1

Bairwa G, Hee Jung W, Kronstad JW (2017) Iron acquisition in fungal pathogens of humans. Metallomics 9:215–227. https://doi.org/10.1039/C6MT00301J

Banerjee M, Pal R, Bhadada SK (2021) Intercepting the deadly trinity of mucormycosis, diabetes and COVID-19 in India. Postgrad Med J. https://doi.org/10.1136/POSTGRADMEDJ-2021-140537

Barros SS, Londero AT (1972) Cardiac zygomycosis in an Australian parakeet. Sabouraudia 10:189–190. https://doi.org/10.1080/00362177285190361

Beigelman PM, Warner NE (1973) Thirty-two fatal cases of severe diabetic ketoacidosis, including a case of mucormycosis. Diabetes 22:847–850. https://doi.org/10.2337/DIAB.22.11.847

Belic S, Page L, Lazariotou M et al (2019) Comparative analysis of inflammatory cytokine release and alveolar epithelial barrier invasion in a transwell ® bilayer model of mucormycosis. Front Microbiol 9. https://doi.org/10.3389/FMICB.2018.03204

Berbudi A, Rahmadika N, Tjahjadi AI, Ruslami R (2020) Type 2 diabetes and its impact on the immune system. Curr Diabetes Rev 16:442. https://doi.org/10.2174/1573399815666191024085838

Bigland CH, Graesser FE, Pennifold KS (1961) An osteolytic mucor mycosis in a penguin. Avian Dis 5(4):367–370

Biswal M, Gupta P, Kanaujia R et al (2022) Evaluation of hospital environment for presence of Mucorales during COVID-19-associated mucormycosis outbreak in India e a multi-centre study. J Hosp Infect 122:173–179. https://doi.org/10.1016/j.jhin.2022.01.016

Bitar D, Van Cauteren D, Lanternier F et al (2009) Increasing incidence of Zygomycosis (Mucormycosis), France, 1997–2006. Emerg Infect Dis 15:1395. https://doi.org/10.3201/EID1509.090334

Boelaert JR, Fenves AZ, Coburn JW (1991) Deferoxamine therapy and mucormycosis in dialysis patients: report of an international registry. Am J Kidney Dis 18:660–667. https://doi.org/10.1016/S0272-6386(12)80606-8

Boelaert JR, De Locht M, Van Cutsem J et al (1993) Mucormycosis during deferoxamine therapy is a siderophore-mediated infection. In vitro and in vivo animal studies. J Clin Invest 91:1979–1986. https://doi.org/10.1172/JCI116419

Bouchara JP, Oumeziane N, Lissitzky J et al (1996) Attachment of spores of the human pathogenic fungus Rhizopus oryzae to extracellular matrix components. Eur J Cell Biol 70:76–83

Caffrey AK, Obar JJ (2016) Alarmin(g) the innate immune system to invasive fungal infections. Curr Opin Microbiol 32:135–143. https://doi.org/10.1016/J.MIB.2016.06.002

Calo S, Shertz-Wall C, Lee SC et al (2014) Antifungal drug resistance evoked via RNAi-dependent epimutations. Nature 513:555–558. https://doi.org/10.1038/nature13575

Caramalho R, Tyndall JDA, Monk BC et al (2017) Intrinsic short-tailed azole resistance in mucormycetes is due to an evolutionary conserved aminoacid substitution of the lanosterol 14α-demethylase. Sci Rep 7. https://doi.org/10.1038/S41598-017-16123-9

Carballeira JD, Quezada MA, Hoyos P et al (2009) Microbial cells as catalysts for stereoselective red–ox reactions. Biotechnol Adv 27:686–714. https://doi.org/10.1016/J.BIOTECHADV.2009.05.001

Carrasco et al. 1993. https://www.jstor.org/stable/1592056?origin=crossref&seq=2

Carroll CS, Grieve CL, Murugathasan I et al (2017) The rhizoferrin biosynthetic gene in the fungal pathogen Rhizopus delemar is a novel member of the NIS gene family. Int J Biochem Cell Biol 89:136–146. https://doi.org/10.1016/J.BIOCEL.2017.06.005

Castillo P, Wright KE, Kontoyiannis DP et al (2018) A new method for reactivating and expanding T cells specific for Rhizopus oryzae. Mol Ther Methods Clin Dev 9:305–312. https://doi.org/10.1016/J.OMTM.2018.03.003/ATTACHMENT/C33056B1-C81C-4569-B038-F57179ACB362/MMC1.PDF

Cervantes M, Vila A, Nicolás FE et al (2013) A single argonaute gene participates in exogenous and endogenous RNAi and controls cellular functions in the basal fungus Mucor circinelloides. PLoS One 8:e69283. https://doi.org/10.1371/JOURNAL.PONE.0069283

Chakrabarti A, Sood P, Denning D (2023) Estimating fungal infection burden in India using computational mod Mucormycosis burden as a case study

Chamilos G, Lewis RE, Lamaris G et al (2008) Zygomycetes hyphae trigger an early, robust proinflammatory response in human polymorphonuclear neutrophils through toll-like receptor 2 induction but display relative resistance to oxidative damage. Antimicrob Agents Chemother 52:722–724. https://doi.org/10.1128/AAC.01136-07/ASSET/93860CAF-C9FB-4C48-AF9B-7E17A159362F/ASSETS/GRAPHIC/ZAC0020870580003.JPEG

Chamilos G, Ganguly D, Lande R et al (2010) Generation of IL-23 producing dendritic cells (DCs) by airborne fungi regulates fungal pathogenicity via the induction of T(H)-17 responses. PLoS One 5:e12955. https://doi.org/10.1371/JOURNAL.PONE.0012955

Chang L, Qi Y, Wang Y et al (2021) Systemic mucormycosis caused by Rhizopus microsporus in a captive bottlenose dolphin. Vet Med Sci 7:2404–2409. https://doi.org/10.1002/VMS3.602

Cherney CL, Chutuape A, Fikrig MK (1999) Fatal invasive gastric mucormycosis occurring with emphysematous gastritis: case report and literature review. Am J Gastroenterol 94:252–256. https://doi.org/10.1111/J.1572-0241.1999.00809.X

Chibucos MC, Soliman S, Gebremariam T et al (2016) An integrated genomic and transcriptomic survey of mucormycosis-causing fungi. Nat Commun 7:1–11. https://doi.org/10.1038/ncomms12218

Chinn RYW, Diamond RD (1982) Generation of chemotactic factors by Rhizopus oryzae in the presence and absence of serum: Relationship to hyphal damage mediated by human neutrophils and effects of hyperglycemia and ketoacidosis. Infect Immun 38:1123–1129. https://doi.org/10.1128/IAI.38.3.1123-1129.1982

Connolly JE, McAdams HP, Erasmus JJ, Rosado-De-Christenson ML (1999) Opportunistic fungal pneumonia. J Thorac Imaging 14:51–62. https://doi.org/10.1097/00005382-199901000-00005

Connor B, Anderson R, Smith J (1979) Mucor mediastinitis. Chest 75:530. https://doi.org/10.1378/CHEST.75.4.525

Cornely OA, Alastruey-Izquierdo A, Arenz D et al (2019) Global guideline for the diagnosis and management of mucormycosis: an initiative of the European Confederation of Medical Mycology in cooperation with the Mycoses Study Group Education and Research Consortium. Lancet Infect Dis 19:e405–e421. https://doi.org/10.1016/S1473-3099(19)30312-3

Dawson CO, Wheeldon EB, McNeil PE (1976) Air sac and renal mucormycosis in an African gray parrot (Psittacus erithacus). Avian Dis:593–600

de Haro JP, Calo S, Cervantes M et al (2009) A single dicer gene is required for efficient gene silencing associated with two classes of small antisense RNAs in mucor circinelloides. Eukaryot Cell 8:1486–1497. https://doi.org/10.1128/EC.00191-09/SUPPL_FILE/SUPPLEMENTAL_MATERIAL.DOC

De Locht M, Boelaert JR, Schneider YJ (1994) Iron uptake from ferrioxamine and from ferrirhizoferrin by germinating spores of rhizopus microsporus. Biochem Pharmacol 47:1843–1850. https://doi.org/10.1016/0006-2952(94)90314-X

De Los E, Monteros A, Carrasco L, King JM, Jensen HE (1999) Nasal zygomycosis and pulmonary aspergillosis in an American bison. J Wildl Dis 35:790–795. https://doi.org/10.7589/0090-3558-35.4.790

de Souza CAF, Lima DX, Gurgel LMS, de Santiago ALCMA (2017) Coprophilous Mucorales (ex Zygomycota) from three areas in the semi-arid of Pernambuco, Brazil. Braz J Microbiol 48:79–86. https://doi.org/10.1016/J.BJM.2016.09.008

Demoulin JB, Essaghir A (2014) PDGF receptor signaling networks in normal and cancer cells. Cytokine Growth Factor Rev 25:273–283. https://doi.org/10.1016/J.CYTOGFR.2014.03.003

Echols RM, Selinger DS, Hallowell C et al (1979) Rhizopus osteomyelitis. A case report and review. Am J Med 66:141–145. https://doi.org/10.1016/0002-9343(79)90505-9

Fazili ABA, Shah AM, Zan X et al (2022) Mucor circinelloides: a model organism for oleaginous fungi and its potential applications in bioactive lipid production. Microb Cell Factories 21:1–19. https://doi.org/10.1186/S12934-022-01758-9/TABLES/4

Garg PK, Gupta N, Gautam V, Hadke NS (2008) Gastric zygomycosis: unusual cause of gastric perforation in an immunocompetent patient. South Med J 101:449–450. https://doi.org/10.1097/SMJ.0B013E318167BB31

Gebremariam T, Liu M, Luo G et al (2014) CotH3 mediates fungal invasion of host cells during mucormycosis. J Clin Invest 124. https://doi.org/10.1172/JCI71349

Geethanjali PA, Hittanahallikoppal GG, Jayashankar M (2020) Optimization of culture conditions for hyper-production of laccase from an indigenous litter dwelling fungus Mucor circinelloides GL1. Environ Sustain 3:481–495. https://doi.org/10.1007/S42398-020-00137-7

Gest H (2004) The discovery of microorganisms by Robert Hooke and Antoni Van Leeuwenhoek, fellows of the Royal Society. Notes Rec R Soc Lond 58:187–201. https://doi.org/10.1098/RSNR.2004.0055

Ghuman H, Voelz K (2017) Innate and adaptive immunity to mucorales. J Fungi 3(3):48. https://doi.org/10.3390/JOF3030048

Gil-Lamaignere C, Simitsopoulou M, Roilides E et al (2005) Interferon-γ and granulocyte-macrophage colony-stimulating factor augment the activity of polymorphonuclear leukocytes against medically important zygomycetes. J Infect Dis 191:1180–1187. https://doi.org/10.1086/428503/2/191-7-1180-FIG006.GIF

Gulbahar MY, Agaoglu Z, Biyik H, Yuksek N (2000) Zygomycotic proventriculitis and ventriculitis in ostriches (Struthio camelus) with impaction. Aust Vet J 78:247–249. https://doi.org/10.1111/J.1751-0813.2000.TB11744.X

Guillot J, Collobert C, Jensen HE et al (2000) Two cases of equine mucormycosis caused by Absidia corymbifera. Equine Vet J 32:453–456. https://doi.org/10.2746/042516400777591039

Han BZ, Rombouts FM, Nout MJR (2001) A Chinese fermented soybean food. Int J Food Microbiol 65:1–10. https://doi.org/10.1016/S0168-1605(00)00523-7

Hassan MIA, Voigt K (2019) Pathogenicity patterns of mucormycosis: epidemiology, interaction with immune cells and virulence factors. Med Mycol 57:S245–S256. https://doi.org/10.1093/MMY/MYZ011

Heung LJ (2020) Monocytes and the host response to fungal pathogens. Front Cell Infect Microbiol 10:34. https://doi.org/10.3389/FCIMB.2020.00034/BIBTEX

Hibbett DS, Binder M, Bischoff JF et al (2007) A higher-level phylogenetic classification of the Fungi. Mycol Res 111:509–547. https://doi.org/10.1016/J.MYCRES.2007.03.004

Hoenigl M, Seidel D, Carvalho A et al (2022) The emergence of COVID-19 associated mucormycosis: a review of cases from 18 countries. Lancet Microb 3:e543–e552. https://doi.org/10.1016/S2666-5247(21)00237-8

Hoffmann K, Pawłowska J, Walther G et al (2013) The family structure of the Mucorales: a synoptic revision based on comprehensive multigene-genealogies. Persoonia 30:57. https://doi.org/10.3767/003158513X666259

Hosseini SMS, Borghei P (2005) Rhinocerebral mucormycosis: pathways of spread. Eur Arch Otorrinolaringol 262:932–938. https://doi.org/10.1007/S00405-005-0919-0/TABLES/1

Howard DH (1999) Acquisition, transport, and storage of iron by pathogenic fungi. Clin Microbiol Rev 12:394–404. https://doi.org/10.1128/CMR.12.3.394/ASSET/251D5669-2C79-49D8-9144-147766F08F87/ASSETS/GRAPHIC/CM0390020002.JPEG
Huggins JL, Garner MM, Raverty SA et al (2020) The emergence of mucormycosis in free-ranging marine mammals of the Pacific Northwest. Front Mar Sci 7:555. https://doi.org/10.3389/FMARS.2020.00555/BIBTEX
Ibrahim AS, Kontoyiannis DP (2013) Update on mucormycosis pathogenesis. Curr Opin Infect Dis 26:508–515. https://doi.org/10.1097/QCO.0000000000000008
Ibrahim AS, Gebremariam T, Lin L et al (2010) The high affinity iron permease is a key virulence factor required for Rhizopus oryzae pathogenesis. Mol Microbiol 77:587–604. https://doi.org/10.1111/J.1365-2958.2010.07234.X
Ibrahim AS, Spellberg B, Walsh TJ, Kontoyiannis DP (2012) Pathogenesis of Mucormycosis. Clin Infect Dis 54:S16–S22. https://doi.org/10.1093/CID/CIR865
Index Fungorum (2022) Mucoromycotina. http://www.indexfungorum.org/names/NamesRecord.asp?RecordID=90756. Accessed 31 Jul 2022
Ingold CT, Zoberi MH (1963) The asexual apparatus of mucorales in relation to spore liberation. Trans Br Mycol Soc 46:115–134, IN10–IN11. doi:https://doi.org/10.1016/S0007-1536(63)80014-5
Jacobsen ID (2019) Animal models to study mucormycosis. J Fungi 5. https://doi.org/10.3390/JOF5020027
Jensen HE, Espinosa de los Monteros A, Carrasco L (1996) Caprine mastitis due to aspergillosis and zygomycosis: a pathological and immunohistochemical study. J Comp Pathol 114:183–191. https://doi.org/10.1016/S0021-9975(96)80007-7
Jeong W, Keighley C, Wolfe R et al (2019) Systematic review The epidemiology and clinical manifestations of mucormycosis: a systematic review and meta-analysis of case reports. Clin Microbiol Infect 25:26. https://doi.org/10.1016/j.cmi.2018.07.011
Karimi K, Zamani A (2013) Mucor indicus: biology and industrial application perspectives: a review. Biotechnol Adv 31:466–481. https://doi.org/10.1016/J.BIOTECHADV.2013.01.009
Karimi K, Brandberg T, Edebo L, Taherzadeh MJ (2005) Fed-batch cultivation of Mucor indicus in dilute-acid lignocellulosic hydrolyzate for ethanol production. Biotechnol Lett 27:1395–1400. https://doi.org/10.1007/S10529-005-0688-2
Kobayashi SD, DeLeo FR (2009) Role of neutrophils in innate immunity: a systems biology-level approach. Wiley Interdiscip Rev Syst Biol Med 1:309–333. https://doi.org/10.1002/WSBM.32
Kordy FN, Al-Mohsen IZ, Hashem F et al (2004) Successful treatment of a child with posttraumatic necrotizing fasciitis caused by Apophysomyces elegans: case report and review of literature. Pediatr Infect Dis J 23:877–879. https://doi.org/10.1097/01.INF.0000136870.17071.FD
Kraibooj K, Park HR, Dahse HM et al (2014) Virulent strain of Lichtheimia corymbifera shows increased phagocytosis by macrophages as revealed by automated microscopy image analysis. Mycoses 57:56–66. https://doi.org/10.1111/MYC.12237
Kumar BV, Connors TJ, Farber DL (2018) Human T cell development, localization, and function throughout life. Immunity 48:202–213. https://doi.org/10.1016/J.IMMUNI.2018.01.007
Lambrecht BN, Prins J-B, Hoogsteden HC (2001) Lung dendritic cells and host immunity to infection. Eur Respir J 18:962–704
Lax C, Pérez-arques C, Navarro-mendoza MI et al (2020) Genes, pathways, and mechanisms involved in the virulence of mucorales. Genes (Basel) 11. https://doi.org/10.3390/GENES11030317
Lee SC, Li A, Calo S, Heitman J (2013) Calcineurin plays key roles in the dimorphic transition and virulence of the human pathogenic zygomycete Mucor circinelloides. PLoS Pathog 9:e1003625. https://doi.org/10.1371/JOURNAL.PPAT.1003625
Lelievre L, Garcia-Hermoso D, Abdoul H et al (2014) Posttraumatic mucormycosis: a nationwide study in France and review of the literature. Medicine (United States) 93:395–404. https://doi.org/10.1097/MD.0000000000000221
Lennartsson PR, Niklasson C, Taherzadeh MJ (2011) A pilot study on lignocelluloses to ethanol and fish feed using NMMO pretreatment and cultivation with zygomycetes in an air-lift reactor. Bioresour Technol 102:4425–4432. https://doi.org/10.1016/J.BIORTECH.2010.12.089
Levitz SM, Selsted ME, Ganz T et al (1986) In vitro killing of spores and hyphae of Aspergillus fumigatus and Rhizopus oryzae by rabbit neutrophil cationic peptides and bronchoalveolar macrophages. J Infect Dis 154:483–489. https://doi.org/10.1093/INFDIS/154.3.483
Liles WC, Huang JE, Van Burik JAH et al (1997) Granulocyte colony-stimulating factor administered in vivo augments neutrophil-mediated activity against opportunistic fungal pathogens. J Infect Dis 175:1012–1015. https://doi.org/10.1086/513961
Liu M, Spellberg B, Phan QT et al (2010) The endothelial cell receptor GRP78 is required for mucormycosis pathogenesis in diabetic mice. J Clin Invest 120:1914–1924. https://doi.org/10.1172/JCI42164
Londoño-Hernández L, Ramírez-Toro C, Ruiz HA et al (2017) Rhizopus oryzae – Ancient microbial resource with importance in modern food industry. Int J Food Microbiol 257:110–127. https://doi.org/10.1016/J.IJFOODMICRO.2017.06.012
López-Fernández L, Sanchis M, Navarro-Rodríguez P et al (2018) Understanding Mucor circinelloides pathogenesis by comparative genomics and phenotypical studies. Virulence 9:707–720. https://doi.org/10.1080/21505594.2018.1435249

López-Muñoz A, Nicolás FE, García-Moreno D et al (2018) An adult zebrafish model reveals that mucormycosis induces apoptosis of infected macrophages. Sci Rep 8. https://doi.org/10.1038/S41598-018-30754-6

López-Romero T, Ramírez-Bravo OE, Camargo-Rivera EE et al (2021) Detection of mucormycosis caused by Apophysomyces elegans in a Lesser Long-nosed Bat (Leptonycteris yerbabuenae) in Central Mexico. Neotropical Biology and Conservation 16(4):493–499

Ma LJ, Ibrahim AS, Skory C et al (2009) Genomic analysis of the basal lineage fungus Rhizopus oryzae reveals a whole-genome duplication. PLoS Genet 5:e1000549. https://doi.org/10.1371/JOURNAL.PGEN.1000549

McCaskey PC, Langheinrich KA (1984) Zygomycosis in the duck. Avian Dis:791–798

Michalak DM, Cooney DR, Rhodes KH et al (1980) Gastrointestinal mucormycoses in infants and children: a cause of gangrenous intestinal cellulitis and perforation. J Pediatr Surg 15:320–324. https://doi.org/10.1016/S0022-3468(80)80147-3

Millon L, Herbrecht R, Grenouillet F et al (2016) Early diagnosis and monitoring of mucormycosis by detection of circulating DNA in serum: retrospective analysis of 44 cases collected through the French Surveillance Network of Invasive Fungal Infections (RESSIF). Clin Microbiol Infect 22:810.e1–810.e8. https://doi.org/10.1016/J.CMI.2015.12.006

Mitchell SJ, Gray J, Morgan MEI et al (1996) Nosocomial infection with Rhizopus microsporus in preterm infants: association with wooden tongue depressors. Lancet 348:441–443. https://doi.org/10.1016/S0140-6736(96)05059-3

Monod M, Capoccia S, Léchenne B et al (2002) Secreted proteases from pathogenic fungi. Int J Med Microbiol 292:405–419. https://doi.org/10.1078/1438-4221-00223

Montaño DE, Voigt K (2020) Host Immune Defense upon Fungal Infections with Mucorales: Pathogen-Immune Cell Interactions as Drivers of Inflammatory Responses. Journal of Fungi 2020, Vol 6, Page 173 6:173. https://doi.org/10.3390/JOF6030173

Nakagun S, Okazaki M, Toyotome T et al (2018) Fatal Pulmonary and Cerebellar Zygomycosis due to Rhizomucor pusillus in a Ringed Seal (Pusa hispida). Mycopathologia 183:979–985. https://doi.org/10.1007/S11046-018-0268-3

Nicolas FE, Moxon S, de Haro JP et al (2010) Endogenous short RNAs generated by Dicer 2 and RNA-dependent RNA polymerase 1 regulate mRNAs in the basal fungus Mucor circinelloides. Nucleic Acids Res 38:5535–5541. https://doi.org/10.1093/NAR/GKQ301

Nicolás FE, Murcia L, Navarro E et al (2020) Mucorales species and macrophages. J Fungi 6:94. https://doi.org/10.3390/JOF6020094

Nosanchuk JD, Casadevall A (2006) Impact of melanin on microbial virulence and clinical resistance to antimicrobial compounds. Antimicrob Agents Chemother 50:3519–3528. https://doi.org/10.1128/AAC.00545-06/ASSET/695D039E-FE28-4D88-B04F-B82FAED56A1A/ASSETS/GRAPHIC/ZAC0110660880002.JPEG

Novilla MN, Flyger V, Jacobson ER et al (1981) Systemic phycomycosis and multiple fibromas in a gray squirrel (Sciurus carolinensis). J Wildl Dis 17:89–95. https://doi.org/10.7589/0090-3558-17.1.89

Obendorf DL, Peel BF, Munday BL (1993) Mucor amphibiorum Infection in Platypus (Ornithorhynchus anatinus) from Tasmania. J Wildl Dis 29:485–487. https://doi.org/10.7589/0090-3558-29.3.485

Orguc S, Yücetürk AV, Demir MA, Goktan C (2005) Rhinocerebral mucormycosis: Perineural spread via the trigeminal nerve. J Clin Neurosci 12:484–486. https://doi.org/10.1016/J.JOCN.2004.07.015

Paes De Oliveira-Neto M, Da Silva M, Cezar Fialho Monteiro P et al (2006) Cutaneous mucormycosis in a young, immunocompetent girl. Med Mycol 44:567–570. https://doi.org/10.1080/13693780600622411

Pagano L, Ricci P, Tonso A et al (1997) Mucormycosis in patients with haematological malignancies: a retrospective clinical study of 37 cases. Br J Haematol 99:331–336. https://doi.org/10.1046/J.1365-2141.1997.3983214.X

Page L, Weis P, Müller T et al (2018) Evaluation of Aspergillus and Mucorales specific T-cells and peripheral blood mononuclear cell cytokine signatures as biomarkers of environmental mold exposure. Int J Med Microbiol 308:1018–1026. https://doi.org/10.1016/J.IJMM.2018.09.002

Pal R, Singh B, Bhadada SK et al (2021) COVID-19-associated mucormycosis: an updated systematic review of literature. Mycoses 64:1452–1459. https://doi.org/10.1111/MYC.13338

Paltauf A (1885) Mycosis mucorina. Archiv für pathologische Anatomie und Physiologie und für klinische Medicin 102(3):543–564. https://doi.org/10.1007/BF01932420

Pana ZD, Farmaki E, Roilides E (2014) Host genetics and opportunistic fungal infections. Clin Microbiol Infect 20:1254–1264. https://doi.org/10.1111/1469-0691.12800

Parapouli M, Vasileiadis A, Afendra A-S, Hatziloukas E (2020) Saccharomyces cerevisiae and its industrial applications. AIMS Microbiol 6:1–31. https://doi.org/10.3934/microbiol.2020001

Park M, Do E, Jung WH (2013) Lipolytic enzymes involved in the virulence of human pathogenic fungi. Mycobiology 41:67. https://doi.org/10.5941/MYCO.2013.41.2.67

Passamonte PM, Dix JD (1985) Nosocomial pulmonary mucormycosis with fatal massive hemoptysis. Am J Med Sci 289:65–67. https://doi.org/10.1097/00000441-198502000-00005

Patel A, Kaur H, Xess I et al (2020) A multicentre observational study on the epidemiology, risk factors, management and outcomes of mucormycosis in India. Clin Microbiol Infect 26:944.e9–944.e15. https://doi.org/10.1016/J.CMI.2019.11.021/

ATTACHMENT/6050CEF9-56E4-43D3-8053-2A746BE27604/MMC3.DOCX

Patil A, Mohanty HS, Kumar S et al (2016) Angioinvasive rhinocerebral mucormycosis with complete unilateral thrombosis of internal carotid artery—case report and review of literature. BJR Case Rep 2:20150448. https://doi.org/10.1259/BJRCR.20150448

Patiño-Medina JA, Maldonado-Herrera G, Pérez-Arques C et al (2018) Control of morphology and virulence by ADP-ribosylation factors (Arf) in Mucor circinelloides. Curr Genet 64:853–869. https://doi.org/10.1007/S00294-017-0798-0

Peters M, Eikelberg D, Jongmans V et al (2014) Rhinoorbitocerebral zygomycosis caused by Rhizopus microsporus in a roe deer (Capreolus capreolus). Berl Munch Tierarztl Wochenschr 127:301–304

Petrikkos G, Skiada A, Lortholary O et al (2012) Epidemiology and clinical manifestations of mucormycosis. Clin Infect Dis 54:S23–S34. https://doi.org/10.1093/CID/CIR866

Polacheck I, Hearing VJ, Kwon-Chung KJ (1982) Biochemical studies of phenoloxidase and utilization of catecholamines in Cryptococcus neoformans. J Bacteriol 150:1212–1220. https://doi.org/10.1128/JB.150.3.1212-1220.1982

Potenza L, Vallerini D, Barozzi P et al (2011) Mucorales-specific T cells emerge in the course of invasive mucormycosis and may be used as a surrogate diagnostic marker in high-risk patients. Blood 118:5416–5419. https://doi.org/10.1182/BLOOD-2011-07-366526

Potenza L, Vallerini D, Barozzi P et al (2016) Mucorales-specific T cells in patients with hematologic malignancies. PLoS One 11:e0149108. https://doi.org/10.1371/JOURNAL.PONE.0149108

Prakash H, Chakrabarti A (2019) Global epidemiology of mucormycosis. J Fungi 5(1):26. https://doi.org/10.3390/jof5010026

Prakash H, Chakrabarti A (2021) Epidemiology of mucormycosis in India. Microorganisms 9:1–12. https://doi.org/10.3390/MICROORGANISMS9030523

Quesada Ó, Rodríguez F, Herráez P, et al (2007) Mucor ramosissimus Associated with Feather Loss in Canaries (Serinus canarius). 101637/0005-2086(2007)51[643:MRAWFL]20CO;2 51:643–645. https://doi.org/10.1637/0005-2086(2007)51

Ram R, Swarnalatha G, Prasad N, Dakshinamurty KV (2007) Exit site infection due to Zygomycosis resulting in abdominal wall necrosis in a continuous ambulatory peritoneal dialysis patient. Nephrol Dial Transplant 22:266–267. https://doi.org/10.1093/NDT/GFL535

Rees JR, Pinner RW, Hajjeh RA et al (1998) The epidemiological features of invasive mycotic infections in the San Francisco Bay Area, 1992–1993: results of population-based laboratory active surveillance. Clin Infect Dis 27:1138–1147

Rehman MEU, Saeed S, Franco D et al (2022) Mucormycosis in hematopoetic stem cell transplant recipients: A systematic review 40:e19043. https://doi.org/10.1200/JCO.2022.40.16_SUPPL.E19043

Reynaldi FJ, Giacoboni G, Córdoba SB et al (2017) Mucormicosys due to Saksenaea vasiformis in a dog. Med Mycol Case Rep 16:4–7

Ribes JA, Vanover-Sams CL, Baker DJ (2000) Zygomycetes in human disease. Clin Microbiol Rev 13:236–301. https://doi.org/10.1128/CMR.13.2.236

Richardson M (2009) The ecology of the Zygomycetes and its impact on environmental exposure. Clin Microbiol Infect 15:2–9. https://doi.org/10.1111/J.1469-0691.2009.02972.X

Richardson JP, Brown R, Kichik N et al (2022) Candidalysins are a new family of cytolytic fungal peptide toxins. MBio 13. https://doi.org/10.1128/MBIO.03510-21

Rigby C (1976) Natural infections of guinea-pigs. Lab Anim 10:119–142. https://doi.org/10.1258/002367776781071503/ASSET/002367776781071503.FP.PNG_V03

Roden MM, Zaoutis TE, Buchanan WL et al (2005) Epidemiology and outcome of zygomycosis: a review of 929 reported cases. Clin Infect Dis 41:634–653. https://doi.org/10.1086/432579

Roilides E, Antachopoulos C, Simitsopoulou M (2014) Pathogenesis and host defence against Mucorales: the role of cytokines and interaction with antifungal drugs. Mycoses 57:40–47. https://doi.org/10.1111/MYC.12236

Ross R, Glomset J, Kariya B, Harker L (1974) A platelet dependent serum factor that stimulates the proliferation of arterial smooth muscle cells in vitro. Proc Natl Acad Sci USA 71:1207–1210. https://doi.org/10.1073/PNAS.71.4.1207

Rubin AI, Grossman ME (2004) Bull's-eye cutaneous infarct of zygomycosis: a bedside diagnosis confirmed by touch preparation. J Am Acad Dermatol 51:996–1001. https://doi.org/10.1016/J.JAAD.2004.07.027

Satari B, Karimi K, Zamani A (2016) Oil, chitosan, and ethanol production by dimorphic fungus Mucor indicus from different lignocelluloses. J Chem Technol Biotechnol 91:1835–1843. https://doi.org/10.1002/JCTB.4776

Schmidt S, Tramsen L, Perkhofer S et al (2012) Characterization of the cellular immune responses to Rhizopus oryzae with potential impact on immunotherapeutic strategies in hematopoietic stem cell transplantation. J Infect Dis 206:135–139. https://doi.org/10.1093/INFDIS/JIS308

Schmidt S, Schneider A, Demir A et al (2016) Natural killer cell-mediated damage of clinical isolates of mucormycetes. Mycoses 59:34–38. https://doi.org/10.1111/MYC.12431

Schmitz A, Korbel R, Rinder M (2018) Zygomycosis due to Mucor racemosus Associated with a Circovirus Infection in a Blue Tit (Cyanistes caeruleus). J Comp Pathol 165:67–71. https://doi.org/10.1016/J.JCPA.2018.09.011

Shelton BG, Kirkland KH, Dana Flanders W, Morris GK (2002) Profiles of airborne fungi in buildings and outdoor environments in the United States. Appl Environ Microbiol 68:1743–1753. https://doi.org/10.1128/AEM.68.4.1743-1753.2002

Singh N, Aguado JM, Bonatti H et al (2009) Zygomycosis in solid organ transplant recipients: a prospective, matched case-control study to assess risks for disease and outcome. J Infect Dis 200:1002–1011. https://doi.org/10.1086/605445

Skaria J, John TM, Varkey S, Kontoyiannis DP (2022) Are unique regional factors the missing link in India's COVID- 19-associated mucormycosis crisis? mBio 13. https://doi.org/10.1128/MBIO.00473-22/ASSET/2EE25D57-1461-4851-9F23-285F0F99A774/ASSETS/IMAGES/MEDIUM/MBIO.00473-22-F001.GIF

Skiada A, Pagano L, Groll A et al (2011) Zygomycosis in Europe: analysis of 230 cases accrued by the registry of the European Confederation of Medical Mycology (ECMM) Working Group on Zygomycosis between 2005 and 2007. Clin Microbiol Infect 17:1859–1867. https://doi.org/10.1111/J.1469-0691.2010.03456.X

Skiada A, Lass-Floerl C, Klimko N et al (2018) Challenges in the diagnosis and treatment of mucormycosis. Med Mycol 56. https://doi.org/10.1093/mmy/myx101

Soliman SSM, Hamdy R, Elseginy SA, Gebremariam T, Hamoda AM, Madkour M, Venkatachalam T, Ershaid MN, Mohammad MG, Chamilos G, Ibrahim AS (2020) Selective inhibition of *Rhizopus* eumelanin biosynthesis by novel natural product scaffold-based designs caused significant inhibition of fungal pathogenesis. Biochem J 477(13):2489–2507. https://doi.org/10.1042/BCJ20200310

Soliman SSM, Baldin C, Gu Y et al (2021) Mucoricin is a ricin-like toxin that is critical for the pathogenesis of mucormycosis. Nat Microbiol 6:313. https://doi.org/10.1038/S41564-020-00837-0

Spatafora JW, Chang Y, Benny GL et al (2016) A phylum-level phylogenetic classification of zygomycete fungi based on genome-scale data. Mycologia 108:1028. https://doi.org/10.3852/16-042

Spatafora JW, Chang Y, Benny GL, et al (2017) A phylum-level phylogenetic classification of zygomycete fungi based on genome-scale data. 108:1028–1046. https://doi.org/10.3852/16-042

Speare R, Berger L, O'Shea P et al (1997) Pathology of mucormycosis of cane toads in Australia. J Wildl Dis 33:105–111. https://doi.org/10.7589/0090-3558-33.1.105

Spellberg B (2012) Gastrointestinal mucormycosis: an evolving disease. Gastroenterol Hepatol (N Y) 8:140

Spellberg B, Edwards J, Ibrahim A (2005) Novel perspectives on mucormycosis: Pathophysiology, presentation, and management. Clin Microbiol Rev 18:556–569. https://doi.org/10.1128/CMR.18.3.556-569.2005/ASSET/80400028-3FD6-4594-B2F6-40DA534ADE4E/ASSETS/GRAPHIC/ZCM0030521440001.JPEG

Spreer A, Rüchel R, Reichard U (2006) Characterization of an extracellular subtilisin protease of Rhizopus microsporus and evidence for its expression during invasive rhinoorbital mycosis. Med Mycol 44:723–731. https://doi.org/10.1080/13693780600936399

Stanford FA, Voigt K (2020) Iron assimilation during emerging infections caused by opportunistic fungi with emphasis on mucorales and the development of antifungal resistance. Genes (Basel) 11:1–30. https://doi.org/10.3390/GENES11111296

Stanford FA, Matthies N, Cseresnyés Z et al (2021) Expression patterns in reductive iron assimilation and functional consequences during phagocytosis of Lichtheimia corymbifera, an emerging cause of mucormycosis. J Fungi (Basel) 7. https://doi.org/10.3390/JOF7040272

Steinlage SJT, Sander JE, Brown TP, et al (2003) Disseminated Mycosis in Layer Cockerels and Pullets. https://doi.org/101637/0005-2086(2003)047[0229:DMILCA]20CO;2 47:229–233. https://doi.org/10.1637/0005-2086(2003)047

Stenkat J, Krautwald-Junghanns ME, Schmidt V (2013) Causes of morbidity and mortality in free-living birds in an Urban environment in Germany. EcoHealth 10:352–365. https://doi.org/10.1007/S10393-013-0868-9/TABLES/5

Sun HY, Aguado JM, Bonatti H et al (2009) Pulmonary zygomycosis in solid organ transplant recipients in the current era. Am J Transplant 9:2166–2171. https://doi.org/10.1111/J.1600-6143.2009.02754.X

Sun HY, Forrest G, Gupta KL et al (2010) Rhino-orbital-cerebral zygomycosis in solid organ transplant recipients. Transplantation 90:85–92. https://doi.org/10.1097/TP.0B013E3181DDE8FC

Takahashi T, Asai T (1988) The products of fermentation by the Mucor group. Zentr Bakt Parasitenk, pp 276–284

Thanh NV, Rombouts FM, Nout MJR (2005) Effect of individual amino acids and glucose on activation and germination of Rhizopus oligosporus sporangiospores in tempe starter. J Appl Microbiol 99:1204–1214. https://doi.org/10.1111/J.1365-2672.2005.02692.X

Therakathu J, Prabhu S, Irodi A et al (2018) Imaging features of rhinocerebral mucormycosis: a study of 43 patients. Egypt J Radiol Nucl Med 49:447–452. https://doi.org/10.1016/J.EJRNM.2018.01.001

Thieken A, Winkelmann G (1992a) Rhizoferrin: a complexone type siderophore of the mocorales and entomophthorales (Zygomycetes). FEMS Microbiol Lett 94:37–41. https://doi.org/10.1111/J.1574-6968.1992.TB05285.X

Thieken A, Winkelmann G (1992b) Rhizoferrin: a complexone type siderophore of the Mucorales and entomophthorales (Zygomycetes). FEMS Microbiol Lett 73:37–41. https://doi.org/10.1016/0378-1097(92)90579-D

Thomas M, Varghese A, Samuel KA, Kurian P (2008) Zygomycosis in captive Asian Elephant (Elephas maximus). Indian J Med Microbiol 26:269–271

Torres-Narbona M, Guinea J, Martínez-Alarcón J et al (2007) Impact of zygomycosis on microbiology workload: a survey study in Spain. J Clin Microbiol 45:2051–2053. https://doi.org/10.1128/JCM.02473-06

Torres-Urbano CJ, Rose RE, Walters SL (2010) Disseminated Zygomycosis in a Cynomolgus Monkey (Macaca fascicularis). J Am Assoc Lab Anim Sci 49:75

Urban CF, Nett JE (2019) Neutrophil extracellular traps in fungal infection. Semin Cell Dev Biol 89:47–57. https://doi.org/10.1016/J.SEMCDB.2018.03.020

Uribe-Quero E, Rosales C (2017) Control of phagocytosis by microbial pathogens. Front Immunol 8:1368. https://doi.org/10.3389/FIMMU.2017.01368/BIBTEX

Virmani R, Connor DH, McAllister HA (1982) Cardiac mucormycosis. A report of five patients and review of 14 previously reported cases. Am J Clin Pathol 78:42–47. https://doi.org/10.1093/AJCP/78.1.42

Vítovec J, Vladík P, Proks C, Fragner P (1975) Mucormycosis (Mucor pusillus) with asteroids in a young bull. Acta Vet Acad Sci Hung 25:31–35

Vivier E, Tomasello E, Baratin M et al (2008) Functions of natural killer cells. Nat Immunol 9:503–510. https://doi.org/10.1038/ni1582

Voigt K (2012) Zygomycota. In: Frey W (ed) Syllabus of plant families – A. Engler's Syllabus der Pflanzenfamilien. Part 1/1: Blue-green algae, Myxomycetes and Myxomycete-like organisms, Phytoparasitic protists, Heterotrophic Heterokontobionta and Fungi. Borntraeger Verlag, pp 130–162

Waldorf AR, Diamond RD (1985) Neutrophil chemotactic responses induced by fresh and swollen Rhizopus oryzae spores and Aspergillus fumigatus conidia. Infect Immun 48:458–463. https://doi.org/10.1128/IAI.48.2.458-463.1985

Waldorf AR, Ruderman N, Diamond RD (1984) Specific susceptibility to mucormycosis in murine diabetes and bronchoalveolar macrophage defense against Rhizopus. J Clin Invest 74:150–160. https://doi.org/10.1172/JCI111395

Walsh TJ, Gamaletsou MN (2013) Treatment of fungal disease in the setting of neutropenia. Hematology 2013:423–427. https://doi.org/10.1182/ASHEDUCATION-2013.1.423

Walther G, Wagner L, Kurzai O (2019) Updates on the taxonomy of mucorales with an emphasis on clinically important taxa. J Fungi 5:106. https://doi.org/10.3390/JOF5040106

Wang XM, Guo LC, Xue SL, Bin CY (2016) Pulmonary mucormycosis: a case report and review of the literature. Oncol Lett 11:3049. https://doi.org/10.3892/OL.2016.4370

Wilson W, Ali-Osman F, Sucher J et al (2019) Invasive fungal wound infection in an otherwise healthy trauma patient (Mucor Trauma). Trauma Case Rep 24:100251. https://doi.org/10.1016/J.TCR.2019.100251

Wray JD, Sparkes AH, Johnson EM (2008) Infection of the subcutis of the nose in a cat caused by Mucor species: successful treatment using posaconazole. J Feline Med Surg 10:523–527. https://doi.org/10.1016/J.JFMS.2008.06.001

Wurster S, Thielen V, Weis P et al (2017) Mucorales spores induce a proinflammatory cytokine response in human mononuclear phagocytes and harbor no rodlet hydrophobins. Virulence 8:1708. https://doi.org/10.1080/21505594.2017.1342920

Yona S, Kim KW, Wolf Y et al (2013) Fate mapping reveals origins and dynamics of monocytes and tissue macrophages under homeostasis. Immunity 38:79–91. https://doi.org/10.1016/J.IMMUNI.2012.12.001/ATTACHMENT/537C07DF-073D-4D07-A933-CF3660CE1793/MMC1.PDF

Pathogenicity Strategies of *Candida* Species During Interactions with Epithelial Cells

2

Marina Pekmezovic, Bernhard Hube, and Mark S. Gresnigt

Abstract

Epithelial cells present an interface between potentially harmful microorganisms on the one side and innate immune cells on the other side. Therefore, their role is to tolerate commensal microbes and initiate host defense upon infection. The commensal lifestyle of *Candida* species on epithelial surfaces is finely conditioned by the bacterial microbiota and the immune system, but if these protective mechanisms are compromised, *Candida* spp. can undergo a commensal-to-pathogen shift and cause infection. In that case, fungal proliferation on the mucosal surfaces of the gut, oral cavity, or vagina, leads to epithelial damage and the initiation of pro-inflammatory immune responses. *C. albicans, C. tropicalis, C. glabrata,* and *C. parapsilosis* are responsible for more than 90% of *Candida* infections, but their pathogenicity strategies are remarkably different. Furthermore, each mucosal niche has unique properties, where both host and *Candida* spp. have to sense, adapt, and respond to the dynamics of specific microenvironments, depending on the nature of their host-pathogen interaction.

Keywords

Candida · Host-pathogen interaction · Epithelial cells · Mucosa · Pathogenicity strategies · Oral cavity · Vaginal mucosa · Intestinal tract · Commensalism

2.1 Introduction

The skin and mucosal surfaces are the main interfaces between the host and its external environment. Therefore, epithelial cells are an immunologically responsive physical barrier between potentially harmful microorganisms and the innate immune system. In the majority of the human population, *Candida* species are constituents of the normal mucosal microbiota (Moyes and Naglik 2011; Naglik and Moyes 2011; Hallen-Adams and Suhr 2017; Nash et al. 2017). The commensalism of *Candida* species is maintained by the bacterial microbiota and the host immune system (Moyes and Naglik 2011; Naglik and Moyes 2011). Consequently, when protective

M. Pekmezovic · M. S. Gresnigt (✉)
Junior Research Group Adaptive Pathogenicity Strategies, Leibniz Institute for Natural Product Research and Infection Biology – Hans Knöll Institute (HKI), Jena, Germany
e-mail: Mark.Gresnigt@leibniz-hki.de

B. Hube
Department of Microbial Pathogenicity Mechanisms, Leibniz Institute for Natural Product Research and Infection Biology – Hans Knöll Institute (HKI), Jena, Germany

Institute of Microbiology, Friedrich-Schiller University, Jena, Germany

A. A. Brakhage et al. (eds.), *Human and Animal Relationships*, The Mycota 6,
https://doi.org/10.1007/978-3-031-64853-3_2

bacterial species are removed by antibiotic therapy and/or the epithelial-immune barrier gets compromised, colonizing *Candida* spp. can undergo a commensal-to-pathogen shift and cause an infection. In this case, *Candida* spp. can proliferate on mucosal surfaces of the gut, oral cavity, or vagina, potentially causing epithelial damage and initiating inflammatory responses that are not necessarily beneficial to clear the infection (Naglik and Moyes 2011; Richardson et al. 2018b). This clearly points toward a role of epithelial cells in shaping innate immune response toward tolerance (commensal state) or initiation of host defense (upon the infection). In this chapter, we will introduce the four most common *Candida* species causing human infections and discuss the epithelium-*Candida* interactions in different mucosal niches, during commensalism and infection.

2.2 Pathogenicity Strategies of *Candida* Species

The majority of *Candida* infections are caused by *Candida albicans,* followed by *C. glabrata* (*Nakaseomyces glabratus*), *C. parapsilosis,* and *C. tropicalis.* Despite belonging to the same genus, these species are phylogenetically remarkably diverse, with *C. albicans, C. parapsilosis,* and *C. tropicalis* being diploid fungi belonging to the CTG clade and *C. glabrata* being haploid and is closer associated to the bakers yeast *Saccharomyces cerevisiae.* These different *Candida* species behave differently during infection (Gabaldon et al. 2016; Pekmezovic et al. 2021a). Moreover, their adaptive commensal and pathogenic strategies are different, indicating that these strategies have evolved independently (Gabaldon et al. 2016).

2.2.1 *Candida albicans*

C. albicans is the most common cause of *Candida* infections. Therefore, it is the best-studied organism of the genus and the *Candida* literature is mostly focused on this species. The *C. albicans* genome is diploid with 6419 mapped open reading frames (ORFs) (Jones et al. 2004). Several gene families encoding virulence-associated proteins have been described. This includes the *ALS* (agglutinin-like sequence) gene family, required for the adhesion to host surfaces (Hoyer 2001), and the secreted aspartyl protease (*SAP*), lipase (*LIP*), and phospholipase B (*PLB*) gene families, which can hydrolyze surrounding proteins, lipids, or phospholipids (Ghannoum 2000; Calderone and Fonzi 2001). Some of these activities can contribute to degradation of host molecules and support invasion of surrounding tissue (Ghannoum 2000; Calderone and Fonzi 2001; Hoyer 2001). A crucial virulence attribute is the morphological transition from yeast to hyphal growth, which can be initiated by various conditions present in the host environment (Sudbery 2011; Jacobsen et al. 2012; Henry et al. 2021). Moreover, several virulence factors are exclusively associated with the hyphal morphology, such as the toxin candidalysin (Moyes et al. 2016), specific proteases such as Sap4-6 (Naglik et al. 2004), and the oxidative stress resistance mediator superoxide dismutase 5 (Sod5) (Martchenko et al. 2004). Finally, *C. albicans* can form biofilms, microbial communities embedded in a structural matrix made of extracellular polymers. Such biofilms are highly resistant to antifungal drugs, immune recognition and clearance, and enable pathogen dissemination (Potera 1999; Cavalheiro and Teixeira 2018). Biofilms can be formed on all mucosal surfaces, but also on abiotic surfaces, such as medical devices, further highlighting the importance of biofilms in clinical settings (Auler et al. 2010; Cavalheiro and Teixeira 2018).

2.2.2 *C. tropicalis*

C. tropicalis is a species closely related to *C. albicans* and considered to be the second most virulent *Candida* species. Specifically in Latin-American countries, but also in Asia *C. tropicalis* is the second or even most frequent species causing candidemia (Nucci and Colombo 2007; Adhikary and Joshi 2011; Rajni et al. 2022). *C.*

tropicalis can form hyphae and biofilms, and expresses a range of virulence factors similar to *C. albicans*, including Als adhesins (Als1-3, Hwp1), Sap proteases (Sapt1-4), and phospholipases (Plb1, Plc1) (reviewed in: Zuza-Alves et al. (2017)). Interestingly, *C. tropicalis* has an *ECE1* gene encoding a candidalysin-like peptide, but the gene expression is low both *in vitro* and *in vivo* and it is unknown whether it plays any role during infection (Willems et al. 2018; Pekmezovic et al. 2021a; Richardson et al. 2022). *C. tropicalis* contains many hydrophobic proteins in the cell wall that can facilitate adherence to plastic, yet also host proteins such as laminin, fibrinogen, and fibronectin (Tronchin et al. 2008). Therefore, *C. tropicalis* isolates exhibit the highest biofilm-forming ability compared to different *Candida* species obtained from patients with systemic candidiasis (Marcos-Zambrano et al. 2014; Atiencia-Carrera et al. 2022). Finally, several studies have reported high levels of *C. tropicalis* resistance to antifungal drugs, specifically azoles (Marie and White 2009; Morschhäuser 2010; Eddouzi et al. 2013; Forastiero et al. 2013; Vincent et al. 2013; Arastehfar et al. 2020).

2.2.3 *C. glabrata* (*Nakaseomyces glabratus*)

C. glabrata is widely reported as the second most common cause of *Candida* infections in many studies (Vijaya et al. 2014; Reyes-Montes et al. 2017; Sharifynia et al. 2017; Makanjuola et al. 2018). Its haploid genome contains 5294 ORFs, of which just over 10% have a verified function [www.candidagenome.org]. *C. glabrata* is very distant from most other common *Candida* species on the evolutionary scale (Kurtzman 2003), which is supported by both genetic and phenotypic differences compared to *C. albicans* (Brunke and Hube 2013). For example, true hyphal formation, toxin production, or active invasion into host tissues were, so far, not shown for *C. glabrata*, and its exact pathogenicity mechanisms by which it causes infection and tissue damage are largely unknown.

Despite being unable to invade or produce toxins, *C. glabrata* has other important virulence attributes: adhesion, a remarkable stress tolerance, the ability to persist for long periods in hosts, and an intrinsic antifungal resistance (Kumar et al. 2019). Compared to close relatives that are non-pathogenic, *C. glabrata* expresses an increased number of adhesins, which is considered a key evolutionary step toward the ability to infect humans (Gabaldón et al. 2013). Although little is known about their function, there are more than 60 putative adhesins identified, classified into several families: epithelial adhesins (Epa), PA14 domain-containing cell wall proteins (Pwp), adherence-to-endothelial cell proteins (Aed), and adhesin-like wall proteins (Awp) (De Groot et al. 2008; Gómez-Molero et al. 2015). Furthermore, analysis of clinical *C. glabrata* isolates showed variations in the number of adhesin-encoding genes among them (De Groot et al. 2008; Salazar et al. 2018; Kumar et al. 2019), and the presence of significant genetic variability (Carreté et al. 2018). Therefore, the adaptative ability of *C. glabrata* to colonize different niches could be based on the presence of diverse adhesin genes and their complex differential regulation (Vale-Silva et al. 2017; Kumar et al. 2019). Although not expressing Saps, *C. glabrata* does have 11 glycosylphosphatidylinositol (GPI)-linked aspartyl proteases (yapsins) on the cell surface, but how they contribute to virulence remains unclear (Kaur et al. 2007; Galocha et al. 2019).

C. glabrata is highly tolerant to a variety of stresses, including oxidative, cell wall, osmotic, and endoplasmic reticulum stress (Nikolaou et al. 2009). This permits not only persistence on mucosal surfaces, but also within the phagosome of immune cells, such as macrophages (Seider et al. 2011). Finally, the reduced susceptibility to azoles and emerging resistance to echinocandins is characteristic for *C. glabrata*, which represents an important clinical problem (Whaley and Rogers 2016; Kumar et al. 2019).

2.2.4 *C. parapsilosis*

C. parapsilosis has dramatically increased in prevalence over the past two decades, especially in neonates, elderly, and intensive care unit patients (Trofa et al. 2008; Lamoth et al. 2018). Prior gastrointestinal colonization with *C. parapsilosis* was found to be associated with its translocation into the bloodstream in recipients of allogeneic hematopoietic cell transplant that received echinocandin prophylaxis (Forrest et al. 2008; Zhai et al. 2020). Interestingly, invasive disease caused by *C. parapsilosis* does not necessarily require prior colonization (Trofa et al. 2008). Apart from being a human commensal, *C. parapsilosis* can also be isolated from the environment, including different surfaces, water, and soil (Brandão et al. 2010). Therefore, contaminated external sources such as medical devices or fluids, the hands of medical staff, prosthetic devices, catheters, and air, represent a risk for horizontal transmission (Sabino et al. 2011). In fact, environmental strains are often more virulent compared to clinical strains (Sabino et al. 2011). Even though the genome of *C. parapsilosis* contains 5837 ORFs, to date, only 112 are verified [www.candidagenome.org], indicating that, despite its clinical importance, this species is highly understudied. Knowledge about virulence traits of *C. parapsilosis* is limited, but the adherence is known to be mediated by two Als-like proteins: CPAR2_404800 and CPAR2_404780. They are important for virulence in animal models (Bertini et al. 2016), but also for binding human fibronectin, vitronectin, and/or laminin (Kozik et al. 2015). *C. parapsilosis* can secrete aspartyl proteases Sapp1 and Sapp2 that influence phagocytosis, killing capacity, and cytokine secretion by human macrophages (Singh et al. 2019). Finally, biofilm formation enables an increased resistance to fluconazole, voriconazole, ravuconazole, amphotericin B, nystatin, chlorhexidine, and terbinafine (Kuhn et al. 2002).

2.3 Commensalism *Versus* Infection at the Epithelial Barriers

The barriers formed by epithelial cells represent a first line of defense against potential pathogens and are essential in sensing and responding to early infection, and subsequent initiation and modulation of innate immune responses (Moyes and Naglik 2011; Naglik and Moyes 2011; Graf et al. 2019). The recognition of different fungal determinants (pathogen-associated molecular patterns; PAMPs) is mediated by pattern recognition receptors (PRRs), primarily Toll-like receptors (TLRs), and C-type lectin receptors (CLRs). Furthermore, exposed β-glucans on *C. albicans* surface can also trigger an epithelial cell response *via* the ephrin type-A receptor 2 (EphA2) (Swidergall et al. 2018). All known PRRs activate mitogen-activated protein kinase (MAPK) and nuclear factor kappa B (NFκB) signaling pathways, although the extent of activation or combination of these pathways, which modulates the generated response, can differ (Naglik and Moyes 2011).

As hyphal formation is in general associated with invasion of *C. albicans*, differential recognition of yeast and hyphae could be a way of discrimination between commensal and pathogenic *C. albicans* cells. Such a differential recognition was shown for various host immune cells (Gantner et al. 2005; van der Graaf et al. 2005; Bi et al. 2010; Gow et al. 2012). For epithelial cells, it is known that the toxin candidalysin, produced exclusively during hyphal growth, can activate epithelial cells *via* the epithelial growth factor receptor (EGFR) (Ho et al. 2019). *C. albicans* hyphae were shown to activate the NLRP3 inflammasome and the subsequent release of pro-inflammatory cytokines *in vitro* and in *in vivo* murine models of oral and vaginal candidiasis (Hise et al. 2009; Cheng et al. 2011; Bruno et al. 2015; Roselletti et al. 2017, 2019). More specifically, several hypha-derived factors, such as Saps and candidalysin, are potent inducers of the immune response, as observed in human vaginal samples (Roselletti et al. 2017) and upon the interaction with mononuclear phagocytes

(Kasper et al. 2018). Therefore, all these virulence factors could be particularly relevant for the ability of epithelial cells to sense the commensal-to-pathogen shift of *C. albicans* (Moyes et al. 2016; d'Enfert et al. 2020). It was shown that MAPK/MKP1/c-Fos activation is a "danger response" pathway in epithelial cells that is dependent on hyphal formation and increased fungal burdens, which is critical for identifying and responding to the commensal-pathogen switch (Moyes et al. 2010). However, the morphology *per se* is probably only partially contributing to this discrimination, as it was shown that *C. albicans* colonizes the intestinal tract as a mixed population of yeast and hyphae (Witchley et al. 2019). Furthermore, hyphae are also prominent in the oral mucosa, even in the absence of an infection (Schonherr et al. 2017; Perez 2019). In addition, PI3K/Akt signaling is not associated with discrimination of *C. albicans* morphology, but it activates processes associated with the protection against damage upon *C. albicans* infection (Moyes et al. 2014).

On the pathogen side, the commensal-to-pathogen shift is still not fully understood, but *C. albicans* has specialized programs optimized for commensalism or pathogenesis in different niches, mostly elucidated on a transcriptional level (Hube 2004; White et al. 2007; Pande et al. 2013; Neville et al. 2015; Noble et al. 2017; Witchley et al. 2019). The knowledge about the "commensal-to-pathogen" shift for other *Candida* species, especially those that do not form hyphae or do not damage *in vitro*, remains limited, yet interactions with antagonistic bacteria in the bacterial microbiota are believed to play a crucial role in driving commensalism (Allonsius et al. 2017; Graf et al. 2019; d'Enfert et al. 2021; MacAlpine et al. 2021; Alonso-Roman et al. 2022).

2.3.1 Stages of *Candida*-Epithelia Interaction: Adhesion, Invasion, Damage

The ability to adhere to biotic and abiotic surfaces is an essential feature for both commensal and pathogenic behavior. *Adhesion* is a complex, dynamic, and multifactorial process that can be unspecific (hydrophobic and electrostatic interactions) or based on specific covalent interactions (Hazen 1989; Richardson et al. 2018a; Pekmezovic et al. 2019). *Candida* species have proteins—adhesins, that bind to different host proteins, or attach indirectly, for example *via* other microorganisms (Wächtler et al. 2012).

The attachment to epithelial cells can facilitate the next step: *invasion* into host tissues (Mogavero and Hube 2021). *C. albicans* can invade host tissues using two mechanisms characterized so far: induced endocytosis and active penetration. Induced endocytosis is the ability of a pathogen to induce its own uptake by non-professional phagocytic host cells, such as epithelial cells (Pizarro-Cerdá and Cossart 2006). For example, the *C. albicans* hyphal-associated protein Als3 mimics host cell cadherins and binds to E-cadherin on oral epithelial cells and N-cadherin on endothelial cells, which induces endocytosis (Phan et al. 2007). In the intestine, M-cells have been proposed to be a target of induced endocytosis (Albac et al. 2016). On the other hand, active penetration is the process that requires hyphal extension and physical forces at the hyphal tip, potentially supported by the hydrolytic activity of enzymes, such as proteases, lipases, and phospholipases (Schaller et al. 2005). Active penetration is an entire fungal-driven mechanism and fully relies on fungal viability (Zakikhany et al. 2007; Dalle et al. 2010), while induced endocytosis can occur even with dead *C. albicans* hyphae (Park et al. 2005). Invading hyphae of *C. albicans* secrete candidalysin into an invasion pocket, so epithelial cells initiate at least two repair mechanisms to maintain epithelial integrity and prevent mucosal damage (Westman et al. 2022). Of note, *C. albicans* can invade epithelial cells and tissues even without causing cellular damage, by a mechanism called trans-cellular tunneling (Lachat et al. 2022). Furthermore, paracellular translocation has been proposed as an additional mechanism of invasion; yet this route seems to play a less significant role than active penetration during interaction with intestinal cells (Allert et al. 2018).

While invading host tissues, hyphae of *C. albicans* are producing the toxin candidalysin, an Ece1-derived peptide toxin encoded by the gene *ECE1* (extent of cell elongation 1) (Moyes et al. 2016). Candidalysin intercalates into host membranes, thereby causing host cell *damage* and release of damage-associated molecular patterns (DAMPs) that potently induce immune activation (Moyes et al. 2016; Richardson et al. 2018b). Hypha-mediated invasion into host cells by induced endocytosis or active penetration does not necessarily inflict host cell damage (Wächtler et al. 2011). For example, a *C. albicans* mutant lacking the *ECE1* gene is still able to invade host cells but causes no damage (Moyes et al. 2016). Finally, for the pathogenic *Candida* species that do not form true hyphae, such as *C. glabrata* and *C. parapsilosis*, the knowledge is limited to how host cell damage is induced. However, *C. glabrata* was observed to cause high epithelial damage in the presence of human albumin, indicating the existence of candidalysin-independent mechanism of causing epithelial damage, but more studies are needed to fully understand this process (Pekmezovic et al. 2021b). For this reason, the integration of host factors in *in vitro* models that are specific for the respective *in vivo* niche may be a crucial step in fully uncovering the pathogenicity strategies of *Candida* species.

2.3.2 Mucosal *Candida* Infections

Oral, vaginal, and intestinal mucosae are very complex environments that are harboring very diverse microbial communities (Pekmezovic et al. 2019; Vila et al. 2020). Therefore, any disturbances can support the dominance of a specific pathogenic species (dysbiosis) and development of disease (Pekmezovic et al. 2019, Vila et al. 2020). Although dysbiosis in all three niches can lead to overgrowth of *Candida spp.* and disease, these infections have also unique characteristics, mostly due to specificities of the anatomical site and local immune responses (Fig. 2.1). For example, in the mouth, candidiasis is manifested by formation of white plaques consisting of desquamated epithelial cells, fibrin, and fungal hyphae (McCullough and Savage 2005). Vaginal infections are actually an immunopathological condition, resulting from overstimulated neutrophil response, characterized by very high morbidity and recurrence (Yano et al. 2019; Rosati et al. 2020; Ardizzoni et al. 2021). Finally, in case of compromised host defenses, such as in corticosteroid-treated patients or patients treated with cytostatic drugs, fungal outgrowth and dysbiosis in the gut can lead to translocation of *Candida spp* from the intestine into the bloodstream causing a fungal sepsis also called candidemia (Allert et al. 2018; Zhai et al. 2020). This disseminated candidiasis leads to fungal invasion of vital organs and is a life-threatening condition associated with a dramatically high mortality (Kullberg and Arendrup 2015).

2.3.2.1 Oral Cavity

Salivary flow and swallowing are important factors in host defense against *Candida* spp. overgrowth, as it removes loosely attached *Candida* cells from the oral surface (Lewis and Williams 2017). Therefore, adhesion of *Candida* spp., independently or as co-adhesion with bacteria, is a prerequisite for fungal persistence and infection (reviewed in: Vila et al. (2020)). Apart from the constant flow, saliva is highly enriched in antimicrobial peptides (AMPs), which are an essential part of host defense against pathogenic microbes (Bahar and Ren 2013). Salivatory glands are secreting histatins, a family of 12 histidine-rich cationic peptides with broad-spectrum antimicrobial activity, of which histatin-5 (Hst-5) specifically has potent antifungal activity (Edgerton et al. 1998). Hst-5 causes localized disruptions on the plasma membrane of *C. albicans* and therefore provides species-specific killing with minimal damage to the host and the rest of the microbiota (Mochon and Liu 2008). Furthermore, Hst-5 is actively taken up by the transporter protein Dur31 of *C. albicans* mediating fungal killing (Mayer et al. 2012).

Within the oral cavity polymicrobial biofilms of *Candida spp* and bacteria are formed (O'Donnell et al. 2015). These complex communities are highly organized structures that are difficult to be removed, very resistant to host

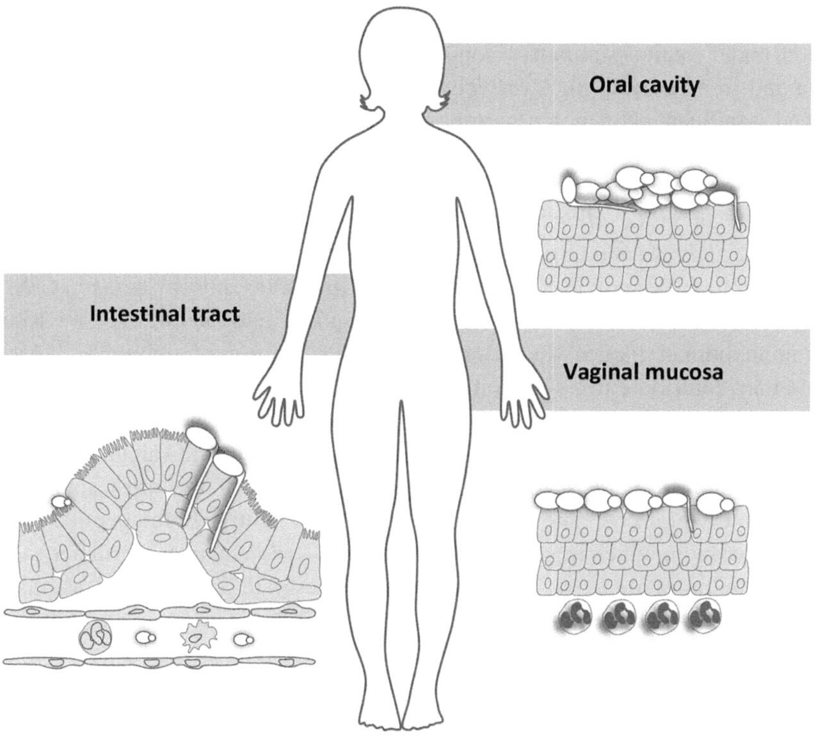

Fig. 2.1 Manifestation of *Candida albicans* infections on mucosal surface. *C. albicans* is a normal commensal, but can overgrow and cause infections on (**a**) oral, (**b**) vaginal, and (**c**) intestinal lining. In general, predisposing factors for mucosal candidiasis are immunosuppression and long-term antibiotic therapy. However, each type of infection has additional specific risk factors, such as human immunodeficiency virus infection, wearing dentures or dry mouth (oral), pregnancy and elevated estrogen levels (vaginal) or microbiota dysbiosis, compromised innate immunity and changes in epithelial barrier function (intestinal infection) (Jensena and Barkvoll 1998; Kumamoto 2011; Aguin and Sobel 2015; Patil et al. 2015; Kushnirenko 2019)

defenses, and serve as a reservoir for infection (Gheorghe et al. 2021). *C. albicans* also has been isolated from dental plaque and may play a role in caries development (Xiang et al. 2023). As mentioned above, biofilms can also be formed on abiotic surfaces, such as implants and prosthetic dental materials (Eguia et al. 2020). Therefore, one current research focus is to develop novel biomaterials with inherent antifungal properties to prevent biofilm formation and oral candidiasis development (Gheorghe et al. 2021).

Oropharyngeal candidiasis (OPC) affects the intraoral, pharyngeal, and perioral regions, which causes discomfort, pain, and loss of taste (Siqueira et al. 2015). Increased risk for the development of OPC is associated with immune dysfunctions, use of immune suppressant or long-term antibiotic therapy, chemotherapy, radiotherapy, underlying diseases (e.g., diabetes or human immunodeficiency virus (HIV) infection), and use of alcohol, tobacco, and dental prostheses (reviewed in: Gheorghe et al. (2021)). In fact, OPC occurs in ~80–90% of untreated HIV-positive individuals, therefore being the most prevalent and recurrent opportunistic infection in these patients (Vazquez 2000; Li et al. 2013).

2.3.2.2 Vaginal Mucosa

The human vaginal niche is characterized by acidic pH (4.5 or less), the presence of lactic-acid-producing bacteria, mucus secretion, variable oxygen availability, and epithelial shedding due to hormonal changes (Amabebe and Anumba 2018; Pekmezovic et al. 2019). These conditions require specific phenotypic and metabolic adaptations in bacteria and fungi colonizing this niche. Once cultivated under vagina-simulative conditions (hypoxia and iron restriction), *C. albicans* cell wall showed significant change in its composition (Sosinska et al. 2008). Moreover, an acidic pH causes cell wall remodeling and consequent unmasking of β-glucan in *C. albicans* and *C. tropicalis*, but not *C. glabrata* and *C. parapsilosis* (Sherrington et al. 2017; Pericolini et al. 2018). Unmasked β-glucan induces NLRP3 inflammasome activation in human monocytes (Camilli et al. 2018) and may contribute to a symptomatic vaginal infection (Sherrington et al. 2017, Pericolini et al. 2018). In contrast, *C. albicans*, *C. dubliniensis*, *C. lusitaniae*, *C. parapsilosis*, *C. orthopsilosis,* and *C. glabrata* show lactate-induced β-glucan masking, which might promote immune evasion (Ballou et al. 2016), indicating the dynamics of host-pathogen interactions. Furthermore, low pH increases the resistance of *C. albicans* and *C. glabrata* to antifungal drugs, which highlights the limitations of *in vitro* susceptibility testing at pH 7.0 and can be a reason for ineffective therapy and recurrence of VVC (Danby et al. 2012; Sobel and Akins 2022).

Lactic acid bacteria (*Lactobacillus* spp.) dominate the vaginal microbiota of most women, and maintain a healthy vaginal environment through the production of antimicrobial metabolites and host immune modulators (Valenti et al. 2018). Furthermore, they can prevent the adhesion and invasion of vaginal epithelial cells by pathogenic microorganisms (Valenti et al. 2018). *Lactobacillus crispatus,* for example, significantly reduces adhesion and induced endocytosis of *C. albicans in vitro,* but also enhances the local immune response of vaginal epithelial cells (Niu et al. 2017). In addition, supernatants of *L. crispatus* and *L. fermentum* inhibit the growth and hyphal induction of *C. albicans* (Jang et al. 2019). In contrast, the antifungal activity of vaginal lactobacilli against *C. glabrata*, *C. krusei*, *C. tropicalis*, and *C. parapsilosis* appears to be low or non-existent compared to *C. albicans* (Parolin et al. 2015).

Finally, the vaginal mucosal surface is coated with a mixture of cervical mucus and vaginal exudate, cervicovaginal secretion (CVS), which acts as a barrier between potential pathogens and vaginal epithelial cells (Amabebe and Anumba 2018; Pekmezovic et al. 2019). It is a mucous layer composed primarily of mucins, large and highly glycosylated proteins, that adhere directly to *C. albicans* (Domino et al. 2009) and inhibit the yeast-to-hyphae transition, required for epithelial invasion (Kavanaugh et al. 2014). In addition to mucins, host proteins in CVS proteins can recognize and neutralize microbial products [AMPs, immunoglobulins (Amabebe and Anumba 2018)] or bind iron, which is an essential metal necessary for microbial survival and growth [lactoferrin (Valenti et al. 2018)]. Finally, albumin, one of the most abundant proteins of CVS, can neutralize the peptide toxin candidalysin of *C. albicans* (Austermeier et al. 2021), but also facilitate fungal iron acquisition (Pinsky et al. 2020) and increase the pathogenicity potential of *C. glabrata* (Pekmezovic et al. 2021b).

Infection of the vaginal mucosa is called vulvovaginal candidiasis (VVC). During VVC, vaginal epithelial cells secrete antimicrobial peptides and pro-inflammatory signaling molecules, which should normally induce the recruitment of both innate and adaptive immune cells to the site of infection, ultimately leading to pathogen clearance (Höfs et al. 2016). However, symptomatic VVC is characterized by an infiltration of neutrophils into the vaginal lumen that are dysfunctional in their capacity to reduce the fungal burden leading to a persistent hyperinflammatory state (Pekmezovic et al. 2019; Yano et al. 2019; Rosati et al. 2020). Therefore, VVC is rather an immunopathology caused by an aggressive overreaction of the innate immune system than an outcome of a deficient adaptive immune response to *Candida* (Fidel 2007; Rosati et al. 2020). Although the pathogenesis of VVC seems to depend on both the host and fungal pathogenicity (Ardizzoni et al. 2021), the knowledge on precise fungal factors that play a role in epithelial-immune communication remains to a large extent limited to candidalysin (Richardson et al. 2018b). Expression of the *C. albicans* zinc-capturing protein Pra1 has been identified as a crucial driver of rapid neutrophil recruitment (Roselletti et al. 2023).

2.3.2.3 Intestinal Tract

The commensal-to-pathogen shift of *Candida* at the intestinal lining can occur upon a dysbalance of the microbiota, disrupted epithelial barrier, and when the immune system is compromised (Koh et al. 2008; Kumamoto et al. 2020). In that case, *C. albicans* has an opportunity to translocate through the gut barrier, reach the bloodstream, and cause disseminated candidiasis associated with high mortality rates (Koehler et al. 2019). Accordingly, the gut is considered a main reservoir of *Candida* species causing systemic candidiasis (Nucci and Anaissie 2001).

Similar to vaginal environment, members of the intestinal microbiota exhibit antagonistic effects that keep *Candida* spp. commensal by suppressing its overgrowth and pathogenicity (Förster et al. 2016; Perez 2021). Therefore, long-term antibiotic exposure impairs the control of *Candida* spp. in the intestine and can promote systemic infections (Drummond et al. 2022). Particularly lactic acid bacteria such as *Lactobacillus* species have been characterized as highly efficient antagonists of *C. albicans* pathogenicity. In fact, the integration of *L. rhamnosus* as a representative member in *in vitro* models alone can antagonize *C. albicans* pathogenicity in a multifactorial manner (Graf et al. 2019; Alonso-Roman et al. 2022). Particularly, the competition for favored nutrient sources as well as the production of metabolites that compromise virulence were characterized as mechanisms by which antagonistic bacteria such as lactobacilli can antagonize *C. albicans* pathogenicity and may reinforce commensalism (Alonso-Roman et al. 2022).

Prior to bloodstream infection, patients had intestinal expansion of pathogenic *Candida* species, associated with reduced bacterial burden and diversity (Zhai et al. 2020). This confirms that the population of *Candida spp* in the gut is the main reservoir for systemic infections and highlights the importance of a microbial balance and the immune system at the intestinal barrier.

2.4 Conclusion

To fully understand host-pathogen interactions, these interactions should be studied from both the microbial and host perspectives. Furthermore, it is crucial to consider the nature of these interactions, as commensalism and infection can be associated with very different phenotypic programs for the two players. As shown in this chapter, our knowledge of candidiasis so far has been largely limited to *C. albicans*, the most common causative agent of *Candida* infections, and particularly the reference laboratory strain (SC5314) used in most studies. Other clinical isolates seem to act far less aggressive as compared to this strain. The increasing prevalence of other *Candida* species prompts the need for a better understanding of the pathogenic mechanisms of non-*albicans Candida* species and their interaction with the host at the mucosal linings. Moreover, variability in these interactions depending on specific *C. albicans* isolates requires in-depth characterization to get a more comprehensive picture of the pathogenicity strategies that facilitate *Candida* spp. infections in susceptible hosts.

References

Adhikary R, Joshi S (2011) Species distribution and anti-fungal susceptibility of Candidaemia at a multi super-specialty center in Southern India. Indian J Med Microbiol 29(3):309–311

Aguin TJ, Sobel JD (2015) Vulvovaginal candidiasis in pregnancy. Curr Med Group LLC 1:17

Albac S, Schmitz A, Lopez-Alayon C, d'Enfert C, Sautour M, Ducreux A, Labruere-Chazal C, Laue M, Holland G, Bonnin A, Dalle F (2016) Candida albicans is able to use M cells as a portal of entry across the intestinal barrier in vitro. Cell Microbiol 18(2):195–210

Allert S, Förster TM, Svensson CM, Richardson JP, Pawlik T, Hebecker B, Rudolphi S, Juraschitz M, Schaller M, Blagojevic M, Morschhäuser J, Figge MT, Jacobsen ID, Naglik JR, Kasper L, Mogavero S, Hube B (2018) Candida albicans-induced epithelial damage mediates translocation through intestinal barriers. MBio 9(3)

Allonsius CN, van den Broek MFL, De Boeck I, Kiekens S, Oerlemans EFM, Kiekens F, Foubert K, Vandenheuvel D, Cos P, Delputte P, Lebeer S (2017) Interplay between Lactobacillus rhamnosus GG and Candida and the involvement of exopolysaccharides. Microb Biotechnol 10(6):1753–1763

Alonso-Roman R, Last A, Mirhakkak MH, Sprague JL, Moller L, Grossmann P, Graf K, Gratz R, Mogavero S, Vylkova S, Panagiotou G, Schauble S, Hube B, Gresnigt MS (2022) Lactobacillus rhamnosus colonisation antagonizes Candida albicans by forcing metabolic adaptations that compromise pathogenicity. Nat Commun 13(1):3192

Amabebe E, Anumba DOC (2018) The vaginal microenvironment: the physiologic role of Lactobacilli. Front Media S.A. 5:181–181

Arastehfar A, Daneshnia F, Hafez A, Khodavaisy S, Najafzadeh MJ, Charsizadeh A, Zarrinfar H, Salehi M, Shahrabadi ZZ, Sasani E, Zomorodian K, Pan W, Hagen F, Ilkit M, Kostrzewa M, Boekhout T (2020) Antifungal susceptibility, genotyping, resistance mechanism, and clinical profile of Candida tropicalis blood isolates. Med Mycol 58(6):766–773

Ardizzoni A, Wheeler RT, Pericolini E (2021) It takes two to tango: how a dysregulation of the innate immunity, coupled with Candida virulence. Triggers VVC Onset Front Microbiol 12:692491

Atiencia-Carrera MB, Cabezas-Mera FS, Tejera E, Machado A (2022) Prevalence of biofilms in Candida spp. bloodstream infections: a meta-analysis. PLoS One 17(2):e0263522

Auler ME, Morreira D, Rodrigues FFO, Abrão MS, Margarido PFR, Matsumoto FE, Silva EG, Silva BCM, Schneider RP, Paula CR (2010) Biofilm formation on intrauterine devices in patients with recurrent vulvovaginal candidiasis Biofilm formation on IUDs in patients with vulvovaginal candidiasis Auler et al. Med Mycol 48(1):211–216

Austermeier S, Pekmezovic M, Porschitz P, Lee S, Kichik N, Moyes DL, Ho J, Kotowicz NK, Naglik JR, Hube B, Gresnigt MS (2021) Albumin neutralizes hydrophobic toxins and modulates Candida albicans pathogenicity. MBio 12(3):e0053121

Bahar AA, Ren D (2013) Antimicrobial peptides. Pharmaceuticals (Basel) 6(12):1543–1575

Ballou ER, Avelar GM, Childers DS, Mackie J, Bain JM, Wagener J, Kastora SL, Panea MD, Hardison SE, Walker LA, Erwig LP, Munro CA, Gow NAR, Brown GD, MacCallum DM, Brown AJP (2016) Lactate signalling regulates fungal β-glucan masking and immune evasion. Nat Microbiol 2

Bertini A, Zoppo M, Lombardi L, Rizzato C, De Carolis E, Vella A, Torelli R, Sanguinetti M, Tavanti A (2016) Targeted gene disruption in Candida parapsilosis demonstrates a role for CPAR2_404800 in adhesion to a biotic surface and in a murine model of ascending urinary tract infection, vol 7. Taylor and Francis Inc, pp 85–97

Bi L, Gojestani S, Wu W, Hsu YM, Zhu J, Ariizumi K, Lin X (2010) CARD9 mediates dectin-2-induced IkappaBalpha kinase ubiquitination leading to activation of NF-kappaB in response to stimulation by the hyphal form of Candida albicans. J Biol Chem 285(34):25969–25977

Brandão LR, Medeiros AO, Duarte MC, Barbosa AC, Rosa CA (2010) Diversity and antifungal susceptibility of yeasts isolated by multiple-tube fermentation from three freshwater lakes in Brazil. J Water Health 8(2):279–289

Brunke S, Hube B (2013) Two unlike cousins: Candida albicans and C.glabrata infection strategies. Cell Microbiol 15:701–708

Bruno VM, Shetty AC, Yano J, Fidel PL Jr, Noverr MC, Peters BM (2015) Transcriptomic analysis of vulvovaginal candidiasis identifies a role for the NLRP3 inflammasome. MBio 6(2)

Calderone RA, Fonzi WA (2001) Virulence factors of Candida albicans. Trends Microbiol 9:327–335

Camilli G, Eren E, Williams DL, Aimanianda V, Meunier E, Quintin J (2018) Impaired phagocytosis directs human monocyte activation in response to fungal derived β-glucan particles. Eur J Immunol 48(5):757–770

Carreté L, Ksiezopolska E, Pegueroles C, Gómez-Molero E, Saus E, Iraola-Guzmán S, Loska D, Bader O, Fairhead C, Gabaldón T (2018) Patterns of genomic variation in the opportunistic pathogen Candida glabrata suggest the existence of mating and a secondary association with humans. Curr Biol 28(1):15–27.e17

Cavalheiro M, Teixeira MC (2018) Candida biofilms: threats, challenges, and promising strategies. Front Med 5:28

Cheng S-C, van de Veerdonk FL, Lenardon M, Stoffels M, Plantinga T, Smeekens S, Rizzetto L, Mukaremera L, Preechasuth K, Cavalieri D, Kanneganti TD, van der Meer JWM, Kullberg BJ, Joosten LAB, Gow NAR, Netea MG (2011) The dectin-1/inflammasome pathway is responsible for the induction of protective T-helper 17 responses that discriminate between yeasts and hyphae of Candida albicans. J Leukoc Biol 90(2):357–366

d'Enfert C, Kaune A-K, Alaban L-R, Chakraborty S, Cole N, Delavy M, Kosmala D, Marsaux B, Fróis-Martins R, Morelli M, Rosati D, Valentine M, Xie Z, Emritloll Y, Warn PA, Bequet F, Bougnoux M-E, Bornes S, Gresnigt MS, Hube B, Jacobsen ID, Legrand M, Leibundgut-Landmann S, Manichanh C, Munro CA, Netea MG, Queiroz K, Roget K, Thomas V, Thoral C, Van den Abbeele P, Walker AW, Brown AJP (2020) The impact of the Fungus-Host-Microbiota interplay upon Candida albicans infections: current knowledge and new perspectives. FEMS Microbiol Rev

d'Enfert C, Kaune AK, Alaban LR, Chakraborty S, Cole N, Delavy M, Kosmala D, Marsaux B, Frois-Martins R, Morelli M, Rosati D, Valentine M, Xie Z, Emritloll Y, Warn PA, Bequet F, Bougnoux ME, Bornes S, Gresnigt MS, Hube B, Jacobsen ID, Legrand M, Leibundgut-Landmann S, Manichanh C, Munro CA, Netea MG, Queiroz K, Roget K, Thomas V, Thoral C, Van den Abbeele P, Walker AW, Brown AJP (2021)

The impact of the Fungus-Host-Microbiota interplay upon Candida albicans infections: current knowledge and new perspectives. FEMS Microbiol Rev 45(3)

Dalle F, Wächtler B, L'Ollivier C, Holland G, Bannert N, Wilson D, Labruère C, Bonnin A, Hube B (2010) Cellular interactions of *Candida albicans* with human oral epithelial cells and enterocytes. Cell Microbiol 12(2):248–271

Danby CS, Boikov D, Rautemaa-Richardson R, Sobel JD (2012) Effect of pH on in vitro susceptibility of Candida glabrata and Candida albicans to 11 antifungal agents and implications for clinical use. Antimicrob Agents Chemother 56(3):1403–1406

De Groot PWJ, Kraneveld EA, Qing YY, Dekker HL, Groß U, Crielaard W, De Koster CG, Bader O, Klis FM, Weig M (2008) The cell wall of the human pathogen Candida glabrata: differential incorporation of novel adhesin-like wall proteins. Eukaryot Cell 7(11):1951–1964

Domino SE, Hurd EA, Thomsson KA, Karnak DM, Holmén Larsson JM, Thomsson E, Bäckström M, Hansson GC (2009) Cervical mucins carry α(1,2) fucosylated glycans that partly protect from experimental vaginal candidiasis. Glycoconj J 26(9):1125–1134

Drummond RA, Desai JV, Ricotta EE, Swamydas M, Deming C, Conlan S, Quinones M, Matei-Rascu V, Sherif L, Lecky D, Lee CR, Green NM, Collins N, Zelazny AM, Prevots DR, Bending D, Withers D, Belkaid Y, Segre JA, Lionakis MS (2022) Long-term antibiotic exposure promotes mortality after systemic fungal infection by driving lymphocyte dysfunction and systemic escape of commensal bacteria. Cell Host Microbe

Eddouzi J, Parker JE, Vale-Silva LA, Coste A, Ischer F, Kelly S, Manai M, Sanglard D (2013) Molecular mechanisms of drug resistance in clinical Candida species isolated from tunisian hospitals. Antimicrob Agents Chemother 57(7):3182–3193

Edgerton M, Koshlukova SE, Lo TE, Chrzan BG, Straubinger RM, Raj PA (1998) Candidacidal activity of salivary histatins. Identification of a histatin 5-binding protein on Candida albicans. J Biol Chem 273(32):20438–20447

Eguia A, Arakistain A, De-la-Pinta I, Lopez-Vicente J, Sevillano E, Quindos G, Eraso E (2020) Candida albicans biofilms on different materials for manufacturing implant abutments and prostheses. Med Oral Patol Oral Cir Bucal 25(1):e13–e20

Fidel PL (2007) History and update on host defense against vaginal candidiasis. Am J Reprod Immunol 57:2–12

Forastiero A, Mesa-Arango AC, Alastruey-Izquierdo A, Alcazar-Fuoli L, Bernal-Martinez L, Pelaez T, Lopez JF, Grimalt JO, Gomez-Lopez A, Cuesta I, Zaragoza O, Mellado E (2013) Candida tropicalis antifungal cross-resistance is related to different azole target (Erg11p) modifications. Antimicrob Agents Chemother 57(10):4769–4781

Forrest GN, Weekes E, Johnson JK (2008) Increasing incidence of Candida parapsilosis candidemia with caspofungin usage. J Infect 56(2):126–129

Förster TM, Mogavero S, Dräger A, Graf K, Polke M, Jacobsen ID, Hube B (2016) Enemies and brothers in arms: *Candida albicans* and gram-positive bacteria. Cell Microbiol 18(12):1709–1715

Gabaldón T, Martin T, Marcet-Houben M, Durrens P, Bolotin-Fukuhara M, Lespinet O, Arnaise S, Boisnard S, Aguileta G, Atanasova R, Bouchier C, Couloux A, Creno S, Almeida Cruz J, Devillers H, Enache-Angoulvant A, Guitard J, Jaouen L, Ma L, Marck C, Neuvéglise C, Pelletier E, Pinard A, Poulain J, Recoquillay J, Westhof E, Wincker P, Dujon B, Hennequin C, Fairhead C (2013) Comparative genomics of emerging pathogens in the Candida glabrata clade. BMC Genomics 14(1)

Gabaldon T, Naranjo-Ortiz MA, Marcet-Houben M (2016) Evolutionary genomics of yeast pathogens in the Saccharomycotina. FEMS Yeast Res 16(6)

Galocha M, Pais P, Cavalheiro M, Pereira D, Viana R, Teixeira MC (2019) Divergent approaches to virulence in C. Albicans and C. Glabrata: two sides of the same coin. Int J Mol Sci 20(9)

Gantner BN, Simmons RM, Underhill DM (2005) Dectin-1 mediates macrophage recognition of Candida albicans yeast but not filaments. EMBO J 24(6):1277–1286

Ghannoum MA (2000) Potential role of phospholipases in virulence and fungal pathogenesis. Clin Microbiol Rev 13(1):122–143

Gheorghe DC, Niculescu AG, Birca AC, Grumezescu AM (2021) Biomaterials for the prevention of oral candidiasis development. Pharmaceutics 13(6)

Gómez-Molero E, de Boer AD, Dekker HL, Moreno-Martínez A, Kraneveld EA, Ichsan N, Chauhan M, Weig JJ, de Soet CG, de Koster OB, de Groot PWJ (2015) Proteomic analysis of hyperadhesive candida glabrata clinical isolates reveals a core wall proteome and differential incorporation of adhesins. FEMS Yeast Res 15(8)

Gow NAR, Van De Veerdonk FL, Brown AJP, Netea MG (2012) Candida albicans morphogenesis and host defence: discriminating invasion from colonization. Nat Publ Group 10:112–122

Graf K, Last A, Gratz R, Allert S, Linde S, Westermann M, Gröger M, Mosig AS, Gresnigt MS, Hube B (2019) Keeping Candida commensal: how lactobacilli antagonize pathogenicity of Candida albicans in an in vitro gut model. DMM Dis Models Mech 12(9)

Hallen-Adams HE, Suhr MJ (2017) Fungi in the healthy human gastrointestinal tract. Virulence 8(3):352–358

Hazen KC (1989) Participation of yeast cell surface hydrophobicity in adherence of Candida albicans to human epithelial cells. Infect Immun 57(7):1894–1900

Henry M, Burgain A, Tebbji F, Sellam A (2021) Transcriptional control of hypoxic hyphal growth in the fungal pathogen Candida albicans. Front Cell Infect Microbiol 11:770478

Hise AG, Tomalka J, Ganesan S, Patel K, Hall BA, Brown GD, Fitzgerald KA (2009) An essential role for the NLRP3 inflammasome in host defense against the human fungal pathogen Candida albicans. Cell Host Microb 5(5):487–497

Ho J, Yang X, Nikou SA, Kichik N, Donkin A, Ponde NO, Richardson JP, Gratacap RL, Archambault LS, Zwirner CP, Murciano C, Henley-Smith R, Thavaraj S, Tynan CJ, Gaffen SL, Hube B, Wheeler RT, Moyes DL, Naglik JR (2019) Candidalysin activates innate epithelial immune responses via epidermal growth factor receptor. Nat Commun 10(1):1–12

Höfs S, Mogavero S, Hube B (2016) Interaction of Candida albicans with host cells: virulence factors, host defense, escape strategies, and the microbiota. Microbiol Soc Korea 54:149–169

Hoyer LL (2001) The ALS gene family of Candida albicans. Trends Microbiol 9:176–180

Hube B (2004) From commensal to pathogen: stage- and tissue-specific gene expression of Candida albicans. Curr Opin Microbiol 7:336–341

Jacobsen ID, Wilson D, Wachtler B, Brunke S, Naglik JR, Hube B (2012) Candida albicans dimorphism as a therapeutic target. Expert Rev Anti-Infect Ther 10(1):85–93

Jang SJ, Lee K, Kwon B, You HJ, Ko GP (2019) Vaginal lactobacilli inhibit growth and hyphae formation of Candida albicans. Sci Rep 9(1):1–9

Jensena JL, Barkvoll P (1998) Clinical implications of the dry mouth: oral mucosal diseases. Ann N Y Acad Sci 842(1 SALIVARY GLAN):156–162

Jones T, Federspiel NA, Chibana H, Dungan J, Kalman S, Magee BB, Newport G, Thorstenson YR, Agabian N, Magee PT, Davis RW, Scherer S (2004) The diploid genome sequence of Candida albicans. Proc Natl Acad Sci USA 101(19):7329–7334

Kasper L, König A, Koenig PA, Gresnigt MS, Westman J, Drummond RA, Lionakis MS, Groß O, Ruland J, Naglik JR, Hube B (2018) The fungal peptide toxin Candidalysin activates the NLRP3 inflammasome and causes cytolysis in mononuclear phagocytes. Nat Commun 9(1):1–20

Kaur R, Ma B, Cormack BP (2007) A family of glycosylphosphatidylinositol-linked aspartyl proteases is required for virulence of Candida glabrata. Proc Natl Acad Sci USA 104(18):7628–7633

Kavanaugh NL, Zhang AQ, Nobile CJ, Johnson AD, Ribbeck K (2014) Mucins suppress virulence traits of Candida albicans. MBio 5(6)

Koehler P, Stecher M, Cornely OA, Koehler D, Vehreschild M, Bohlius J, Wisplinghoff H, Vehreschild JJ (2019) Morbidity and mortality of candidaemia in Europe: an epidemiologic meta-analysis. Clin Microbiol Infect 25(10):1200–1212

Koh AY, Kohler JR, Coggshall KT, Van Rooijen N, Pier GB (2008) Mucosal damage and neutropenia are required for Candida albicans dissemination. PLoS Pathog 4(2):e35

Kozik A, Karkowska-Kuleta J, Zajac D, Bochenska O, Kedracka-Krok S, Jankowska U, Rapala-Kozik M (2015) Fibronectin-, vitronectin- and laminin-binding proteins at the cell walls of Candida parapsilosis and Candida tropicalis pathogenic yeasts. BMC Microbiol 15(1):197–197

Kuhn DM, George T, Chandra J, Mukherjee PK, Ghannoum MA (2002) Antifungal susceptibility of Candida biofilms: unique efficacy of amphotericin B lipid formulations and echinocandins. Antimicrob Agents Chemother 46(6):1773–1780

Kullberg BJ, Arendrup MC (2015) Invasive candidiasis. N Engl J Med 373(15):1445–1456

Kumamoto CA (2011) Inflammation and gastrointestinal Candida colonization. NIH Public Access 14:386–391

Kumamoto CA, Gresnigt MS, Hube B (2020) The gut, the bad and the harmless: Candida albicans as a commensal and opportunistic pathogen in the intestine. Curr Opin Microbiol 56:7–15

Kumar K, Askari F, Sahu M, Kaur R (2019) Candida glabrata: a lot more than meets the eye. Microorganisms 7(2):39–39

Kurtzman C (2003) Phylogenetic circumscription of, and other members of the Saccharomycetaceae, and the proposal of the new genera Lachancea, Nakaseomyces, Naumovia, Vanderwaltozyma and Zygotorulaspora. FEMS Yeast Res 4(3):233–245

Kushnirenko IV (2019) Risk factors for developing mucosal candidiasis of the upper gastrointestinal tract. The problem of antibiotic therapy. Gastroenterology 53(2):77–85

Lachat J, Pascault A, Thibaut D, Le Borgne R, Verbavatz JM, Weiner A (2022) Trans-cellular tunnels induced by the fungal pathogen Candida albicans facilitate invasion through successive epithelial cells without host damage. Nat Commun 13(1):3781

Lamoth F, Lockhart SR, Berkow EL, Calandra T (2018) Changes in the epidemiological landscape of invasive candidiasis. J Antimicrob Chemother 73(suppl_1):i4–i13

Lewis MAO, Williams DW (2017) Diagnosis and management of oral candidosis. Br Dent J 223(9):675–681

Li X, Lei L, Tan D, Jiang L, Zeng X, Dan H, Liao G, Chen Q (2013) Oropharyngeal Candida colonization in human immunodeficiency virus infected patients. APMIS 121(5):375–402

MacAlpine J, Daniel-Ivad M, Liu Z, Yano J, Revie NM, Todd RT, Stogios PJ, Sanchez H, O'Meara TR, Tompkins TA, Savchenko A, Selmecki A, Veri AO, Andes DR, Fidel PL Jr, Robbins N, Nodwell J, Whitesell L, Cowen LE (2021) A small molecule produced by Lactobacillus species blocks Candida albicans filamentation by inhibiting a DYRK1-family kinase. Nat Commun 12(1):6151

Makanjuola O, Bongomin F, Fayemiwo SA (2018) An update on the roles of non-albicans Candida Species in vulvovaginitis. J Fungi (Basel) 4(4)

Marcos-Zambrano LJ, Escribano P, Bouza E, Guinea J (2014) Production of biofilm by Candida and non-Candida spp. isolates causing fungemia: Comparison of biomass production and metabolic activity and development of cut-off points. Int J Med Microbiol 304(8):1192–1198

Marie C, White TC (2009) Genetic basis of antifungal drug resistance. NIH Public Access 3:163–169

Martchenko M, Alarco AM, Harcus D, Whiteway M (2004) Superoxide dismutases in Candida albicans: transcriptional regulation and functional characterization of the hyphal-induced SOD5 gene. Mol Biol Cell 15(2):456–467

Mayer FL, Wilson D, Jacobsen ID, Miramon P, Grosse K, Hube B (2012) The novel Candida albicans transporter Dur31 Is a multi-stage pathogenicity factor. PLoS Pathog 8(3):e1002592

McCullough MJ, Savage NW (2005) Oral candidosis and the therapeutic use of antifungal agents in dentistry. Aust Dent J 50:4–4

Mochon AB, Liu H (2008) The antimicrobial peptide histatin-5 causes a spatially restricted disruption on the Candida albicans surface, allowing rapid entry of the peptide into the cytoplasm. PLoS Pathog 4(10):e1000190

Mogavero S, Hube B (2021) Candida albicans interaction with oral epithelial cells: adhesion, invasion, and damage assays. Methods Mol Biol 2260:133–143

Morschhäuser J (2010) Regulation of multidrug resistance in pathogenic fungi. Fungal Genet Biol 47(2):94–106

Moyes DL, Naglik JR (2011) Mucosal immunity and Candida albicans infection. Clin Dev Immunol 2011:346307

Moyes DL, Runglall M, Murciano C, Shen C, Nayar D, Thavaraj S, Kohli A, Islam A, Mora-Montes H, Challacombe SJ, Naglik JR (2010) A biphasic innate immune MAPK response discriminates between the yeast and hyphal forms of Candida albicans in epithelial cells. Cell Host Microbe 8(3):225–235

Moyes DL, Shen C, Murciano C, Runglall M, Richardson JP, Arno M, Aldecoa-Otalora E, Naglik JR (2014) Protection against epithelial damage during Candida albicans infection is mediated by PI3K/Akt and mammalian target of rapamycin signaling. J Infect Dis 209(11):1816–1826

Moyes DL, Wilson D, Richardson JP, Mogavero S, Tang SX, Wernecke J, Hofs S, Gratacap RL, Robbins J, Runglall M, Murciano C, Blagojevic M, Thavaraj S, Forster TM, Hebecker B, Kasper L, Vizcay G, Iancu SI, Kichik N, Hader A, Kurzai O, Luo T, Kruger T, Kniemeyer O, Cota E, Bader O, Wheeler RT, Gutsmann T, Hube B, Naglik JR (2016) Candidalysin is a fungal peptide toxin critical for mucosal infection. Nature 532(7597):64–68

Naglik JR, Moyes D (2011) Epithelial cell innate response to Candida albicans. Adv Dent Res 23(1):50–55

Naglik J, Albrecht A, Bader O, Hube B (2004) Candida albicans proteinases and host/pathogen interactions. Cell Microbiol 6(10):915–926

Nash AK, Auchtung TA, Wong MC, Smith DP, Gesell JR, Ross MC, Stewart CJ, Metcalf GA, Muzny DM, Gibbs RA, Ajami NJ, Petrosino JF (2017) The gut mycobiome of the Human Microbiome Project healthy cohort. Microbiome 5(1):153

Neville BA, d'Enfert C, Bougnoux ME (2015) Candida albicans commensalism in the gastrointestinal tract. FEMS Yeast Res 15

Nikolaou E, Agrafioti I, Stumpf M, Quinn J, Stansfield I, Brown AJ (2009) Phylogenetic diversity of stress signalling pathways in fungi. BMC Evol Biol 9(1):44–44

Niu XX, Li T, Zhang X, Wang SX, Liu ZH (2017) Lactobacillus crispatus modulates vaginal epithelial cell innate response to Candida albicans. Chin Med J 130(3):273–279

Noble SM, Gianetti BA, Witchley JN (2017) Candida albicans cell-type switching and functional plasticity in the mammalian host. Nat Rev Microbiol 15(2):96–108

Nucci M, Anaissie E (2001) Revisiting the source of candidemia: skin or gut? Clin Infect Dis 33(12):1959–1967

Nucci M, Colombo AL (2007) Candidemia due to Candida tropicalis: clinical, epidemiologic, and microbiologic characteristics of 188 episodes occurring in tertiary care hospitals. Diagn Microbiol Infect Dis 58(1):77–82

O'Donnell LE, Millhouse E, Sherry L, Kean R, Malcolm J, Nile CJ, Ramage G (2015) Polymicrobial Candida biofilms: friends and foe in the oral cavity. FEMS Yeast Res 15(7)

Pande K, Chen C, Noble SM (2013) Passage through the mammalian gut triggers a phenotypic switch that promotes Candida albicans commensalism. Nat Genet 45(9):1088–1091

Park H, Myers CL, Sheppard DC, Phan QT, Sanchez AA, Edwards JE, Filler SG (2005) Role of the fungal Ras-protein kinase A pathway in governing epithelial cell interactions during oropharyngeal candidiasis. Cell Microbiol 7(4):499–510

Parolin C, Marangoni A, Laghi L, Foschi C, Palomino RAÑ, Calonghi N, Cevenini R, Vitali B (2015) Isolation of vaginal lactobacilli and characterization of anti-candida activity. PLoS One 10(6)

Patil S, Rao RS, Majumdar B, Anil S (2015) Clinical appearance of oral candida infection and therapeutic strategies. Front Microbiol 6:1391–1391

Pekmezovic M, Mogavero S, Naglik JR, Hube B (2019) Host-pathogen interactions during female genital tract infections. Trends Microbiol 27(12):982–996

Pekmezovic M, Hovhannisyan H, Gresnigt MS, Iracane E, Oliveira-Pacheco J, Siscar-Lewin S, Seemann E, Qualmann B, Kalkreuter T, Müller S, Kamradt T, Mogavero S, Brunke S, Butler G, Gabaldón T, Hube B (2021a) Candida pathogens induce protective mitochondria-associated type I interferon signalling and a damage-driven response in vaginal epithelial cells. Nat Microbiol

Pekmezovic M, Kaune AK, Austermeier S, Hitzler SUJ, Mogavero S, Hovhannisyan H, Gabaldon T, Gresnigt MS, Hube B (2021b) Human albumin enhances the pathogenic potential of Candida glabrata on vaginal epithelial cells. PLoS Pathog 17(10):e1010037

Perez JC (2019) Candida albicans dwelling in the mammalian gut. Curr Opin Microbiol 52:41–46

Perez JC (2021) The interplay between gut bacteria and the yeast Candida albicans. Gut Microbes 13(1):1979877

Pericolini E, Perito S, Castagnoli A, Gabrielli E, Mencacci A, Blasi E, Vecchiarelli A, Wheeler RT (2018) Epitope unmasking in vulvovaginal candidiasis is associated with hyphal growth and neutrophilic infiltration. PLoS One 13(7)

Phan QT, Myers CL, Fu Y, Sheppard DC, Yeaman MR, Welch WH, Ibrahim AS, Edwards JE, Filler SG (2007) Als3 is a Candida albicans invasin that binds to cadherins and induces endocytosis by host cells. PLoS Biol 5(3):0543–0557

Pinsky M, Roy U, Moshe S, Weissman Z, Kornitzer D (2020) Human serum albumin facilitates heme-iron utilization by fungi. MBio 11(2)

Pizarro-Cerdá J, Cossart P (2006) Bacterial adhesion and entry into host cells. Cell 124:715–727

Potera C (1999) Forging a link between biofilms and disease. Science 283:1837–1839

Rajni E, Chaudhary P, Garg VK, Sharma R, Malik M (2022) A complete clinico-epidemiological and microbiological profile of candidemia cases in a tertiary-care hospital in Western India. Antimicrob Steward Healthc Epidemiol 2(1):e37

Reyes-Montes MDR, Duarte-Escalante E, Martínez-Herrera E, Acosta-Altamirano G, León MGF-D (2017) Current status of the etiology of candidiasis in Mexico. Asociacion Espanola de Micologia 34:203–210

Richardson JP, Ho J, Naglik JR (2018a) Candida–epithelial interactions. MDPI AG 4

Richardson JP, Willems HME, Moyes DL, Shoaie S, Barker KS, Tan SL, Palmer GE, Hube B, Naglik JR, Peters BM (2018b) Candidalysin drives epithelial signaling, neutrophil recruitment, and immunopathology at the vaginal mucosa. Infect Immun 86(2)

Richardson JP, Brown R, Kichik N, Lee S, Priest E, Mogavero S, Maufrais C, Wickramasinghe DN, Tsavou A, Kotowicz NK, Hepworth OW, Gallego-Cortes A, Ponde NO, Ho J, Moyes DL, Wilson D, d'Enfert C, Hube B, Naglik JR (2022) Candidalysins are a new family of cytolytic fungal peptide toxins. mBio:e0351021

Rosati D, Bruno M, Jaeger M, Ten Oever J, Netea MG (2020) Recurrent vulvovaginal candidiasis: an immunological perspective. Microorganisms 8(2)

Roselletti E, Perito S, Gabrielli E, Mencacci A, Pericolini E, Sabbatini S, Cassone A, Vecchiarelli A (2017) NLRP3 inflammasome is a key player in human vulvovaginal disease caused by Candida albicans. Sci Rep 7(1):17877–17877

Roselletti E, Monari C, Sabbatini S, Perito S, Vecchiarelli A, Sobel JD, Cassone A (2019) A role for yeast/pseudohyphal cells of Candida albicans in the correlated expression of NLRP3 inflammasome inducers in women with acute vulvovaginal candidiasis. Front Microbiol 10:2669–2669

Roselletti E, Pericolini E, Nore A, Takacs P, Kozma B, Sala A, De Seta F, Comar M, Usher J, Brown GD, Wilson D (2023) Zinc prevents vaginal candidiasis by inhibiting expression of an inflammatory fungal protein. Sci Transl Med 15(725):eadi3363

Sabino R, Sampaio P, Carneiro C, Rosado L, Pais C (2011) Isolates from hospital environments are the most virulent of the Candida parapsilosis complex. BMC Microbiol 11:180–180

Salazar SB, Wang C, Münsterkötter M, Okamoto M, Takahashi-Nakaguchi A, Chibana H, Lopes MM, Güldener U, Butler G, Mira NP (2018) Comparative genomic and transcriptomic analyses unveil novel features of azole resistance and adaptation to the human host in Candida glabrata. FEMS Yeast Res 18(1)

Schaller M, Borelli C, Korting HC, Hube B (2005) Hydrolytic enzymes as virulence factors of Candida albicans. Mycoses 48:365–377

Schonherr FA, Sparber F, Kirchner FR, Guiducci E, Trautwein-Weidner K, Gladiator A, Sertour N, Hetzel U, Le GTT, Pavelka N, d'Enfert C, Bougnoux ME, Corti CF, LeibundGut-Landmann S (2017) The intraspecies diversity of C. albicans triggers qualitatively and temporally distinct host responses that determine the balance between commensalism and pathogenicity. Mucosal Immunol 10(5):1335–1350

Seider K, Brunke S, Schild L, Jablonowski N, Wilson D, Majer O, Barz D, Haas A, Kuchler K, Schaller M, Hube B (2011) The facultative intracellular pathogen Candida glabrata subverts macrophage cytokine production and phagolysosome maturation. J Immunol 187(6):3072–3086

Sharifynia S, Falahati M, Akhlaghi L, Foroumadi A, Fateh R (2017) Molecular identification and antifungal susceptibility profile of Candida species isolated from patients with vulvovaginitis in Tehran. Iran J Res Med Sci 22(12)

Sherrington SL, Sorsby E, Mahtey N, Kumwenda P, Lenardon MD, Brown I, Ballou ER, MacCallum DM, Hall RA (2017) Adaptation of Candida albicans to environmental pH induces cell wall remodelling and enhances innate immune recognition. PLoS Pathog 13(5):e1006403

Singh DK, Németh T, Papp A, Tóth R, Lukácsi S, Heidingsfeld O, Dostal J, Vágvölgyi C, Bajtay Z, Józsi M, Gácser A (2019) Functional Characterization of Secreted Aspartyl Proteases in Candida parapsilosis. mSphere 4(4)

Siqueira AB, Rodriguez LR, Santos RK, Marinho RR, Abreu S, Peixoto RF, Gurgel BC (2015) Antifungal activity of propolis against Candida species isolated from cases of chronic periodontitis. Braz Oral Res 29

Sobel JD, Akins R (2022) Determining susceptibility in Candida vaginal isolates. Antimicrob Agents Chemother 66(6):e0236621

Sosinska GJ, de Groot PWJ, de Mattos MJT, Dekker HL, de Koster CG, Hellingwerf KJ, Klis FM (2008) Hypoxic conditions and iron restriction affect the cell-wall proteome of Candida albicans grown under vagina-simulative conditions. Microbiology 154(2):510–520

Sudbery PE (2011) Growth of Candida albicans hyphae. Nat Rev Microbiol 9(10):737–748

Swidergall M, Solis NV, Lionakis MS, Filler SG (2018) EphA2 is an epithelial cell pattern recognition receptor for fungal β-glucans. Nat Microbiol 3(1):53–61

Trofa D, Gácser A, Nosanchuk JD (2008) Candida parapsilosis, an emerging fungal pathogen. Clin Microbiol Rev 21:606–625

Tronchin G, Pihet M, Lopes-Bezerra L, Bouchara JP (2008) Adherence mechanisms in human pathogenic fungi. Med Mycol 46:749–772
Valenti P, Rosa L, Capobianco D, Lepanto MS, Schiavi E, Cutone A, Paesano R, Mastromarino P (2018) Role of lactobacilli and lactoferrin in the mucosal cervicovaginal defense. Front Immunol 9
Vale-Silva L, Beaudoing E, Tran VDT, Sanglard D (2017) Comparative genomics of two sequential Candida glabrata clinical isolates. G3: Genes, Genomes. Genetics 7(8):2413–2426
van der Graaf CA, Netea MG, Verschueren I, van der Meer JW, Kullberg BJ (2005) Differential cytokine production and Toll-like receptor signaling pathways by Candida albicans blastoconidia and hyphae. Infect Immun 73(11):7458–7464
Vazquez JA (2000) Therapeutic options for the management of oropharyngeal and esophageal candidiasis in HIV/AIDS patients. HIV Clin Trials 1(1):47–59
Vijaya D, Dhanalakshmi TA, Kulkarni S (2014) Changing trends of vulvovaginal candidiasis. J Lab Physicians 6(01):028–030
Vila T, Sultan AS, Montelongo-Jauregui D, Jabra-Rizk MA (2020) Oral candidiasis: a disease of opportunity. J Fungi (Basel) 6(1)
Vincent BM, Lancaster AK, Scherz-Shouval R, Whitesell L, Lindquist S (2013) Fitness trade-offs restrict the evolution of resistance to amphotericin B. PLoS Biol 11(10):e1001692
Wächtler B, Wilson D, Haedicke K, Dalle F, Hube B (2011) From attachment to damage: defined genes of Candida albicans mediate adhesion, invasion and damage during interaction with oral epithelial cells. PLoS One 6(2)
Wächtler B, Citiulo F, Jablonowski N, Förster S, Dalle F, Schaller M, Wilson D, Hube B (2012) Candida albicans-epithelial interactions: Dissecting the roles of active penetration, induced endocytosis and host factors on the infection process. PLoS One 7(5)
Westman J, Plumb J, Licht A, Yang M, Allert S, Naglik JR, Hube B, Grinstein S, Maxson ME (2022) Calcium-dependent ESCRT recruitment and lysosome exocytosis maintain epithelial integrity during Candida albicans invasion. Cell Rep 38(1):110187
Whaley SG, Rogers PD (2016) Azole resistance in Candida glabrata. Curr Med Group LLC 1:18
White SJ, Rosenbach A, Lephart P, Nguyen D, Benjamin A, Tzipori S, Whiteway M, Mecsas J, Kumamoto CA (2007) Self-regulation of Candida albicans population size during GI colonization. PLoS Pathog 3(12):1866–1878
Willems HME, Lowes DJ, Barker KS, Palmer GE, Peters BM (2018) Comparative analysis of the capacity of the candida species to elicit vaginal immunopathology. Infect Immun 86(12)
Witchley JN, Penumetcha P, Abon NV, Woolford CA, Mitchell AP, Noble SM (2019) Candida albicans morphogenesis programs control the balance between gut commensalism and invasive infection. Cell Host Microbe 25(3):432–443 e436
Xiang Z, Wakade RS, Ribeiro AA, Hu W, Bittinger K, Simon-Soro A, Kim D, Li J, Krysan DJ, Liu Y, Koo H (2023) Human tooth as a fungal niche: Candida albicans traits in dental plaque isolates. MBio 14(1):e0276922
Yano J, Sobel JD, Nyirjesy P, Sobel R, Williams VL, Yu Q, Noverr MC, Fidel PL Jr (2019) Current patient perspectives of vulvovaginal candidiasis: incidence, symptoms, management and post-treatment outcomes. BMC Womens Health 19(1):48
Zakikhany K, Naglik JR, Schmidt-Westhausen A, Holland G, Schaller M, Hube B (2007) In vivo transcript profiling of Candida albicans identifies a gene essential for interepithelial dissemination. Cell Microbiol 9(12):2938–2954
Zhai B, Ola M, Rolling T, Tosini NL, Joshowitz S, Littmann ER, Amoretti LA, Fontana E, Wright RJ, Miranda E, Veelken CA, Morjaria SM, Peled JU, van den Brink MRM, Babady NE, Butler G, Taur Y, Hohl TM (2020) High-resolution mycobiota analysis reveals dynamic intestinal translocation preceding invasive candidiasis. Nat Res 26:59–64
Zuza-Alves DL, Silva-Rocha WP, Chaves GM (2017) An update on Candida tropicalis based on basic and clinical approaches. Front Microbiol 8:1927

Malassezia Yeasts in Animals in the Next-Generation Sequencing Era

3

Leyna Díaz, Gemma Castellá, M. Rosa Bragulat, and F. Javier Cabañes

Abstract

The yeasts of the genus *Malassezia* are part of the normal skin microbiota of a wide range of warm-blooded animals, including humans. Currently, this genus includes 18 species that have been only isolated from different wild and domestic animals and humans. A main characteristic of these yeasts is their lipid dependency, which is caused by the lack of the gene encoding the fatty acid synthase. Thus, *Malassezia* yeasts have a predilection for sebaceous skin sites. Despite being part of the normal skin microbiota, the population of *Malassezia* can overgrow under certain circumstances and these yeasts can become opportunistic pathogens causing *Malassezia*-related diseases such as dermatitis, otitis, and other skin diseases. In dogs, *Malassezia pachydermatis* is considered a common etiological agent causing otitis and dermatitis. Although azole antifungals are usually effective in resolving these diseases, some cases of treatment failure have recently been reported.

Keywords

Animals · Antifungal resistance · Dermatitis *ERG11* gene · *Malassezia* · Metagenomics Mycobiome · NGS · Otitis · Speciation

3.1 Introduction

The yeasts of the genus *Malassezia* are part of the normal skin microbiota of a wide range of warm-blooded animals (Sugita et al. 2010). Although next-generation sequencing (NGS) techniques have made it possible to detect the presence of *Malassezia* DNA sequences in other animals and ecosystems (e.g., marine corals, lobsters) (Amend 2014), to date these yeasts have only been isolated from humans and various wild and domestic animals.

Some species appear to have a broad host range, while others are more host-specific and closely adapted to the cutaneous ecosystem of a single animal species or a group of phylogenetically related animals (Guillot and Bond 2020). A key characteristic of these yeasts is their lipid dependence, which is caused by the absence of the gene encoding fatty acid synthase. As a result, they are unable to synthetize long-chain fatty acids *de novo* and they rely on the host as an

L. Díaz · G. Castellá · M. R. Bragulat
Department of Animal Health and Anatomy, Veterinary Mycology Group, Universitat Autònoma de Barcelona, Bellaterra, Catalonia, Spain

F. J. Cabañes (✉)
Facultat de Veterinària, Departament de Sanitat i d'Anatomia Animals, Grup de Micologia Veterinària, Universitat Autònoma de Barcelona, Bellaterra (Barcelona), Spain
e-mail: javier.cabanes@uab.cat

A. A. Brakhage et al. (eds.), *Human and Animal Relationships*, The Mycota 6, https://doi.org/10.1007/978-3-031-64853-3_3

exogenous source of fatty acids; thus, *Malassezia* yeasts have a preference for sebaceous skin sites (Wu et al. 2015). Nevertheless, some of these yeasts have recently been detected for the first time in the canine duodenum but it is unknown whether they are resident or transient fungi at this site (Kathrani et al. 2023).

Despite being part of the normal skin microbiota, under certain circumstances the population of *Malassezia* can overgrow and the yeasts can act as an opportunistic pathogen causing *Malassezia*-related diseases such as dermatitis, otitis, and other skin diseases (Sugita et al. 2010). Besides, some *Malassezia* species, among other microorganisms of the microbiota, have been associated with tumorigenesis in humans and mice (Elaskandrany et al. 2021).

In dogs and cats, *Malassezia pachydermatis* is considered a common etiological agent causing otitis and dermatitis. In the case of the aforementioned diseases, it has evolved from a controversial to a routine diagnosis in small animal practice, with significant benefits for many animals (Bond et al. 2020). Although azole antifungals are usually effective in resolving *Malassezia* diseases, some cases of treatment failure have been recently reported (Peano et al. 2020). This chapter will mainly focus on the major advances in the field that have occurred during the last 15 years, as new molecular techniques such as NGS have become more popular and affordable.

3.2 Current Perspective of the Genus *Malassezia*

The genus *Malassezia* (Baillon) was described in 1889 and consisted of a single species, *Malassezia furfur*, that was detected in human skin lesions. This genus remained limited to only two species for over a century: *M. furfur* (Robin) Baillon 1889 and *M. pachydermatis* (Weidman) Dodge 1925 (Cabañes 2014). Traditionally, the lipid-dependent species *M. furfur* (*sensu lato*) was thought to occur only on human skin, being the causal agent of distinct skin disorders. On the contrary, the classically considered non–lipid-dependent species *M. pachydermatis*, the only species in the genus that does not usually require lipid supplementation for development in Sabouraud glucose agar (SGA), was restricted to animal skin (Gueho-Kellerman et al. 2010). However, different names have been proposed for these yeasts in different genera (e.g., *Pityrosporum*, *Malassezia*) and those belonging to the genus *Malassezia* can be consulted chronologically in Fig. 3.1.

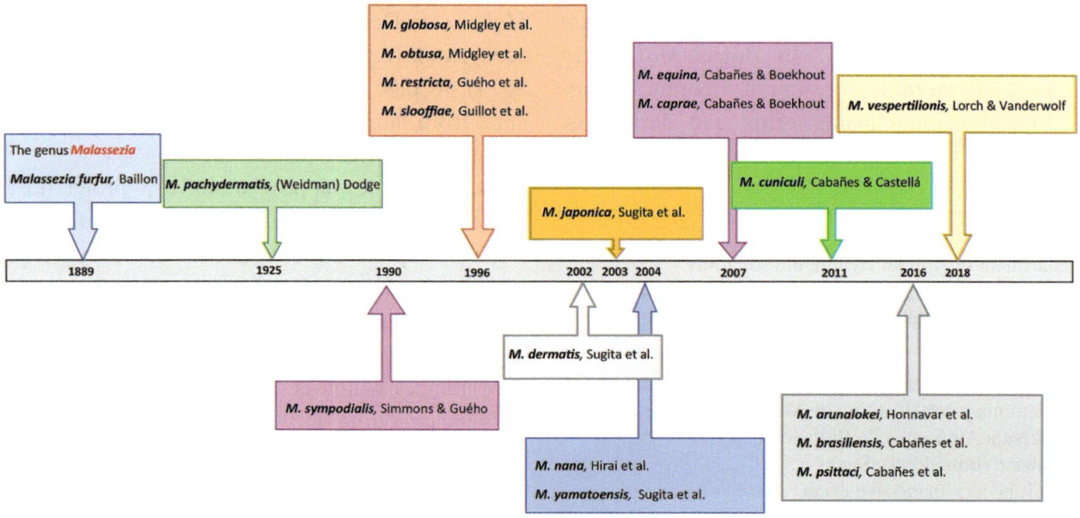

Fig. 3.1 Current accepted *Malassezia* species in chronological order

Today, according to current taxonomy the genus *Malassezia* consists of 18 species, placed in the phylum *Basidiomycota*, subphylum *Ustilaginomycotina*, class *Malasseziomycetes*, order *Malasseziales* and family *Malasseziaceae* (Wang et al. 2014). New species associated with the skin of animals were described, such as *Malassezia nana* from cats and cows (Hirai et al. 2004), *Malassezia caprae* and *Malassezia equina* from goats and horses, respectively (Cabañes et al. 2007), *Malassezia cuniculi* from rabbits (Cabañes et al. 2011), *Malassezia psittaci* and *Malassezia brasiliensis* from birds (Cabañes et al. 2016), and *Malassezia vespertilionis* from bats (Lorch et al. 2018).

Phylogeny studies using the D1/D2 sequences of the 18 *Malassezia* species allow to group the species into four main groups (Fig. 3.2). One group includes species isolated from human skin (*M. sympodialis* and *M. dermatis*) and animal skin (*M. equina*, *M. caprae*, and *M. nana*). A second group includes human skin species (*M. restricta*, *M. anuralokei*, and *M. globosa*) and *M. pachydermatis* which is found nearly exclusively on animal skin. A third group includes *M. furfur*, three other species rarely found on human skin (*M. yamatoensis*, *M. obtusa*, and *M. japonica*), and two species isolated from parrots (*M. brasiliensis* and *M. psittaci*). And finally, a basal group consists of more divergent species (*M. cuniculi*, *M. slooffiae*, and *M. vespertilionis*). Except for the branch position of *M. pachydermatis* in the cluster of the *M. sympodialis* group, a similar topology was observed when the genomes of some of these species were used to construct the phylogenetic tree (Wu et al. 2015).

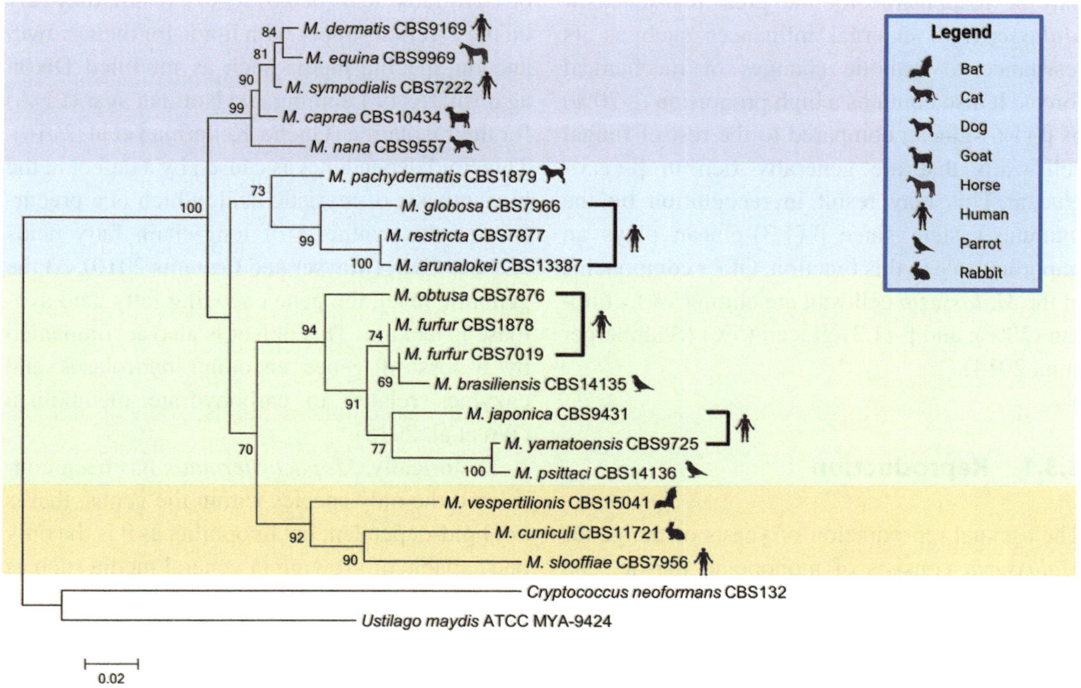

Fig. 3.2 Phylogeny of 18 *Malassezia* species using sequences of the D1/D2 region of the 26S rRNA gene inferred by using the maximum likelihood method based on the Tamura-Nei model. Bootstrap values in 1000 replications are shown at the nodes. Sequences of *Ustilago maydis* ATCC MYA-4924 and *Cryptococcus neoformans* CBS 132 were selected as outgroup for the tree construction. The phylogenetic tree was generated using W-IQ-TREE (Trifinopoulos et al. 2016)

3.3 Main Characteristics of *Malassezia* Yeasts

At the microscopic level, *Malassezia* yeast cells could be spherical, ovoid, ellipsoidal, or cylindrical. Cells bud in a monopolar pattern on a narrow to broad base. The size of the cells is also variable depending on the species and additionally, several species can form filaments or pseudo-hyphae. The *Malassezia* cell wall presents a series of characteristics that make it unique among the rest of the fungal genera. Its wall is thick and formed by various layers including an outer lamella, a multilayer wall, and a plasma membrane with a corrugate pattern and surrounded by an electro-lucent band (Gueho-Kellerman et al. 2010).

The cell wall of *Malassezia* spp. is composed of a high percentage of lipids (15%) compared to other genera such as *Saccharomyces* spp. (1–2%) (Mayser and Gaitanis 2010). This high lipid ratio may be responsible for the great resistance of *Malassezia* to external influences such as its resistance to osmotic changes or mechanical forces. It also contains a high proportion (~70%) of β-(1,6)-glucan compared to the rest of fungal cell walls that are generally rich in β-(1,3)-glucan. This may result in recognition by the immune system since β-(1,3)-glucan plays an important role in this function. Other components of the *Malassezia* cell wall are chitin (5%), chitosan (20%), and β-(1,3)-glucan (5%) (Stalhberger et al. 2014).

3.3.1 Reproduction

The asexual reproduction of yeasts of the genus *Malassezia* consists of monopolar, blastic, and percurrent budding. This budding can have a wider or narrower base depending on the species. It is hypothesized that sexual reproduction in fungi can contribute to gene reorganization through recombination, which can be associated with an increased virulence (Gioti et al. 2013; Nielsen and Heitman 2007). In *Malassezia*, the high genetic variability between species suggests the existence of genomic recombination although its sexual cycle is still unknown (Xu et al. 2007; Coelho et al. 2013). Genome analyses have shown the presence of MAT genes in most *Malassezia* species. Thus, if *Malassezia* has a sexual cycle, it is likely to be bipolar or pseudo-bipolar with two types of alleles of the MAT gene rather than tetrapolar, which is more common in other basidiomycetes (Coelho et al. 2010; Heitman et al. 2014; Wu et al. 2015). These MAT genes are related to sexual reproduction in other basidiomycetous species. It is still unknown whether these genes are associated with the yeast-hyphal transition, which has rarely been observed in *Malassezia* species (Ianiri et al. 2022).

3.3.2 Lipid Dependency

Yeasts of the genus *Malassezia* are characterized by their inability to synthesize long-chain (C14 or C16) fatty acids *de novo*. As a result, they rely on the external supply with lipids for their growth and on specific media such as modified Dixon agar (mDA) or Leeming and Notman agar (LNA) for their isolation (Gueho-Kellerman et al. 2010). This lipid dependency is caused by a defect in the biosynthesis of myristic acid, which is a precursor for the synthesis of long-chain fatty acids (C14 or C16) (Mayser and Gaitanis 2010). At the genomic level, the gene encoding fatty acid synthase is lacking. This defect is also accompanied by a loss of genes encoding hydrolases and enzymes related to carbohydrate metabolism (Wu et al. 2015).

Historically, *M. pachydermatis* has been considered the only species within the genus, that is not lipid-dependent but lipophilic, as it is the only one capable of growing in general media such as SGA (Cabañes 2020). Despite this, genome sequencing of *M. pachydermatis* showed that it also lacks the enzyme responsible for synthesizing long-chain fatty acids but instead, it can use lipid fractions of the peptone of SGA. Therefore, *M. pachydermatis* is unable to grow in lipid-free synthetic culture media and should also be considered a lipid-dependent species (Wu et al. 2015). In fact, there is a very low percentage of strains of *M. pachydermatis* that poorly grow in

SGA or are unable to grow in this medium (Puig et al. 2017). Most *M. pachydermatis* isolates only need culture media that contain peptone, such as SGA, which provides a sufficient amount of short-chain fatty acids.

3.3.3 Genome

The genome of *Malassezia* is small and compact (~7–9 Mb) and encodes a very low number of genes (~4400) (Wu et al. 2015; Sankaranarayanan et al. 2020). Genes are heavily packed throughout the genome and are interrupted by only few introns (present in ~27% of genes). Furthermore, the genome contains few repetitive elements (~0.78% of the genome) (Xu et al. 2007). Comparative gene analysis revealed extensive turnover events characterized by gene gain and loss, probably a result of evolutionary adaptation to colonize a specific niche such as animal skin. The loss of the fatty acid synthase and the δ-9 desaturase in all known *Malassezia* species may be due to the abundance of lipids in their natural habitat (Wu et al. 2015). However, the genomes of *Malassezia* contain an abundance of genes encoding hydrolases to utilize the proteins, lipids, and glycerol of the media (Ianiri et al. 2016). Haploid *Malassezia* species contain six to nine chromosomes. The existence of an ancestral karyotype with nine chromosomes is hypothesized to be conserved in *M. slooffiae*, *M. globosa*, and *M. restricta*. Chromosome breakage phenomena and subsequent centromere inactivation resulted in eight chromosomes in *M. sympodialis*, *M. equina*, *M. caprae*, *M. nana*, *M. dermatis*, and *M. furfur*. In the case of *M. pachydermatis* and other *Malassezia* species, the evolutionary selection pressure on their chromosomes is still unclear (Ianiri et al. 2022; Sankaranarayanan et al. 2020).

Very recently, the genomes of mitochondria of all described *Malassezia* spp. have been studied (Christinaki et al. 2023). In this study, the comparative analysis of these *Malassezia* mitogenomes contributed to phylogenetic tree construction for all species and their mitogenomes presented significant size and gene order diversity which correlates to their phylogeny.

3.4 Epidemiology and Ecology of *Malassezia* Yeasts

Malassezia yeasts are part of the microbiota of the skin and mucous membranes of a wide range of warm-blooded animals, including humans. Some *Malassezia* species can be detected in multiple hosts while others have a narrow host specificity and have undergone a series of adaptations to the skin ecosystem of an animal species or a group of phylogenetically close animals (Cabañes 2014).

The predominant species identified by non-culture-based studies in human skin are *M. globosa* and *M. restricta* (Gaitanis et al. 2012; Ianiri et al. 2022) but other species such as *M. furfur*, *M. sympodialis*, *M. obtusa*, *M. dermatis*, *M. japonica*, and *M. arunalokei* have also been isolated from this organ (Table 3.1). Despite being part of the microbiota, under certain circumstances the population of these yeasts can proliferate excessively and the fungi can become opportunistic pathogens, causing various skin pathologies (Bond et al. 2020). The factors involved in the transformation of these yeasts from commensals to pathogens are still unknown. *Malassezia* species are associated with different diseases in human skin. The most common ones are pityriasis versicolor, seborrheic dermatitis, atopic dermatitis, and folliculitis (Sugita et al. 2010). Cases of iatrogenic fungemia in hospitals caused by *M. pachydermatis*, *M. furfur*, and *M. sympodialis* have been reported. These infections usually occur in neonatal or immunocompromised patients and are related to parenteral nutrition (Rhimi et al. 2020).

3.4.1 *Malassezia* in Animals

Malassezia is part of the normal microbiota of a wide range of warm-blooded vertebrate animals (Cabañes 2014). *Malassezia* yeasts have been isolated from almost every domestic animal species, but also from wild animals (Table 3.1). *Malassezia pachydermatis* has been isolated from healthy dogs and cats but also from dogs and cats with otitis externa. However, this species

Table 3.1 Current described *Malassezia* species and main hosts

Malassezia species	Main host/others[a]
M. furfur, (Robin) Baillon (1889)	Human/pig, elephant, cattle, monkey, sheep, horse, goat, cat, dog, ibis, ostrich, parrot, spoonbill
M. pachydermatis, (Weidman) Dodge (1925)	Dog, cat/carnivores, birds, pig, bear
M. sympodialis, Simmons and Guého (1990)	Human/pig, sheep, horse
M. globosa, Midgley et al. (1996)	Human/cheetah, cattle
M. obtusa, Midgley et al. (1996)	Human
M. restricta, Guého et al. (1996)	Human
M. slooffiae, Guillot et al. (1996)	Human, pig/cat, sheep, goat
M. dermatis, Sugita et al. (2002)	Human
M. japonica, Sugita et al. (2003)	Human
M. nana, Hirai et al. (2004)	Cat, cattle/dog
M. yamatoensis, Sugita et al. (2004)	Human
M. caprae, Cabañes and Boekhout (2007)	Goat/horse
M. equina, Cabañes and Boekhout (2007)	Horse/cattle
M. cuniculi, Cabañes and Castellá (2011)	Rabbit
M. arunalokei, Honnavar et al. (2016)	Human
M. brasiliensis, Cabañes et al. (2016)	Parrot
M. psittaci, Cabañes et al. (2016)	Parrot
M. vespertilionis, Lorch and Vanderwolf (2018)	Bat

[a]Cited only hosts from which these species were isolated and their identification was confirmed by rDNA sequencing methods

is more frequently isolated from dogs than from cats (Bond et al. 2010). From dogs, the lipid-dependent species *M. nana* and *M. furfur* have also been isolated. On cats, *M. nana* was found to be the most common lipid-dependent species but other lipid-dependent species such as *M. furfur* and *M. slooffiae* have also been isolated (Bond et al. 2020; Puig et al. 2018).

Lipid-dependent species, other than *M. pachydermatis* constitute the major population of *Malassezia* spp. in herbivores, such as horses, sheep, cows, and goats (Cabañes 2014; Crespo et al. 2002). *Malassezia furfur* has been found on cattle, sheep, goat, and horses (Puig et al. 2018) while *M. globosa* and *M. nana* have only been isolated from cattle (Hirai et al. 2004). Until now, *M. caprae* and *M. equina* have only been isolated from herbivores (Cabañes et al. 2007).

The skin of pigs is colonized by a range of *Malassezia* species such as *M. pachydermatis*, *M. furfur*, *M. sympodialis*, and *M. slooffiae* that have been isolated from the external ear canal and the skin of healthy pigs (Garau et al. 2005; Nardoni et al. 2010; Puig et al. 2018). Also, *M. pachydermatis* has been isolated from the external ear canal of pigs with otitis externa (Pinter et al. 2002). Recently, this species has been also found in the external ear canal and the anus of healthy brown bears (Hađina et al. 2023).

In birds, different *Malassezia* species have been isolated such as *M. brasiliensis*, *M. psittaci* (Cabañes et al. 2016), *M. furfur*, and *M. pachydermatis* (Puig et al. 2018). *Malassezia* species are apparently rarely isolated from lagomorphs. Only the lipid-dependent species *M. cuniculi* has been described in the skin and the external ear canal of healthy rabbits (Cabañes et al. 2011). In a recent NGS study of these animals (Díaz et al. 2021), *Malassezia* phylotype 131, a non-described *Malassezia* yeast detected in the external auditory ear canal of humans (Zhang et al. 2012), was the predominant taxon in all samples. However, *M. cuniculi* was also detected in all studied samples. Although no growth was obtained in any of the culture media used, the presence of spherical yeast cells was observed in more than 50% of the samples (Fig. 3.3). The last *Malassezia* species described is *M. vespertilionis* from bats (Lorch et al. 2018). The mitogenome of a putative new species isolated from bats has been recently published (Christinaki et al. 2023).

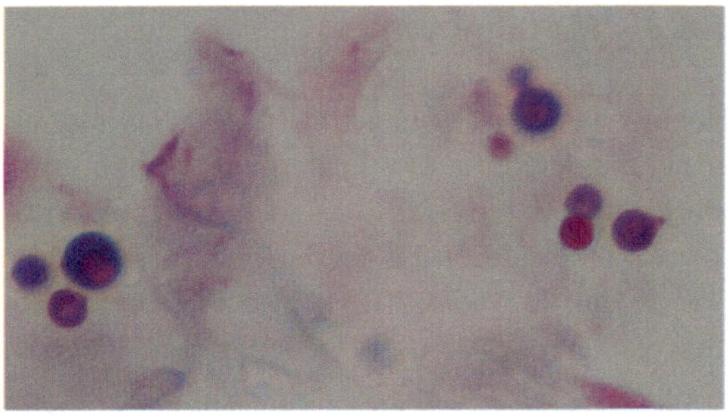

Fig. 3.3 Diff-Quick stain of a smear from an otic swab of a rabbit showing the presence of spherical yeasts cells, possibly of *Malassezia* phylotype 131 or *M. cuniculi* yeasts cells

3.5 *Malassezia*-Related Diseases in Animals

Many factors may be associated with an overgrowth of *Malassezia* yeasts in animals and with their transformation from commensals to opportunistic pathogens (Sugita et al. 2010). The presence of *Malassezia* yeasts within the stratum corneum of the skin exposes the host to an array of allergens, immunogens, and chemicals such as cell-wall carbohydrates, proteins and lipids, and metabolic byproducts (Ashbee and Bond 2010). *Malassezia* antigens can stimulate the host's immune response as well as trigger hypersensitivity reactions. Although the activation of the immune response offers a degree of protection against the over-proliferation of the yeasts, it may also be harmful causing skin inflammation leading to pruritus and dermatitis (Bond et al. 2020; Goh et al. 2022). *Malassezia pachydermatis* is now recognized as an important cause of dermatitis and otitis externa in dogs, however, appears to be a relatively uncommon pathogen in other animal hosts (Bond et al. 2010).

3.5.1 Dermatitis

In *M. pachydermatis* dermatitis, the yeast population can be increased up to 100-10.000-fold above the counts found on a healthy skin. Clinical signs are associated with a hypersensitivity reaction against the yeasts, and, however, also low yeast counts can potentially produce severe clinical signs. Thus, any number of *Malassezia* yeasts may be considered relevant when associated with clinical signs (Bond et al. 2020). Canine pododermatitis caused by *M. pachydermatis* is a common disorder in the general veterinary practice, especially in atopic animals or animals diagnosed with endocrine disease. It may affect the interdigital spaces, nail folds, and nails, and usually more than one paw is affected (Duclos 2013). In dogs with *Malassezia* pododermatitis, pruritus is a major and constant sign often associated with erythema and/or greasy exudate.

3.5.2 Otitis Externa

Otitis externa is the most common disorder of the canine external ear canal, affecting approximately 15% of the dogs presented at the veterinary clinic. *Malassezia pachydermatis* is the most common etiological agent associated with canine otitis externa and is involved in up to 57% of all infections (Scott et al. 2001). It is also the most common *Malassezia* yeast isolated from cats' ears, and the isolation of cells increased in cats with otitis externa (Bond et al. 2020).

3.6 Laboratory Diagnosis and Detection Techniques for *Malassezia* Yeasts

Diagnosis of *Malassezia* otitis externa and dermatitis is commonly based on clinical signs, response of clinical signs to antifungal therapy, and the presence of an elevated number of the yeasts by direct microscopy. Culturing of *Malassezia* is not usually performed in the clinical setting (Bond et al. 2020). However, high cell counts may not necessarily be clinically relevant in each case when detected. By contrast, normal or at least lower cell counts may be sufficient to exacerbate cutaneous inflammation in patients with immediate or delayed hypersensitivity responses (Bond et al. 2020; Cabañes 2021). In addition, we have different molecular techniques that allow detecting these yeasts directly from animal samples.

3.6.1 Cytology

Different cytological techniques have been developed including scraping methods, tape stripping, direct impression with glass slides or smears on a slide from a swab previously rubbed onto the lesional skin. Tape stripping has been widely accepted in the veterinary clinical practice as a rapid and versatile method for recovering *Malassezia* yeasts from the stratum corneum of the skin (Bond et al. 2020).

3.6.2 Real-Time PCR

Molecular techniques, especially PCR approaches such as real-time PCR, namely also quantitative real-time PCR or qPCR, are useful and reliable to detect the presence of *Malassezia* species without the need to culture the yeasts (Sugita et al. 2010). In the animal skin, some qPCR techniques have been developed for *Malassezia* detection and quantification. A study developed a qPCR technique based on the amplification of the β-tubulin gene. This technique allowed to quantify *M. pachydermatis* in the external ear canal of dogs (Puig et al. 2019). Also, multicopy ITS has been used as a target to study the population of *Malassezia* species in samples from dog skin using qPCR (Meason-Smith et al. 2020; Díaz et al. 2023b).

3.6.3 Metagenomics

Next-generation sequencing (NGS) techniques allow the detection of microorganisms in a sample by sequencing millions of DNA fragments in parallel, without the need of prior cultivation (Forde and O'Toole 2013). Two different NGS approaches can be used to detect and identify species within a sample, i.e., shotgun metagenomic sequencing and amplicon-based metagenomic sequencing. In amplicon-based analyses, several regions of fungal ribosomal genes can be used as target to study taxonomy and diversity within samples (Ianiri et al. 2022). Most studies selected the internal transcribed spacer regions 1 and 2 of the rRNA. However, the length of these regions varies between different fungal genera and species which could lead to preferential amplification of distinct regions and thus a biased assessment of abundance. Therefore, the use of alternative target genes has been promoted, such as portions of the genes coding for 26S and 18S rRNA (De Filippis et al. 2017).

At the moment, only a few mycobiome studies have been conducted in dogs and cats. These studies included healthy animals and clinically affected animals with otitis, dermatitis, pododermatitis, and/or allergy (Rodrigues Hoffmann et al. 2014; Meason-Smith et al. 2015, 2017, 2020; Korbelik et al. 2018; Chermprapai et al. 2019; Bradley et al. 2020; Tang et al. 2020; Díaz et al. 2023b). In all these studies, the main phylum identified in healthy animals was *Ascomycota* and the main genera were *Cladosporium* and *Alternaria*, considered transient mycobiota. In clinically affected animals

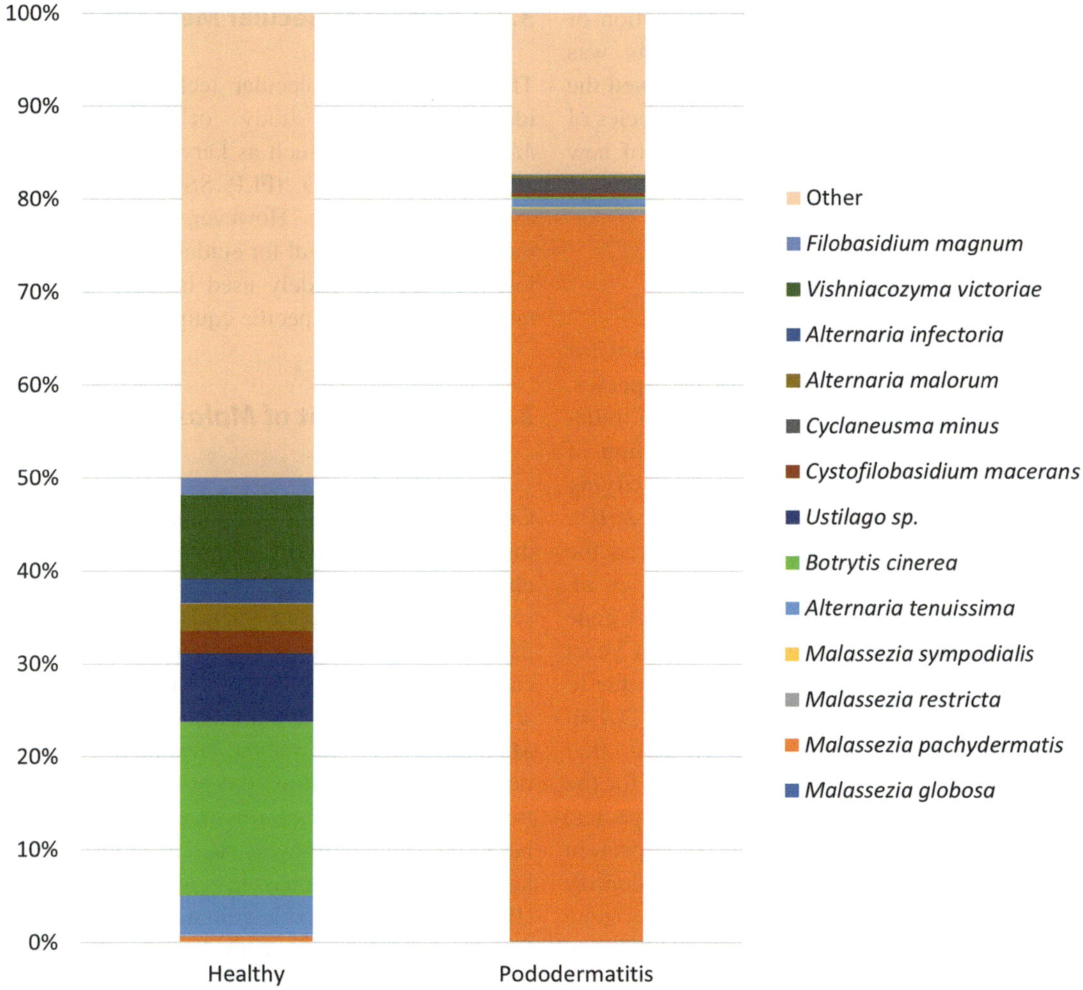

Fig. 3.4 Average relative abundance of fungal species in the interdigital fold of healthy dogs and dogs with pododermatitis obtained in a recent NGS study (Díaz et al. 2023b)

with dermatitis, otitis, and/or allergy, *Basidiomycota* was the main phylum and *Malassezia* the main genus, with *M. pachydermatis* as the main species (Fig. 3.4).

3.7 Identification Methods for *Malassezia* Yeasts

Due to the dependency of *Malassezia* on external lipids, conventional laboratory techniques used to identify yeasts are not applicable to this genus. The main phenotypic *Malassezia* identification tests are based on catalase activity, β-glucosidase activity, growth at different temperatures (32, 37, and 40 °C), assimilation of Tweens 20, 40, 60, 80, and Cremophor EL. Furthermore, *M. pachydermatis* is probably identifiable by its unique ability to grow on SGA (Gueho-Kellerman et al. 2010).

To date, current molecular techniques are most reliable to identify *Malassezia* species, since diagnostic results solely obtained through phenotypic characteristics of cultures can be ambiguous and do not allow the description of new species (Sugita et al. 2010). This was achieved by molecular methods that allowed the unequivocal separation of the different species of *Malassezia* as well as the identification of new species.

3.7.1 DNA Sequencing

DNA sequencing is the most reliable tool for the identification of *Malassezia* species. Ribosomal gene sequencing has made an indispensable contribution to the understanding of the biodiversity of *Malassezia* by identifying all the different species (Sugita et al. 2010). Even though the ITS region is considered the universal barcode for fungi (Schoch et al. 2012), the D1/D2 region of the 26S rRNA gene is more suitable for the identification of yeast species, especially in *Malassezia* yeasts due to its lower variability (Castellá et al. 2014). Scorzetti et al. (2002) demonstrated that sequencing the D1/D2 region is reliable for the identification of basidiomycetous species. Apart from the ribosomal genes, some protein coding genes have been applied to identify *Malassezia* species, such as the β-tubulin gene (Castellá et al. 2011, 2014) and CHS2 gene (Kano et al. 1999).

3.7.2 Genome Sequencing

Whole genome sequencing has made it possible to obtain the genomes of different species within the genus *Malassezia*, thus allowing for a better understanding of the genus. In 2007, the first genome of a species of the genus *Malassezia*, *M. globosa*, obtained by whole genome shotgun sequencing was published (Xu et al. 2007). Then, the genomes of *M. sympodialis* (Gioti et al. 2013) and *M. pachydermatis* (Triana et al. 2015) were published. In 2015, a study reported the genome sequences of 14 *Malassezia* species (Wu et al. 2015).

3.7.3 Other Molecular Methods

There are other molecular techniques for the identification and study of the different *Malassezia* species such as karyotyping, PFGE, AFLP, RFLP, RAPD, tFLP, SSCP, and DGGE (Sugita et al. 2010). However, although these methods may be useful for epidemiological studies, they are not widely used because they are tedious and require specific equipment.

3.8 Treatment of *Malassezia* Diseases

Current options for the treatment of *Malassezia* diseases include various antiseptics such as chlorhexidine, and for systemic and topic therapy several antifungal agents (Bond et al. 2020). The used antifungal agents belong to different chemical classes and have different mechanisms of action. Due to the intrinsic resistance of *Malassezia* species to some antifungals, in particular to echinocandins, the primary antifungal treatment for *Malassezia*-associated diseases is based on azole antifungals such as clotrimazole, ketoconazole, itraconazole, and miconazole (Bond et al. 2010). Azole-containing compounds function by interacting with the lanosterol-14α-demethylase that is involved in the biosynthesis of ergosterol (Vanden Bossche et al. 2003). Also, polyene macrolides like nystatin and natamycin, which interact with membrane sterols, are also effective against *M. pachydermatis*, but at higher concentrations *in vitro* (Bond et al. 2010). The use of amphotericin B, a polyene macrolide, although very active against *M. pachydermatis*, is not recommended for dogs due to its cost, difficulty of usage, and toxicity. Nevertheless, thiabendazole is useful to control *Malassezia* overgrowth. It inhibits the assembly of microtubules in yeast cells (Peano et al. 2020). Antifungal treatments are generally successful in controlling and resolving *Malassezia* yeast overgrowth and disease, but failure or rapid recurrence occasionally occurs. Usually, the reason for failure is the lack of identifying and resolving predisposing factors and primary causes (Miller 2013). In fact, some predisposing breeds, the animal's external environment, skin folds, food allergies, and hypersensitivity disorders, especially in atopic dermatitis, primary cornification defects of the skin and endocrinopathies are common underlying factors and need to be managed together with the antifungal treatment (Bond et al. 2010).

At the moment, most wild-type *Malassezia* yeast are susceptible to the main used azole antifungals such as itraconazole, ketoconazole, and miconazole (Brilhante et al. 2018; Schlemmer et al. 2019), although the efficacy of fluconazole is variable (Velegraki et al. 2004; Weiler et al. 2013; Díaz et al. 2023a).

3.9 Antifungal Resistance in *M. pachydermatis*

Some years ago, Jesus et al. (2011) observed that some *M. pachydermatis* strains can become resistant to azoles by inducing resistance *in vitro*. Recently, some sporadic cases of treatment failure with azoles in canine *M. pachydermatis* dermatitis associated with an increased azole tolerance *in vitro* have been reported (Chiavassa et al. 2014; Watanabe et al. 2014). However, to date only two *M. pachydermatis* strains have been found to be resistant to azoles (Angileri et al. 2019; Kano et al. 2019).

3.9.1 Susceptibility Testing

Establishing a set of standardized criteria for *in vitro* susceptibility testing is essential, since different testing variables are known to impact the *in vitro* determinations. The last edition of the Clinical and Laboratory Standards Institute (2017) developed a reference broth microdilution method for susceptibility testing of yeasts. However, it is universally accepted that these general conditions designed for testing yeasts are not suitable for susceptibility testing of *Malassezia* species due to their unique physiologic features such as their lipid dependency, slower growth rate, and tendency to form clusters. Therefore, different adjustments have been made to susceptibility testing of *Malassezia* yeast, such as modifying the media, time of incubation, and inoculum size but official guidelines are not available yet. Also, other susceptibility testing methods such as disk diffusion method and E-test gradient have been applied to *Malassezia* yeasts (Peano et al. 2017, 2020; Díaz et al. 2023a). However, officially recognized interpretive criteria of *in vitro* results, clinical breakpoints, and epidemiological cut-off values are not available for *Malassezia* spp. While *in vitro* susceptibility results are indicators of the relative susceptibility of a strain to antifungals, they are insufficient to predict the clinical outcome *in vivo* (Peano et al. 2020).

3.9.2 Mechanisms of Resistance

In *Malassezia*, reduced azole susceptibility may be caused by a reduced affinity of the target for the azole (Kano et al. 2019) or by increased expression of the *ERG11* gene encoding the target enzyme of azole (Kim et al. 2018). Also, an increased efflux by overexpression of genes encoding membrane transporters of the ABC transporter family (*CDR1*/*CDR2*) or the major facilitator (*MDR1*) family has been suggested (Bond et al. 2020).

3.9.2.1 Alterations of the *ERG11* Gene

Azole antifungals act by inhibiting a cytochrome P450 enzyme named lanosterol-14-α-demethylase, which participates in the biosynthesis of ergosterol, a major constituent of the fungal cell membrane (Peano et al. 2020). This enzyme is encoded by the *ERG11* gene that belongs to the cytochrome P450 monooxygenase (CYP) superfamily (Zhang et al. 2019). The inactivation of this enzyme causes the demethylation of lanosterol, which inhibits the biosynthesis of ergosterol and leads to the accumulation of bypass products as toxic methylated sterols in the fungal membrane, defects in fungal cytoplasmic membrane integrity, and inhibition of cell growth (Park et al. 2020). Because azole antifungals are fungistatic, their prolonged use can cause resistance.

It is interesting to note that there is a high variability of *ERG11* gene sequences within *M. pachydermatis* strains (Díaz et al. 2023a). Mutations in the coding region of the *ERG11* gene encode amino acid substitutions which alter the structure of the enzyme and thus, reduce its affinity to the antifungal drug (Sanglard et al. 1998; Warrilow et al. 2010). Accordingly, some studies have reported the isolation of strains of different *Malassezia* spp. with tolerance to different azoles associated with mutations in the *ERG11* gene (Kim et al. 2011; Kano et al. 2019, 2020; Leong et al. 2021; Díaz et al. 2023a).

Also, overexpression of the *ERG11* gene results in increased levels of the antifungal target protein, and thus to azole resistance (Flowers et al. 2015; Feng et al. 2017). A tandem quadruplication of a region in chromosome 4, which includes the genes *ERG11* and *ERG4* whose overexpression is associated with azole resistance, was detected in *M. pachydermatis* isolates that exhibited a high minimum inhibitory concentration (MIC) to ketoconazole (Kim et al. 2018).

3.9.2.2 Efflux Pumps

The increased expression of drug efflux pumps is considered the major cause of azole resistance in *Candida* spp. There are two classes of efflux pumps, i.e., the ATP-binding cassette (ABC) transporters (encoded by *CDR1/CDR2* genes) that use as energy source the hydrolysis of ATP, and the major facilitator superfamily (encoded by *MDR1* gene) that use the membrane potential to drive the efflux. The up-regulation of the genes encoding these transporters induces resistance in *Candida* spp. (Cannon et al. 2009). Accordingly, it has been reported that efflux pumps, especially the major facilitator superfamily might be a defense mechanism against azoles in *M. restricta* (Park et al. 2020), *M. pachydermatis*, and *M. furfur* (Iatta et al. 2017).

3.10 Conclusion

Knowledge about the presence of *Malassezia* yeasts and their significance in domestic or wild animals is currently scarce. These yeasts are isolated from the skin of animals including humans, although they have also been detected in other organs. However, currently it is unknown whether their presence in the microbiota in these internal locations is purely transient.

By using classic microbiology techniques, such as direct microscopic observation and culture, and the most innovative molecular techniques, such as NGS, we begin to understand the role of *M. pachydermatis* in different skin pathologies in different animal species, mainly in dogs. In these animals, pathologies as common as otitis and dermatitis usually respond well to current antifungals. However, although most strains of *M. pachydermatis* tend to be mostly sensitive, some strains that are resistant or have high MICs to some antifungals have already been detected. Therefore, standardized and validated techniques are necessary to evaluate antifungal susceptibility *in vitro* of these yeasts.

The use of culture-independent techniques, such as NGS, has allowed the detection of *Malassezia* DNA sequences in animal species other than mammals and birds, such as nematodes (Renker et al. 2003), coral colonies, lobsters (Amend 2014), and marine sponges (Gao et al. 2008). However, the presence of cells from these *Malassezia* yeasts in these animals has yet to be confirmed. Thus, the significance of the presence of these DNA sequences in these unusual locations remains to be proven.

References

Amend A (2014) From dandruff to deep-sea vents: *Malassezia*-like fungi are ecologically hyper-diverse. PLoS Pathog 10(8):e1004277

Angileri M, Pasquetti M, De Lucia M, Peano A (2019) Azole resistance of *Malassezia pachydermatis* causing treatment failure in a dog. Med Mycol Case Rep 23:58–61

Ashbee HR, Bond R (2010) *Malassezia* species and immunity: host-pathogen interactions. In: Boekhout T, Gueho E, Mayser P et al (eds) *Malassezia* and the skin. Science and clinical practice. Springer, Berlin, Heidelberg, pp 139–174

Bond R, Guillot J, Cabañes FJ (2010) *Malassezia* yeast in animal disease. In: Boekhout T, Guého-Kellermann E, Mayser P, Velegraki A (eds) *Malassezia* and the Skin. Science and clinical practice. Springer, Berlin, Heidelberg, pp 271–299

Bond R, Morris DO, Guillot J et al (2020) Biology, diagnosis and treatment of *Malassezia* dermatitis in dogs and cats Clinical Consensus Guidelines of the World Association for Veterinary Dermatology. Vet Dermatol 31:28–74

Bradley CW, Lee FF, Rankin SC et al (2020) The otic microbiota and mycobiota in a referral population of dogs in eastern USA with otitis externa. Vet Dermatol 31:225–249

Brilhante RSN, Rocha MGD, Guedes GMM et al (2018) *Malassezia pachydermatis* from animals: planktonic and biofilm antifungal susceptibility and its virulence arsenal. Vet Microbiol 220:47–52

Cabañes FJ (2014) *Malassezia* yeasts: how many species infect humans and animals? PLoS Pathog 10(2):e1003892

Cabañes FJ (2020) *Malassezia pachydermatis*: to be, or not to be lipid-dependent. Rev Iberoam Micol 37:3–4

Cabañes FJ (2021) Diagnosis of *Malassezia* dermatitis and otitis in dogs and cats, is it just a matter of counting? Rev Iberoam Micol 38:3–4

Cabañes FJ, Theelen B, Castellá G, Boekhout T (2007) Two new lipid-dependent *Malassezia* species from domestic animals. FEMS Yeast Res 7:1064–1076

Cabañes FJ, Vega S, Castellá G (2011) *Malassezia cuniculi* sp. nov., a novel yeast species isolated from rabbit skin. Med Mycol 49:40–48

Cabañes FJ, Coutinho SD, Puig L et al (2016) New lipid-dependent *Malassezia* species from parrots. Rev Iberoam Micol 33:92–99

Cannon RD, Lamping E, Holmes AR et al (2009) Efflux-mediated antifungal drug resistance. Clin Microbiol Rev 22:291–321

Castellá G, De Bellis F, Bond R, Cabañes FJ (2011) Molecular characterization of *Malassezia nana* isolates from cats. Vet Microbiol 148:363–367

Castellá G, Coutinho SD, Cabañes FJ (2014) Phylogenetic relationships of *Malassezia* species based on multilocus sequence analysis. Med Mycol 52:99–105

Chermprapai S, Ederveen THA, Broere F et al (2019) The bacterial and fungal microbiome of the skin of healthy dogs and dogs with atopic dermatitis and the impact of topical antimicrobial therapy, an exploratory study. Vet Microbiol 229:90–99

Chiavassa E, Tizzani P, Peano A (2014) *In vitro* antifungal susceptibility of *Malassezia pachydermatis* strains isolated from dogs with chronic and acute otitis externa. Mycopathologia 178:315–319

Christinaki AC, Theelen B, Zania A et al (2023) Co-evolution of large inverted repeats and G-quadruplex DNA in fungal mitochondria may facilitate mitogenome stability: the case of *Malassezia*. Sci Rep 13(1):6308

CLSI. Performance Standards for Antifungal Susceptibility Testing of Yeasts (2017) 1st ed. In: CLSI supplement M60. Clinical and Laboratory Standards Institute, Wayne, PA

Coelho MA, Sampaio JP, Goncalves P (2010) A deviation from the bipolar-tetrapolar mating paradigm in an early diverged basidiomycete. PLoS Genet 6(8):e1001052

Coelho MA, Sampaio JP, Gonçalves P (2013) Living and thriving on the skin: *Malassezia* genomes tell the story. mBio 4(2):e00117–e00113

Crespo MJ, Abarca ML, Cabañes FJ (2002) Occurrence of *Malassezia* spp. in horses and domestic ruminants. Mycoses 45:333–337

De Filippis F, Laiola M, Blaiotta G, Ercolini D (2017) Different amplicon targets for fungal sequencing-based studies of fungal diversity. Appl Environ Microbiol 83(17): e00905-17

Díaz L, Castellá G, Bragulat MR et al (2021) External ear canal mycobiome of some rabbit breeds. Med Mycol 59:683–693

Díaz L, Castellá G, Bragulat MR, Cabañes FJ (2023a) *ERG11* gene variability and azole susceptibility in *Malassezia pachydermatis*. Mycopathologia 188:21–34

Díaz L, Castellá G, Bragulat MR et al (2023b) Study of the variation of the *Malassezia* load in the interdigital fold of dogs with pododermatitis. Vet Res Commun 47:385–396

Duclos D (2013) Canine pododermatitis. Vet Clin North Am Small Anim Pract 43:69–70

Elaskandrany M, Patel R, Patel M et al (2021) Fungi, host immune response, and tumorigenesis. Am J Physiol Gastrointest Liver Physiol 321:G213–G222

Feng W, Yang J, Xi Z et al (2017) Mutations and/or overexpressions of *ERG4* and *ERG11* genes in clinical azoles-resistant isolates of *Candida albicans*. Microb Drug Resist 2:563–570

Flowers SA, Colón B, Whaley SG et al (2015) Contribution of clinically derived mutations in *ERG11* to azole resistance in *Candida albicans*. Antimicrob Agents Chemother 59:450–460

Forde BM, O'Toole PW (2013) Next-generation sequencing technologies and their impact on microbial genomics. Brief Funct Genomics 12:440–453

Gaitanis G, Magiatis P, Hantschke M et al (2012) The *Malassezia* genus in skin and systemic diseases. Clin Microbiol Rev 25:106–141

Gao Z, Li B, Zheng C, Wang G (2008) Molecular detection of fungal communities in the Hawaiian marine sponges *Suberites zeteki* and *Mycale armata*. Appl Environ Microbiol 74:6091–6101

Garau M, del Palacio A, García J (2005) Prevalence of *Malassezia* spp. in healthy pigs. Mycoses 48:17–20

Gioti A, Nystedt B, Li W et al (2013) Genomic insights into the atopic eczema-associated skin commensal yeast *Malassezia sympodialis*. mBio 4(1):e00572–e00512

Goh JPZ, Ruchti F, Poh SE et al (2022) The human pathobiont *Malassezia furfur* secreted protease Mfsap1 regulates cell dispersal and exacerbates skin inflammation. Proc Natl Acad Sci USA 119(49):e2212533119

Gueho-Kellerman E, Boekhout T, Begerow D et al (2010) Biodiversity, phylogeny and ultrastructure. In: Boekhout T, Gueho-Kellerman E, Mayser P et al (eds) *Malassezia* and the skin. Science and clinical practice. Springer, Berlin, Heidelberg, pp 17–63

Guillot J, Bond R (2020) *Malassezia* yeasts in veterinary dermatology: an updated overview. Front Cell Infect Microbiol 10:79. https://doi.org/10.3389/fcimb.2020.00079

Hađina S, Bruvo Mađarić B, Kazazić S et al (2023) *Malassezia pachydermatis* from brown bear: a comprehensive analysis reveals novel genotypes and distribution of all detected variants in domestic and wild animals. Front Microbiol 14:1151107

Heitman J, Carter DA, Dyer PS, Soll DR (2014) Sexual reproduction of human fungal pathogens. Cold Spring Harb Perspect Med 4(8):a019281

Hirai A, Kano R, Makimura K et al (2004) *Malassezia nana* sp. nov., a novel lipid-dependent yeast species isolated from animals. Int J Syst Evol Microbiol 54:623–627

Ianiri G, Averette AF, Kingsbury JM, Heitman J, Idnurm A (2016) Gene function analysis in the ubiquitous human commensal and pathogen *Malassezia* genus. mBio 7(6):e01853–e01816

Ianiri G, LeibundGut-Landmann S, Dawson TL Jr (2022) *Malassezia:* a commensal, pathogen, and mutualist of human and animal skin. Ann Rev Microbiol 76:757–782

Iatta R, Puttilli MR, Immediato D et al (2017) The role of drug efflux pumps in *Malassezia pachydermatis* and *Malassezia furfur* defense against azoles. Mycoses 60:178–182

Jesus FP, Lautert C, Zanette RA et al (2011) *In vitro* susceptibility of fluconazole-susceptible and -resistant isolates of *Malassezia pachydermatis* against azoles. Vet Microbiol 152:161–164

Kano R, Aizawa T, Nakamura Y et al (1999) Chitin synthase 2 gene sequence of *Malassezia* species. Microbiol Immunol 43:813–815

Kano R, Yokoi S, Kariya N et al (2019) Multi-azole-resistant strain of *Malassezia pachydermatis* isolated from a canine *Malassezia* dermatitis. Med Mycol 57:346–350

Kano R, Aramaki C, Murayama N et al (2020) High multi-azoleresistant *Malassezia pachydermatis* clinical isolates from canine *Malassezia* dermatitis. Med Mycol 58:197–200

Kathrani A, Theelen B, Bond R (2023) Isolation of *Malassezia* yeasts from dogs with gastrointestinal disease undergoing duodenal endoscopy. J Small Anim Pract 1–8. https://doi.org/10.1111/jsap.13649

Kim D, Lim YR, Ohk SO et al (2011) Functional expression and characterization of CYP51 from dandruff-causing *Malassezia globosa*. FEMS Yeast Res 11:80–87

Kim M, Cho Y, Park M et al (2018) Genomic tandem quadruplication is associated with ketoconazole resistance in *Malassezia pachydermatis*. J Microbiol Biotechnol 28:1937–1945

Korbelik J, Singh A, Rosseau J, Weese JS (2018) Analysis of the otic mycobiota in dogs with otitis externa compared to healthy individuals. Vet Dermatol 29(5):417–e138

Leong C, Kit JCW, Lee SM et al (2021) Azole resistance mechanisms in pathogenic *M. furfur*. Antimicrob Agents Chemother 65(5):e01975–ee2020

Lorch JM, Palmer JM, Vanderwolf KJ et al (2018) *Malassezia vespertilionis* sp. nov.: a new cold-tolerant species of yeast isolated from bats. Persoonia 41:56–70

Mayser P, Gaitanis G (2010) Physiology and biochemistry. In: Boekhout T, Guého-Kellermann E, Mayser P, Velegraki A (eds) *Malassezia* and the Skin. Science and clinical practice. Springer, Berlin, Heidelberg, pp 271–299

Meason-Smith C, Diesel A, Patterson AP et al (2015) What is living on your dog's skin? Characterization of the canine cutaneous mycobiota and fungal dysbiosis in canine allergic dermatitis. FEMS Microbiol Ecol 91(12):fiv139

Meason-Smith C, Diesel A, Patterson AP et al (2017) Characterization of the cutaneous mycobiota in healthy and allergic cats using next generation sequencing. Vet Dermatol 28(1):71e-17

Meason-Smith C, Olivry T, Lawhon SD, Hoffmann AR (2020) *Malassezia* species dysbiosis in natural and allergen-induced atopic dermatitis in dogs. Med Mycol 58:756–765

Miller WH (ed) (2013) Muller and Kirk's small animal dermatology, 7th edn. Saunders WB, Philadelphia

Nardoni S, Merildi V, Frangioni S et al (2010) Isolation and characterization of *Malassezia* spp. in healthy swine of different breeds. Vet Microbiol 141:155–158

Nielsen K, Heitman J (2007) Sex and virulence of human pathogenic fungi. Adv Genet 57:143–173

Park M, Cho Y, Lee YW, Jung WH (2020) Genomic multiplication and drug efflux influence ketoconazole resistance in *Malassezia restricta*. Front Cell Infect Microbiol 10:191

Peano A, Pasquetti M, Tizzani P et al (2017) Methodological issues in antifungal susceptibility testing of *Malassezia pachydermatis*. J Fungi 3(3):37

Peano A, Johnson E, Chiavassa E et al (2020) Antifungal resistance regarding *Malassezia pachydermatis*: where are we now? J Fungi 6(2):93

Pinter L, Anthony RM, Glumac N et al (2002) Apparent cross-infection with a single strain of *Malassezia pachydermatis* on a pig farm. Acta Vet Hung 50:151–156

Puig L, Bragulat MR, Castellá G, Cabañes FJ (2017) Characterization of the species *Malassezia pachydermatis* and re-evaluation of its lipid dependence using a synthetic agar medium. PLoS One 12(6):e0179148

Puig L, Bragulat MR, Castellá G, Cabañes FJ (2018) Phenotypic and genetic diversity of *Malassezia furfur* from domestic and zoo animals. Med Mycol 56:941–949

Puig L, Castellá G, Cabañes FJ (2019) Quantification of *Malassezia pachydermatis* by real-time PCR un swabs from the external ear canal of dogs. J Vet Diagn Invest 31:440–447

Renker C, Alphei J, Buscot F (2003) Soil nematodes associated with the mammal pathogenic fungal genus Malassezia (Basidiomycota: Ustilaginomycetes) in Central European forests. Biol Fertil Soils 37:70–72

Rhimi W, Theelen B, Boekhout T, Otranto D, Cafarchia C (2020) *Malassezia* spp. yeasts of emerging concern in fungemia. Front Cell Infect Microbiol 10:370

Rodrigues Hoffmann A, Patterson AP, Diesel A et al (2014) The skin microbiome in healthy and allergic dogs. PLoS One 9(1):e83197

Sanglard D, Ischer F, Koymans L, Bille J (1998) Amino acid substitutions in the cytochrome P-450 lanosterol 14alpha-demethylase (CYP51A1) from azole-resistant *Candida albicans* clinical isolates contribute to resistance to azole antifungal agents. Antimicrob Agents Chemother 42:241–253

Sankaranarayanan SR, IaniriG CMA et al (2020) Loss of centromere function drives karyotype evolution in closely related *Malassezia* species. elife 9:e53944

Schlemmer KB, de Jesus FPK, Loreto ES et al (2019) *In vitro* combination of antifungal agents against *Malassezia pachydermatis*. Med Mycol 57:324–327

Schoch CL, Seifert KA, Hunhdorf S et al (2012) Nuclear ribosomal trancribed spacer (ITS) region as a universal DNA barcode marker for Fungi. Proc Natl Acad Sci USA 109:6241–6246

Scorzetti G, Fell JW, Fonseca A, Statzell-Tallman A (2002) Systematics of basidiomycetous yeasts: a comparison of large subunit D1/D2 and internal transcribed spacer rDNA regions. FEMS Yeast Res 2:495–517

Scott DW, Miller WH, Griffin CE (2001) Muller & Kirk's small animal dermatology. WB Saunders, Philadelphia

Stalhberger T, Simenel C, Clavaud C et al (2014) Chemical organization of the cell wall polysaccharide core of *Malassezia restricta*. J Biol Chem 289:12647–12656

Sugita T, Boekhout T, Velegraki A et al (2010) Epidemiology of *Malassezia*-related skin diseases. In: Boekhout T, Gueho-Kellerman E, Mayser P et al (eds) *Malassezia* and the skin. Science and clinical practice. Springer, Berlin, Heidelberg, pp 65–119

Tang S, Prem A, Tjokrosurjo J et al (2020) The canine skin and ear microbiome: a comprehensive survey of pathogens implicated in canine skin and ear infections using a novel next-generation-sequencing-based assay. Vet Microbiol 247:108764

Triana S, Gonzalez A, Ohm RA et al (2015) Draft genome sequence of the animal and human pathogen *Malassezia pachydermatis* Strain CBS 1879. Genome Announc 3(5):e01197–e01115

Trifinopoulos J, Nguyen LT, von Haeseler A, Minh BQ (2016) W-IQ-TREE: a fast online phylogenetic tool for maximum likelihood analysis. Nucleic Acids Res 44:W232–W235

Vanden Bossche H, Engelen M, Rochette F (2003) Antifungal agents of use in animal health-chemical, biochemical and pharmacological aspects. J Vet Pharmacol Ther 26:5–29

Velegraki A, Alexopoulos EC, Kritikou S, Gaitanis G (2004) Use of fatty acid RPMI 1640 media for testing susceptibilities of eight *Malassezia* species to the new triazole Posaconazole and to six established antifungal agents by a modified NCCLS M27–A2 microdilution method and Etest. J Clin Microbiol 42:3589–3593

Wang QM, Theelen B, Groenewald M, Bai FY, Boekhout T (2014) *Moniliellomycetes* and *Malasseziomycetes*, two new classes in *Ustilaginomycotina*. Persoonia 33:41–47

Warrilow AG, Martel CM, Parker JE et al (2010) Azole binding properties of *Candida albicans* sterol 14-a demethylase (CaCYP51). Antimicrob Agents Chemother 54:4235–4245

Watanabe S, Koike A, Kano R et al (2014) In vitro susceptibility of *Malassezia pachydermatis* isolates from canine skin with atopic dermatitis to ketoconazole and itraconazole in East Asia. J Vet Med Sci 76:579–581

Weiler CB, de Jesus FP, Nardi GH et al (2013) Susceptibility variation of *Malassezia pachydermatis* to antifungal agents according to isolate source. Braz J Microbiol 44:174–178

Wu G, Zhao H, Li C et al (2015) Genus-wide comparative genomics of *Malassezi*a delineates its phylogeny, physiology, and niche adaptation on human skin. PLoS Genet 11(11):e1005614

Xu J, Saunders CW, Hu P et al (2007) Dandruff-associated *Malassezia* genomes reveal convergent and divergent virulence traits shared with plant and human fungal pathogens. Proc Natl Acad Sci USA 104:18730–18735

Zhang E, Tanaka T, Tsuboi R et al (2012) Characterization of *Malassezia* microbiota in the human external auditory canal and on the sole of the foot. Microbiol Immunol 56:238–244

Zhang J, Li L, Lv Q et al (2019) The fungal CYP51s: their functions, structures, related drug resistance, and inhibitors. Front Microbiol 10:691

Extracellular Proteins and Their Roles in *Aspergillus fumigatus* Pathogenesis

4

Lei-Jie Jia and Axel A. Brakhage

Abstract

Aspergillus fumigatus is the most significant airborne human-pathogenic fungus capable of causing infections and allergies in humans. Diseases associated with *A. fumigatus* are initiated by inhalation of its conidia. Surface proteins on these conidia can mediate direct contact with host immune system, facilitate adhesion to host cells, but also shield conidia from phagosomal elimination. Conidia, germlings, and outgrowing hyphae not only damage host cells and tissues, resulting in infections, but also expose and release antigens that provoke allergic reactions. Despite the identification of nearly two thousand surface-exposed or secreted proteins through bioinformatics and/or proteomics studies, only a few of these proteins have been functionally evaluated for their roles in pathogenesis. In this chapter, we will provide a summary of the current findings regarding *A. fumigatus* extracellular proteins and for their contributions to the development of diseases.

Keywords

Aspergillus fumigatus · Surface protein Adhesin · Phagocytosis · Phagosome Complement · Allergen

4.1 Introduction

Aspergillus fumigatus is an opportunistic human pathogen that causes disseminated infections in immunocompromised patients, chronic colonization in immunocompetent individuals, and allergic responses in atopic patients (Brakhage 2005; Dagenais and Keller 2009; Feys et al. 2022; Fisher et al. 2022; Kousha et al. 2011; Latgé and Chamilos 2019; Lionakis et al. 2023; Taccone et al. 2015). During the coronavirus disease 2019 (COVID-19) pandemic, pulmonary aspergillosis has occurred in about 10% of patients with severe respiratory virus infections who required admis-

L.-J. Jia
Junior Research Group Phagosome Biology and Engineering, Leibniz Institute for Natural Product Research and Infection Biology – Hans Knöll Institute (Leibniz-HKI), Jena, Germany

Department of Molecular and Applied Microbiology, Leibniz Institute for Natural Product Research and Infection Biology – Hans Knöll Institute (Leibniz-HKI), Jena, Germany
e-mail: leijie.jia@leibniz-hki.de

A. A. Brakhage (✉)
Department of Molecular and Applied Microbiology, Leibniz Institute for Natural Product Research and Infection Biology – Hans Knöll Institute (Leibniz-HKI), Jena, Germany

Department of Microbiology and Molecular Biology, Institute of Microbiology, Friedrich Schiller University, Jena, Germany
e-mail: axel.brakhage@leibniz-hki.de

A. A. Brakhage et al. (eds.), *Human and Animal Relationships*, The Mycota 6,
https://doi.org/10.1007/978-3-031-64853-3_4

sion to the intensive care unit, regardless of their immune status, and coinfection has been associated with a high (>50%) mortality rate (Feys et al. 2022; Kariyawasam et al. 2022; Schauwvlieghe et al. 2018). Because of these reasons, *A. fumigatus* has been listed as one of the particularly critical fungal pathogens by the World Health Organization, alongside *Candida albicans*, *Candida auris*, and *Cryptococcus neoformans*, due to their significant impact on human health and their ability to cause life-threatening infections (WHO 2022).

A. fumigatus produces conidia, asexually produced spores, that are released into the air and are continuously inhaled by human individuals. Because of their small size of 2–3 μm, conidia can easily reach the lung alveoli and cause a wide spectrum of diseases in humans, depending on their immune status (Brakhage and Langfelder 2002; Latgé and Chamilos 2019). Despite this ongoing challenge, healthy individuals rarely develop *Aspergillus*-associated diseases. However, the fungus can cause life-threatening invasive aspergillosis in immunocompromised individuals (Fig. 4.1a) caused by neutropenia, corticosteroid therapy, or by haematopoietic stem cell transplantation, and in patients with severe respiratory virus infections (Brakhage 2005; Earle et al. 2023; Hoenigl et al. 2022; Lionakis et al. 2023). In patients with lung cavities, commonly due to pulmonary tuberculosis, the inhalation of conidia can lead to chronic pulmonary aspergillosis (Bongomin 2020; Denning et al. 2016). For the atopic patients, especially those with cystic fibrosis, inhalation of conidia may cause allergic bronchopulmonary aspergillosis (ABPA) or severe asthma with fungal sensitization (Agarwal et al. 2013). In all the aforementioned *Aspergillus*-associated diseases (Fig. 4.1a), fungal surface cell wall components as well as secreted molecules mediate the first contact with the host immune system and contribute to immune stimulation, immune evasion, adherence, and virulence (Fig. 4.1b) (Blango et al. 2019; Gow et al. 2017; Heinekamp et al. 2015; Latgé and Chamilos 2019; Schwarz et al. 2023; Voltersen et al. 2018; Wartenberg et al. 2011). Therefore, these proteins could serve as possible biomarkers for the diagnosis and as valuable targets for the treatment of *Aspergillus* infections.

The dynamic cell wall structures of *A. fumigatus* conidial and hyphal morphotypes are well reviewed (Latgé et al. 2017). The innermost layer is composed of a conserved structural

Fig. 4.1 *A. fumigatus* pathogenesis, and fungal extracellular proteins influencing host-pathogen interaction. (**a**) Diseases caused by *A. fumigatus* and the estimated number of individuals affected. ABPA, allergic bronchopulmonary aspergillosis; CPA, chronic pulmonary aspergillosis; IPA, invasive pulmonary aspergillosis. (**b**) *A. fumigatus* pathogenesis at the cellular level. Upon inhalation by the human host, conidia can reach the lung alveoli, where they come into contact with and are phagocytosed by host immune cells, such as macrophages and neutrophils, as well as epithelial cells. As an opportunistic pathogen, conidia of *A. fumigatus* can escape from phagosomal killing by interfering with phagosome maturation and redirecting phagosomes to the non-degradative pathway. The formed hyphae invade host tissues in immunocompromised patients or cause allergic responses in atopic patients by exposing and releasing allergens. (**c**) The roles of extracellular proteins in *A. fumigatus* pathogenesis include their interference with phagocytosis, disarming complement, and triggering allergic responses in the host. To escape from professional phagocytes, the fungus covers PAMPs through hydrophobin RodA and the CcpA protein, as well as DHN-melanin on the surface of dormant conidia. Surface adhesins and DHN-melanin also contribute to inducing phagocytosis of conidia by epithelial cells. *A. fumigatus* prevents opsonization by complement through the recruitment of complement regulators and the secretion of complement-degrading proteases. Inside phagosomes, *A. fumigatus* anchors p11 to the phagosome membrane, thereby redirecting phagosomes to the non-degradative pathway. Additionally, the fungus secretes enzymes like CAT and SOD to counteract the oxidants within the phagosomal lumen. The germination of conidia into hyphae exposes and releases more antigens, triggering allergic responses in atopic patients. A2t, annexin A2-p11 heterotetramer; Alp1, alkaline protease 1; Aspf2, allergen Asp f 2; CAT, catalase; CD4, cluster of differentiation 4; C3b, complement component 3b; C4, complement component 4; C5, complement component 5; CR, complement receptor; EP, early phagosome; FHL-1, factor H-like protein 1; HscA, heat shock cognate A; iC3b, inactivated C3b; LP, late phagosome; Mep1, metalloprotease 1; SOD, superoxide dismutase; TCR, Tcell receptor; TRPV4, transient receptor potential cation channel subfamily V member 4

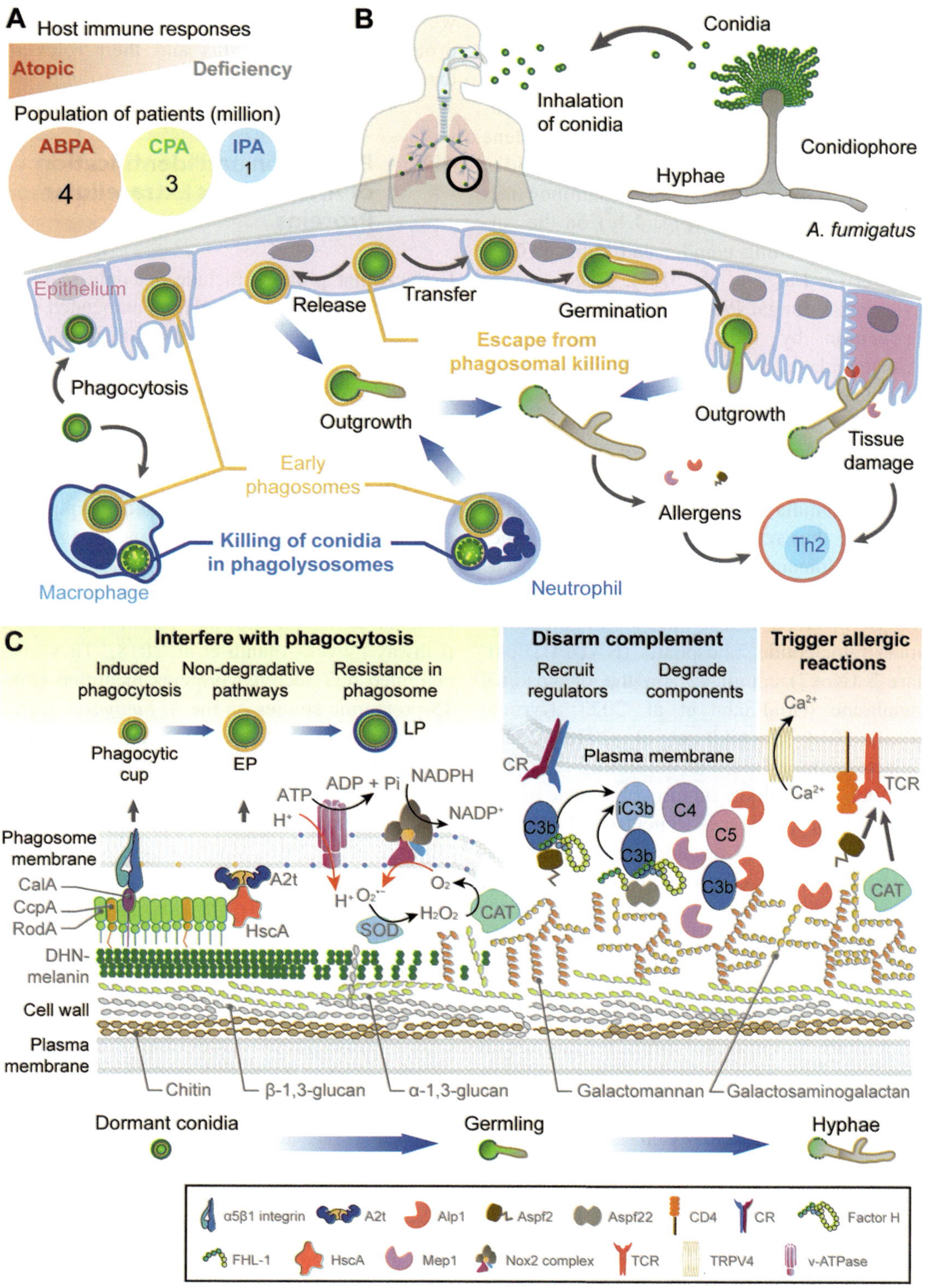
A
Host immune responses
Atopic
Deficiency
Population of patients (million)
ABPA 4
CPA 3
IPA 1
B
Conidia
Inhalation of conidia
Conidiophore
Hyphae
A. fumigatus
Epithelium
Release
Transfer
Germination
Escape from phagosomal killing
Phagocytosis
Outgrowth
Outgrowth
Tissue damage
Early phagosomes
Allergens
Killing of conidia in phagolysosomes
Th2
Macrophage
Neutrophil
C
Interfere with phagocytosis
Induced phagocytosis
Non-degradative pathways
Resistance in phagosome
Disarm complement
Recruit regulators
Degrade components
Trigger allergic reactions
Phagocytic cup
EP
LP
CR
Plasma membrane
TCR
Phagosome membrane
CalA
CcpA
RodA
A2t
HscA
ATP
ADP + Pi
NADPH
NADP+
SOD
CAT
C3b
iC3b
C4
C5
DHN-melanin
Cell wall
Plasma membrane
Chitin
β-1,3-glucan
α-1,3-glucan
Galactomannan
Galactosaminogalactan
Dormant conidia
Germling
Hyphae
α5β1 integrin
A2t
Alp1
Aspf2
Aspf22
CD4
CR
Factor H
FHL-1
HscA
Mep1
Nox2 complex
TCR
TRPV4
v-ATPase

skeletal layer of β-1,3-glucan and chitin (Fig. 4.1c), which are critical pathogen-associated molecular patterns (PAMPs) for fungal pathogens. The outer layers consist of a proteinaceous rodlet layer and a dihydroxynaphthalene (DHN)-melanin layer found on dormant conidia, as well as polysaccharidic galactosaminogalactan and galactomannan (Fig. 4.1c) on the surface of germinating conidia and hyphae (Gow et al. 2017; Latgé et al. 2017). These surface layers mask fungal PAMPs to prevent immune recognition by the host immune system (Aimanianda et al. 2009; Akoumianaki et al. 2016; Carrion et al. 2013; Gravelat et al. 2013; Langfelder et al. 2003). Additional virulence attributes were associated with DHN-melanin and galactosaminogalactan. DHN-melanin on dormant conidia sequesters Ca^{2+} inside phagosome and thus reduces the formation of lipid-raft microdomains on the phagosomal membrane thereby reducing the assembly of vacuolar-type ATPase (v-ATPase) and the nicotinamide adenine dinucleotide phosphate (NADPH) oxidase-2 (Nox2) complexes on the phagosomal membrane (Goldmann et al. 2023; Kyrmizi et al. 2018; Schmidt et al. 2018, 2020). Galactosaminogalactan on the surface of hyphae was found to facilitate binding of hyphae to host cells, enhance resistance to neutrophil extracellular traps, induce neutrophil apoptosis, dysregulate interleukin (IL)-1 signalling, and the formation of biofilm (Fontaine et al. 2011; Gresnigt et al. 2014; Lee et al. 2015; Rambach et al. 2015; Robinet et al. 2014).

Although polysaccharides are much more abundant in the cell wall than proteins (Latgé et al. 2017), hundreds of different surface and secreted proteins have been detected in several proteomics studies (Blango et al. 2020; Jia et al. 2020; Valsecchi et al. 2019a; Vivek-Ananth et al. 2018; Voltersen et al. 2018). Many surface proteins or secreted proteins of *A. fumigatus* have recently been found to play important roles in fungal infection or as allergens, causing allergic diseases. In this chapter, we provide a summary of these extracellular proteins of *A. fumigatus* and their roles in diseases.

4.2 Prediction and Identification of *A. fumigatus* Extracellular Proteins

Extracellular proteins include not only secreted but also cell wall- and membrane-associated proteins and cell membrane proteins that could be exposed on the cell surface. These proteins can be predicted based on the presence of signal peptides, glycosylphosphatidylinositol anchor domains, or transmembrane domains in the deduced protein sequences. This way, Vivek-Ananth et al. employed a computational pipeline to predict extracellular proteins, including those secreted through the classical and non-classical secretion pathways, as well as cell membrane proteins delivered *via* the classical secretion pathway (Vivek-Ananth et al. 2018). They also compiled 437 secreted proteins identified from 15 proteomic studies on the *A. fumigatus* secretome (Adav et al. 2013, 2015; Asif et al. 2006; Champer et al. 2016; Farnell et al. 2012; Kim et al. 2007; Kumar et al. 2011; Liu et al. 2013; Neustadt et al. 2009; Ouyang et al. 2010; Sharma Ghimire et al. 2016; Sharma et al. 2011; Singh et al. 2010; Sriranganadane et al. 2010; Wartenberg et al. 2011). This comprehensive approach led to the identification of an extracellular proteome containing 1791 *A. fumigatus* proteins (Vivek-Ananth et al. 2018).

However, numerous proteins of predominantly cytoplasmic origin, such as translation elongation factors and heat shock proteins, have also been detected on the surface of *A. fumigatus* (Blango et al. 2020; Jia et al. 2020, 2023). Through surface proteome methods, e.g., trypsin shaving and surface biotinylation coupled with liquid chromatography-tandem mass spectrometry (LC-MS/MS) analysis (Bigalke et al. 2023), 946 different proteins were detected as surface-exposed proteins on different *A. fumigatus* morphotypes (Blango et al. 2020; Jia et al. 2020; Valsecchi et al. 2019a; Voltersen et al. 2018). Valsecchi et al. identified 1045 to 2326 surface proteins from conidia lacking either a single or multiple components of rodlet layer, melanin layer, and α-1,3-glucan (Valsecchi et al. 2019a). These findings indicate a high number of extracellular proteins reflecting the complex cell wall biosynthesis but likely also the complex interaction between *A. fumigatus* and host cells.

4.3 Extracellular Proteins and Their Roles in Pathogenicity of *A. fumigatus*

After inhalation, dormant conidia of *A. fumigatus* encounter host macrophages, airway epithelial cells, and neutrophils in the lung alveoli and are engulfed and eliminated by these cells. However, several studies have shown that *A. fumigatus* conidia can survive from phagosomal killing by masking PAMPs, interfering with phagosomal acidification, preventing LC-3 associated phagocytosis and phagosome maturation, and inducing non-lytic expulsion (Akoumianaki et al. 2016; Amin et al. 2014; Brakhage et al. 2010; Jia et al. 2023; Kyrmizi et al. 2018; Liu et al. 2016; Pazhakh et al. 2019; Schmidt et al. 2020; Seidel et al. 2020; Thywißen et al. 2011; Wasylnka and Moore 2003). Surface molecules, including DHN-melanin as well as surface proteins, play an essential role in these immune evasion strategies.

4.3.1 Roles of Hydrophobins and DHN-Melanin in Masking PAMPs

The outer layer of the *A. fumigatus* dormant conidium is composed of DHN-melanin interwoven with a proteinaceous rodlet layer that confers hydrophobic properties to conidia (Latgé et al. 2017). Proteomics studies have shown that the hydrophobin protein rodlet protein A (RodA), encoded by the *rodA* gene (Afu5g09580, as indicated in Table 4.1), is the most abundant surface protein found on dormant conidia of *A. fumigatus* (Blango et al. 2020; Valsecchi et al. 2019a; Voltersen et al. 2018). Both DHN-melanin and rodlet layers prevent conidia from environmental insults and mask PAMPs, such as β-glucan, thereby preventing immune recognition (Latgé and Chamilos 2019).

The *A. fumigatus* genome harbours seven genes supposed to encode hydrophobins, with two of them displaying high expression levels under laboratory conditions (Valsecchi et al. 2017). Specifically, the *rodA* gene exhibits the highest expression levels among hydrophobin genes during sporulation. Additionally, a homologous gene, *rodB* (Afu1g17250), encoding rodlet protein B (RodB), is highly expressed during biofilm formation and in the context of mice infection. The remaining hydrophobin genes are expressed at considerably lower levels (Valsecchi et al. 2017). *A. fumigatus* mutant strains lacking single or multiple hydrophobin genes further showed that RodA is the main hydrophobin protein of the rodlet layer (Fig. 4.1c), as an amorphous surface was observed on Δ*rodA* conidia. Additionally, Δ*rodA* conidia were more sensitive to physical insult than wild-type conidia, indicating a notable weakening of their cell wall compared to the parental strain (Valsecchi et al. 2017).

RodA exhibits an amphiphilic character, featuring a hydrophobic face with hydrophobic residues and compensated charges, along with a more hydrophilic surface containing clusters of net charges on the opposite face (Valsecchi et al. 2019b). The coverage of conidia with RodA masks PAMPs and makes dormant conidia immunologically inert (Aimanianda et al. 2009; Carrion et al. 2013). Conidia lacking RodA, either by chemical removal using hydrofluoric acid, or by deletion of the *rodA* gene, or by conidial germination, trigger a higher immune activation (Aimanianda et al. 2009). However, the absence of RodA does not affect phagolysosomal acidification in mouse macrophages (Thywißen et al. 2011). The *rodA* deletion mutants, including Δ*rodA* and Δ*rodBCDEFGA* (a mutant strain lacking seven hydrophobin genes), have been found to exhibit the same level of virulence as

Table 4.1 *A. fumigatus* surface proteins mediate fungal pathogenesis

Protein	Description	Gene ID	Protein length (aa)	Roles of protein in *A. fumigatus* pathogenesis	References
Complement evasion					
Alp1 (or Aspf13)	Alkaline protease 1 or Asp f 13	Afu4g11800	403	Degradation of complement components; allergens	Behnsen et al. (2010), Monod et al. (1993), Smith et al. (1994), Wiesner et al. (2020), Namvar et al. (2015)
Aspf2	Allergen Asp f 2	Afu4g09580	314	Recruitment of complement regulators; allergen	Banerjee et al. (1998), Dasari et al. (2018), Schwarz et al. (2023)
Aspf22 (or Eno1)	Allergen Asp f 22 or enolase 1	Afu6g06770	438	Recruitment of complement regulators; allergen	Dasari et al. (2019), Schwarz et al. (2023)
Mep1 (or Aspf5)	Secreted metalloprotease or allergen Asp f 5	Afu8g07080	634	Degradation of complement components; allergen	Jaton-Ogay et al. (1994), Shende et al. (2018)
Masking of PAMPs or adherence					
CcpA	Conidial cell wall protein A	Afu1g13670	235	Masking of PAMPs; virulence factor in mouse infection model	Voltersen et al. (2018)
RodA	Conidial hydrophobin	Afu5g09580	159	Masking of PAMPs; virulence unclear	Aimanianda et al. (2009), Carrion et al. (2013), Valsecchi et al. (2019b)
CspA	Glycophosphatidylinositol-anchored protein	Afu3g08990	430	Surface adhesion	Levdansky et al. (2007)
FleA	L-fucose-specific lectin	Afu5g14740	315	Role in fucose-dependent mucin binding and macrophage recognition	Kerr et al. (2016)
Interference with phagocytosis					
CalA	Laminin-binding protein with extracellular thaumatin domain	Afu3g09690	177	Binding to integrin $\alpha5\beta1$ and mediation of invasion	Liu et al. (2016), Upadhyay et al. (2009)
HscA	HSP70 family chaperone	Afu8g03930	614	Redirection of phagosomes to non-degradative pathway by hijacking p11	Jia et al. (2023)

Table 4.1 (continued)

Protein	Description	Gene ID	Protein length (aa)	Roles of protein in *A. fumigatus* pathogenesis	References
Stress resistance					
CatA	Catalase A	Afu6g03890	750	Resistance to ROI	Paris et al. (2003)
CatB (or Cat1)	Catalase B	Afu3g02270	728	Resistance to ROI	Paris et al. (2003)
Cat2	Catalase-peroxidase	Afu8g01670	759	Resistance to ROI	Paris et al. (2003)
Sod1	Superoxide dismutase [Cu-Zn]	Afu5g09240	154	Resistance to ROI	Lambou et al. (2010)
Sod2	Superoxide dismutase [Mn]	Afu4g11580	229	Resistance to ROI	Lambou et al. (2010)
Sod3 (or Aspf6)	Superoxide dismutase [Mn]	Afu1g14550	210	Resistance to ROI; allergen	Lambou et al. (2010)
Aspf1	Allergen Asp f 1; ribonuclease mitogillin	Afu5g02330	176	IgE-binding ribotoxin; allergen	Garciá-Ortega et al. (2005), Liu et al. (2018, 2021)
Aspf3	Allergen Asp f 3; peroxiredoxin family reductase	Afu6g02280	168	Resistance to ROI; allergen	Brantl et al. (2021), Hillmann et al. (2016)
Grx1	Glutaredoxin	Afu1g06100	102	Resistance to ROI	
Afu3g12270	Glutathione peroxidase	Afu3g12270	235	Resistance to ROI	Kurucz et al. (2018)
Afu5g01440	Thioredoxin peroxidase	Afu5g01440	202	Resistance to ROI	
GstC	Glutathione S-transferase	Afu4g14530	225	Resistance to ROI	Burns et al. (2005)
Afu1g15960	Glutathione reductase	Afu1g15960	554	Resistance to ROI	

their parental strain in mouse infection models (Thau et al. 1994; Valsecchi et al. 2017). These data indicate that the function of RodA for the virulence of *A. fumigatus* appears to be more complex.

In addition to RodA, another protein known for masking PAMPs, conidial cell wall protein A (CcpA), has been found to be highly abundant on the surface of *A. fumigatus* dormant conidia (Blango et al. 2020; Voltersen et al. 2018). The predicted CcpA protein, encoded by the *ccpA gene (*Afu1g13670, Table 4.1), consists of 235 amino acids. It is likely also a conidial surface hydrophobin because it shares similarities with RodA as an amphiphilic protein, characterized by a hydrophobic N-terminal region and a hydrophilic C-terminus. Considering that RodA serves as the main rodlet layer protein, the deletion of *ccpA* (Δ*ccpA*) did not cause substantial changes in the surface characteristics of resting conidia when compared to those of wild-type conidia (Voltersen et al. 2018). Compared to the wild-type strain, the Δ*ccpA* mutant strain did not exhibit any defects in growth on agar plates, even under various stress conditions, including temperature stress, oxidative stress, or exposure to cell wall stressors. Notably, the dormant conidia of the Δ*ccpA* mutant strain exposed more surface proteins and were more readily recognized by human polymorphonuclear leukocytes and dendritic cells than those of the wild-type strain (Voltersen et al. 2018). This enhanced recognition is likely attributed to the absence of the amphiphilic CcpA protein, which aids in concealing surface epitopes. In virulence assays, Δ*ccpA* showed attenuated virulence in mice that were immunosuppressed with cortisone acetate and caused less damage to A549 pulmonary epithelial cells (Voltersen et al. 2018). This indicates that the masking of PAMPs with CcpA is required for virulence although other effects of CcpA on host cells cannot be excluded yet.

4.3.2 Surface Proteins as Adhesins

After inhalation, dormant conidia adhere to host airway epithelia, and this adhesion is essential for the pathogenicity of *A. fumigatus* (Sheppard 2011). Cell wall carbohydrates, such as sialic acid and galactosaminogalactan, have been implicated in the adherence of *A. fumigatus* conidia and hyphae to host cells (Gravelat et al. 2013; Warwas et al. 2007). Pre-treatment of *A. fumigatus* conidia with trypsin reduced the binding of conidia with fibronectin and laminin, suggesting that fungal proteins could also mediate adhesion (Gil et al. 1996; Penalver et al. 1996).

There are a few *A. fumigatus* proteins, including RodA, Asp f 2 (Aspf2), CalA, CspA, and FleA, that have been shown to adhere to basement membrane macromolecules such as laminin, collagen, and fibronectin (Banerjee et al. 1998; Kerr et al. 2016; Levdansky et al. 2007; Thau et al. 1994; Upadhyay et al. 2009). Although RodA is a hydrophobin, conidia of the rodletless Δ*rodA* strain display a decreased capability of adherence to collagen and bovine serum albumin compared to that of the wild type, indicating an adhesion role of this rodlet protein (Thau et al. 1994). Consistently, *A. fumigatus* allergen Aspf2 (Aspf2, see also Table 4.1) was found to bind laminin *in vitro* (Banerjee et al. 1998). More than 82 candidate adhesins were predicted by an *in silico* screening of the proteome of *A. fumigatus*, including RodA, several glycosylphosphatidylinositol (GPI)-anchored proteins, and an extracellular thaumatin domain protein named CalA (Upadhyay et al. 2009). In the same study, CalA (Afu3g09690, Table 4.1) was shown to bind to laminin and adhere to murine lung and spleen cells, suggesting its significance in the adhesion of fungal conidia to host cells. In another study, Levdansky et al. identified a cell wall localized GPI-anchored protein CspA (for cell surface protein A) that also contributes to the adhesion of conidia to the extracellular matrix. However, deletion of the *cspA* gene did not alter fungal virulence in a murine model of systemic aspergillosis (Levdansky et al. 2007). The FleA (for L-fucose-specific lectin) serves as another example of a conidial surface protein that mediates adhesion to the host (Kerr et al. 2016). Detection of FleA by the host is essential for proper clearance of conidia, as macrophages exhibited reduced binding and phagocytosis of Δ*fleA* conidia compared to wild-type conidia. Consequently, Δ*fleA* conidia caused more severe pneumonia and invasive aspergillosis than wild-type conidia in mice (Kerr et al. 2016).

4.3.3 Surface Proteins Interfering with Phagocytosis

Phagocytosis serves as a crucial mechanism for pathogen clearance (Flannagan et al. 2012). Once engulfed, the cargo is embedded in phagosomes, which then mature into phagolysosomes where killing and digestion occur. However, as a pathogen, *A. fumigatus* can interfere with the progress of phagocytosis, hindering phagosome maturation and can then initiate the infection process (Akoumianaki et al. 2016; Goldmann et al. 2023; Heinekamp et al. 2015; Jia et al. 2023; Kyrmizi et al. 2018; Liu et al. 2016; Pazhakh et al. 2019; Schmidt et al. 2020).

4.3.3.1 Induction of Phagocytosis of Conidia by Epithelial Cells

In addition to adhering to epithelia, conidia of *A. fumigatus* are also internalized by these cells through a process designated as induced phagocytosis (DeHart et al. 1997; Liu et al. 2016; Wasylnka and Moore 2003). Most of these conidia are subsequently eliminated through phagosomal killing (Amin et al. 2014; Bertuzzi et al. 2022; Ewald et al. 2021). However, survival and germination of phagocytosed conidia inside epithelial cells have been observed (Fig. 4.1b), and these abilities are associated with the pathogen's invasive capability (Jia et al. 2023; Seidel et al. 2020).

Phagocytosis of pathogens begins with the direct recognition of pathogen-associated molecular patterns (PAMPs) by phagocytic receptors or indirectly through opsonization of pathogens with opsonins (Hatinguais et al. 2020). Since the yeast *Saccharomyces cerevisiae* does not adhere to or is not phagocytosed by lung epithelial cells (Liu et al. 2016), there must be surface factors targeting host cell receptors and inducing phagocytosis.

One of these factors is the above-mentioned adhesin CalA (Table 4.1). CalA binds to epithelial and endothelial cells through $\alpha5\beta1$ integrin (Fig. 4.1c), thereby inducing the phagocytosis of *A. fumigatus* conidia (Liu et al. 2016). Deletion of the fungal *calA* gene or blocking the host integrin $\beta1$ or $\alpha5$ reduces the phagocytosis of *A. fumigatus* conidia by host cells. *S. cerevisiae* cells expressing both CalA and the adhesin Als1 from *C. albicans* can induce their phagocytosis of epithelial cells. Therefore, CalA is a surface effector protein directly inducing host cell phagocytosis. Importantly, Δ*calA* mutant conidia exhibited reduced virulence compared to the wild-type strain in a mouse model of invasive aspergillosis. In agreement, antibody-mediated blockade of CalA reduced the internalization of wild-type conidia by host cells and protected mice from invasive aspergillosis (Liu et al. 2016). Collectively, these results support the finding of CalA as an important invasin and as a potential target for passive immunotherapy against invasive pulmonary aspergillosis.

Another surface protein that interferes with epithelial cell phagocytosis is the surface-exposed heat shock protein HscA (Jia et al. 2020, 2023). Latex beads coated with recombinant HscA protein were observed to attach to and be internalized by epithelial cells more efficiently than beads coated with another heat shock protein, i.e., Hsp70 (Jia et al. 2023). Mechanistically, HscA anchors the human p11 (also called S100A10) protein on the phagocytic cup and the phagosomal membrane (Fig. 4.1c). This prevents the maturation of phagosomes and triggers the recruitment of exocytosis mediators. As a consequence, conidia in such phagosomes with p11 on the surface of the membrane could escape phagosomal killing by germinating inside a non-mature phagosome, being released to the medium through exocytosis, or even being transferred *via* whole conidia-containing phagosomes from one cell to another (Jia et al. 2023).

In addition to the protein factors CalA and HscA, the surface pigment DHN-melanin also plays a role in inducing phagocytosis by lung epithelial cells. Melanized conidia were recognized and taken up better compared to non-melanized conidia by epithelial cells (Amin et al. 2014). Despite the melanin-sensing C-type lectin receptor (MelLec) has been shown to recognize DHN-melanin (Stappers et al. 2018), its involvement in the phagocytosis of *A. fumigatus* conidia by epithelial cells remains unclear. On the contrary, non-melanized conidia were phagocytosed more efficiently by macrophages than melanized conidia (Amin et al. 2014; Thywißen et al. 2011). This is likely due to the exposure of more PAMPs in the DHN-melanin deletion strain that can be detected by the professional phagocytes.

4.3.3.2 Resistance of *A. fumigatus* Against Phagosomal Killing

After phagocytosis, phagosomes gradually accumulate protons and reactive oxygen intermediates (ROI) within the phagosomal lumen by assembling the v-ATPase and Nox2 complex at the phagosomal membrane (Fig. 4.1c). The acidic, oxidative phagosomal lumen directly restricts the growth of pathogens and indirectly activates hydrolases that contribute to the killing of engulfed pathogens (Flannagan et al. 2012). In neutrophils, macrophages, as well as epithelial cells, however, internalized *A. fumigatus* conidia are capable of interfering with phagosomal acidification to facilitate their germination and outgrowth (Fig. 4.1b) (Akoumianaki et al. 2016; Goldmann et al. 2023; Jia et al. 2023; Kyrmizi et al. 2018; Pazhakh et al. 2019; Rafiq et al. 2022; Schmidt et al. 2020; Thywißen et al. 2011). Compared to bacterial pathogens, our understanding of how fungal pathogens manipulate phagosome maturation is limited. The tolerance of *A. fumigatus* to ROI (Fig. 4.1c) and low pH, along with specific mechanisms that modulate phagosome maturation, contribute to the survival of this pathogen within phagosomes.

Catalase (CAT), superoxide dismutase (SOD), and glutathione peroxidase are discussed as defence against oxidants (Ighodaro and Akinloye 2019). On the surface of *A. fumigatus* conidia and hyphae three CATs (CatA, CatB, and Cat2), three SODs (Sod1, Sod2, and Sod3), five putative glutathione-associated enzymes (Grx1, GstC, Afu3g12270, Afu5g01440, and Afu1g15960), and the peroxiredoxin Aspf3 were detected (Jia et al. 2020). The triple *SOD* deletion mutant Δ*sod1*/Δ*sod2*/Δ*sod3*, but not the Δ*catA* or Δ*cat1*/Δ*cat2* strains, was shown to have reduced hyphal survival when exposed to human neutrophils and in a mouse corneal infection model (Leal Jr. et al. 2012). Surprisingly, the deletion of many ROI-detoxifying enzymes or regulators, such as the triple mutant Δ*sod1*/Δ*sod2*/Δ*sod3*, did not result in a significant reduction of virulence in invasive aspergillosis mouse models (Lambou et al. 2010; Latgé and Chamilos 2019; Lessing et al. 2007). However, the *aspf3*-deletion strain also sensitive to ROI was avirulent in a cortisone acetate-induced mouse model of pulmonary aspergillosis (Hillmann et al. 2016). Subsequently, it was found that iron supplementation could restore the virulence of the Δ*aspf3* mutant in the same mouse infection model (Brantl et al. 2021). While the expression of *aspf3* is responsive to iron availability, it has been observed that Aspf3 does not affect iron homeostasis. Instead, the protein plays a role in overcoming iron limitation by compensating for the loss of iron-dependent antioxidant enzymes (Brantl et al. 2021).

In addition to the antioxidant proteins, the multifunctional spore pigment DHN-melanin has been implicated in protecting the fungus against ROI and has also been intensely investigated for its role in inhibiting phagosome maturation (Heinekamp et al. 2013). Notably, it was shown to sequester Ca^{2+} inside phagosome, thus preventing LC3-associated phagocytosis by interference with calcium-calmodulin-dependent signalling pathways (Kyrmizi et al. 2018). Additionally, the presence of DHN-melanin on conidia diminishes the formation of lipid-raft microdomains in the phagosomal membrane. These microdomains are crucial for the maturation of phagolysosomes as they provide platforms for the assembly of v-ATPase and Nox2 complexes (Goldmann et al. 2023; Schmidt et al. 2020).

Recently, we discovered that *A. fumigatus* utilizes an additional mechanism, i.e., the surface protein HscA hijacks the host p11 protein allowing conidia to escape phagosome killing (Jia et al. 2023). Phagosomes containing wild-type *A. fumigatus* conidia were less acidified compared to those containing Δ*hscA* mutant conidia. There were also less Nox2 complexes assembled on phagosomes containing wild-type conidia. The HscA protein present on the conidial surface of *A. fumigatus* anchors the human p11 protein on conidia-containing phagosomes (Fig. 4.1b, c), excludes the recruitment of Ras-related protein 7 (Rab7), and triggers the recruitment of exocytosis mediators such as Rab11. Consequently, conidia in such phagosomes could germinate, or were released into the medium through exocytosis, or even were transferred from one cell to another cell (Fig. 4.1b) (Jia et al. 2023). HscA was not found in secretome studies of *A. fumigatus* swollen conidia cultivated in RPMI medium (Blango et al. 2020) but traces were detectable in the supernatant of cultures grown with collagen (Shemesh et al. 2017). This finding suggests that the majority of extracellular HscA is bound on the surface of conidia, while a minority may be shed from the surface. Therefore, the defined HscA protein represents an effector protein of *A. fumigatus*.

4.3.3.3 Interaction of *A. fumigatus* with Phagocytes

After sensing of *A. fumigatus*, lung epithelial cells produce chemokines to recruit dendritic cells and neutrophils the latter are essential for fungal clearance (Guo et al. 2020; Jhingran et al. 2015; Mircescu et al. 2009). Neutrophils eliminate conidia through phagocytosis and kill hyphae by releasing granules, neutrophil extracellular traps (NETs), and extracellular vesicles whereby extracellular vesicles play the major role (Brakhage et al. 2010; Bruns et al. 2010; Rafiq et al. 2022; Shopova et al. 2020). Alveolar macrophages are also professional phagocytes that are involved in the early recognition of inhaled conidia (Ewald et al. 2021).

A. fumigatus also produces a number of toxic secondary metabolites, such as fumagillin, gliotoxin, and helvolic acid, that are known to interfere with the function of neutrophils and macrophages (Fallon et al. 2010; Mitchell et al. 1997; Scharf et al. 2012). However, secreted effector proteins are rarely identified, even in other human fungal pathogens. An example is Cpl1 (capsule formation 1), a secreted cysteine-rich protein from the fungal pathogen *C. neoformans.* It triggers the alternative activation of macrophages, also known as M2 polarization, by directly activating Toll-like receptor 4 (TLR4), leading to the creation of a favourable niche for the fungus (Dang et al. 2022). Similarly, after infection of mice with *A. fumigatus*, the mouse alveolar macrophages predominantly express Arginase 1, a key marker of alternatively activated macrophages (Bhatia et al. 2011). Given the absence of a Cpl1 homologue in *S. cerevisiae* (Liu et al. 2008) or *A. fumigatus* (by protein BLAST, https://blast.ncbi.nlm.nih.gov/Blast.cgi), the existence of a small effector protein with a similar function in *A. fumigatus* remains to be shown.

Along with secondary metabolite toxins, *A. fumigatus* produces the secreted allergen Aspf1 for cell toxicity. Aspf1 is a ribotoxin with cytotoxic ribonuclease activity that degrades ribosomal RNA, thereby impairing protein biosynthesis (Garciá-Ortega et al. 2005). In a non-neutropenic mouse infection model, an increased expression of *aspf1* in the gliotoxin-deficient Δ*gliP* mutant was found (Liu et al. 2018). An *in vitro* cell toxicity assay showed that a Δ*aspf1* mutant strain caused less damage to RAW 264.7 mouse macrophages than that of the wild-type strain. Cortisone acetate-treated mice infected with the Δ*aspf1* mutant had prolonged survival, suggesting a role of Aspf1 for *A. fumigatus* virulence (Liu et al. 2018, 2021). However, Aspf1 was not required for virulence in a cyclophosphamide-induced neutropenic mouse infection model (Smith et al. 1994).

4.3.4 Surface Proteins and Secreted Proteins Disarm the Complement System

Upon inhalation, *A. fumigatus* triggers the activation of all three complement pathways: classical, alternative, and lectin pathways, which converge at the formation of convertases that cleave the C3 and C5 proteins (Dellière and Aimanianda 2023; Reis et al. 2019). Activation of complement leads to the formation of membrane attack complexes and the generation of active complement proteins, such as C3a, C3b, iC3b, C3dg, C4a, C4b, and C5a (Zipfel and Skerka 2009). Activated C3b opsonizes *A. fumigatus* by binding to surface molecules, such as RodA, cell wall β-glucan, and galactomannan (Wong et al. 2020). The opsonization process enhances the phagocytosis of *A. fumigatus* conidia by phagocytes. Additionally, C5a induces cellular inflammation, while C5b initiates the formation of the terminal complement complex, leading to the lysis of target cells (Zipfel and Skerka 2009). Consequently, the complement system plays a pivotal role in defence against *A. fumigatus* infection, as mice deficient in complement components exhibit a heightened mortality rate and more severe pulmonary aspergillosis (Garlanda et al. 2002; Shende et al. 2022).

A successful establishment of an *A. fumigatus* infection, therefore, requires the efficient evasion of the complement system in the lung. In line with this notion, this fungal pathogen has developed a variety of strategies to disarm the host complement. *A. fumigatus* can recruit host complement regulators that reduce opsonization (Behnsen et al. 2008; Dasari et al. 2018, 2019; Vogl et al. 2008) and secret proteases to degrade host complement components (Behnsen et al. 2010; Shende et al. 2018) (see next chapters).

4.3.4.1 Surface Proteins Recruit Complement Regulators

A. fumigatus has been shown to evade the host complement attack by binding to complement regulators, including factor H, factor H-like protein 1 (FHL-1), factor H-related protein 1 (FHR-1), and plasminogen, which all help the fungus to evade the host complement attack (Behnsen et al. 2008). Both factor H and FHL-1 exhibit cofactor activity for factor I-mediated inactivation of C3b and accelerate the dissociation of C3 convertase to C3bBb (Skerka et al. 2013). FHR-1 was found to inhibit complement C5 convertase activity and the formation of the lytic membrane attack complex (Heinen et al. 2009). Active plasmin is generated from plasminogen and can cleave and inactivate human complement proteins, such as C1, C3, and C5 (Barthel et al. 2012). Under homeostatic conditions, these proteins contribute to the prevention of uncontrolled complement activation and accelerate the decay of convertases or mediate the cleavage of activated fragments (Reis et al. 2019). However, they are also hijacked by *A. fumigatus* (Fig. 4.1c) to hide from the complement system (Dasari et al. 2018, 2019).

The strongest binding of complement regulators was observed for dormant conidia, which represent the first morphotype to encounter the host immune system, while lower binding was detected for swollen conidia and hyphae (Behnsen et al. 2008). Subsequent studies have identified Aspf2 and Aspf22 (Table 4.1 and Fig. 4.1c) as surface-exposed proteins that directly recruit complement regulators (Dasari et al. 2018, 2019). Aspf2 was shown to recruit complement regulators H, FHL-1, FHR-1, and plasminogen (Dasari et al. 2018). Factor H and FHL-1 bound to Aspf2 retained cofactor activity for factor I-mediated cleavage of C3b. Attachment of plasmin to *A. fumigatus* conidia resulted in damage of human lung epithelial cells, induced cell retraction, and exposed the extracellular matrix. Deletion of the *aspf2* gene in *A. fumigatus* led to increased C3 deposition on conidia of the mutant strain and enhanced their phagocytosis and killing by human neutrophils compared to wild-type conidia (Dasari et al. 2018). Hence, Aspf2 plays an important role in immune evasion and contributes to the disruption of human lung epithelial cell layers. Similarly, Aspf22, also known as *A. fumigatus* enolase, was bound by factor H, FHL-1, C4b binding protein (C4BP), and plasminogen (Dasari et al. 2019).

4.3.4.2 Secreted Proteases Cleave Complement Components

In addition to the virulence factors binding to complement inhibitors, *A. fumigatus* also releases proteases that cleave host complement components (Fig. 4.1c) (Behnsen et al. 2010; Shende et al. 2018). Bioinformatic analyses suggest that *A. fumigatus* encodes over 100 proteases (Rawlings et al. 2010). To identify the complement degrading proteases, Behnsen et al. employed two-dimensional gel analysis of the *A. fumigatus* hyphal secretome and found four abundant proteases, including dipeptidyl peptidase (Afu2g09030), the aspartic-type endoprotease AP3 (Afu4g09400), the aminopeptidase Y (Afu3g00650), and the alkaline serine protease Alp1 (also known as Aspf13, Table 4.1) (Behnsen et al. 2010). By using different protease inhibitors and by generating an *alp1* deletion mutant, the authors confirmed that Alp1 is the major protease responsible for cleaving C3b, C4, C5, and C1q. However, in both a cortisone acetate murine infection model and a cyclophosphamide-based murine infection model, the Δ*alp1* mutant strain exhibited the same level of virulence as the wild-type strain (Behnsen et al. 2010; Monod et al. 1993; Smith et al. 1994). This may be attributed to the fact that Alp1 is secreted by hyphae, which are only produced in the later stages of infection. In another study, *A. fumigatus* conidia were cultured in a medium designed to mimic the lung environment. In this medium, the activity of extracellular proteases against human complement proteins was assessed (Shende et al. 2018). The inhibition of protease activity by EDTA helped to identify the metalloprotease Mep1 (Table 4.1), which cleaves complement components C3, C4, and C5. Mep1 is stored in the conidial cell wall and released into the medium containing albumin or collagen. The presence of Mep1 inhibits C3b deposition on conidia, consequently reducing their phagocytosis by macrophages (Shende et al. 2018).

Gene redundancy and functional complementation in many fungal pathogens pose challenges when assigning definitive roles to genes in virulence (Jia and Tang 2015). Although the complement-disarming factors mentioned above clearly contribute to the inhibition of complement activation, both the Δ*mep1* mutant and Δ*alp1*Δ*mep1* double mutant exhibited virulence levels similar to that of the wild-type strain (Jaton-Ogay et al. 1994; Shende et al. 2018). Similarly, an *A. fumigatus* Δ*prtT* (Afu4g10120) mutant, which lacks a conserved positive regulator of secreted proteases, exhibited a 70% reduction in casein proteolytic activity but remains fully virulent in a leukopenic mouse infection model (Bergmann et al. 2009; Sharon et al. 2009). Because the complement system is important for defence against *A. fumigatus*, it is conceivable that the fungus has developed complement-disarming redundancy as an adaptive strategy in response to adverse changes in host conditions.

4.3.5 *A. fumigatus* Proteins as Allergens

The above-mentioned complement-disarming factors, i.e., Aspf2, Aspf22, and Alp1, while not required for the virulence of *A. fumigatus*, have been reported to possess antigenic properties that trigger a spectrum of different diseases, including immunoglobulin E (IgE)-mediated sensitization to ABPA (Agarwal et al. 2013; Banerjee et al. 1998; Namvar et al. 2015; Schwarz et al. 2023). The development of ABPA can be induced by *Aspergillus* sensitization, particularly in patients with chronic obstructive pulmonary disease (COPD), asthma, or cystic fibrosis who cannot efficiently clear inhaled *Aspergillus* conidia. In such cases, the conidia germinate and produce hyphae, thereby exposing and releasing allergens that activate the adaptive immune responses and usually exacerbate the diseases (Agarwal et al. 2013; Bacher et al. 2014b; Fairs et al. 2010; Goh et al. 2017; Schwarz et al. 2023; Tiew et al. 2023).

CD4^{+} T cells, including type 1 T helper (Th1) cells, Th2 cells, and Th17 cells, are central organizers of adaptive immunity against fungal pathogens and play important roles in both the resolution and worsening of fungal infections (Speakman et al. 2020; van de Veerdonk et al. 2017). Th1 cells produce interferon-γ, which augments phagocytosis and killing of pathogens, whereas Th2 cells release interleukin-4 (IL-4), IL-5, and IL-13, which induce the production of IgE (van de Veerdonk et al. 2017). ABPA is characterized by uncontrolled Th2 responses to *A. fumigatus* (Bacher et al. 2014a, 2016; Schwarz et al. 2023). In addition, although *A. fumigatus* is not an inducer of human Th17 responses (Chai et al. 2010), Th17 cells induced by *C. albicans* in the gut can migrate to the lung. Because these cells can react to *A. fumigatus*, they may contribute to *A. fumigatus*-driven airway inflammation (Bacher et al. 2019). Therefore, identifying *A. fumigatus* allergens could help to develop improved diagnostic assays and facilitate the precise targeting of cross-reactive Th2 cells as a therapeutic approach.

4.3.5.1 Identification of *A. fumigatus* Extracellular Allergens

Allergens from *A. fumigatus*, also called "Asp f" proteins, are recognized by serum IgE antibodies or Th2 cells in patients with ABPA or *A. fumigatus*-sensitized individuals (Kurup 2005; Singh et al. 2010). According to the Allergome database (https://www.allergome.org/index.php), there are 48 allergens listed. Proteomics studies have identified at least 32 of them (Table 4.2) to be exposed on the surface or secreted by various morphotypes of *A. fumigatus* (Jia et al. 2020; Singh et al. 2010; Wartenberg et al. 2011). The representativeness of crude allergen extracts, which usually lack specificity and are prone to contamination, for screening for sensitization responses has faced challenges (Curin et al. 2017). Therefore, most of the allergens have been produced as recombinant proteins and have been used in various clinical and immunological investigations (Bacher et al. 2014b; Chaudhary et al. 2010; Schwarz et al. 2023; Tiew et al. 2023). For example, 20 recombinant *A. fumigatus* allergens were tested for their association with clinical outcomes in COPD. It was found that sensitization to Aspf1, Aspf3, Aspf5, and Aspf6 holds critical clinical relevance in COPD, as these allergens are associated with poor clinical outcomes (Tiew et al. 2023). The levels of IgE antibodies against Aspf1, Aspf2, and Aspf4 have also been suggested to be useful for the diagnosis of ABPA

(Muthu et al. 2020). Schwarz et al. utilized 22 recombinant *A. fumigatus* proteins to enrich antigen-reactive T cells and identified Aspf2, Aspf3, CatB, CpcB, CsnB, Fg-gap, and GliT that exhibited reactivity with Th2 cells (Schwarz et al. 2023).

Despite the identification of many allergens from *A. fumigatus*, our understanding of the function of each of them remains limited. Only a few of them have been implicated in roles as allergic initiators, such as proteases that can lead to damage in the airway epithelium, thereby triggering the release of danger signals (Komlósi et al. 2022; Wiesner et al. 2020), or as direct targets of Th2 cells in those atopic patients who have already established type 2 immune responses (Bacher et al. 2014b; Schwarz et al. 2023).

4.3.5.2 Epithelial Cell Barrier Damage Leads to Th2 Cell Sensitization

The epithelial barrier serves as the initial line of both physical and immunologic defence against environmental threats (Komlósi et al. 2022). Consequently, damage to the airway epithelium caused by *A. fumigatus* may act as a trigger for allergic sensitization (Fig. 4.1b). Convergent findings have highlighted the pivotal roles of allergens with proteolytic activity in both the initiation and progression of allergic responses (Soh et al. 2023).

Four protease allergens, namely Alp1, Mep1, Aspf10, and Aspf18 (Table 4.2), have been found to be secreted or present on the surface of *A. fumigatus* (Behnsen et al. 2010; Jia et al. 2020; Wartenberg et al. 2011). The culture filtrates from Δ*mep1* and wild-type strains exhibited higher protease activity compared to that of the Δ*alp1* strain, suggesting that Alp1 is the primary secreted protease (Namvar et al. 2015). In mouse models of airway allergy, it has been demonstrated that exposure to Alp1 damages the epithelial barrier (Balenga et al. 2015; Wiesner et al. 2020). Alp1 promotes airway hyper-responsiveness by infiltrating the bronchial submucosa. The protein disrupts interactions between the airway smooth muscle cell and the extracellular matrix. Thereby a pathophysiological RhoA-dependent Ca^{2+} sensitivity and bronchoconstriction are evoked (Balenga et al. 2015). Alp1 also damages the junctions of bronchiolar club cells by cleaving E-cadherin, a key component of adherent junctions. The damage is detected as a danger signal by a mechanosensitive calcium channel, known as transient receptor potential cation channel subfamily V member 4 (TRPV4). This recognition triggers a calcium influx (Fig. 4.1c), ultimately resulting in the upregulation of several proinflammatory genes *via* the calcineurin-signalling pathway (Wiesner et al. 2020). However, chronic intranasal exposure of mice to filtrates of the Δ*alp1* mutant strain and the wild-type strain resulted in comparable airway hyper-reactivity and Th2 responses (Namvar et al. 2015). This finding indicates that multiple allergens contribute to the development of an allergy against *A. fumigatus*.

4.3.5.3 Extracellular Proteins Detected by Th2 and Th17 Cells

Th2 and Th17 responses are particularly associated with pulmonary inflammatory diseases (Bacher et al. 2019; Iwanaga and Kolls 2019). Studying the allergens that activate these T cell populations is crucial for advancing our understanding of allergies and for improving diagnosis and treatment. *Ex vivo* characterization of reactive T cells responding to specific allergens provides opportunities to classify these allergens and to delineate the triggered T cell responses (Bacher et al. 2019; Schwarz et al. 2023).

For patients allergic to *A. fumigatus*, specific Th2 cells could be enriched from their peripheral blood (Bacher et al. 2016; Bacher et al. 2014a; Schwarz et al. 2023). By applying a sensitive antigen-reactive T cell enrichment method, rare $CD4^+$ T cells in human peripheral blood that react against *A. fumigatus* allergens were systematically quantified and characterized (Bacher et al. 2014a, b, 2016; Schwarz et al. 2023). This way, fungus-reactive Th cells in the peripheral blood of patients with cystic fibrosis who are allergic to *A. fumigatus* and healthy controls were investigated and Th2 target proteins were identified that included Aspf2, Aspf3, CatB, CpcB, CsnB, Fg-gap, and GliT, as well as Th17 target proteins Aspf22 and Scw4 (Table 4.2) (Schwarz et al.

Table 4.2 Exposed or secreted allergens of *A. fumigatus*

Allergens	Description	Gene ID	Protein length (aa)	References
Aspf1	Allergen Aspf1; ribonuclease mitogillin	Afu5g02330	176	Garciá-Ortega et al. (2005), Muthu et al. (2020), Tiew et al. (2023)
Aspf2	Allergen Aspf2	Afu4g09580	314	Banerjee et al. (1998), Muthu et al. (2020), Schwarz et al. (2023)
Aspf3	Allergen Aspf3; peroxiredoxin family reductase	Afu6g02280	168	Chaudhary et al. (2010), Schwarz et al. (2023), Tiew et al. (2023)
Aspf4	Allergen Aspf4	Afu2g03830	322	Muthu et al. (2020), Ramachandran et al. (2004)
Mep1 (or Asp f 5)	Allergen Aspf5; secreted metalloprotease	Afu8g07080	634	Namvar et al. (2015), Tiew et al. (2023)
Aspf6 (or Sod3)	Allergen Aspf4; superoxide dismutase [Mn]	Afu1g14550	210	Schwienbacher et al. (2005), Tiew et al. (2023)
Aspf7	Allergen Aspf7	Afu4g06670	270	Scharrenberg et al. (2010)
Aspf8	Allergen Aspf8; large ribosomal subunit protein P2	Afu2g10100	111	Rhyner et al. (2011)
Aspf9 (or Crf1)	Allergen Aspf9; probable glycosidase crf1	Afu1g16190	395	Bacher et al. (2014b), Chaudhary et al. (2010), Singh et al. (2010)
Aspf10	Allergen Aspf10: putative extracellular aspartic endopeptidase Pep1	Afu5g13300	395	Singh et al. (2010)
Aspf11	Allergen Aspf11; putative cyclophilin	Afu2g03720	205	Chaudhary et al. (2010), Fluckiger et al. (2002)
Aspf12 (or Hsp90)	Allergen Aspf12; heat shock protein 90 kDa	Afu5g04170	706	Kumar et al. (1993), Singh et al. (2010)
Alp1 (or Aspf13)	Alkaline protease 1 or Aspf13	Afu4g11800	403	Balenga et al. (2015), Namvar et al. (2015), Wiesner et al. (2020)
Aspf15	Allergen Aspf15	Afu2g12630	152	Singh et al. (2010)
Aspf17	Allergen Aspf17	Afu4g03240	284	
Aspf18 (or Alp2)	Allergen Aspf18; alkaline protease 2	Afu5g09210	495	Poll et al. (2009)
Aspf22 (or Eno1)	Allergen Aspf22; enolase 1	Afu6g06770	438	Bacher et al. (2014b). Schwarz et al. (2023)
Aspf23 (or Rpl3)	Allergen Aspf23; ribosomal L3 protein	Afu2g11850	392	Saxena et al. (2003)
Aspf27	Allergen Aspf27; putative peptidyl-prolyl *cis-trans* isomerase	Afu3g07430	163	Glaser et al. (2006)
Aspf28	Allergen Aspf28	Afu6g10300	171	Limacher et al. (2007)
Aspf29	Allergen Aspf29; thioredoxin	Afu5g11320	110	Limacher et al. (2007)
Aspf34	Allergen Aspf34; cell wall protein PhiA	Afu3g03060	185	Glaser et al. (2009), Tiew et al. (2023)
Asp f FDH	Putative NAD-dependent formate dehydrogenase	Afu6g04920	418	Singh et al. (2010)
Asp f MDH	Putative NAD-dependent malate dehydrogenase	Afu7g05740	340	Singh et al. (2010)
Asp f gamma_actin	Actin	Afu6g04740	393	Singh et al. (2010)

(continued)

Table 4.2 (continued)

Allergens	Description	Gene ID	Protein length (aa)	References
Asp f RPS3	40S ribosomal protein S3	Afu1g05630	266	Singh et al. (2010)
CatB	Asp f catalase	Afu3g02270	728	Bacher et al. (2014b), Schwarz et al. (2023)
CpcB	G-protein complex beta subunit	Afu4g13170	316	Bacher et al. (2014b), Schwarz et al. (2023)
CsnB	Asp f chitosanase	Afu4g01290	238	Schwarz et al. (2023)
Exg12 (or Asp f glucosidase)	Secreted beta-glucosidase	Afu1g05770	863	Singh et al. (2010)
Fg-gap (or Aspf39)	FG-gap repeat protein	Afu1g04130	307	Bacher et al. (2014b), Schwarz et al. (2023)
Gel2 (or Asp f GT)	GPI-anchored 1,3-beta-glucanosyltransferase	Afu6g11390	475	Singh et al. (2010)
GliT	Gliotoxin sulfhydryl oxidase required for gliotoxin biosynthesis	Afu6g09740	334	Bacher et al. (2014b), Schwarz et al. (2023)
MsdS (or Asp f Mannosidase)	Putative 1,2-alpha-mannosidase	Afu1g14560	503	Singh et al. (2010)

2023). Interestingly, *A. fumigatus* itself does not generate Th17 responses in healthy individuals (Bacher et al. 2019; Chaudhary et al. 2010). The Th17 cells targeted to Scw4 or Aspf22 are probably cross-reactive and were programmed by homologous proteins of *C. albicans* (Bacher et al. 2019). Hence, the modulation of specific Th2 cells may offer a potential therapeutic approach for the treatment of *A. fumigatus*-induced allergic diseases. Furthermore, the generation of *A. fumigatus*-specific T cells holds the potential to safeguard immunosuppressed patients against invasive infections. In a recent study, researchers generated T cells expressing a gene-engineered *A. fumigatus*-specific chimeric antigen receptor (Af-CAR) (Seif et al. 2022). Af-CAR has been found to be reactive to a conserved protein antigen in the cell wall of *A. fumigatus* hyphae, although this antigen has not yet been identified. The study demonstrated that Af-CAR T cells provide protective effects in mouse infection models (Seif et al. 2022).

4.4 Conclusions and Perspectives

Research on the pathogenicity of *A. fumigatus* and host immune responses has progressed rapidly, resulting in significant breakthroughs in our fundamental comprehension of the cellular and molecular factors governing fungal-host interaction mechanisms. An important role is played by surface proteins of the pathogen that form the first contact with the host. Although these proteins are also interesting targets for the development of diagnosis and therapy, a specific and sensitive diagnosis of *A. fumigatus* infections is still needed, and its development remains a major challenge. This lack of specific diagnostic markers, together with resistance against the few antimycotics, and the complexities of host immune status are still major factors contributing to the high mortality of *A. fumigatus* infections. In this chapter, we have reported on the functions of surface-exposed or secreted proteins produced by *A. fumigatus*, shedding light on their contributions as adhesins, virulence factors, stress resistance factors, and/or allergens to disease development. This knowledge could potentially facilitate personalized risk assessment and prognosis for individuals who are susceptible to fungal infections. Additionally, these proteins may also serve as promising candidate targets for the development of diagnostic and immunotherapy strategies.

Acknowledgements We thank all our co-workers for their valuable discussions. Work in the here summarized area in the authors' laboratory is funded by the Deutsche Forschungsgemeinschaft (DFG) cluster of excellence

Balance of the Microverse (Project-ID 390713860, Gepris 2051), the DFG-ANR project "Afulnf (316898429)", DFG CRC/Transregio 124 "FungiNet" (A1, Z2; 210879364), DFG CRC 1278 "PolyTarget" (Z01; 316213987), the Leibniz project (K217/2016), and the Free State of Thuringia and the European Social Fund Plus project (PhagoInf, 2023FGR0043).

References

Adav SS, Ravindran A, Sze SK (2013) Proteomic analysis of temperature dependent extracellular proteins from *Aspergillus fumigatus* grown under solid-state culture condition. J Proteome Res 12:2715–2731

Adav SS, Ravindran A, Sze SK (2015) Quantitative proteomic study of *Aspergillus fumigatus* secretome revealed deamidation of secretory enzymes. J Proteome 119:154–168

Agarwal R, Chakrabarti A, Shah A, Gupta D, Meis JF, Guleria R, Moss R, Denning DW, A.c.a.I.w. group (2013) Allergic bronchopulmonary aspergillosis: review of literature and proposal of new diagnostic and classification criteria. Clin Exp Allergy 43:850–873

Aimanianda V, Bayry J, Bozza S, Kniemeyer O, Perruccio K, Elluru SR, Clavaud C, Paris S, Brakhage AA, Kaveri SV, Romani L, Latgé JP (2009) Surface hydrophobin prevents immune recognition of airborne fungal spores. Nature 460:1117–1121

Akoumianaki T, Kyrmizi I, Valsecchi I, Gresnigt MS, Samonis G, Drakos E, Boumpas D, Muszkieta L, Prevost MC, Kontoyiannis DP, Chavakis T, Netea MG, van de Veerdonk FL, Brakhage AA, El-Benna J, Beauvais A, Latgé JP, Chamilos G (2016) *Aspergillus* cell wall melanin blocks LC3-associated phagocytosis to promote pathogenicity. Cell Host Microbe 19:79–90

Amin S, Thywissen A, Heinekamp T, Saluz HP, Brakhage AA (2014) Melanin dependent survival of *Apergillus fumigatus* conidia in lung epithelial cells. Int J Med Microbiol 304:626–636

Asif AR, Oellerich M, Amstrong VW, Riemenschneider B, Monod M, Reichard U (2006) Proteome of conidial surface associated proteins of *Aspergillus fumigatus* reflecting potential vaccine candidates and allergens. J Proteome Res 5:954–962

Bacher P, Kniemeyer O, Schonbrunn A, Sawitzki B, Assenmacher M, Rietschel E, Steinbach A, Cornely OA, Brakhage AA, Thiel A, Scheffold A (2014a) Antigen-specific expansion of human regulatory T cells as a major tolerance mechanism against mucosal fungi. Mucosal Immunol 7:916–928

Bacher P, Kniemeyer O, Teutschbein J, Thon M, Vodisch M, Wartenberg D, Scharf DH, Koester-Eiserfunke N, Schutte M, Dubel S, Assenmacher M, Brakhage AA, Scheffold A (2014b) Identification of immunogenic antigens from *Aspergillus fumigatus* by direct multiparameter characterization of specific conventional and regulatory CD4+ T cells. J Immunol 193:3332–3343

Bacher P, Heinrich F, Stervbo U, Nienen M, Vahldieck M, Iwert C, Vogt K, Kollet J, Babel N, Sawitzki B, Schwarz C, Bereswill S, Heimesaat MM, Heine G, Gadermaier G, Asam C, Assenmacher M, Kniemeyer O, Brakhage AA, Ferreira F, Wallner M, Worm M, Scheffold A (2016) Regulatory T cell specificity directs tolerance versus allergy against aeroantigens in humans. Cell 167:1067–1078. e1016

Bacher P, Hohnstein T, Beerbaum E, Röcker M, Blango MG, Kaufmann S, Röhmel J, Eschenhagen P, Grehn C, Seidel K, Rickerts V, Lozza L, Stervbo U, Nienen M, Babel N, Milleck J, Assenmacher M, Cornely OA, Ziegler M, Wisplinghoff H, Heine G, Worm M, Siegmund B, Maul J, Creutz P, Tabeling C, Ruwwe-Glösenkamp C, Sander LE, Knosalla C, Brunke S, Hube B, Kniemeyer O, Brakhage AA, Schwarz C, Scheffold A (2019) Human anti-fungal Th17 immunity and pathology rely on cross-reactivity against *Candida albicans*. Cell 176:1340–1355

Balenga NA, Klichinsky M, Xie Z, Chan EC, Zhao M, Jude J, Laviolette M, Panettieri RA Jr, Druey KM (2015) A fungal protease allergen provokes airway hyper-responsiveness in asthma. Nat Commun 6:6763

Banerjee B, Greenberger PA, Fink JN, Kurup VP (1998) Immunological characterization of Asp f 2, a major allergen from *Aspergillus fumigatus* associated with allergic bronchopulmonary aspergillosis. Infect Immun 66:5175–5182

Barthel D, Schindler S, Zipfel PF (2012) Plasminogen is a complement inhibitor. J Biol Chem 287:18831–18842

Behnsen J, Hartmann A, Schmaler J, Gehrke A, Brakhage AA, Zipfel PF (2008) The opportunistic human pathogenic fungus *Aspergillus fumigatus* evades the host complement system. Infect Immun 76:820–827

Behnsen J, Lessing F, Schindler S, Wartenberg D, Jacobsen ID, Thoen M, Zipfel PF, Brakhage AA (2010) Secreted *Aspergillus fumigatus* protease Alp1 degrades human complement proteins C3, C4, and C5. Infect Immun 78:3585–3594

Bergmann A, Hartmann T, Cairns T, Bignell EM, Krappmann S (2009) A regulator of *Aspergillus fumigatus* extracellular proteolytic activity is dispensable for virulence. Infect Immun 77:4041–4050

Bertuzzi M, Howell GJ, Thomson DD, Fortune-Grant R, Möslinger A, Dancer P, Van Rhijn N, Motsi N, Du X, Codling A, Sash R, Demirbag M, Bignell EM (2022) Epithelial uptake of *Aspergillus fumigatus* drives efficient fungal clearance *in vivo* and is aberrant in Chronic Obstructive Pulmonary Disease (COPD). *bio Rxiv*:2022.2002.2001.478664

Bhatia S, Fei M, Yarlagadda M, Qi Z, Akira S, Saijo S, Iwakura Y, van Rooijen N, Gibson GA, St Croix CM, Ray A, Ray P (2011) Rapid host defense against *Aspergillus fumigatus* involves alveolar macrophages with a predominance of alternatively activated phenotype. PLoS One 6:e15943

Bigalke A, Krüger T, Jia L-J, Kniemeyer O, Brakhage AA (2023) Pathoproteomik des humanpathogenen Pilzes *Aspergillus fumigatus*. BIOspektrum 29:269–272

Blango MG, Kniemeyer O, Brakhage AA (2019) Conidial surface proteins at the interface of fungal infections. PLoS Pathog 15:e1007939

Blango MG, Pschibul A, Rivieccio F, Krüger T, Rafiq M, Jia L-J, Zheng T, Goldmann M, Voltersen V, Li J, Panagiotou G, Kniemeyer O, Brakhage AA (2020) The dynamic surface proteomes of allergenic fungal conidia. J Proteome Res 19:2092–2104

Bongomin F (2020) Post-tuberculosis chronic pulmonary aspergillosis: an emerging public health concern. PLoS Pathog 16:e1008742

Brakhage A (2005) Systemic fungal infections caused by *Aspergillus* species: epidemiology, infection process and virulence determinants. Curr Drug Targets 6:875–886

Brakhage A, Langfelder K (2002) Menacing mold: the molecular biology of *Aspergillus fumigatus*. Ann Rev Microbiol 56:433–455

Brakhage AA, Bruns S, Thywissen A, Zipfel PF, Behnsen J (2010) Interaction of phagocytes with filamentous fungi. Curr Opin Microbiol 13:409–415

Brantl V, Boysen JM, Yap A, Golubtsov E, Ruf D, Heinekamp T, Strassburger M, Dichtl K, Haas H, Hillmann F, Wagener J (2021) Peroxiredoxin Asp f3 is essential for *Aspergillus fumigatus* to overcome iron limitation during infection. MBio 12:e00976-21

Bruns S, Kniemeyer O, Hasenberg M, Aimanianda V, Nietzsche S, Thywißen A, Jeron A, Latgé J-P, Brakhage AA, Gunzer M (2010) Production of extracellular traps against *Aspergillus fumigatus* in vitro and in infected lung tissue is dependent on invading neutrophils and influenced by hydrophobin RodA. PLoS Pathog 6:e1000873

Burns C, Geraghty R, Neville C, Murphy A, Kavanagh K, Doyle S (2005) Identification, cloning, and functional expression of three glutathione transferase genes from *Aspergillus fumigatus*. Fungal Genet Biol 42:319–327

Carrion SDJ, Leal SM, Ghannoum MA, Aimanianda V, Latgé JP, Pearlman E (2013) The RodA hydrophobin on *Aspergillus fumigatus* spores masks dectin-1- and dectin-2-dependent responses and enhances fungal survival in vivo. J Immunol 191:2581–2588

Chai LY, van de Veerdonk F, Marijnissen RJ, Cheng SC, Khoo AL, Hectors M, Lagrou K, Vonk AG, Maertens J, Joosten LA, Kullberg BJ, Netea MG (2010) Anti-*Aspergillus* human host defence relies on type 1 T helper (Th1), rather than type 17 T helper (Th17), cellular immunity. Immunology 130:46–54

Champer J, Ito JI, Clemons KV, Stevens DA, Kalkum M (2016) Proteomic analysis of pathogenic fungi reveals highly expressed conserved cell wall proteins. J Fungi 2:6

Chaudhary N, Staab JF, Marr KA (2010) Healthy human T-cell responses to *Aspergillus fumigatus* antigens. PLoS One 5:e9036

Curin M, Garib V, Valenta R (2017) Single recombinant and purified major allergens and peptides: how they are made and how they change allergy diagnosis and treatment. Ann Allergy Asthma Immunol 119:201–209

Dagenais TR, Keller NP (2009) Pathogenesis of *Aspergillus fumigatus* in invasive Aspergillosis. Clin Microbiol Rev 22:447–465

Dang EV, Lei S, Radkov A, Volk RF, Zaro BW, Madhani HD (2022) Secreted fungal virulence effector triggers allergic inflammation via TLR4. Nature 608:161–167

Dasari P, Shopova IA, Stroe M, Wartenberg D, Martin-Dahse H, Beyersdorf N, Hortschansky P, Dietrich S, Cseresnyés Z, Figge MT, Westermann M, Skerka C, Brakhage AA, Zipfel PF (2018) Aspf2 from *Aspergillus fumigatus* recruits human immune regulators for immune evasion and cell damage. Front Immunol 9:1635

Dasari P, Koleci N, Shopova IA, Wartenberg D, Beyersdorf N, Dietrich S, Sahagún-Ruiz A, Figge MT, Skerka C, Brakhage AA, Zipfel PF (2019) Enolase from *Aspergillus fumigatus* is a moonlighting protein that binds the human plasma complement proteins factor H, FHL-1, C4BP, and plasminogen. Front Immunol 10:2573

DeHart DJ, Agwu DE, Julian NC, Washburn RG (1997) Binding and germination of *Aspergillus fumigatus* conidia on cultured A549 pneumocytes. J Infect Dis 175:146–150

Dellière S, Aimanianda V (2023) Humoral immunity against *Aspergillus fumigatus*. Mycopathologia 188:603–621

Denning DW, Cadranel J, Beigelman-Aubry C, Ader F, Chakrabarti A, Blot S, Ullmann AJ, Dimopoulos G, Lange C, M. European Society for Clinical Microbiology and Infectious Diseases and European Respiratory Society (2016) Chronic pulmonary aspergillosis: rationale and clinical guidelines for diagnosis and management. Eur Respir J 47:45–68

Earle K, Valero C, Conn DP, Vere G, Cook P, Bromley M, Bowyer P, Gago S (2023) Pathogenicity and virulence of *Aspergillus fumigatus*. Virulence 14:2172264

Ewald J, Rivieccio F, Radosa L, Schuster S, Brakhage AA, Kaleta C (2021) Dynamic optimization reveals alveolar epithelial cells as key mediators of host defense in invasive aspergillosis. PLoS Comput Biol 17:e1009645

Fairs A, Agbetile J, Hargadon B, Bourne M, Monteiro WR, Brightling CE, Bradding P, Green RH, Mutalithas K, Desai D, Pavord ID, Wardlaw AJ, Pashley CH (2010) IgE sensitization to *Aspergillus fumigatus* is associated with reduced lung function in asthma. Am J Respir Crit Care Med 182:1362–1368

Fallon JP, Reeves EP, Kavanagh K (2010) Inhibition of neutrophil function following exposure to the *Aspergillus fumigatus* toxin fumagillin. J Med Microbiol 59:625–633

Farnell E, Rousseau K, Thornton DJ, Bowyer P, Herrick SE (2012) Expression and secretion of *Aspergillus fumigatus* proteases are regulated in response to different protein substrates. Fungal Biol 116:1003–1012

Feys S, Gonçalves SM, Khan M, Choi S, Boeckx B, Chatelain D, Cunha C, Debaveye Y, Hermans G, Hertoghs M, Humblet-Baron S, Jacobs C, Lagrou K, Marcelis L, Maizel J, Meersseman P, Nyga R,

Seldeslachts L, Starick MR, Thevissen K, Vandenbriele C, Vanderbeke L, Vande Velde G, Van Regenmortel N, Vanstapel A, Vanmassenhove S, Wilmer A, Van De Veerdonk FL, De Hertogh G, Mombaerts P, Lambrechts D, Carvalho A, Van Weyenbergh J, Wauters J (2022) Lung epithelial and myeloid innate immunity in influenza-associated or COVID-19-associated pulmonary aspergillosis: an observational study. Lancet Respir Med 10:1147–1159

Fisher MC, Alastruey-Izquierdo A, Berman J, Bicanic T, Bignell EM, Bowyer P, Bromley M, Brüggemann R, Garber G, Cornely OA, Gurr SJ, Harrison TS, Kuijper E, Rhodes J, Sheppard DC, Warris A, White PL, Xu J, Zwaan B, Verweij PE (2022) Tackling the emerging threat of antifungal resistance to human health. Nat Rev Microbiol 20:557–571

Flannagan RS, Jaumouillé V, Grinstein S (2012) The cell biology of phagocytosis. Annu Rev Pathol Mech Dis 7:61–98

Fluckiger S, Fijten H, Whitley P, Blaser K, Crameri R (2002) Cyclophilins, a new family of cross-reactive allergens. Eur J Immunol 32:10–17

Fontaine T, Delangle A, Simenel C, Coddeville B, van Vliet SJ, van Kooyk Y, Bozza S, Moretti S, Schwarz F, Trichot C, Aebi M, Delepierre M, Elbim C, Romani L, Latgé JP (2011) Galactosaminogalactan, a new immunosuppressive polysaccharide of *Aspergillus fumigatus*. PLoS Pathog 7:e1002372

Garciá-Ortega L, Lacadena J, Villalba M, Rodríguez R, Crespo JF, Rodríguez J, Pascual C, Olmo N, Oñaderra M, del Pozo AM, Gavilanes JG (2005) Production and characterization of a noncytotoxic deletion variant of the *Aspergillus fumigatus* allergen Aspf1 displaying reduced IgE binding. FEBS J 272:2536–2544

Garlanda C, Hirsch E, Bozza S, Salustri A, De Acetis M, Nota R, Maccagno A, Riva F, Bottazzi B, Peri G, Doni A, Vago L, Botto M, De Santis R, Carminati P, Siracusa G, Altruda F, Vecchi A, Romani L, Mantovani A (2002) Non-redundant role of the long pentraxin PTX3 in anti-fungal innate immune response. Nature 420:182–186

Gil ML, Penalver MC, Lopez-Ribot JL, O'Connor JE, Martinez JP (1996) Binding of extracellular matrix proteins to *Aspergillus fumigatus* conidia. Infect Immun 64:5239–5147

Glaser AG, Limacher A, Fluckiger S, Scheynius A, Scapozza L, Crameri R (2006) Analysis of the cross-reactivity and of the 1.5 A crystal structure of the *Malassezia sympodialis* Mala s 6 allergen, a member of the cyclophilin pan-allergen family. Biochem J 396:41–49

Glaser AG, Kirsch AI, Zeller S, Menz G, Rhyner C, Crameri R (2009) Molecular and immunological characterization of Asp f 34, a novel major cell wall allergen of *Aspergillus fumigatus*. Allergy 64:1144–1151

Goh KJ, Yii ACA, Lapperre TS, Chan AK, Chew FT, Chotirmall SH, Koh MS (2017) Sensitization to *Aspergillus* species is associated with frequent exacerbations in severe asthma. J Asthma Allergy 10:131–140

Goldmann M, Schmidt F, Cseresnyés Z, Orasch T, Jahreis S, Hartung S, Figge MT, von Lilienfeld-Toal M, Heinekamp T, Brakhage AA (2023) The lipid raft-associated protein stomatin is required for accumulation of dectin-1 in the phagosomal membrane and for full activity of macrophages against *Aspergillus fumigatus*. *mSphere* 8:e00523-22

Gow NAR, Latgé JP, Munro CA (2017) The fungal cell wall: structure, biosynthesis, and function. Microbiol Spectr 5

Gravelat FN, Beauvais A, Liu H, Lee MJ, Snarr BD, Chen D, Xu W, Kravtsov I, Hoareau CM, Vanier G, Urb M, Campoli P, Al Abdallah Q, Lehoux M, Chabot JC, Ouimet MC, Baptista SD, Fritz JH, Nierman WC, Latge JP, Mitchell AP, Filler SG, Fontaine T, Sheppard DC (2013) *Aspergillus* galactosaminogalactan mediates adherence to host constituents and conceals hyphal β-glucan from the immune system. PLoS Pathog 9:e1003575

Gresnigt MS, Bozza S, Becker KL, Joosten LA, Abdollahi-Roodsaz S, van der Berg WB, Dinarello CA, Netea MG, Fontaine T, De Luca A, Moretti S, Romani L, Latge JP, van de Veerdonk FL (2014) A polysaccharide virulence factor from *Aspergillus fumigatus* elicits anti-inflammatory effects through induction of Interleukin-1 receptor antagonist. PLoS Pathog 10:e1003936

Guo Y, Kasahara S, Jhingran A, Tosini NL, Zhai B, Aufiero MA, Mills KAM, Gjonbalaj M, Espinosa V, Rivera A, Luster AD, Hohl TM (2020) During *Aspergillus* infection, monocyte-derived DCs, neutrophils, and plasmacytoid DCs enhance innate Immune defense through CXCR3-dependent cross-talk. Cell Host Microbe 28:104–116

Hatinguais R, Willment JA, Brown GD (2020) PAMPs of the Fungal Cell Wall and Mammalian PRRs. Curr Top Microbiol Immunol 425:187–223. https://doi.org/10.1007/82_2020_201. PMID: 32180018

Heinekamp T, Thywissen A, Macheleidt J, Keller S, Valiante V, Brakhage AA (2013) *Aspergillus fumigatus* melanins: interference with the host endocytosis pathway and impact on virulence. Front Microbiol 3:440

Heinekamp T, Schmidt H, Lapp K, Pähtz V, Shopova I, Köster-Eiserfunke N, Krüger T, Kniemeyer O, Brakhage AA (2015) Interference of *Aspergillus fumigatus* with the immune response. Semin Immunopathol 37:141–152

Heinen S, Hartmann A, Lauer N, Wiehl U, Dahse HM, Schirmer S, Gropp K, Enghardt T, Wallich R, Halbich S, Mihlan M, Schlotzer-Schrehardt U, Zipfel PF, Skerka C (2009) Factor H-related protein 1 (CFHR-1) inhibits complement C5 convertase activity and terminal complex formation. Blood 114:2439–2447

Hillmann F, Bagramyan K, Straßburger M, Heinekamp T, Hong TB, Bzymek KP, Williams JC, Brakhage AA, Kalkum M (2016) The crystal structure of peroxiredoxin Asp f3 provides mechanistic insight into oxidative stress resistance and virulence of *Aspergillus fumigatus*. Sci Rep 6:33396

Hoenigl M, Seidel D, Sprute R, Cunha C, Oliverio M, Goldman GH, Ibrahim AS, Carvalho A (2022) COVID-19-associated fungal infections. Nat Microbiol 7:1127–1140

Ighodaro OM, Akinloye OA (2019) First line defence antioxidants-superoxide dismutase (SOD), catalase (CAT) and glutathione peroxidase (GPX): Their fundamental role in the entire antioxidant defence grid. Alexandria J Med 54:287–293

Iwanaga N, Kolls JK (2019) Updates on T helper type 17 immunity in respiratory disease. Immunology 156:3–8

Jaton-Ogay K, Paris S, Huerre M, Quadroni M, Falchetto R, Togni G, Latgé JP, Monod M (1994) Cloning and disruption of the gene encoding an extracellular metalloprotease of *Aspergillus fumigatus*. Mol Microbiol 14:917–928

Jhingran A, Kasahara S, Shepardson KM, Junecko BA, Heung LJ, Kumasaka DK, Knoblaugh SE, Lin X, Kazmierczak BI, Reinhart TA, Cramer RA, Hohl TM (2015) Compartment-specific and sequential role of MyD88 and CARD9 in chemokine induction and innate defense during respiratory fungal infection. PLoS Pathog 11:e1004589

Jia LJ, Tang WH (2015) The omics era of *Fusarium graminearum*: opportunities and challenges. New Phytol 207:1–3

Jia L-J, Krüger T, Blango MG, von Eggeling F, Kniemeyer O, Brakhage AA (2020) Biotinylated surfome profiling identifies potential biomarkers for diagnosis and therapy of *Aspergillus fumigatus* infection. mSphere 5:e00535–20

Jia L-J, Rafiq M, Radosa L, Hortschansky P, Cunha C, Cseresnyés Z, Krüger T, Schmidt F, Heinekamp T, Straßburger M, Löffler B, Doenst T, Lacerda JF, Campos A, Figge MT, Carvalho A, Kniemeyer O, Brakhage AA (2023) *Aspergillus fumigatus* hijacks human p11 to redirect fungal-containing phagosomes to non-degradative pathway. Cell Host Microbe 31:373–388

Kariyawasam RM, Dingle TC, Kula BE, Vandermeer B, Sligl WI, Schwartz IS (2022) Defining COVID-19-associated pulmonary aspergillosis: systematic review and meta-analysis. Clin Microbiol Infect 28:920–927

Kerr SC, Fischer GJ, Sinha M, McCabe O, Palmer JM, Choera T, Lim FY, Wimmerova M, Carrington SD, Yuan S, Lowell CA, Oscarson S, Keller NP, Fahy JV (2016) FleA expression in *Aspergillus fumigatus* is recognized by fucosylated structures on mucins and macrophages to prevent lung infection. PLoS Pathog 12:e1005555

Kim KH, Brown KM, Harris PV, Langston JA, Cherry JR (2007) A proteomics strategy to discover beta-glucosidases from *Aspergillus fumigatus* with two-dimensional page in-gel activity assay and tandem mass spectrometry. J Proteome Res 6:4749–4757

Komlósi ZI, van de Veen W, Kovacs N, Szűcs G, Sokolowska M, O'Mahony L, Akdis M, Akdis CA (2022) Cellular and molecular mechanisms of allergic asthma. Mol Asp Med 85:100995

Kousha M, Tadi R, Soubani AO (2011) Pulmonary aspergillosis: a clinical review. Eur Respir Rev 20:156–174

Kumar A, Reddy LV, Sochanik A, Kurup VP (1993) Isolation and characterization of a recombinant heat shock protein of *Aspergillus fumigatus*. J Allergy Clin Immunol 91:1024–1030

Kumar A, Ahmed R, Singh PK, Shukla PK (2011) Identification of virulence factors and diagnostic markers using immunosecretome of *Aspergillus fumigatus*. J Proteome 74:1104–1112

Kurucz V, Krüger T, Antal K, Dietl AM, Haas H, Pócsi I, Kniemeyer O, Emri T (2018) Additional oxidative stress reroutes the global response of *Aspergillus fumigatus* to iron depletion. BMC Genomics 19:357

Kurup VP (2005) *Aspergillus* antigens: which are important? Med Mycol 43(Suppl 1):S189–S196

Kyrmizi I, Ferreira H, Carvalho A, Figueroa JAL, Zarmpas P, Cunha C, Akoumianaki T, Stylianou K, Deepe GS Jr, Samonis G, Lacerda JF, Campos A Jr, Kontoyiannis DP, Mihalopoulos N, Kwon-Chung KJ, El-Benna J, Valsecchi I, Beauvais A, Brakhage AA, Neves NM, Latgé JP, Chamilos G (2018) Calcium sequestration by fungal melanin inhibits calcium-calmodulin signalling to prevent LC3-associated phagocytosis. Nat Microbiol 3:791–803

Lambou K, Lamarre C, Beau R, Dufour N, Latge J-P (2010) Functional analysis of the superoxide dismutase family in *Aspergillus fumigatus*. Mol Microbiol 75:910–923

Langfelder K, Streibel M, Jahn B, Haase G, Brakhage AA (2003) Biosynthesis of fungal melanins and their importance for human pathogenic fungi. Fungal Genet Biol 38:143–158

Latgé JP, Chamilos G (2019) *Aspergillus fumigatus* and Aspergillosis in 2019. Clin Microbiol Rev 33:e00140–e00118

Latgé JP, Beauvais A, Chamilos G (2017) The cell wall of the human fungal pathogen *Aspergillus fumigatus*: biosynthesis, organization, immune response, and virulence. Ann Rev Microbiol 71:99–116

Leal SM Jr, Vareechon C, Cowden S, Cobb BA, Latge JP, Momany M, Pearlman E (2012) Fungal antioxidant pathways promote survival against neutrophils during infection. J Clin Invest 122:2482–2498

Lee MJ, Liu H, Barker BM, Snarr BD, Gravelat FN, Al Abdallah Q, Gavino C, Baistrocchi SR, Ostapska H, Xiao T, Ralph B, Solis NV, Lehoux M, Baptista SD, Thammahong A, Cerone RP, Kaminskyj SG, Guiot MC, Latge JP, Fontaine T, Vinh DC, Filler SG, Sheppard DC (2015) The fungal exopolysaccharide galactosaminogalactan mediates virulence by enhancing resistance to neutrophil extracellular traps. PLoS Pathog 11:e1005187

Lessing F, Kniemeyer O, Wozniok I, Loeffler J, Kurzai O, Haertl A, Brakhage AA (2007) The *Aspergillus fumigatus* transcriptional regulator AfYap1 represents the major regulator for defense against reactive oxygen intermediates but is dispensable for pathogenicity in an intranasal mouse infection model. Eukaryot Cell 6:2290–2302

Levdansky E, Romano J, Shadkchan Y, Sharon H, Verstrepen KJ, Fink GR, Osherov N (2007) Coding tandem repeats generate diversity in *Aspergillus fumigatus* genes. Eukaryot Cell 6:1380–1391

Limacher A, Glaser AG, Meier C, Schmid-Grendelmeier P, Zeller S, Scapozza L, Crameri R (2007) Cross-reactivity and 1.4-A crystal structure of *Malassezia sympodialis* thioredoxin (Mala s 13), a member of a new pan-allergen family. J Immunol 178:389–396

Lionakis MS, Drummond RA, Hohl TM (2023) Immune responses to human fungal pathogens and therapeutic prospects. Nat Rev Immunol 23:433–452

Liu OW, Chun CD, Chow ED, Chen C, Madhani HD, Noble SM (2008) Systematic genetic analysis of virulence in the human fungal pathogen *Cryptococcus neoformans*. Cell 135:174–188

Liu D, Li J, Zhao S, Zhang R, Wang M, Miao Y, Shen Y, Shen Q (2013) Secretome diversity and quantitative analysis of cellulolytic *Aspergillus fumigatus* Z5 in the presence of different carbon sources. Biotechnol Biofuels 6:149

Liu H, Lee MJ, Solis NV, Phan QT, Swidergall M, Ralph B, Ibrahim AS, Sheppard DC, Filler SG (2016) *Aspergillus fumigatus* CalA binds to integrin α5β1 and mediates host cell invasion. Nat Microbiol 2:16211

Liu H, Xu W, Solis NV, Woolford C, Mitchell AP, Filler SG (2018) Functional convergence of *gliP* and *aspf1* in *Aspergillus fumigatus* pathogenicity. Virulence 9:1062–1073

Liu H, Xu W, Bruno VM, Phan QT, Solis NV, Woolford CA, Ehrlich RL, Shetty AC, McCraken C, Lin J, Bromley MJ, Mitchell AP, Filler SG (2021) Determining *Aspergillus fumigatus* transcription factor expression and function during invasion of the mammalian lung. PLoS Pathog 17:e1009235

Mircescu MM, Lipuma L, van Rooijen N, Pamer EG, Hohl TM (2009) Essential role for neutrophils but not alveolar macrophages at early time points following *Aspergillus fumigatus* infection. J Infect Dis 200:647–656

Mitchell CG, Slight J, Donaldson K (1997) Diffusible component from the spore surface of the fungus *Aspergillus fumigatus* which inhibits the macrophage oxidative burst is distinct from gliotoxin and other hyphal toxins. Thorax 52:796–801

Monod M, Paris S, Sarfati J, Jaton-Ogay K, Ave P, Latgé JP (1993) Virulence of alkaline protease-deficient mutants of *Aspergillus fumigatus*. FEMS Microbiol Lett 106:39–46

Muthu V, Singh P, Choudhary H, Sehgal IS, Dhooria S, Prasad KT, Aggarwal AN, Garg M, Chakrabarti A, Agarwal R (2020) Diagnostic cutoffs and clinical utility of recombinant *Aspergillus fumigatus* antigens in the diagnosis of allergic bronchopulmonary aspergillosis. J Allergy Clin Immunol Pract 8:579–587

Namvar S, Warn P, Farnell E, Bromley M, Fraczek M, Bowyer P, Herrick S (2015) *Aspergillus fumigatus* proteases, Asp f 5 and Asp f 13, are essential for airway inflammation and remodelling in a murine inhalation model. Clin Exp Allergy 45:982–993

Neustadt M, Costina V, Kupfahl C, Buchheidt D, Eckerskorn C, Neumaier M, Findeisen P (2009) Characterization and identification of proteases secreted by *Aspergillus fumigatus* using free flow electrophoresis and MS. Electrophoresis 30:2142–2150

Ouyang H, Luo Y, Zhang L, Li Y, Jin C (2010) Proteome analysis of *Aspergillus fumigatus* total membrane proteins identifies proteins associated with the glycoconjugates and cell wall biosynthesis using 2D LC-MS/MS. Mol Biotechnol 44:177–189

Paris S, Wysong D, Debeaupuis JP, Shibuya K, Philippe B, Diamond RD, Latgé JP (2003) Catalases of *Aspergillus fumigatus*. Infect Immun 71:3551–3562

Pazhakh V, Ellett F, Croker BA, O'Donnell JA, Pase L, Schulze KE, Greulich RS, Gupta A, Reyes-Aldasoro CC, Andrianopoulos A, Lieschke GJ (2019) β-glucan–dependent shuttling of conidia from neutrophils to macrophages occurs during fungal infection establishment. PLoS Biol 17:e3000113

Penalver MC, O'Connor JE, Martinez JP, Gil ML (1996) Binding of human fibronectin to *Aspergillus fumigatus* conidia. Infect Immun 64:1146–1153

Poll V, Denk U, Shen HD, Panzani RC, Dissertori O, Lackner P, Hemmer W, Mari A, Crameri R, Lottspeich F, Rid R, Richter K, Breitenbach M, Simon-Nobbe B (2009) The vacuolar serine protease, a cross-reactive allergen from *Cladosporium herbarum*. Mol Immunol 46:1360–1373

Rafiq M, Rivieccio F, Zimmermann A-K, Visser C, Bruch A, Krüger T, González Rojas K, Kniemeyer O, Blango MG, Brakhage AA (2022) PLB-985 neutrophil-like cells as a model to study *Aspergillus fumigatus* pathogenesis. mSphere 7:e0094021

Ramachandran H, Banerjee B, Greenberger PA, Kelly KJ, Fink JN, Kurup VP (2004) Role of C-terminal cysteine residues of *Aspergillus fumigatus* allergen Asp f 4 in immunoglobulin E binding. Clin Diagn Lab Immunol 11:261–265

Rambach G, Blum G, Latgé JP, Fontaine T, Heinekamp T, Hagleitner M, Jeckström H, Weigel G, Würtinger P, Pfaller K, Krappmann S, Löffler J, Lass-Flörl C, Speth C (2015) Identification of *Aspergillus fumigatus* surface components that mediate interaction of conidia and hyphae with human platelets. J Infect Dis 212:1140–1149

Rawlings ND, Barrett AJ, Bateman A (2010) MEROPS: the peptidase database. Nucleic Acids Res 38:D227–D233

Reis ES, Mastellos DC, Hajishengallis G, Lambris JD (2019) New insights into the immune functions of complement. Nat Rev Immunol 19:503–516

Rhyner C, Daigle I, Crameri R (2011) Auto-reactive IgE responses to acidic ribosomal P(2) protein in systemic lupus erythematosus. Allergy 66:1127–1129

Robinet P, Baychelier F, Fontaine T, Picard C, Debré P, Vieillard V, Latgé JP, Elbim C (2014) A polysaccharide virulence factor of a human fungal pathogen induces neutrophil apoptosis via NK cells. J Immunol 192:5332–5342

Saxena S, Madan T, Muralidhar K, Sarma PU (2003) cDNA cloning, expression and characterization of an allergenic L3 ribosomal protein of *Aspergillus fumigatus*. Clin Exp Immunol 134:86–91

Scharf DH, Heinekamp T, Remme N, Hortschansky P, Brakhage AA, Hertweck C (2012) Biosynthesis and function of gliotoxin in *Aspergillus fumigatus*. Appl Microbiol Biotechnol 93:467–472

Scharrenberg A, Gerber V, Swinburne JE, Wilson AD, Klukowska-Rotzler J, Laumen E, Marti E (2010) IgE, IgGa, IgGb and IgG(T) serum antibody levels in offspring of two sires affected with equine recurrent airway obstruction. Anim Genet 41(Suppl 2):131–137

Schauwvlieghe A, Rijnders BJA, Philips N, Verwijs R, Vanderbeke L, Van Tienen C, Lagrou K, Verweij PE, Van de Veerdonk FL, Gommers D, Spronk P, Bergmans D, Hoedemaekers A, Andrinopoulou ER, van den Berg C, Juffermans NP, Hodiamont CJ, Vonk AG, Depuydt P, Boelens J, Wauters J, g. Dutch-Belgian Mycosis study. (2018) Invasive aspergillosis in patients admitted to the intensive care unit with severe influenza: a retrospective cohort study. Lancet Respir Med 6:782–792

Schmidt H, Vlaic S, Krüger T, Schmidt F, Balkenhol J, Dandekar T, Guthke R, Kniemeyer O, Heinekamp T, Brakhage AA (2018) Proteomics of *Aspergillus fumigatus* conidia-containing phagolysosomes identifies processes governing immune evasion. Mol Cell Proteomics 17:1084–1096

Schmidt F, Thywißen A, Goldmann M, Cunha C, Cseresnyés Z, Schmidt H, Rafiq M, Galiani S, Gräler MH, Chamilos G, Lacerda JF, Campos A, Eggeling C, Figge MT, Heinekamp T, Filler SG, Carvalho A, Brakhage AA (2020) Flotillin-dependent membrane microdomains are required for functional phagolysosomes against fungal infections. Cell Rep 32:108017

Schwarz C, Eschenhagen P, Schmidt H, Hohnstein T, Iwert C, Grehn C, Roehmel J, Steinke E, Stahl M, Lozza L, Tikhonova E, Rosati E, Stervbo U, Babel N, Mainz JG, Wisplinghoff H, Ebel F, Jia LJ, Blango MG, Hortschansky P, Brunke S, Hube B, Brakhage AA, Kniemeyer O, Scheffold A, Bacher P (2023) Antigen specificity and cross-reactivity drive functionally diverse anti-*Aspergillus fumigatus* T cell responses in cystic fibrosis. J Clin Invest 133:e161593

Schwienbacher M, Israel L, Heesemann J, Ebel F (2005) Asp f6, an *Aspergillus* allergen specifically recognized by IgE from patients with allergic bronchopulmonary aspergillosis, is differentially expressed during germination. Allergy 60:1430–1435

Seidel C, Moreno-Velásquez SD, Ben-Ghazzi N, Gago S, Read ND, Bowyer P (2020) Phagolysosomal survival enables non-lytic hyphal escape and ramification through lung epithelium during *Aspergillus fumigatus* infection. Front Microbiol 11:1955

Seif M, Kakoschke TK, Ebel F, Bellet MM, Trinks N, Renga G, Pariano M, Romani L, Tappe B, Espie D, Donnadieu E, Hünniger K, Häder A, Sauer M, Damotte D, Alifano M, White PL, Backx M, Nerreter T, Machwirth M, Kurzai O, Prommersberger S, Einsele H, Hudecek M, Löffler J (2022) CAR T cells targeting *Aspergillus fumigatus* are effective at treating invasive pulmonary aspergillosis in preclinical models. Sci Transl Med 14:eabh1209

Sharma Ghimire P, Ouyang H, Wang Q, Luo Y, Shi B, Yang J, Lu Y, Jin C (2016) Insight into enzymatic degradation of corn, wheat, and soybean cell wall cellulose using quantitative secretome analysis of *Aspergillus fumigatus*. J Proteome Res 15:4387–4402

Sharma M, Soni R, Nazir A, Oberoi HS, Chadha BS (2011) Evaluation of glycosyl hydrolases in the secretome of *Aspergillus fumigatus* and saccharification of alkali-treated rice straw. Appl Biochem Biotechnol 163:577–591

Sharon H, Hagag S, Osherov N (2009) Transcription factor PrtT controls expression of multiple secreted proteases in the human pathogenic mold *Aspergillus fumigatus*. Infect Immun 77:4051–4060

Shemesh E, Hanf B, Hagag S, Attias S, Shadkchan Y, Fichtman B, Harel A, Krüger T, Brakhage AA, Kniemeyer O, Osherov N (2017) Phenotypic and proteomic analysis of the *Aspergillus fumigatus* Δ*PrtT*, Δ*XprG* and Δ*XprG*/Δ*PrtT* protease-deficient mutants. Front Microbiol 8:2490

Shende R, Wong SSW, Rapole S, Beau R, Ibrahim-Granet O, Monod M, Guhrs KH, Pal JK, Latge JP, Madan T, Aimanianda V, Sahu A (2018) *Aspergillus fumigatus* conidial metalloprotease Mep1p cleaves host complement proteins. J Biol Chem 293:15538–15555

Shende R, Wong SSW, Meitei HT, Lal G, Madan T, Aimanianda V, Pal JK, Sahu A (2022) Protective role of host complement system in *Aspergillus fumigatus* infection. Front Immunol 13:978152

Sheppard DC (2011) Molecular mechanism of *Aspergillus fumigatus* adherence to host constituents. Curr Opin Microbiol 14:375–379

Shopova IA, Belyaev I, Dasari P, Jahreis S, Stroe MC, Cseresnyes Z, Zimmermann AK, Medyukhina A, Svensson CM, Kruger T, Szeifert V, Nietzsche S, Conrad T, Blango MG, Kniemeyer O, von Lilienfeld-Toal M, Zipfel PF, Ligeti E, Figge MT, Brakhage AA (2020) Human neutrophils produce antifungal extracellular vesicles against *Aspergillus fumigatus*. MBio 11:e00596–20

Singh B, Oellerich M, Kumar R, Kumar M, Bhadoria DP, Reichard U, Gupta VK, Sharma GL, Asif AR (2010) Immuno-reactive molecules identified from the secreted proteome of *Aspergillus fumigatus*. J Proteome Res 9:5517–5529

Skerka C, Chen Q, Fremeaux-Bacchi V, Roumenina LT (2013) Complement factor H related proteins (CFHRs). Mol Immunol 56:170–180

Smith JM, Tang CM, Van Noorden S, Holden DW (1994) Virulence of *Aspergillus fumigatus* double mutants lacking restriction and an alkaline protease in a low-dose model of invasive pulmonary aspergillosis. Infect Immun 62:5247–5254

Soh WT, Zhang J, Hollenberg MD, Vliagoftis H, Rothenberg ME, Sokol CL, Robinson C, Jacquet A (2023) Protease allergens as initiators-regulators of allergic inflammation. Allergy 78:1148–1168

Speakman EA, Dambuza IM, Salazar F, Brown GD (2020) T cell antifungal immunity and the role of C-type lectin receptors. Trends Immunol 41:61–76

Sriranganadane D, Waridel P, Salamin K, Reichard U, Grouzmann E, Neuhaus JM, Quadroni M, Monod M (2010) *Aspergillus* protein degradation pathways with different secreted protease sets at neutral and acidic pH. J Proteome Res 9:3511–3519

Stappers MHT, Clark AE, Aimanianda V, Bidula S, Reid DM, Asamaphan P, Hardison SE, Dambuza IM, Valsecchi I, Kerscher B, Plato A, Wallace CA, Yuecel R, Hebecker B, da Glória Teixeira M, Sousa C, Cunha Y, Liu T, Feizi AA, Brakhage KJ, Kwon-Chung NAR, Gow M, Zanda M, Piras C, Zanato M, Jaeger MG, Netea FL, van de Veerdonk JF, Lacerda A, Campos A, Carvalho JA, Willment JPL, Brown GD (2018) Recognition of DHN-melanin by a C-type lectin receptor is required for immunity to *Aspergillus*. Nature 555:382–386

Taccone FS, Van den Abeele A-M, Bulpa P, Misset B, Meersseman W, Cardoso T, Paiva J-A, Blasco-Navalpotro M, De Laere E, Dimopoulos G, Rello J, Vogelaers D, Blot SI (2015) Epidemiology of invasive aspergillosis in critically ill patients: clinical presentation, underlying conditions, and outcomes. Crit Care 19:7

Thau N, Monod M, Crestani B, Rolland C, Tronchin G, Latgé JP, Paris S (1994) *Rodletless* mutants of *Aspergillus fumigatus*. Infect Immun 62:4380–4388

Thywißen A, Heinekamp T, Dahse H-M, Schmaler-Ripcke J, Nietzsche S, Zipfel PF, Brakhage AA (2011) Conidial dihydroxynaphthalene melanin of the human pathogenic fungus *Aspergillus fumigatus* interferes with the host endocytosis pathway. Front Microbiol 2:96

Tiew PY, Narayana JK, Quek MSL, Ang YY, Ko FWS, Poh ME, Jaggi TK, Xu H, Thng KX, Koh MS, Tee A, Hui DSC, Abisheganaden JA, Tsaneva-Atanasova K, Chew FT, Chotirmall SH (2023) Sensitisation to recombinant *Aspergillus fumigatus* allergens and clinical outcomes in COPD. Eur Respir J 61:2200507

Upadhyay SK, Mahajan L, Ramjee S, Singh Y, Basir SF, Madan T (2009) Identification and characterization of a laminin-binding protein of *Aspergillus fumigatus*: extracellular thaumatin domain protein (AfCalAp). J Med Microbiol 58:714–722

Valsecchi I, Dupres V, Stephen-Victor E, Guijarro JI, Gibbons J, Beau R, Bayry J, Coppee JY, Lafont F, Latgé JP, Beauvais A (2017) Role of hydrophobins in *Aspergillus fumigatus*. J Fungi 4:2

Valsecchi I, Dupres V, Michel JP, Duchateau M, Matondo M, Chamilos G, Saveanu C, Guijarro JI, Aimanianda V, Lafont F, Latgé JP, Beauvais A (2019a) The puzzling construction of the conidial outer layer of *Aspergillus fumigatus*. Cell Microbiol 21:e12994

Valsecchi I, Lai JI, Stephen-Victor E, Pillé A, Beaussart A, Lo V, Pham CLL, Aimanianda V, Kwan AH, Duchateau M, Gianetto QG, Matondo M, Lehoux M, Sheppard DC, Dufrene YF, Bayry J, Guijarro JI, Sunde M, Latgé J-P (2019b) Assembly and disassembly of *Aspergillus fumigatus* conidial rodlets. Cell Surf 5:100023

van de Veerdonk FL, Gresnigt MS, Romani L, Netea MG, Latge JP (2017) *Aspergillus fumigatus* morphology and dynamic host interactions. Nat Rev Microbiol 15:661–674

Vivek-Ananth RP, Mohanraj K, Vandanashree M, Jhingran A, Craig JP, Samal A (2018) Comparative systems analysis of the secretome of the opportunistic pathogen *Aspergillus fumigatus* and other *Aspergillus* species. Sci Rep 8:6617

Vogl G, Lesiak I, Jensen DB, Perkhofer S, Eck R, Speth C, Lass-Florl C, Zipfel PF, Blom AM, Dierich MP, Wurzner R (2008) Immune evasion by acquisition of complement inhibitors: the mould *Aspergillus* binds both factor H and C4b binding protein. Mol Immunol 45:1485–1493

Voltersen V, Blango MG, Herrmann S, Schmidt F, Heinekamp T, Strassburger M, Kruger T, Bacher P, Lother J, Weiss E, Hunniger K, Liu H, Hortschansky P, Scheffold A, Loffler J, Krappmann S, Nietzsche S, Kurzai O, Einsele H, Kniemeyer O, Filler SG, Reichard U, Brakhage AA (2018) Proteome analysis reveals the conidial surface protein CcpA essential for virulence of the pathogenic fungus *Aspergillus fumigatus*. MBio 9:e01557–18

Wartenberg D, Lapp K, Jacobsen ID, Dahse HM, Kniemeyer O, Heinekamp T, Brakhage AA (2011) Secretome analysis of *Aspergillus fumigatus* reveals Asp-hemolysin as a major secreted protein. Int J Med Microbiol 301:602–611

Warwas ML, Watson JN, Bennet AJ, Moore MM (2007) Structure and role of sialic acids on the surface of *Aspergillus fumigatus* conidiospores. Glycobiology 17:401–410

Wasylnka JA, Moore MM (2003) *Aspergillus fumigatus* conidia survive and germinate in acidic organelles of A549 epithelial cells. J Cell Sci 116:1579–1587

WHO (2022) WHO fungal priority pathogens list to guide research, development and public health action

Wiesner DL, Merkhofer RM, Ober C, Kujoth GC, Niu M, Keller NP, Gern JE, Brockman-Schneider RA, Evans MD, Jackson DJ, Warner T, Jarjour NN, Esnault SJ, Feldman MB, Freeman M, Mou H, Vyas JM, Klein BS (2020) Club cell TRPV4 serves as a damage sensor driving lung allergic inflammation. Cell Host Microbe 27:614–628. e616

Wong SSW, Daniel I, Gangneux J-P, Jayapal JM, Guegan H, Dellière S, Lalitha P, Shende R, Madan T, Bayry J, Guijarro JI, Kuppamuthu D, Aimanianda V (2020) Differential interactions of serum and bronchoalveolar lavage fluid complement proteins with conidia of airborne fungal pathogen *Aspergillus fumigatus*. Infect Immun 88:e00212–20

Zipfel PF, Skerka C (2009) Complement regulators and inhibitory proteins. Nat Rev Immunol 9:729–740

Part II

Host-Pathogen Interaction

RNA as a Mediator of Host-Fungal Pathogenesis

5

Alexander Bruch and Matthew G. Blango

Abstract

In this chapter, we explore the wide range of functions RNA plays in mediating interactions between fungi and their hosts in the animal kingdom. We cover examples of RNA in the regulation of gene expression, novel RNA structures and non-coding RNAs with links to fungal biology, as well as the emerging fields of epimutation and extracellular RNA. Next, the host response to fungal infection is investigated by considering cases of RNA recognition by host pattern recognition receptors, together with host regulatory mechanisms dependent on non-coding RNAs. Many of these themes are then revisited in the discussion of interkingdom communication, where RNA is sent between fungi and animals to influence the outcome of a given infection. We end by discussing the recent progress in RNA-based therapeutics against human fungal pathogens, where this foundation of basic research is leading to potential improvements in the way we treat human diseases.

A. Bruch · M. G. Blango (✉)
Leibniz Institute for Natural Product Research and Infection Biology–Hans Knöll Institute (Leibniz-HKI), Jena, Germany
e-mail: alexander.bruch@leibniz-hki.de; matthew.blango@leibniz-hki.de

Keywords

RNA · Fungal pathogenesis · *Aspergillus fumigatus* · *Candida albicans* · *Cryptococcus neoformans* · Host-pathogen · Fungi · Mycota
Pattern recognition receptor

5.1 Introduction

When discovered, nucleic acids were first associated with the source of their isolation. What was initially known as "thymus nucleic acid" would eventually be defined as a specific macromolecule, deoxyribonucleic acid (DNA); the mysterious "yeast nucleic acid" was in due course classified as ribonucleic acid (RNA). In this chapter we will focus on the "yeast nucleic acids" and explore the contribution of the RNA macromolecule to fungal-animal relationships, with a particular emphasis on human fungal pathogens. We will examine far-reaching examples of RNA regulation in a variety of systems and explore concepts that will likely prove key in host-fungal interactions as this field further develops. In the conclusion section, we will provide an update on recent progress in using RNA more widely in therapeutic intervention, both as a target and as a therapeutic itself.

Before revealing the complexity of RNA regulatory pathways in the host-fungal interaction, we first must know a bit more about RNA biochem-

A. A. Brakhage et al. (eds.), *Human and Animal Relationships*, The Mycota 6,
https://doi.org/10.1007/978-3-031-64853-3_5

istry. The core structure of the RNA macromolecule is composed of alternating phosphate groups and ribose sugars. The ribose sugars of RNA carry, attached at their 1′ position, the nitrogenous bases adenine (A), guanine (G), cytosine (C), and uracil (U)—notably not thymine (T) as in DNA—and a hydroxyl group at the 2′ position, which helps distinguish RNA from DNA (Fig. 5.1). Unlike DNA, RNA is primarily found to be single-stranded, but this does not mean the molecule is wholly unstructured. In fact, the flexibility of the molecule and the incorporation of >170 modifications (Frye et al. 2018; Zhang et al. 2023) means that RNA is frequently folded into complicated secondary, and even tertiary structures (Väre et al. 2017). The structure of the tRNA is often used to illustrate the importance of RNA folding, with its numerous hairpins forming into a cloverleaf structure that finally twists into a more compact L-shaped structure via tertiary interactions between the helices.

RNA nucleotides can base pair in a manner similar to DNA, with guanine pairing with cytosine, and adenine pairing with uracil according to the standard Watson and Crick base pairing rules. One of the features that makes RNA so special is the ability of the bases to also pair in unconventional ways using the Hoogsteen edge or even the sugar edge for instance. Modified bases like pseudouridine or inosine inject even more possibilities into RNA folding and are likely only the tip of the iceberg of RNA folding complexity found throughout the kingdoms of life. Recent efforts in modeling RNA folding have revealed a highly dynamic process that can be facilitated by binding to proteins or other RNA molecules to form intermolecular structures (Yu et al. 2021).

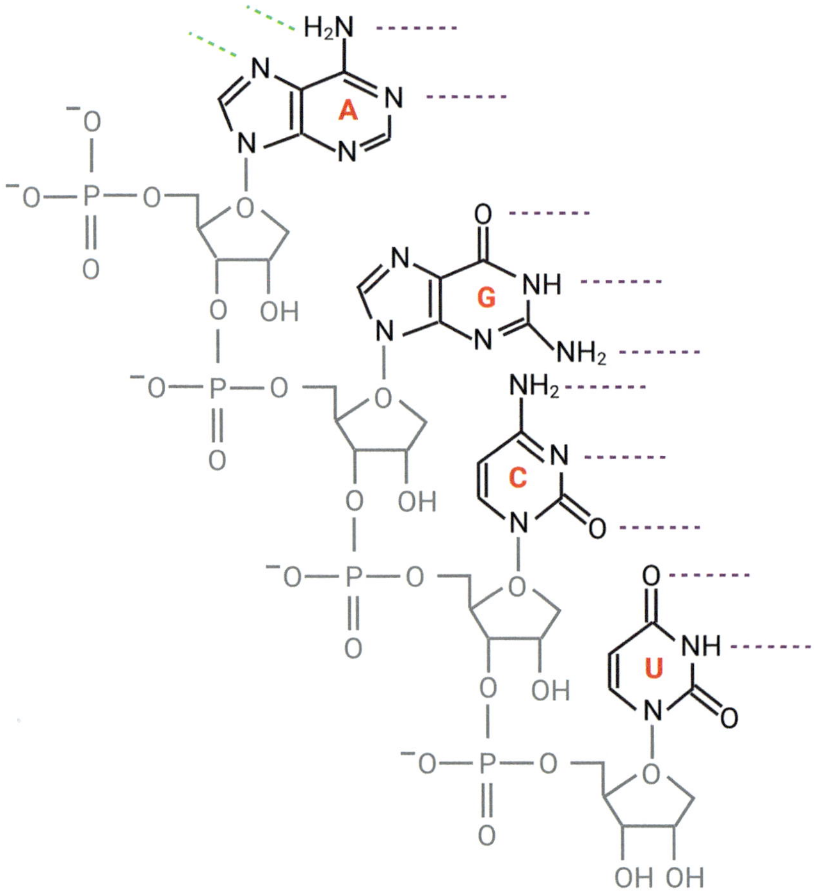

Fig. 5.1 Chemical structure of RNA showing the sugar phosphate backbone in gray and the nitrogenous bases adenine (A), guanine (G), cytosine (C), and uracil (U) in black. The Watson-Crick base-pairing interactions are highlighted in purple with an example of Hoogsteen pairing shown in green. Created with BioRender.com

With the discovery of messenger RNA (mRNA) in the early 1960s, RNA was divided into coding and non-coding sequences. In eukaryotes, RNA polymerase II typically transcribes mRNA, a process that is coupled with splicing, polyadenylation, and the addition of a 5′ protective cap structure. A combination of proteins and non-coding RNAs including transfer RNAs (tRNA) and ribosomal RNAs (rRNA) then help to decode and translate the mRNA into protein. In subsequent years, numerous additional regulatory RNAs have been described, including long non-coding RNAs (lncRNA) and a variety of subclasses of small RNA molecules like the microRNAs (miRNA) and short interfering RNAs (siRNA) that will be discussed further in this chapter. Collectively, these non-coding RNA species modulate gene expression by aiding in RNA processing, modification, and even trafficking.

The integral nature of RNA throughout gene expression and regulation means that it is also frequently involved in host pathogenesis, both as a modulator of the immune response and as a signal of foreign entities. In looking at the case of human-viral interactions, RNA intermediates like double-stranded RNA (dsRNA) are frequently recognized as a signal of a viral infection by pattern recognition receptors. In bacteria, small RNA species aid in virulence and gene regulation during infection, whereas many hosts send RNA molecules as opposing signals to promote immunity. We have learned time and again that RNA is involved in signaling between organisms and can even play important roles in immune recognition pathways.

The most notable examples of intercellular RNA communication come from plant-fungal interactions, where studies of bidirectional cross-kingdom RNA interference have revealed that small RNAs are produced by both the host and the pathogen to influence the outcome of infection in a sort of molecular arms race (Weiberg et al. 2013; Cai et al. 2018b). Canonically, these small RNAs are produced by dsRNA-precursor cleavage by a dicer protein before loading into an argonaute protein for gene silencing, often as part of a larger RNA-induced silencing complex (Wilson and Doudna 2013). Sequence complementarity between the small RNA and target drives mRNA degradation or translational repression. As an example, the small RNAs of the grey mold *Botrytis cinerea* were shown to be transferred to the plant host and internalized via clathrin-mediated endocytosis to silence plant immunity genes (Weiberg et al. 2013; He et al. 2023). The host plant also uses a similar approach to fight back against fungal infection by delivering effector small RNAs and mRNAs in exosome-like extracellular vesicles into the fungus (Cai et al. 2018b; Wang et al. 2024). The endogenous small RNAs produced by plants like *A. thaliana* are transferred to the fungus to target the mRNAs of conserved vesicle pathways important to virulence (Cai et al. 2018b; He et al. 2021). Other examples support the use of small RNAs by plants to target fungal pathogens as well. Wheat plants can suppress the invasion of *Fusarium graminearum* using miRNAs (Jiao and Peng 2018), and cotton plants export miRNAs to inhibit the virulence of *Verticillium dahliae* (Zhang et al. 2016). These studies offer a clear role for RNA in interkingdom communication in plant-fungal interactions.

In this chapter, we extend these principles to explore the wide range of functions RNA plays in interactions between fungi and their hosts in the kingdom *Animalia*. Most of these cases come from studies of human fungal pathogens due to their medical relevance, but we expect that these mechanisms are by no means unique to human pathogens. Instead, similar processes are surely used during all sorts of host-microbe interactions. The studies presented here will highlight how much we can learn by studying RNA regulation in pathogens and how animal hosts naturally respond to these fungal invaders. We will end by discussing the potential of converting our collective knowledge of RNA regulation in fungal pathogenesis into RNA-centric therapeutics to fight back against fungal pathogens and improve human health.

5.2 RNA Regulation in Fungal Pathogens

5.2.1 RNA in Gene Regulation: Transcription and Translation

The classic, textbook function of RNA is in converting the genetic information of DNA into the functional output of protein, but we now know that RNA serves many more roles beyond just that of an intermediary. In the process of transcription, RNA serves as both the product of the RNA polymerase and as a regulator of the entire process of transcription in the form of non-coding RNAs. We have learned a lot from model organisms about the intricacies of RNA regulation of transcription, but in many cases, we have obtained only glimpses of the complexity surrounding RNA regulation in fungal pathogens. The transcriptional response as assessed by RNA-seq is now frequently the first feature of gene expression considered when investigating a new experimental system.

Cryptococcus neoformans is likely the best-studied of the human pathogenic fungi regarding central questions surrounding gene regulation. It was recently shown that upon entry of *C. neoformans* into the host environment, many gene regulatory systems are employed to facilitate global gene expression reprogramming and adaptation to the local host environment. A major aspect of this adaptation comes in the form of temperature response to adjust to the warm host, an essential attribute of any human fungal pathogen. For *C. neoformans*, it was shown that the mRNA decay pathway plays a pivotal role in managing thermotolerance and growth in the host by degrading messages, in particular those of ribosomal proteins (Bloom et al. 2019). In the absence of the mRNA decay pathway using a *ccr4* knockout strain, cell wall remodeling enzymes were repressed, resulting in exposed cell wall glucans on the fungal surface that could trigger activation of the C-type lectin receptor Dectin-1. Recognition of fungal ligands by Dectin-1 led to an increased host response and clearance of the fungus, highlighting the importance of removing unwanted transcripts to facilitate a more fine-tuned transcriptomic response.

The rapid adjustment of transcriptomes during the interaction of organisms is facilitated by global regulatory proteins like the *C. neoformans* virulence-associated DEAD-box RNA helicase VAD1. This RNA-binding protein functions as a negative regulator of the global transcriptional repressor, NOT1, resulting in upregulation of a laccase important for melanin pigment production during infection (Panepinto et al. 2005). Corroborating that the loss of VAD1 was detrimental to virulence, other virulence determinants, including TUF1, PCK1, and MPF3 were also downregulated. Global regulators like VAD1 play important roles in rapidly shaping the transcriptome during the host-pathogen interaction, where integration of virulence cascades into the global transcriptional response is paramount.

Yet, it is not always the case that pathogens respond so aggressively to the new host environment. In fact, the filamentous fungal pathogen *Aspergillus fumigatus* does not mount a recognizable transcriptional response to human airway epithelial cells at all (Watkins et al. 2018), at least in culture, but instead responds more actively to the growth media of these *in vitro* assays. These findings suggest that understanding the entire host environment, including cells, nutrients, and likely otherwise unanticipated features will be important to eventually appreciate gene regulatory cascades and transcriptional responses employed during interaction.

Once mRNA is produced, it must be translated by the ribosome into protein; a process influenced on multiple levels by features of the mRNA, additional non-coding RNAs, and even catalytic RNA molecules in the supramolecular structure of the ribosome. The ribosome is a dominant feature of life that may superficially seem extremely conserved, yet examples of alternative regulation exist even here. One such story comes from the yeast *Candida albicans*, an organism capable of living as both a commensal and a pathogen in human hosts. Unlike the model yeast *Saccharomyces cerevisiae, C. albicans* employs a ribosomal RNA processing cascade that uses both cotranscriptional and post-transcriptional rRNA processing modes to facilitate better responses to changing environmental conditions like those found in a nutrient restricted host (Pendrak and Roberts 2011). By evolving this layered regulatory network, *C. albicans* has acquired an additional tool to manage stressful situations and survive in a harsh host environment. This finding represents the power of studying non-model organisms to reveal amazing new regulatory mechanisms of core molecular processes.

Translation is greatly influenced by the architecture of the mRNA, including parameters such as length, structure, and regulatory elements. In *Cryptococcus*, mRNAs have long transcriptional leader sequences and appear to employ a large number of upstream open reading frames (uORFs) capable of repressing translation, features that are less prevalent in *S. cerevisiae* (Wallace et al. 2020). The observed uORF repression has a strong dependency on the Kozak sequence context of the associated AUG start codon, and a strong context typically promotes nonsense-mediated decay of these mRNAs. It was also shown that start codon context contributes to regulation of predicted dual-localized proteins, a mechanism that is hypothesized to occur as a leaky upstream AUG is separated from a downstream in-frame AUG with strong Kozak context by a mitochondrial localization signal. The selection of start codons thus dictates the abundance and structure of the proteins produced; a characteristic that is now recognized to be maintained in many fungi across the kingdom.

Technological advances in RNA-seq have facilitated incredible progress in our understanding of gene expression, but in general the tools to study many aspects of gene regulation in fungal pathogens remain missing. In many cases, the complex and impermeable cell wall that can scavenge nucleic acids from the environment has proven to be a significant hurdle (Jochl et al. 2009). Molecular and cellular techniques are improving with time and studies that were difficult a decade ago are now becoming more common. For example, mRNA localization studies were nearly impossible a few decades ago, as high background and low signal made data interpretation difficult. Early experiments were promising, and despite these technical challenges, indicated a regulated trafficking of mRNAs in fungi, specifically in organisms like *C. albicans*, where mRNAs were shown to be trafficked to the hyphal tip to aid hyphal development and even invasion during infection (Elson et al. 2009). More recent advances in fluorescent *in situ* hybridization techniques like hybridization chain reaction using split probes have provided improved signal-to-noise ratios and allowed single molecule visualizations to work in some organisms (Moreno-Velasquez and Perez 2021). In this case, multiple mRNAs were visualized together in a single *C. albicans* cell with minimal loss of signal, paving the way for more mechanistic studies of mRNA trafficking and regulation in this and related organisms. Improvements in localizing and tracking RNAs are now facilitating studies of the full lifecycle of a fungal RNA from transcription to degradation.

5.2.2 RNA Structure and Non-coding RNA

In addition to new variations of canonical processes, research has uncovered a tremendous number of novel RNAs and RNA structural motifs in fungi. Large-scale screening of fungal genomes revealed an abundance of unique structural folds, some of which could be assigned functions (Li and Breaker 2017). However, most of these RNA structural motifs and non-coding RNAs remain completely unstudied, leaving a wealth of untapped information available to the field. The study of both RNA (and DNA) G quadruplexes, noncanonical four-stranded structures formed by a series of G-G base pairs, is a growing area of research in many organisms, including fungi. Recent efforts have revealed widespread examples of these intriguing secondary structures in *C. neoformans* and *A. fumigatus* (Leipheimer et al. 2018; Warner et al. 2021), but the full regulatory potential remains unknown. Fungal riboswitches are rarer, with only a few variations of the thiamine pyrophosphate (TPP) riboswitch described to date (Vargas-Junior et al. 2022; Moldovan et al. 2018; Donovan et al. 2018). It can be guaranteed that novel structures and regulatory mechanisms will continue to be unearthed as the field expands into new transcriptomic territory.

In the genus *Candida*, a recent study uncovered thousands of lncRNAs that show evolutionary conservation above that observed for intergenic regions but less than protein-coding regions (Hovhannisyan and Gabaldon 2021). These moderately conserved lncRNAs were also observed to be differentially expressed during infection of epithelial cells with *Candida* species, suggesting a potential function in the regulation of virulence or stress response; however, more work will need to be done to elucidate the full impact of these lncRNAs on gene regulation. Corroborating the likely importance of these newly identified lncRNAs, the *C. auris* lncRNA *DINOR* was shown to contribute to stress response and virulence, with a genetic deletion exhibiting increased sensitivity to stressors and decreased virulence in a mouse model of infection (Gao et al. 2021). Work in *C. neoformans* has uncovered an estimated 1182 lncRNAs, with diverse predicted functions (Kalem and Panepinto 2022). Concerted effort is now rightly underway to characterize these widespread regulatory RNAs.

The first lncRNA characterized from a human pathogen came from *C. neoformans* and can serve as an example of the regulatory potential waiting to be uncovered in lncRNAs. In this study, the lncRNA *RZE1* was shown to regulate the anti-virulence factor and morphogenesis regulator, Znf2 (Chacko et al. 2015). The lncRNA *RZE1* is restricted to its nucleus of origin, where it works primarily in cis to increase the transcript level of *ZNF2* and alter the ratio of cytoplasmic to nuclear transcripts in favor of the cytoplasm, as determined using deletion studies. In the absence of the zinc finger transcription factor Znf2, *C. neoformans* cells are locked in the more virulent yeast form. Consistent with this regulation, the loss of *RZE1* promoted virulence by decreasing *ZNF2* transcript levels, linking its expression to pathogenicity. Although just a few examples, lncRNAs have a vast potential to control gene regulation, stress response, and virulence in diverse ways during interaction with host organisms.

Non-coding RNAs come in many forms, from the lncRNAs mentioned above to tRNA, rRNAs, or even as introns. In the case of introns, examples are scattered throughout the literature of the complex regulation both in the removal of introns from mRNAs by the splicing machinery, as well as the intricate regulatory pathways that result in intron retention or alternative splicing under different conditions. In *C. neoformans*, introns can be retained for gene regulation in a manner that seems to be independent of nonsense-mediated decay (Gonzalez-Hilarion et al. 2016). This is in stark contrast to other organisms where introns are thought to diversify and provide additional flexibility to the proteome (Chorev and Carmel 2012). Introns are also known to help protect eukaryotic genomes against the genomic instability introduced by transcription (Bonnet et al. 2017). It is then not surprising that there is evidence to support differential regulation of alternative splicing during fungal stress and virulence (Sieber et al. 2018). The picture becomes even more convoluted when additional regulatory RNAs are added into the mix. A clear illustration comes from *A. flavus*, where natural antisense transcripts, another form of poorly understood non-coding RNA, were found to be controlled in part by temperature; however, the full implications of this finding remain unknown (Smith et al. 2008). A more careful and systematic assessment of the many factors involved will likely be required to fully understand the regulatory potential of RNA structure and non-coding RNAs during interactions between fungi and animals.

5.2.3 RNAi and Small RNAs: Regulation of Resistance and Genome Defense

The best-studied RNA regulatory pathway in non-model fungi is likely the RNA interference system. Over the past two decades, many fungi have been shown to contain functional RNAi pathways, with *Cryptococcus* spp. and *A. fumiga-*

tus among the first to be characterized (Janbon et al. 2010; Liu et al. 2002; Mouyna et al. 2004; Hammond et al. 2008). The canonical RNA interference system processes dsRNA to sequence-specifically, posttranscriptionally silence gene expression by mRNA cleavage or blockade of translation machinery binding (Wilson and Doudna 2013). The RNAi machinery, and small RNAs generated by the system, can also promote the formation of a heterochromatic state, resulting in transcriptional repression (Martienssen and Moazed 2015). In fungi, the RNAi pathway frequently plays an important role in genome defense from mycoviruses and selfish genetic elements like transposons (Priest et al. 2022; Navarro-Mendoza et al. 2023), but other varied functions have also been assigned (Torres-Martinez and Ruiz-Vazquez 2017). However, functions are not widely conserved, with extreme variation between systems. Small RNA generation in fungi is therefore an active area of investigation, as few examples of canonical small RNA biogenesis pathways have been clearly defined (Jiang et al. 2012). Instead, small RNAs are generated by diverse pathways utilizing RNAi machinery often in conjunction with cellular enzymes that appear to have moonlighting functions in small RNA biogenesis (Lee et al. 2010).

Early on it became clear that fungi employ variations of the RNAi machinery during different growth phases, for example, the sexual reproductive cycle or during vegetative growth. Pathways like quelling (Romano and Macino 1992) or transgene-induced cosuppression (Wang et al. 2012) are active during vegetative growth, whereas meiotic silencing by unpaired DNA (MSUD) (Shiu et al. 2001), repeat-induced point mutation (RIP) (Gladyshev 2017), and sex-induced silencing (Wang et al. 2010, 2013) reveal just a bit of the complexity that has been unearthed in the many variations of small RNA biogenesis during fungal sexual cycles. In asexual spores (conidia) of *A. fumigatus*, the RNAi pathway appears to play a role in supporting transcription of ribosome biogenesis genes (Kelani et al. 2023), a feature reminiscent of prior work proposing that RNAi is required for release of RNA polymerase in cycling and quiescent cells of *Schizosaccharomyces pombe* (Roche et al. 2016). More details on each of these pathways are unfortunately beyond the scope of this chapter.

The RNAi machinery also contributes to other noncanonical regulatory functions, features which were proposed to have accrued over multiple evolutionary gain and loss events (Bernstein et al. 2012; Billmyre et al. 2013). Remarkably, RNAi can even be lost completely by some organisms. Loss of RNAi in *C. gattii* leads to increased virulence and is thought to have been a factor for the success of the pathogen during the now famous pacific northwest outbreak of the 1990s (Datta et al. 2009). Many other members of the *Cryptococcus* complex have retained RNAi, again highlighting the countless paths to becoming a pathogen (Feretzaki et al. 2016). The loss of argonaute proteins in *C. gattii* does have tradeoffs though, as genome rearrangements and higher levels of retrotransposon expression and small RNA generation are observed (Ferrareze et al. 2017). The full consequences of RNAi loss in this organism are still under investigation, but these data support many other studies showing that the RNAi pathway is a major player in genome defense against selfish genetic elements. *C. gattii* also offers a system to explore novel small RNA pathways that are typically overshadowed by an intact, canonical RNAi system. One example of a dicer-independent pathway of transposon silencing has been described in *Cryptococcus* already (Burke et al. 2019). Instead of dicer, *Cryptococcus* uses *RDE3*, which encodes a protein with similarities to the *S. cerevisiae* RNase III protein Rnt1. Together with several other presumed nuclear factors, RDE3 can aid in the production of endo-siRNAs and suppress transposon mobilization. Finally, RNAi has even been repurposed for entirely new functions. The reference strain of *C. albicans* uses dicer for optimal maturation of ribosomal and spliceosomal RNAs (Bernstein et al. 2012), replacing the functionality of the nuclear Rnt1 protein found in more ancestral strains.

Over the past decade, the role of RNAi in stress response and drug resistance has become more apparent with the description of RNAi-dependent epimutation. The majority of this work has been done in *Mucor circinelloides* and other closely related organisms (Calo et al. 2014), where unstable antifungal drug resistance has been attributed to RNAi-dependent epigenetic regulation. Several examples exist, but the best remains the *M. circinelloides* response to the antifungal drug FK506 (Chang et al. 2019; Calo et al. 2014). In response to the challenge with FK506, *M. circinelloides* silences the expression of the *fkbA* gene to limit the production of the FKBP12 protein. The absence of FKPB12 prevents the formation of a complex with FK506 that can inhibit the protein phosphatase calcineurin, resulting in resistance to FK506. The epimutation pathway requires several other peripheral RNAi factors, including the quelling-induced protein (QIP) and a Sad-3-like helicase (RnhA) (Calo et al. 2017). These drug-resistant epimutants were found to be stable during infection in a mouse model, with clear clinical implications for the application of antifungal therapy (Chang and Heitman 2019). A noncanonical RdRP-dependent, dicer-independent silencing pathway reliant on R3B2 and RdRP3 was then shown to inhibit this RNAi-dependent epimutation pathway via mRNA degradation (Calo et al. 2017). The concept is then that the RNAi-dependent epimutation pathway serves in stress response and is balanced by the mRNA degradation pathway that occurs during normal growth. Collectively, these findings highlight the interplay and repurposing of multiple RNAi components to create a balanced regulatory network in *Cryptococcus*. Future work will inevitably reveal how these, and other pathways have evolved in other fungal pathogens for the purposes of stress response and virulence.

In addition to microRNA-like RNAs (milRNAs) and siRNAs, other small RNA species are also found in fungi. *A. fumigatus* was among the first organisms shown to produce tRNA fragments under stressful environmental conditions, including starvation (Jochl et al. 2008). tRNA fragments are a heterogenous and diverse class of molecules typically produced from mature, full-length tRNAs by cleavage with enzymes like Rny1 or Dicer (Xie et al. 2020). tRNA fragments are known to have a variety of ambiguous functions, but conventional knowledge suggests cleavage of tRNAs frequently limits translation, with the tRNA fragments then capable of repressing gene expression. Despite the identification of tRNA fragments in *A. fumigatus*, little is known about the regulation and functional outcome of the production of tRNA halves and smaller fragments in this organism. The world of small RNAs in fungi remains poorly described, and additional studies will certainly uncover new biology.

5.2.4 Extracellular RNA and Extracellular Vesicles

Decades of work have proven the importance of both coding and non-coding RNAs in the regulation of nearly every process within the cell. In recent years, functions have also been emerging for RNA outside of cells. The extracellular RNA population has proven enigmatic, but advances in tools to study extracellular vesicles have revealed a more complex extracellular environment than previously appreciated. Studies in humans have shown an extracellular RNA environment with RNA even taking on previously unrecognized folds, as observed for protected intermolecular interactions formed by some tRNA fragments (Tosar et al. 2020). Our understanding of extracellular RNA released by fungi has been driven by studies of plant-fungal interactions, as mentioned in the introduction (Weiberg et al. 2013). As early as 2009, evidence was emerging that fungal extracellular vesicles may contain RNA (Nicola et al. 2009). The use of membrane and nucleic acid dyes for the staining of extracellular fractions has proven challenging (De Voogt et al. 2021), but improved extracellular vesicle isolation protocols have facilitated new strategies. Together with the increased sensitivity of RNA-seq technologies, a number of studies have been performed confirming the release of extracellular RNA by fungi, including *C. albicans*, *C. neoformans*, *Histoplasma capsulatum*, and *Paracoccidioides brasiliensis* (Peres Da Silva et al. 2015; Alves et al. 2019). Nearly all RNA species have been found extracellularly, with small RNAs most frequently identified (Peres Da Silva et al. 2015; Munhoz Da Rocha et al. 2021; Rayner et al. 2017). Larger RNAs have also been detected, including mRNAs and lncRNAs, in a variety of experimental systems (Peres Da Silva et al. 2019) with a particularly noteworthy example of complete mRNAs being transported from plant host to fungal pathogen and translated (Wang et al. 2024).

How these RNA species traffic to the extracellular space remains unclear. The current consensus is that extracellular vesicles play a large role in carrying RNA and protecting it from the hostile extracellular environments. In *C. deneoformans*, the Cin1-Wsp1-Cdc42 endocytic pathway was proposed to play a role in RNA export by membrane vesicles, with deletion of *CIN1* resulting in major changes to the RNAs produced and released (Liu et al. 2020). In *C. neoformans* several factors are thought to contribute to RNA loading and export, including the Golgi reassembly and stacking protein GRASP (Peres Da Silva et al. 2018) and the scramblase Aim25 (Reis et al. 2019). It is not yet clear if these preliminary discoveries are more widely applicable across fungi, but again, a lot of work is underway to answer these questions.

The functions of extracellular RNAs are just starting to be dissected. Most studies have again focused on plant-fungal pathogens, but some work has been done on human-fungal interactions too. Work from the important human pathogen *C. gattii* revealed it to use extracellular vesicles loaded with protein and RNA cargo to facilitate virulence (Bielska et al. 2018). During infection, the fungus uses a division of labor strategy driven by sensing of host reactive oxygen species, where a subpopulation of cells become dormant and facilitate the rapid growth of another subset of fungal cells (Voelz et al. 2014). This support is reliant on the transfer of RNA and protein cargo by extracellular vesicles released from the fungus and taken up by host macrophages. Much remains to be discovered in this system, but it seems clear that fungi can use extracellular vesicles to their advantage to deliver cargo molecules like RNA between subpopulations in a host environment.

A particularly compelling example of the influence of extracellular RNA on host-fungal interactions comes from the insect pathogen *Beauveria bassiana*. During infection, *B. bassiana* releases a milRNA termed bba-milR1 that attenuates mosquito immunity by exploiting the insect RNAi machinery (Cui et al. 2019). The bba-milR1 milRNA was identified by searching for small RNAs produced during the infection of mosquitos that were specific to the *B. bassiana* genome. Four small RNAs were identified using this dual RNA-seq approach. Bba-milR1 exhibited Dicer-dependent biogenesis via a Dicer-like 2 protein, and deletion of the milRNA led to increased survival of mosquito hosts during infection. RNA precipitation experiments with antibodies against the *Anopheles stephensi* AGO1 protein revealed binding of milR1–4 to the mosquito protein. These results suggested that *B. bassiana* milRs were exported into the mosquito to hijack the host RNAi machinery. Bba-milR1 shared complementarity with several mosquito immunity genes and was shown using dual luciferase reporter assays to regulate gene expression by decreasing expression of the Spätzle-like cytokine Spz4 and increasing expression of the CLIP serine protease CLIPB9. Antagonistic bba-milR1 agomirs blocked the gene regulatory effects, and *in vivo* studies confirmed that bba-milR1 is capable of silencing *spz4* during the cuticle penetration phase of the infection to limit antifungal activity of the mosquito and give the fungus an edge. In an interesting wrinkle, *B. bassiana* then lowers levels of bba-milR1 during hemocoel invasion late in infection to prevent activation of CLIPB9 and the protective host melanization response. Small RNA trafficking and regulation is tightly controlled during host-fungal pathogenesis, likely offering multiple ingress points for potential biocontrol strategies in the future.

5.2.5 Other Examples of RNA Regulation in Fungi

There are too many studies of RNA regulation in fungi to cover each in depth here, but a few cases can provide evidence of the immense complexity yet to be revealed. The first example comes from a nontraditional RNase III enzyme found in *Mucorales*. Canonical RNase III enzymes cleave dsRNA to smaller precursor molecules that can contribute to gene silencing, as discussed above. However, recently a *Mucorales* RNase III enzyme was identified that evolved to cut single-stranded RNA (ssRNA) (Canovas-Marquez et al. 2021). This enzyme, known as R3B2, is an important regulator in the *Mucorales* noncanonical RNAi pathway mentioned above, which has been linked to regulation of virulence through degradation of specific mRNAs. In this system, the function of dicer is replaced with R3B2 and the system requires RNA-dependent RNA polymerase activity, but not argonaute proteins (Trieu et al. 2015), indicating an RNase III enzyme that displays a noncanonical cleavage profile. In addition to noncanonical roles in RNA cleavage, RNase orthologs also exist that completely lack RNase activity, as was shown for the RNase II enzyme Ssd1 in *S. cerevisiae* (Bayne et al. 2022). Ssd1 lacks RNase activity, but retains its ability to bind RNA, a feature which has been linked to its function of controlling stress response and virulence in a variety of fungi. A novel binding site on the surface of the protein is thought to aid interaction, but it remains to be definitively proven whether the observations from model organisms like *S. cerevisiae* will extend to other fungi as well. Collectively, these studies provide a cautionary tale, highlighting potential difficulties in separating presumed enzymatic activity from unanticipated RNA binding functionalities.

The contribution of RNA modifications to regulation of gene expression during stress and host-microbe interactions is another emerging area of research. The field has largely focused on tRNA modifications, revealing that various important modifiers like Ncs2, the Elongator- and KEOPS complexes, as well as Sua5 contribute to the virulence of fungal pathogens like *C. albicans*, *A. fumigatus*, and *C. neoformans* (Zhang et al. 2022; Alings et al. 2023; Choi et al. 2023). Remarkably, all modifiers target the anticodon loop by either modifying wobble uridines (Ncs2, Elongator complex) or position 37 adenosines (KEOPS complex, Sua5), indicative of their importance for tRNA stability and decoding fidelity. In each case ablation results in decline of pathogenicity and stress resistance. In addition to tRNA modifications, additional work has linked modification of mRNAs to control of gene expression, intriguingly showing a requirement for m^6A methylation in the biosynthesis of aflatoxin in *A. flavus* (Yang et al. 2023). A number of studies have shown similar roles for m^6A in many other fungal systems, implying a vast coordinated regulation in response to changing environmental conditions across fungi (Ren et al. 2022; Kim et al. 2023). Although still a field in its infancy, RNA modifications are likely to be revealed as key regulators of host-fungal interactions.

A final interesting regulatory feature of RNA comes in the form of RNA-directed modification. Many examples exist, but here we will focus on the tRNA-dependent addition of amino acids to glycerolipids. The addition of amino acids to glycerolipids is widespread in bacteria, but until recently no examples were described in fungi. Now we know that aspartate can be added to ergosterol in a tRNA-dependent fashion in a wide range of fungi, including important human pathogens like *A. fumigatus* (Yakobov et al. 2020). Ergosterol is the principal sterol in fungal cell membranes, and enzymes involved in ergosterol biosynthesis are frequent targets of antifungal therapeutics. It is therefore quite intriguing that the fungal enzyme ergosterol-3β-O-L-aspartate synthase (ErdS) transfers aspartate from Asp-$tRNA^{Asp}$ to the 3β-OH group of ergosterol. ErdS is a bifunctional protein that has an aspartyl tRNA synthetase fused to a Domain of Unknown Function 2156 that transfers the aspartate to the sterol. An additional enzyme ErdH is then capable of removing the added aspartate. A fuller description of the biosynthesis and regulation of ergosterol may reveal novel therapeutic targets, and certainly the tRNA-dependent addition of amino acids to glycerolipids represents just one of many fungal RNA modification reactions still hiding from view.

5.3 Host RNA Responses

The human immune system, subdivided into the innate and adaptive immune response, must react and modulate its actions according to microbial threat. Recognition of fungal molecules serving as pathogen-associated molecular patterns (PAMPs) is mediated by pattern recognition receptors (PRR). Among these, the ten human Toll-like receptors (TLR) are a class capable of specifically binding different PAMPs to mediate innate immune reactions, leading to inflammation, immune cell recruitment, and phagocytosis of the respective microbe (Moresco et al. 2011). The first indications that these receptors were essential for antifungal defense came from a *Drosophila-A. fumigatus* infection model (Lemaitre et al. 1996), which were later supported by findings describing TLR2 and TLR4 as integral components in combatting fungal invaders like *C. albicans* and *A. fumigatus* (Bellocchio et al. 2004; Mambula et al. 2002; Meier et al. 2003; Netea et al. 2002). TLR2 binds to the cell wall component zymosan of *S. cerevisiae* with the help of TLR6 and also recognizes β-glucan in cooperation with Dectin-1 (Underhill et al. 1999; Gantner et al. 2003; Brown and Netea 2007). TLR4 is classically known to recognize bacterial lipopolysaccharides (LPS) but was also found to bind different mannan structures in *C. neoformans* and *C. albicans* to drive induction of proinflammatory responses (Netea et al. 2006; Tada et al. 2002; Shoham et al. 2001).

These examples illustrate that a set of receptors are capable of recognizing fungi by the molecules exposed on the cell wall, indicating that these receptors are predominantly located on the plasma membrane of the host cell. Interestingly, a variety of other receptors like TLR3, TLR7/8, TLR9, and the RIG-I-like receptors (RLRs) are found both in the cytoplasm (RLRs) and/or membrane-associated, facing the endosomal/lysosomal lumen (TLRs) to recognize foreign nucleic acids (Uehata and Takeuchi 2020). TLR3 and TLR7/8 specifically bind select RNA species to activate the innate immune response (Jannuzzi et al. 2020). This section will discuss some of the recent data suggesting that PRRs can respond to RNAs associated with human fungal infection and contextualize these intriguing findings with other RNA-dependent regulatory systems activated upon host-fungal interactions in a range of organisms.

5.3.1 RNA Recognition by Toll-Like Receptors in Fungal Immunity

Nucleic acids can be actively secreted by microbes within the host environment or released from dying or degraded microorganisms upon phagocytosis or endocytosis by host immune cells (Lee and Barton 2014; Barton 2007). TLR3 and TLR7/8 (as well as TLR9) are trafficked to either endosomes or lysosomes after synthesis in the endoplasmic reticulum, enabling the interaction and activation of these receptors by foreign nucleic acids (Fig. 5.2) (Lee and Barton 2014; Barton 2007). The RNA-specific TLRs appear to play a major role in the response to fungal infections by either recognizing dsRNA (TLR3) or ssRNA (TLR7/8) (Jannuzzi et al. 2020). This was demonstrated in a murine dendritic cell (DC) model system in which TLR3 was deleted and lung DCs were subjected to *A. fumigatus* infection (Carvalho et al. 2012). Lack of TLR3 affected the priming and MHC-I restricted protective memory response of $CD8^+$ T cells by DCs, which was caused by a decrease of the phagocytic index of conidia, no upregulation of *CCR7*, and finally reduced lung-to-lymph node migration. In essence, no recognition of fungal RNA by TLR3 led to an ablation of the critical *CCR7* induction strictly necessary for DCs to migrate to the lymph nodes and prime $CD8^+$ T cells. This higher susceptibility to *A. fumigatus* infection in mice was corroborated by a higher vulnerability of patients with a single nucleotide polymorphism (SNP) in the *TLR3* gene to develop aspergillosis (Carvalho et al. 2012). These patients exhibited defects in DC function and their capability to activate $CD8^+$ T cells.

Interestingly, recognition of pathogenic fungi by TLR3 is not limited to the detection of fungal RNA but can also be mediated by mycoviral dsRNA. Fungal pathogens from the genus

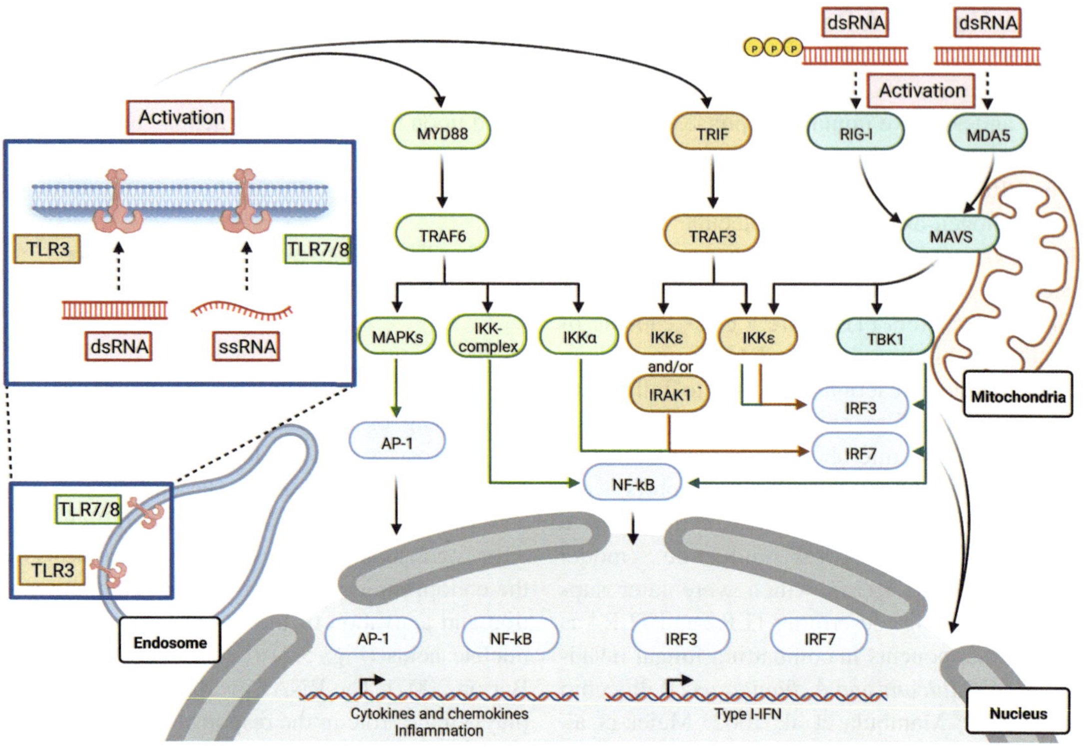

Fig. 5.2 Innate immune signaling cascades of RNA-dependent PRRs activated by fungal pathogens. Sensing of various foreign (fungal) RNAs like dsRNAs or ssRNA is achieved by the endosomal receptors TLR3 or TLR7/8 and cytoplasm-localized RIG-I or MDA5. Signaling is mediated by the TLR3/TRIF axis (orange boxes), TLR7/8/MYD88 axis (green boxes), and/or RIG I/MDA5/MAVS axis (dark green boxes) by employing the transcription factors AP-1, NFκB, IRF3 and, IRF7 for the activation of diverse innate immunity genes. Adapted from Rehwinkel and Gack (2020), Jannuzzi et al. (2020), Uehata and Takeuchi (2020). Created with BioRender.com

Malassezia, known to commonly inhabit the human skin, were found to be infrequently infected by mycoviruses of the family *Totiviridae* (Saunte et al. 2020; Applen Clancey et al. 2020; Park et al. 2020). In general, the mycoviral genomes consisted of dsRNA segments of different sizes packed into viral particles but not excreted to infect other organisms. Instead, the transmission process relies on cell division, mating, or the fusion between individual vegetative cells (hyphal anastomosis) (Ghabrial 2008; Ghabrial et al. 2015). An investigation focusing on *Malassezia restricta* identified a new mycovirus termed MrV40 that was extracted from clinical fungal isolates. The hallmark features of the MrV40-infected strains included an adjustment of the transcriptome to repress genes involved in ribosome biogenesis and an induction of genes participating in energy production and programmed cell death (Park et al. 2020; Applen Clancey et al. 2020). The authors then went one step further and investigated the viral influence on fungal pathogenicity by treating bone marrow-derived dendritic cells (BMDC) with virus-infected and cured *M. restricta* strains or protein components of the viral particles. Surprisingly, the BMDCs displayed an induction of TLR3 expression as soon as they were treated with the infected fungi. Moreover, the cytokines tumor necrosis factor alpha (TNF-α), interleukin 6 (IL-6), IL-10, interferon alpha (IFN-α), and IFN-γ were elevated in a TLR3-dependent manner, underlining the major role of the mycoviral dsRNA as a PAMP recognized by the innate immune system during *M. restricta* infection (Park et al. 2020). A separate study focusing on a related species, *M. sympodialis*, again infected with a dsRNA virus of the *Totiviridae* family termed MsMV1, reported a different outcome regarding fungal pathogenicity. Here, the authors used bone marrow-derived macrophages (BMDM) to test whether virus-infected fungus triggered activation via TLR3. The BMDMs displayed an induction of IFN-β upon infection with the virus-infected fungus in contrast to the virus-cured fungus. But the activation of the macrophages appeared to happen in a TLR3-independent manner, raising the question of what receptor might be mediating the IFN-β induction (Applen Clancey et al. 2020). The authors suggested that other dsRNA-dependent PRRs (e.g., RIG-I or MDA5, see following section) might contribute to interferon expression, although the effect of viral proteins like RNA-dependent RNA polymerases or capsid proteins on macrophage activation were not tested as in the Park study (Park et al. 2020).

The examples discussed intriguingly show how TLR3 can contribute to innate immune activation by the detection of either fungal or mycoviral RNAs, albeit the relevance of TLR3 changes according to the fungal pathogen. Similarly, *P. brasiliensis*, the cause of paracoccidioidomycosis, escapes the immune response by negative regulation of the activity of TLR3 during infection (Jannuzzi et al. 2019). Deletion of TLR3 in murine BMDMs resulted in increased fungicidal activity, reduced fungal burden, and elevated production of nitric oxide. Moreover, intratracheal infection of wild-type or TLR3-deficient mice with *P. brasiliensis* caused an increase in $CD8^+$ T cells in the knockout, which an earlier study showed to be important for resolution of murine paracoccidioidomycosis (Jannuzzi et al. 2015). The increase of these T cells concomitantly occurred with an induction of the cytokines IL-17, IL-1β, IFN-γ, and IL-6, all of which are known to contribute to defense against fungal infections (Jannuzzi et al. 2015; Alegre-Maller et al. 2014; De Souza Silva et al. 2015; Morais et al. 2016). Although this study did not investigate the molecules responsible for the pathogen escape mechanism, it nicely illustrates how PRRs can be exploited during infections to prevent clearance of fungi by the host immune system.

A similar effect was observed in a murine macrophage-infection model investigating the contribution of the ssRNA receptor TLR7 to defense against invasive pulmonary *Aspergillosis* (IPA) caused by *A. fumigatus* (Xu et al. 2021). In contrast to the upregulation of *TLR7* expression during IPA, deletion of the receptor resulted in a reduction of the fungal burden, lung damage, and inflammatory response as well as a decreased mortality of the knockout compared to the wild-type mice, indicating a negative regulatory role of the receptor during *A. fumigatus* infection. Specific activation of TLR7 by the compound R837 increased the mortality of mice infected with *A. fumigatus* in an alveolar macrophage-dependent manner. According to the data, TLR7 activity negatively regulated phagocytosis and the generation of reactive oxygen species leading to a higher susceptibility to invasive pulmonary aspergillosis (Xu et al. 2021).

The outcome of TLR7 activity is highly dependent on the invading fungal pathogen as well as the immune cell responding to those pathogens. An interesting example of this observation was given in a report deciphering the TLRs involved in the innate immune response to *C. albicans* infection (Biondo et al. 2012). As mentioned in the beginning of this chapter, TLR2 and TLR4 play an integral role in the recognition of fungal cell wall components of *C. albicans* (Bellocchio et al. 2004; Mambula et al. 2002; Meier et al. 2003; Netea et al. 2002). Biondo and colleagues aimed to expand these findings by investigating the role of nucleic acid-dependent PRRs by focusing on the role of TLR7 (and TLR9). For this, they utilized mouse BMDMs and BMDCs, which were not only challenged with purified yeast RNA but also with whole yeast cells to reflect a more natural infection situation. The treatment with purified *C. albicans* RNA activated BMDCs but not BMDMs and led to an induction of the cytokines IL-12p70, IL-23, and TNFα in a TLR7-dependent manner. Moreover, downstream effectors like the TLR-hub protein MyD88 (Takeda et al. 2003; Saikh 2021) and transcription factor IRF1 (Feng et al. 2021) were clearly important for the induction of IL-12p70, while IL-23 and TNFα expression seemed to rely on a different transcription factor and even on different receptors when the BMDCs were challenged with whole cells. *TLR7*-deficient BMDCs infected with live *C. albicans* cells only resulted in a depletion of IL-12p70, while IL-23 and TNFα generation were still increased, indicating different receptor dependencies for the cytokine induction and a more concerted inflammatory response of the different PRRs. The appropriate innate immune activation depended on the correct localization of the endosome-situated TLRs since interference with this process by mutating UNC93B1, important for recruitment of TLR3, TLR7/8, and TLR9 to the endosome, ablated cytokine induction, demonstrating that the ssRNA ligand had to be released from internalized/phagocytosed pathogens for proper recognition (Tabeta et al. 2006; Brinkmann et al. 2007; Kim et al. 2008). These findings were recapitulated in mice lacking TLR7, MyD88, IRF1, or functional UNC93B1 strengthening the importance of this inflammatory response axis in candidiasis (Biondo et al. 2012).

Similar results were later obtained regarding the importance of the TLR7/MyD88 axis in the innate immune response upon infection with *H. capsulatum* in a murine infection model utilizing BMDMs and BMDCs (Van Prooyen et al. 2016). Conidia of this dimorphic fungal pathogen enter the human lung and then disseminate into other tissues and organs (Woods 2003; Nosanchuk and Gacser 2008). Controlling *H. capsulatum* infection requires induction of T cells to produce TNFα and IFN-γ, which activate phagocytic cells like macrophages. Macrophages are responsible for phagocytosis and clearance of the pathogen, but are also exploited by the fungus for replication in the phagolysosome (Deepe 1988; Nosanchuk and Gacser 2008). In contrast, DCs are capable of phagocytosing and killing *H. capsulatum* to support an adequate immune response through stimulation of $CD8^+$ T cells (Nosanchuk and Gacser 2008; Gildea et al. 2001; Lin et al. 2005). In line with this notion, the investigation of Van Prooyen et al. focused on the immune responses orchestrated by BMDCs compared to BMDMs. Interestingly, *H. capsulatum* infection of the dendritic cells, but not the macrophages, resulted in an induction of type I IFNs that was dependent on TLR7/TLR9 and MyD88. Combined deletion of both receptors completely ablated IFN-I production *in vitro* in BMDCs and *in vivo* in mice, leading to a higher neutrophil recruitment, lung damage, and mortality rate. The lack of TLR7/TLR9-dependent IFN-I production affected the $CD4^+$ and $CD8^+$ T cell-controlled IFN-γ expression, a known factor in *H. capsulatum* infection control (Wu-Hsieh and Howard 1987; Allendoerfer and Deepe Jr. 1997). These studies clearly show that RNA recognizing-PRRs play an intricate role in both defense against fungal infection and as a point of exploitation by fungal pathogens against host organisms. Additional work will be required to determine how other organisms have evolved to control or even exploit these pathways, and how these signaling cascades might be utilized for antifungal drug development.

5.3.2 RIG-I-Like Receptors and the Antifungal Response

Other RNA-dependent receptors like the retinoic acid-inducible gene I (RIG-I)-like receptors (RLR) are produced and localized to the cytoplasm in many cell types. This class comprises of three proteins, RIG-I, melanoma differentiation-associated protein 5 (MDA5), and laboratory of genetics and physiology 2 (LGP2), which are all characterized by a central helicase domain and a carboxy-terminal domain (Rehwinkel and Gack 2020). RIG-I and MDA5 also contain two N-terminal caspase activation and recruitment domains (CARD) that upon RNA-oligomerization and activation are necessary for the interaction with the CARD of mitochondrial antiviral-signaling protein (MAVS) (Kawai et al. 2005; Meylan et al. 2005; Seth et al. 2005; Xu et al. 2005; Mcwhirter et al. 2005; Hou et al. 2011). MAVS in turn mediates activation of TANK-binding kinase 1 (TBK1), IκB kinase-ε (IKKε), IKKα, and IKKβ, allowing for activation of IFN regulatory factor 3 (IRF3), IRF7, and nuclear kappa-light-chain enhancer of activated B cells (NF-κB). These three transcription factors promote expression of antiviral genes upon translocation to the nucleus (Goubau et al. 2013; Hartmann 2017; Rehwinkel and Gack 2020). The RLR pathway is induced upon interaction of RIG-I and/or MDA5 with RNA molecules possessing specific features that are normally not found on endogenous RNAs. RIG-I recognizes uncapped 5′-di- or triphosphates of blunt-end, dsRNA segments lacking 2′-O-methylations at the 5′-ends (Rehwinkel and Gack 2020). Soon after its description, the RLR pathway was found to be essential in the defense against viral infections (Yoneyama et al. 2004; Gitlin et al. 2006; Kato et al. 2006) and in innate immunity against bacterial infections (Abdullah et al. 2012; Zou et al. 2016; Sahr et al. 2022). The RNA features activating MDA5 remain somewhat ambiguous, but in general very long (>2000 base pair) dsRNA-molecules seem capable of activating the receptor (Takeuchi and Akira 2010; Goubau et al. 2013; Rehwinkel and Gack 2020).

Reports have also revealed a role for these same PRRs in the detection of fungal pathogens. A study from 2015 investigated changes of the transcriptome of macrophages and peripheral blood mononuclear cells to evaluate gene expression in response to *C. albicans* infections (Jaeger et al. 2015). Wild-type or *Δhgc1* (yeast-locked) *C. albicans* cells were used to infect macrophages for 4 or 24 h followed by microarray analysis to assess gene expression during infection. Sixty-two macrophage genes were strongly induced after challenge with wild-type hyphae-forming fungus, predominantly from the IFN signaling pathway, but four genes could be assigned to the RLR signaling cascade, namely *IFIH1* (*MDA5*), *ISG15*, *IL-8*, and *TRIM25* (Jaeger et al. 2015; Loo and Gale Jr. 2011; Rehwinkel and Gack 2020). As a major receptor of RLR signaling, MDA5 was further investigated to determine its role during *Candida* infection. Expression analysis of peripheral blood mononuclear cells stimulated with either heat-killed yeast or hyphae compared to LPS treatment or *Borrelia burgdorferi* infection as controls confirmed a strong induction of *MDA5* in the immune cells (Jaeger et al. 2015), and another study showed that MDA5 enhances invasive *C. albicans* infection by regulating apoptosis and phagocytosis in macrophages (Chen et al. 2023). The importance of *MDA5* was further evaluated in human patients suffering from candidemia by investigation of SNPs occurring mostly in genomic regions involved in immune-mediated diseases (Jaeger et al. 2015). Among other *MDA5*-SNPs associated with increased susceptibility to *Candida* infections, missense mutations of Ala946Thr and His843Arg (rs1990760 and rs3747517) resulted in the clearest dysfunctionalities of RLR signaling, manifesting as dysregulation of cytokine signaling in an *in vitro* model. The authors proposed that these missense SNPs might corrupt the interaction of MDA5 with dsRNA (Ala946Thr and His843Arg) and interfere with the interaction of receptor monomers (His843Arg), which are known to form filaments along the recognized RNA necessary for downstream MAVS interaction (Jaeger et al. 2015; Van Der Lee et al. 2014; Hou et al. 2011; Wu et al. 2013). Recapitulation of *Candida* yeast or hyphae infection with murine splenocytes with or without *MDA5/IFIH1* deletion only hinted to a similar role of the receptor in mice, suggesting other pathways may be required in other organisms. Nevertheless, this data conclusively indicated an important role for MDA5 in the defense against candidiasis. In the future, it will be interesting to see how the presence of fungal dsRNA is influenced by the activity of other host processes, such as modification by host dsRNA-modifying enzymes like the adenosine deaminase that acts on RNA (ADAR) proteins, which were previously shown to be activated during *C. albicans* infection (Huang et al. 2018).

Further investigations of the RLR pathway in response to *A. fumigatus* infections expanded on the importance of MDA5 in antifungal immunity. Deletion of *MDA5* or *MAVS* resulted in a higher susceptibility of mice upon challenge with *A. fumigatus*, which was represented by decreased conidial killing by neutrophils, altered cytokine production, increased lung damage, and ultimately a higher mortality rate. Infection of primary murine fibroblasts supported this by showing that viable conidia or even isolated dsRNA of *A. fumigatus* could lead to the induction of type I and III IFNs, which depended on the action of MDA5 and MAVS (Wang et al. 2020). The importance of both RLR-pathway effectors became even more visible in a follow-up study assessing the risk factors of hematopoietic stem-cell transplantation recipients affected by IPA, where SNPs in MDA5 and MAVS associated with a higher risk of IPA. A clear connection was shown between the missense SNP rs1990760 (Ala946Thr) in *MDA5* and susceptibility to *A. fumigatus* infection, indicating a similar effect of the same mutation not only for IPA but also candidiasis (Wang et al. 2022; Jaeger et al. 2015). Moreover, the SNP rs17857295 in *MAVS* appeared to contribute to a higher IPA-development risk as well. While the identified *MDA5* mutation and increased IPA-risk was associated with the recipient, this was reversed for the *MAVS* SNP displaying a link to the donor. Underlining the importance of MAVS functionality in a hematopoietic stem-cell transplantation context, *A. fumigatus* infection was recapitulated in mice devoid of *MAVS* in CD11c$^+$ cells, revealing alveolar macrophages as largely responsible for antifungal host resistance (Wang et al. 2022). These findings indicate a critical role of the RLR pathway in immunity against some of the most prevalent fungal pathogens, which will likely be expanded to other microbes in future investigations.

5.3.3 Transcriptomic and Non-coding RNA Responses to Fungal Infections

The last sections gave an overview of the role of RNA-dependent PRRs in the regulation of the antifungal immune response, where specific fungal RNA was shown to activate TLR or RLR pathways and induce appropriate innate defense mechanisms. In addition to these immunological studies, several recent reports have investigated how fungi control the coding and non-coding RNA landscape of hosts by affecting expression, functionality, and even host immune regulation. One approach to assess the transcriptional changes induced during fungal infection in diverse immune cells is to perform RNA-seq of the host and pathogen transcriptomes simultaneously, as was reported recently using cytomegalovirus (CMV) together with *A. fumigatus* to infect monocyte-derived dendritic cells (moDCs) (Seelbinder et al. 2020). The authors compared the transcriptomes obtained from untreated moDCs or *A. fumigatus* to infections of the host cells with either the fungus or CMV in co-infection experiments. The transcriptional changes documented in the singular infections of moDCs reflected the previously described transcriptome adjustments induced upon *A. fumigatus* or CMV stimulation. Fungal infection promoted TNF signaling, which induced NF-κB-dependent gene expression in addition to *IL-1B* and *IL-10*, whereas CMV infection strongly elevated signaling of RIG-I, cGAS, and STING pathways as well as *TLR3*, *CXCL10, IFN-B,* and *ZBP1* expression (Seelbinder et al. 2020; Yang et al. 2020; Onomoto et al. 2010). The commonality between all of these signaling responses was that they were strongly downregulated when the moDCs were co-infected with both pathogens, indicating a synergistic effect on host immunity that benefits both the fungus and the virus to the detriment of the host.

Another compelling example of dual RNA-seq usage came from the study of *C. albicans* during infection of BMDMs (Munoz et al. 2019), where the authors aimed to resolve the transcriptional changes occurring during host-pathogen interaction and apply single-cell RNA sequencing to distinguish different subpopulations with unique single-cell transcriptomes. In doing so, they found significant regulatory changes in the transcriptome over time for both the macrophages and the fungus. While the macrophages heavily induced proinflammatory cytokine production and expression of several transmembrane receptors (for fungal recognition) at early timepoints after infection, the internalized live fungal pathogens rapidly adjusted metabolic genes and pathways likely to adapt to the scarce nutrient availability in the phagolysosomal environment. After four hours of infection, repressed fungal gene clusters became more transcriptionally active, including those involved in translation, cell wall remodeling, glucose, and nucleoside metabolism. Interestingly, BMDMs started to repress immune response-relevant genes at later timepoints, including phagocytosis signaling, proinflammatory cytokines, and transmembrane receptors as part of a transcriptional shift toward a reduced immune response. Using single-cell resolution for RNA-seq, the authors correlated late transcriptional changes of the macrophages to the induction of metabolic and filamentation pathways in *C. albicans*, suggesting that the pathogen might drive the transcriptional shift in the host cells. Hence, these data indicated a coordinated transcriptional adjustment during the host-pathogen interaction by both organisms.

Together with the application of RNA-seq to investigate expressional changes of protein-coding RNAs, efforts have also been invested in defining the response of the non-coding transcriptome. The focus of these studies has been the miRNAs due to their defined regulatory role in gene expression via RNAi. In order to investigate the possible functionality of miRNAs during infection, a murine alveolar macrophage cell line model system challenged with *A. fumigatus* was utilized to investigate the influence of a particular miRNA on the expression of TLR2 and how this connects to the induction of autophagy (Wu et al. 2019). As one of the major PRRs responsible to detect *A. fumigatus*, TLR2 contributes to the activation and coordination of the immune response (Bellocchio et al. 2004; Chai et al. 2009; Mambula et al. 2002). Further work established that the regulation of this receptor depended on miRNA-344b-1-3p modulation of TLR2 expression (Xu et al. 2017). This regulatory relationship displayed an inversely proportional expression pattern resulting in the decrease of miRNA levels while the *TLR2*-mRNA abundance increased during *A. fumigatus* infection (Wu et al. 2019). Accordingly, miRNA-dependent regulation of *TLR2* correlated with downstream pathways, including the autophagy response necessary to cope with the fungal pathogen. Similarly, the importance of miRNAs for the antifungal immune response was also shown in a broader experimental set-up investigating small RNA expression patterns in human DCs responding to *A. fumigatus* or *C. albicans* challenge (Dix et al. 2017). By employing small RNA-seq, the authors identified three (after 6 h; miR-132-3p, miR-132-5p, miR-212-5p) and five (after 12 h; miR-132-3p, miR-132-5p, miR-129-5p, miR-212-3p, miR-212-5p) differentially expressed miRNAs that were predicted to target various immune response genes (e.g., *FKBP1B*, *BTN3A2K*). Successful verification of several predicted targets by knockdown of the respective miRNAs underlined the value of the high-throughput findings. Thus, this data supported a scenario in which small RNA expression is adjusted during host-fungal interaction, indicating a more pronounced function as a regulatory element and complementing the growing body of literature reporting similar effects for other cell types [see Boomiraj et al. (2015), Croston et al. (2018)] and non-coding transcripts like lncRNAs [e.g. Riege et al. (2017)].

5.3.4 Interkingdom Communication

As has been mentioned in numerous examples above, the interplay between fungi and animal hosts is highly regulated and complex. In this section, we have focused primarily on the sensor proteins that detect an invading organism and the regulation of RNA in response to activation of these sensors but have barely scratched the surface of the RNA molecules themselves that are transferred between fungi and animals. Although only an emerging field, several examples already exist of RNAs that are trafficked between organisms during infection. These come primarily from studies of plant-fungal pathogens, but other examples do exist in animals. In insects, miRNAs of the let-7 family and others like miR-100 were shown to be transferred to the entomopathogenic fungus *B. bassiana* from the silkworm *Bombyx mori* and the mosquito *A. stephensi* to inhibit expression of the fungal virulence-related genes *sec2p* and *C6TF* (Fig. 5.3) (Wang et al. 2021). The induction of let-7 and miR-100 by the host organism is initiated upon penetration of the hemocoel by the fungus, resulting in decreased production of fungal Sec2p and C6TF and ultimately decreased virulence. Exploiting these findings, the authors then showed that expression

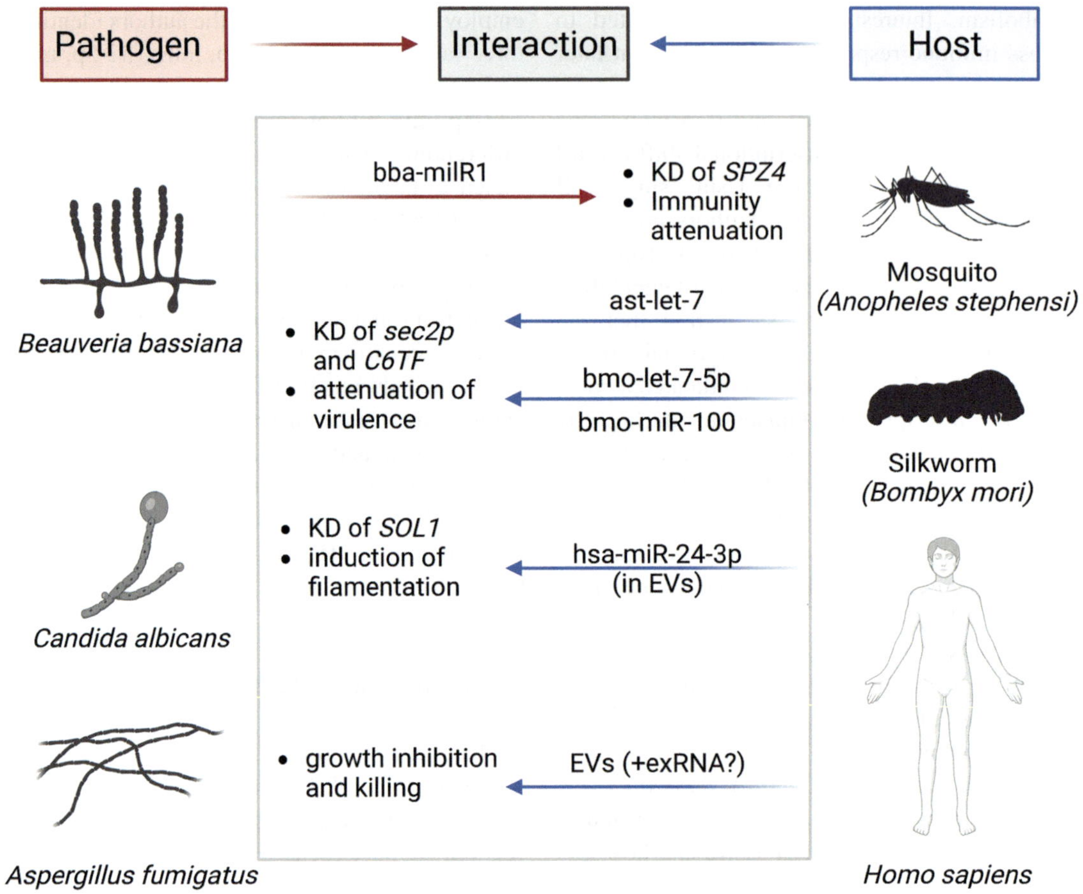

Fig. 5.3 Cross-kingdom RNA trafficking modulates virulence and immunity. The depicted examples show reported cross-kingdom interactions (grey box) of fungal pathogens and their respective hosts. If identified, trafficked small RNAs are given together with the described outcome in the target species (red or blue arrows). *KD* knockdown, *EVs* extracellular vesicles. Created with BioRender.com

of an anti-miR-100 miRNA sponge in the fungus led to increased virulence of the pathogen. The clear implication of this experiment is that a more pathogenic form of *B. bassiana* may hold potential as a biocontrol agent against insect pests in some cases. Of course, additional careful work needs to be done to ensure that such strategies do not have unforeseen consequences.

In humans, *C. albicans* induces miRNA trafficking by host monocytes to influence fungal behavior and elicit a growth advantage (Halder et al. 2020, 2022). In this case, the miRNAs seem to be associated with extracellular vesicles, although the exact mechanism of loading into vesicles and delivery into the fungus remains unclear. Experiments using naked miRNAs also resulted in a growth advantage to the fungus, suggesting that antifungal RNA delivery to *C. albicans* could be a strategy for treatment in the future. Other systems have also investigated the transfer of host cargo molecules by extracellular vesicles, as with human neutrophils and the filamentous pathogen *A. fumigatus* (Shopova et al. 2020; Rafiq et al. 2022). In this case, neutrophils have an inhibitory effect on fungal growth, like that observed in the example from mosquitos above. More work will have to be done to define the RNA cargo trafficked in these vesicles, and how these RNAs might influence the outcome of fungal infection in different patient populations.

Other examples of cross-kingdom RNA interference, or even novel regulatory mechanisms, will most certainly be observed between fungi and their hosts. The large-scale transcriptional responses to most pathogens may hide some of these regulatory features amongst so many changes, but we do know of cases where regulation seems to be manipulated by the fungal pathogen, as with macrophage polarization after *C. neoformans* infection (Subramani et al. 2020). Through an as-yet-unknown mechanism, *C. neoformans* appears to direct macrophages toward an immature, unpolarized state, although more work will be needed to confirm that these regulatory mechanisms also occur *in vivo* during infection. By combining high-quality dual transcriptomic datasets (Niemiec et al. 2017; Munoz et al. 2019) with more rigorous biochemical and genetic characterizations, we are likely to uncover all sorts of new regulatory features previously hidden from view.

5.4 Conclusions: Exploiting RNA Biology for Therapy

Our increased collective knowledge of RNA biochemistry and regulation is now being translated to more applied projects, with the hope of leveraging what we have learned for the treatment of fungal pathogens with novel therapeutics. Although a field in its infancy, the examples highlighted throughout this chapter provide numerous target points for the development of creative, new therapeutics. Arguably one of the first examples of leveraging these basic molecular biological principles for therapeutics came from the nucleoside analog, 5-fluorocytosine, which is converted into 5-fluorouridine in susceptible fungi to inhibit protein and DNA synthesis (Vermes et al. 2000). Unfortunately, high rates of resistance currently limit the overall effectiveness of this class of antifungals (Billmyre et al. 2020), but it remains in use in specific cases.

More recent approaches have tried to leverage chemicals designed to inhibit important RNA molecules, particularly group II intron ribozyme structures found in essential genes of fungi (Liu and Pyle 2021). A small molecule inhibitor was also recently demonstrated to exhibit target specificity and catalytic activity inhibition of the ribozyme against the human pathogen *C. parapsilosis* (Fedorova et al. 2018). Another approach is to design RNA molecules that fit into a particular protein to elicit an effect, so-called RNA aptamers. Although limited examples of RNA aptamers exist in general, they hold therapeutic promise and have already been designed to facilitate the disassembly of *C. albicans* biofilms (Bachtiar et al. 2019). Antisense oligonucleotides have been explored for control of *C. albicans* filamentation with some success (Araujo et al. 2019). For this a 2'-OMethylRNA antisense oligonucleotide (ASO) was used to target the *EFG1* mRNA to limit gene expression and translation of Efgp1, resulting in decreased filamentation and virulence. Previously unexplored therapeutic targets are also being investigated, as is the case with the Tpt1 enzymes found in fungi and plants that are absent from humans (Dantuluri et al. 2021). These essential enzymes are important for tRNA splicing and offer a potential target for novel antifungal therapy building on our improved understanding of basic molecular biological mechanisms in human fungal pathogens. Other approaches could utilize this knowledge by employing other classes of small RNAs besides ASOs to modulate the gene expression patterns of fungal pathogens and influence the outcome of infection. For instance, using small RNA molecules like siRNAs or miRNAs to specifically target certain virulence-relevant genes by hijacking the fungal RNAi machinery could be an elegant way to suppress fungal infections. A proof of concept for this strategy was reported using *A. fumigatus* strains expressing inverted-repeat transgenes as a source for siRNAs targeting mRNAs of either *pksP* or the conditionally essential gene *pabA* (Kelani et al. 2023). Both strategies resulted in efficient knock-down of the respective gene and growth reduction, although decrease of the latter mRNA resulted in a more severe decrease. Thus, targeted RNAi-dependent approaches could be an interesting treatment option for specific fungal infections although efficient uptake of siRNAs still faces challenges (Bruch et al. 2022). In this regard, the agricultural sector might serve as a role model due to recent advances in exploiting the RNAi system and cross-kingdom communication between plant pathogens and the host (Cai et al. 2018a). Strategies involving the use of transgenic plants expressing anti-pathogen small RNAs (host-induced gene silencing) or spray treatment with RNA of the plant host (spray-induced gene silencing) proved to be efficient in infection prevention (Cai et al. 2018a, 2019). Translation of these options into antifungal medication could advance our repertoire tremendously for the treatment of patients suffering from severe fungal infections (Cai et al. 2019).

These approaches offer unique advantages for the treatment of fungal infections over more conventional approaches. The target space over chemical inhibitors is much wider, meaning that RNA drugs can be designed against targets more specifically, even in highly conserved pathways. In addition, RNA is easy to synthesize; fast to adapt to evolving pathogens, and easy to redesign upon development of antifungal resistance. Many difficulties will have to be overcome to employ such drugs against human fungal pathogens, but innovative strategies are already in development. By leveraging the biochemical and biological solutions being employed in related systems, we are poised for rapid advances of these therapeutics against a wide range of pathogens. Advances like those embraced by the mRNA vaccines against SARS-CoV-2 are the most obvious, such as replacement of uridine with modified N1-methyl-pseudouridine to limit recognition and activation of host nucleic acid sensors like those discussed above (Corbett et al. 2020; Kariko et al. 2012; Anderson et al. 2010). Improvements in RNA synthesis and *in vitro* transcription have decreased unwanted immunogenic byproducts like dsRNA (Nelson et al. 2020), further facilitating this new generation of therapeutics.

The largest barrier to the exploitation of RNA regulatory pathways in fungi by RNA therapeutics is likely the formidable fungal cell wall (Bruch et al. 2022). Here, leveraging approaches developed against other human diseases may not be entirely possible, as the fungal cell wall has unique properties compared to a human cell. There is hope that protection of RNA in modified liposomes and nanoparticles or conjugation to aptamers and antibodies may result in improved delivery to cells (Dammes and Peer 2020; Howard et al. 2017; Moustafa et al. 2021), but the utility of these strategies in traversing the fungal cell wall remains understudied. Approaches using engineered extracellular vesicles may hold promise (Zhang et al. 2020), but are far from applicable in the clinic currently. Recent studies showing that host-derived extracellular vesicles respond and influence the course of infection with *A. fumigatus*, *C. albicans*, and *C. neoformans* suggest that modulation of human fungal pathogens is possible (Shopova et al. 2020; Halder et al. 2020; Zhang et al. 2021). Finally, in the absence of direct antifungal strategies, it is possible that more indirect approaches reliant on improving the remaining host immune response may prove valuable (De Candia et al. 2016). Only the future will tell how we can apply our increasing knowledge of RNA in host-fungal pathogenesis for the betterment of human health.

Acknowledgments Work associated with this chapter was supported by the Federal Ministry for Education and Research (BMBF: https://www.bmbf.de/), Germany, Project FKZ 01 K12012 'RFIN – RNA-Biologie von Pilzinfektionen'. The funders had no role in the design, analysis, decision to publish, or preparation of the document.

References

Abdullah Z, Schlee M, Roth S, Mraheil MA, Barchet W, Bottcher J, Hain T, Geiger S, Hayakawa Y, Fritz JH, Civril F, Hopfner KP, Kurts C, Ruland J, Hartmann G, Chakraborty T, Knolle PA (2012) Rig-I detects infection with live *Listeria* by sensing secreted bacterial nucleic acids. EMBO J 31:4153–4164

Alegre-Maller AC, Mendonca FC, Da Silva TA, Oliveira AF, Freitas MS, Hanna ES, Almeida IC, Gay NJ, Roque-Barreira MC (2014) Therapeutic administration of recombinant paracoccin confers protection against *Paracoccidioides brasiliensis* infection: involvement of Tlrs. PLoS Negl Trop Dis 8:E3317

Alings F, Scharmann K, Eggers Aracena CE, Böttcher B, Sokołowski M, Shvetsova E, Sharma P, Roth J, Rashiti L, Glatt S (2023) Ncs2* mediates In vivo virulence of pathogenic yeast through sulphur modification of cytoplasmic transfer RNA. Nucleic Acids Res 51(15):8133–8149

Allendoerfer R, Deepe GS Jr (1997) Intrapulmonary response to *Histoplasma capsulatum* in gamma interferon knockout mice. Infect Immun 65:2564–2569

Alves LR, Peres da Silva R, Sanchez DA, Zamith-Miranda D, Rodrigues ML, Goldenberg S, Puccia R, Nosanchuk JD (2019) Extracellular vesicle-mediated RNA release in *Histoplasma capsulatum*. mSphere 4:e00176–19

Anderson BR, Muramatsu H, Nallagatla SR, Bevilacqua PC, Sansing LH, Weissman D, Kariko K (2010) Incorporation of pseudouridine into mRNA enhances translation by diminishing PKR activation. Nucleic Acids Res 38:5884–5892

Applen Clancey S, Ruchti F, Leibundgut-Landmann S, Heitman J, Ianiri G (2020) A novel mycovirus evokes transcriptional rewiring in the fungus *Malassezia* and stimulates beta interferon production in macrophages. *MBio* 11:e01534–20

Araujo D, Azevedo NM, Barbosa A, Almeida C, Rodrigues ME, Henriques M, Silva S (2019) Application of 2'-OMethylRNA' antisense oligomer to control *Candida albicans* Efg1 virulence determinant. Mol Ther Nucleic Acids 18:508–517

Bachtiar BM, Srisawat C, Bachtiar EW (2019) RNA aptamers selected against yeast cells inhibit *Candida Albicans* biofilm formation in vitro. Microbiology 8:E00812

Barton GM (2007) Viral recognition by toll-like receptors. In: Seminars in immunology. Elsevier, pp 33–40

Bayne RA, Jayachandran U, Kasprowicz A, Bresson S, Tollervey D, Wallace EWJ, Cook AG (2022) Yeast Ssd1 is a non-enzymatic member of the RNase II family with an alternative RNA recognition site. Nucleic Acids Res 50:2923–2937

Bellocchio S, Montagnoli C, Bozza S, Gaziano R, Rossi G, Mambula SS, Vecchi A, Mantovani A, Levitz SM, Romani L (2004) The contribution of the toll-like/Il-1 receptor superfamily to innate and adaptive immunity to fungal pathogens *in vivo*. J Immunol 172:3059–3069

Bernstein DA, Vyas VK, Weinberg DE, Drinnenberg IA, Bartel DP, Fink GR (2012) *Candida Albicans* Dicer (Cadcr1) is required for efficient ribosomal and spliceosomal RNA maturation. Proc Natl Acad Sci USA 109:523–528

Bielska E, Sisquella MA, Aldeieg M, Birch C, O'donoghue EJ, May RC (2018) Pathogen-derived extracellular vesicles mediate virulence in the fatal human pathogen *Cryptococcus Gattii*. Nat Commun 9:1556

Billmyre RB, Calo S, Feretzaki M, Wang X, Heitman J (2013) RNAi function, diversity, and loss in the fungal kingdom. Chromosom Res 21:561–572

Billmyre RB, Applen Clancey S, Li LX, Doering TL, Heitman J (2020) 5-fluorocytosine resistance is associated with hypermutation and alterations in capsule biosynthesis in *Cryptococcus*. Nat Commun 11:127

Biondo C, Malara A, Costa A, Signorino G, Cardile F, Midiri A, Galbo R, Papasergi S, Domina M, Pugliese M (2012) Recognition of fungal RNA By TLR 7 has a nonredundant role in host defense against experimental candidiasis. Eur J Immunol 42:2632–2643

Bloom ALM, Jin RM, Leipheimer J, Bard JE, Yergeau D, Wohlfert EA, Panepinto JC (2019) Thermotolerance in the pathogen *Cryptococcus neoformans* is linked to antigen masking via mRNA decay-dependent reprogramming. Nat Commun 10:4950

Bonnet A, Grosso AR, Elkaoutari A, Coleno E, Presle A, Sridhara SC, Janbon G, Geli V, De Almeida SF, Palancade B (2017) Introns protect eukaryotic genomes from transcription-associated genetic instability. Mol Cell 67:608–621 e6

Boomiraj H, Mohankumar V, Lalitha P, Devarajan B (2015) Human corneal microrna expression profile in fungal keratitis. Invest Ophthalmol Vis Sci 56:7939–7946

Brinkmann MM, Spooner E, Hoebe K, Beutler B, Ploegh HL, Kim Y-M (2007) The interaction between the ER membrane protein UNC93B and TLR3, 7, and 9 is crucial for TLR signaling. J Cell Biol 177:265–275

Brown GD, Netea MG (2007) Immunology of fungal infections. Springer

Bruch A, Kelani AA, Blango MG (2022) RNA-based therapeutics to treat human fungal infections. Trends Microbiol 30:411–420

Burke JE, Longhurst AD, Natarajan P, Rao B, Liu J, Sales-Lee J, Mortensen Y, Moresco JJ, Diedrich JK, Yates JR III, Madhani HD (2019) A non-dicer RNase III and four other novel factors required for RNAi-mediated transposon suppression in the human pathogenic yeast *Cryptococcus neoformans*. *G3 (Bethesda)* 9:2235–2244

Cai Q, He B, Kogel K-H, Jin H (2018a) Cross-kingdom RNA trafficking and environmental RNAi—nature's blueprint for modern crop protection strategies. Curr Opin Microbiol 46:58–64

Cai Q, Qiao L, Wang M, He B, Lin FM, Palmquist J, Huang SD, Jin H (2018b) Plants send small RNAs in extracellular vesicles to fungal pathogen to silence virulence genes. Science 360:1126–1129

Cai Q, He B, Weiberg A, Buck AH, Jin H (2019) Small RNAs and extracellular vesicles: new mechanisms of cross-species communication and innovative tools for disease control. PLoS Pathog 15:e1008090

Calo S, Shertz-Wall C, Lee SC, Bastidas RJ, Nicolas FE, Granek JA, Mieczkowski P, Torres-Martinez S, Ruiz-Vazquez RM, Cardenas ME, Heitman J (2014) Antifungal drug resistance evoked via RNAi-dependent epimutations. Nature 513:555–558

Calo S, Nicolas FE, Lee SC, Vila A, Cervantes M, Torres-Martinez S, Ruiz-Vazquez RM, Cardenas ME, Heitman J (2017) A non-canonical RNA degradation pathway suppresses RNAi-dependent epimutations in the human fungal pathogen *Mucor circinelloides*. PLoS Genet 13:e1006686

Canovas-Marquez JT, Falk S, Nicolas FE, Padmanabhan S, Zapata-Perez R, Sanchez-Ferrer A, Navarro E, Garre V (2021) A ribonuclease III involved in virulence of *Mucorales* fungi has evolved to cut exclusively single-stranded RNA. Nucleic Acids Res 49:5294–5307

Carvalho A, De Luca A, Bozza S, Cunha C, D'angelo C, Moretti S, Perruccio K, Iannitti RG, Fallarino F,

Pierini A (2012) TLR3 essentially promotes protective class I–restricted memory CD8+ T-cell responses to *Aspergillus fumigatus* in hematopoietic transplanted patients. Blood 119:967–977

Chacko N, Zhao Y, Yang E, Wang L, Cai JJ, Lin X (2015) The lncRNA *Rze1* controls Cryptococcal morphological transition. PLoS Genet 11:e1005692

Chai LY, Kullberg BJ, Vonk AG, Warris A, Cambi A, Latge JP, Joosten LA, Van Der Meer JW, Netea MG (2009) Modulation of toll-like receptor 2 (TLR2) and TLR4 responses by *Aspergillus fumigatus*. Infect Immun 77:2184–2192

Chang Z, Heitman J (2019) Drug-resistant epimutants exhibit organ-specific stability and induction during murine infections caused by the human fungal pathogen *Mucor circinelloides*. MBio 10

Chang Z, Billmyre RB, Lee SC, Heitman J (2019) Broad antifungal resistance mediated by RNAi-dependent epimutation in the basal human fungal pathogen *Mucor circinelloides*. PLoS Genet 15:e1007957

Chen Y, Jiang Q, Qing F, Xue J, Xiao Q, He W, Sui L, Liu Z (2023) MDA5 enhances invasive *Candida albicans* infection by regulating macrophage apoptosis and phagocytosis/killing functions. Inflammation:1–18

Choi Y, Hyeon H, Lee K, Bahn Y-S (2023) Sua5 catalyzing universal t6A tRNA modification is responsible for multifaceted functions of the KEOPS complex in *Cryptococcus neoformans*. mSphere:e00557–e00523

Chorev M, Carmel L (2012) The function of introns. *Front Genet* 3:55

Corbett KS, Edwards DK, Leist SR, Abiona OM, Boyoglu-Barnum S, Gillespie RA, Himansu S, Schafer A, Ziwawo CT, Dipiazza AT, Dinnon KH, Elbashir SM, Shaw CA, Woods A, Fritch EJ, Martinez DR, Bock KW, Minai M, Nagata BM, Hutchinson GB, Wu K, Henry C, Bahl K, Garcia-Dominguez D, Ma L, Renzi I, Kong WP, Schmidt SD, Wang L, Zhang Y, Phung E, Chang LA, Loomis RJ, Altaras NE, Narayanan E, Metkar M, Presnyak V, Liu C, Louder MK, Shi W, Leung K, Yang ES, West A, Gully KL, Stevens LJ, Wang N, Wrapp D, Doria-Rose NA, Stewart-Jones G, Bennett H, Alvarado GS, Nason MC, Ruckwardt TJ, Mclellan JS, Denison MR, Chappell JD, Moore IN, Morabito KM, Mascola JR, Baric RS, Carfi A, Graham BS (2020) SARS-Cov-2 mRNA vaccine design enabled by prototype pathogen preparedness. Nature 586:567–571

Croston TL, Lemons AR, Beezhold DH, Green BJ (2018) Microrna regulation of host immune responses following fungal exposure. Front Immunol 9:170

Cui C, Wang Y, Liu J, Zhao J, Sun P, Wang S (2019) A fungal pathogen deploys a small silencing RNA that attenuates mosquito immunity and facilitates infection. Nat Commun 10:4298

Dammes N, Peer D (2020) Paving the road for RNA therapeutics. Trends Pharmacol Sci 41:755–775

Dantuluri S, Schwer B, Abdullahu L, Damha MJ, Shuman S (2021) Activity and substrate specificity of *Candida*, *Aspergillus*, and *Coccidioides* Tpt1: essential tRNA splicing enzymes and potential anti-fungal targets. RNA

Datta K, Bartlett KH, Baer R, Byrnes E, Galanis E, Heitman J, Hoang L, Leslie MJ, Macdougall L, Magill SS, Morshed MG, Marr KA & Cryptococcus Gattii Working Group of the Pacific Northwest (2009) Spread of *Cryptococcus gattii* Into Pacific Northwest region of the United States. Emerg Infect Dis 15:1185–1191

De Candia P, De Rosa V, Casiraghi M, Matarese G (2016) Extracellular RNAs: a secret arm of immune system regulation. J Biol Chem 291:7221–7228

De Souza Silva C, Tavares AH, Sousa Jeronimo M, Soares De Lima Y, Da Silveira Derengowski L, Bocca AL, Silva-Pereira I (2015) The effects of *Paracoccidioides brasiliensis* infection on GM-CSF- and M-CSF-induced mouse bone marrow-derived macrophage from resistant and susceptible mice strains. Mediat Inflamm 2015:605450

De Voogt WS, Tanenbaum ME, Vader P (2021) Illuminating RNA trafficking and functional delivery by extracellular vesicles. Adv Drug Deliv Rev 174:250–264

Deepe G (1988) Histoplasmosis. A granulomatous inflammatory response. In: Inflammation: basic principles and clinical correlates. Raven Press, pp 943–958

Dix A, Czakai K, Leonhardt I, Schaferhoff K, Bonin M, Guthke R, Einsele H, Kurzai O, Loffler J, Linde J (2017) Specific and novel micrornas are regulated as response to fungal infection in human dendritic cells. Front Microbiol 8:270

Donovan PD, Holland LM, Lombardi L, Coughlan AY, Higgins DG, Wolfe KH, Butler G (2018) TPP riboswitch-dependent regulation of an ancient thiamin transporter in *Candida*. PLoS Genet 14:e1007429

Elson SL, Noble SM, Solis NV, Filler SG, Johnson AD (2009) An RNA transport system in *Candida albicans* regulates hyphal morphology and invasive growth. PLoS Genet 5:e1000664

Fedorova O, Jagdmann GE Jr, Adams RL, Yuan L, Van Zandt MC, Pyle AM (2018) Small molecules that target group II introns are potent antifungal agents. Nat Chem Biol 14:1073–1078

Feng H, Zhang YB, Gui JF, Lemon SM, Yamane D (2021) Interferon regulatory factor 1 (IRF1) and anti-pathogen innate immune responses. PLoS Pathog 17:e1009220

Feretzaki M, Billmyre RB, Clancey SA, Wang X, Heitman J (2016) Gene network polymorphism illuminates loss and retention of novel RNAi silencing components in the *Cryptococcus* pathogenic species complex. PLoS Genet 12:e1005868

Ferrareze PA, Streit RS, Dos Santos FM, Schrank A, Kmetzsch L, Vainstein MH, Staats CC (2017) sRNAs as possible regulators of retrotransposon activity in *Cryptococcus gattii* Vgii. BMC Genomics 18:294

Frye M, Harada BT, Behm M, He C (2018) RNA modifications modulate gene expression during development. Science 361:1346–1349

Gantner BN, Simmons RM, Canavera SJ, Akira S, Underhill DM (2003) Collaborative induction of inflammatory responses by dectin-1 and Toll-like receptor 2. J Exp Med 197:1107–1117

Gao J, Chow EWL, Wang H, Xu X, Cai C, Song Y, Wang J, Wang Y (2021) LncRNA *DINOR* is a virulence factor and global regulator of stress responses in *Candida auris*. Nat Microbiol 6:842–851

Ghabrial S (2008) Totiviruses. In: Encyclopedia of virology, pp 163–174

Ghabrial SA, Caston JR, Jiang D, Nibert ML, Suzuki N (2015) 50-plus years of fungal viruses. Virology 479–480:356–368

Gildea LA, Morris RE, Newman SL (2001) *Histoplasma capsulatum* yeasts are phagocytosed via very late antigen-5, killed, and processed for antigen presentation by human dendritic cells. J Immunol 166:1049–1056

Gitlin L, Barchet W, Gilfillan S, Cella M, Beutler B, Flavell RA, Diamond MS, Colonna M (2006) Essential role of mda-5 In type I IFN responses to polyriboinosinic: polyribocytidylic acid and encephalomyocarditis picornavirus. Proc Natl Acad Sci USA 103:8459–8464

Gladyshev E (2017) Repeat-induced point mutation and other genome defense mechanisms in fungi. Microbiol Spectr 5

Gonzalez-Hilarion S, Paulet D, Lee KT, Hon CC, Lechat P, Mogensen E, Moyrand F, Proux C, Barboux R, Bussotti G, Hwang J, Coppee JY, Bahn YS, Janbon G (2016) Intron retention-dependent gene regulation in *Cryptococcus neoformans*. Sci Rep 6:32252

Goubau D, Deddouche S, Reis E, Sousa C (2013) Cytosolic sensing of viruses. Immunity 38:855–869

Halder LD, Jo EAH, Hasan MZ, Ferreira-Gomes M, Kruger T, Westermann M, Palme DI, Rambach G, Beyersdorf N, Speth C, Jacobsen ID, Kniemeyer O, Jungnickel B, Zipfel PF, Skerka C (2020) Immune modulation by complement receptor 3-dependent human monocyte TGF-beta1-transporting vesicles. Nat Commun 11:2331

Halder LD, Babych S, Palme DI, Mansouri-Ghahnavieh E, Ivanov L, Ashonibare V, Langenhorst D, Prusty B, Rambach G, Wich M, Trinks N, Blango MG, Kornitzer D, Terpitz U, Speth C, Jungnickel B, Beyersdorf N, Zipfel PF, Brakhage AA, Skerka C (2022) *Candida albicans* induces cross-kingdom mirna trafficking in human monocytes to promote fungal growth. MBio 13:e0356321

Hammond TM, Bok JW, Andrewski MD, Reyes-Dominguez Y, Scazzocchio C, Keller NP (2008) Rna silencing gene truncation in the filamentous fungus *Aspergillus nidulans*. Eukaryot Cell 7:339–349

Hartmann G (2017) Nucleic acid immunity. Adv Immunol 133:121–169

He B, Cai Q, Qiao L, Huang CY, Wang S, Miao W, Ha T, Wang Y, Jin H (2021) RNA-binding proteins contribute to small rna loading in plant extracellular vesicles. Nat Plants 7:342–352

He B, Wang H, Liu G, Chen A, Calvo A, Cai Q, Jin H (2023) Fungal small RNAs ride in extracellular vesicles to enter plant cells through clathrin-mediated endocytosis. Nat Commun 14:4383

Hou F, Sun L, Zheng H, Skaug B, Jiang QX, Chen ZJ (2011) Mavs forms functional prion-like aggregates to activate and propagate antiviral innate immune response. Cell 146:448–461

Hovhannisyan H, Gabaldon T (2021) The long non-coding rna landscape of *Candida* yeast pathogens. Nat Commun 12:7317

Howard JJ, Sturge CR, Moustafa DA, Daly SM, Marshall-Batty KR, Felder CF, Zamora D, Yabe-Gill M, Labandeira-Rey M, Bailey SM, Wong M, Goldberg JB, Geller BL, Greenberg DE (2017) Inhibition of *Pseudomonas aeruginosa* by peptide-conjugated phosphorodiamidate morpholino oligomers. Antimicrob Agents Chemother 61:e01938-16

Huang Y, Cao Y, Li J, Liu Y, Zhong W, Li X, Chen C, Hao P (2018) A survey on cellular RNA editing activity in response to *Candida albicans* infections. BMC Genomics 19:43

Jaeger M, Van Der Lee R, Cheng SC, Johnson MD, Kumar V, Ng A, Plantinga TS, Smeekens SP, Oosting M, Wang X, Barchet W, Fitzgerald K, Joosten LAB, Perfect JR, Wijmenga C, Van De Veerdonk FL, Huynen MA, Xavier RJ, Kullberg BJ, Netea MG (2015) The Rig-I-like helicase receptor MDA5 (IFIH1) is involved in the host defense against *Candida* infections. Eur J Clin Microbiol Infect Dis 34:963–974

Janbon G, Maeng S, Yang DH, Ko YJ, Jung KW, Moyrand F, Floyd A, Heitman J, Bahn YS (2010) Characterizing the role of RNA silencing components in *Cryptococcus neoformans*. Fungal Genet Biol 47:1070–1080

Jannuzzi GP, Tavares AHF, Kaihami GH, De Almeida JRF, De Almeida SR, Ferreira KS (2015) ScFV from antibody that mimics Gp43 modulates the cellular and humoral immune responses during experimental paracoccidioidomycosis. PLoS One 10:E0129401

Jannuzzi GP, De Almeida JRF, Amarante-Mendes GP, Romera LMDM, Kaihami GH, Vasconcelos JR, Ferreira CP, De Almeida SR, Ferreira KS (2019) TLR3 is a negative regulator of immune responses against *Paracoccidioides brasiliensis*. Front Cell Infect Microbiol 8:426

Jannuzzi GP, De Almeida JRF, Paulo LNM, De Almeida SR, Ferreira KS (2020) Intracellular PRRs activation in targeting the immune response against fungal infections. Front Cell Infect Microbiol 10:591970

Jiang N, Yang Y, Janbon G, Pan J, Zhu X (2012) Identification and functional demonstration of mirnas in the fungus *Cryptococcus neoformans*. PLoS One 7:E52734

Jiao J, Peng D (2018) Wheat microRNA1023 suppresses invasion of *Fusarium graminearum* via targeting and silencing FGSG_03101. J Plant Interact 13:514–521

Jochl C, Rederstorff M, Hertel J, Stadler PF, Hofacker IL, Schrettl M, Haas H, Huttenhofer A (2008) Small ncRNA transcriptome analysis from *Aspergillus fumigatus* suggests a novel mechanism for regulation of protein synthesis. Nucleic Acids Res 36:2677–2689

Jochl C, Loh E, Ploner A, Haas H, Huttenhofer A (2009) Development-dependent scavenging of nucleic acids in the filamentous fungus *Aspergillus fumigatus*. RNA Biol 6:179–186

Kalem MC, Panepinto JC (2022) Long non-coding RNAs in *Cryptococcus neoformans*: insights into fungal pathogenesis. Front Cell Infect Microbiol 12:858317

Kariko K, Muramatsu H, Keller JM, Weissman D (2012) Increased erythropoiesis in mice injected with submicrogram quantities of pseudouridine-containing mRNA encoding erythropoietin. Mol Ther 20:948–953

Kato H, Takeuchi O, Sato S, Yoneyama M, Yamamoto M, Matsui K, Uematsu S, Jung A, Kawai T, Ishii KJ, Yamaguchi O, Otsu K, Tsujimura T, Koh CS, Reis E, Sousa C, Matsuura Y, Fujita T, Akira S (2006) Differential roles of MDA5 and RIG-I helicases in the recognition of RNA viruses. Nature 441:101–105

Kawai T, Takahashi K, Sato S, Coban C, Kumar H, Kato H, Ishii KJ, Takeuchi O, Akira S (2005) IPS-1, an adaptor triggering RIG-I-and Mda5-mediated type I interferon induction. Nat Immunol 6:981–988

Kelani AA, Bruch A, Rivieccio F, Visser C, Krueger T, Weaver D, Pan X, Schäuble S, Panagiotou G, Kniemeyer O (2023) Disruption of the *Aspergillus fumigatus* RNA interference machinery alters the conidial transcriptome. RNA 29:1033–1050

Kim YM, Brinkmann MM, Paquet ME, Ploegh HL (2008) UNC93B1 delivers nucleotide-sensing toll-like receptors to endolysosomes. Nature 452:234–238

Kim H, Hu J, Kang H, Kim W (2023) Phylogenetic and functional analyses of N 6-methyladenosine RNA methylation factors in the wheat scab fungus *Fusarium graminearum*. mSphere 9:E0055223

Lee BL, Barton GM (2014) Trafficking of endosomal toll-like receptors. Trends Cell Biol 24:360–369

Lee HC, Li L, Gu W, Xue Z, Crosthwaite SK, Pertsemlidis A, Lewis ZA, Freitag M, Selker EU, Mello CC, Liu Y (2010) Diverse pathways generate microrna-like RNAs and dicer-independent small interfering RNAs in fungi. Mol Cell 38:803–814

Leipheimer J, Bloom ALM, Baumstark T, Panepinto JC (2018) CNBP homologues Gis2 and Znf9 interact with a putative G-quadruplex-forming 3' untranslated region, altering polysome association and stress tolerance In *Cryptococcus neoformans*. mSphere 3:e0201-18

Lemaitre B, Nicolas E, Michaut L, Reichhart J-M, Hoffmann JA (1996) The dorsoventral regulatory gene cassette spätzle/toll/cactus controls the potent antifungal response in *Drosophila* adults. Cell 86:973–983

Li S, Breaker RR (2017) Identification of 15 candidate structured noncoding RNA motifs in fungi by comparative genomics. BMC Genomics 18:785

Lin JS, Yang CW, Wang DW, Wu-Hsieh BA (2005) Dendritic cells cross-present exogenous fungal antigens to stimulate a protective CD8 T cell response in infection by *Histoplasma capsulatum*. J Immunol 174:6282–6291

Liu T, Pyle AM (2021) Discovery of highly reactive self-splicing group II introns within the mitochondrial genomes of human pathogenic fungi. Nucleic Acids Res 49:12422–12432

Liu H, Cottrell TR, Pierini LM, Goldman WE, Doering TL (2002) RNA interference in the pathogenic fungus *Cryptococcus neoformans*. Genetics 160:463–470

Liu M, Zhang Z, Ding C, Wang T, Kelly B, Wang P (2020) Transcriptomic analysis of extracellular RNA governed by the endocytic adaptor protein Cin1 of *Cryptococcus deneoformans*. Front Cell Infect Microbiol 10:256

Loo YM, Gale M Jr (2011) Immune signaling by Rig-I-like receptors. Immunity 34:680–692

Mambula SS, Sau K, Henneke P, Golenbock DT, Levitz SM (2002) Toll-like receptor (TLR) signaling in response to *Aspergillus fumigatus*. J Biol Chem 277:39320–39326

Martienssen R, Moazed D (2015) RNAi and heterochromatin assembly. Cold Spring Harb Perspect Biol 7:A019323

Mcwhirter SM, Tenoever BR, Maniatis T (2005) Connecting mitochondria and innate immunity. Cell 122:645–647

Meier A, Kirschning CJ, Nikolaus T, Wagner H, Heesemann J, Ebel F (2003) Toll-like receptor (TLR) 2 and TLR4 are essential for *Aspergillus*-induced activation of murine macrophages. Cell Microbiol 5:561–570

Meylan E, Curran J, Hofmann K, Moradpour D, Binder M, Bartenschlager R, Tschopp J (2005) Cardif is an adaptor protein in the Rig-I antiviral pathway and is targeted by hepatitis C virus. Nature 437:1167–1172

Moldovan MA, Petrova SA, Gelfand MS (2018) Comparative genomic analysis of fungal TPP-riboswitches. Fungal Genet Biol 114:34–41

Morais EA, Chame DF, Melo EM, De Carvalho Oliveira JA, De Paula AC, Peixoto AC, Da Silva Santos L, Gomes DA, Russo RC, De Goes AM (2016) TLR 9 involvement in early protection induced by immunization with rPB27 against paracoccidioidomycosis. Microbes Infect 18:137–147

Moreno-Velasquez SD, Perez JC (2021) Imaging and quantification of mRNA molecules at single-cell resolution in the human fungal pathogen *Candida albicans*. mSphere 6:E0041121

Moresco EM, Lavine D, Beutler B (2011) Toll-like receptors. Curr Biol 21:R488–R493

Moustafa DA, Wu AW, Zamora D, Daly SM, Sturge CR, Pybus C, Geller BL, Goldberg JB, Greenberg DE (2021) Peptide-conjugated phosphorodiamidate morpholino oligomers retain activity against multidrug-resistant *Pseudomonas aeruginosa* in vitro and in vivo. mBio 12:e02411-20

Mouyna I, Henry C, Doering TL, Latge JP (2004) Gene silencing with RNA interference in the human pathogenic fungus *Aspergillus fumigatus*. FEMS Microbiol Lett 237:317–324

Munhoz Da Rocha IF, Martins ST, Amatuzzi RF, Zamith-Miranda D, Nosanchuk JD, Rodrigues ML, Alves LR (2021) Cellular and extracellular vesicle RNA analysis in the global threat fungus *Candida auris*. Microbiol Spectr 9:e0153821

Munoz JF, Delorey T, Ford CB, Li BY, Thompson DA, Rao RP, Cuomo CA (2019) Coordinated host-pathogen transcriptional dynamics revealed using sorted subpopulations and single macrophages infected with *Candida albicans*. Nat Commun 10:1607

Navarro-Mendoza MI, Pérez-Arques C, Heitman J (2023) Heterochromatin and RNAi act independently to ensure genome stability in mucorales human fungal pathogens. Proc Natl Acad Sci U S A 120:e2220475120

Nelson J, Sorensen EW, Mintri S, Rabideau AE, Zheng W, Besin G, Khatwani N, Su SV, Miracco EJ, Issa WJ, Hoge S, Stanton MG, Joyal JL (2020) Impact of mRNA chemistry and manufacturing process on innate immune activation. Sci Adv 6:eaaz6893

Netea MG, Van Der Graaf CA, Vonk AG, Verschueren I, Van Der Meer JW, Kullberg BJ (2002) The role of toll-like receptor (TLR) 2 and TLR4 in the host defense against disseminated candidiasis. J Infect Dis 185:1483–1489

Netea MG, Gow NA, Munro CA, Bates S, Collins C, Ferwerda G, Hobson RP, Bertram G, Hughes HB, Jansen T, Jacobs L, Buurman ET, Gijzen K, Williams DL, Torensma R, Mckinnon A, Maccallum DM, Odds FC, Van Der Meer JW, Brown AJ, Kullberg BJ (2006) Immune sensing of *Candida albicans* requires cooperative recognition of mannans and glucans by lectin and toll-like receptors. J Clin Invest 116:1642–1650

Nicola AM, Frases S, Casadevall A (2009) Lipophilic dye staining of *Cryptococcus neoformans* extracellular vesicles and capsule. Eukaryot Cell 8:1373–1380

Niemiec MJ, Grumaz C, Ermert D, Desel C, Shankar M, Lopes JP, Mills IG, Stevens P, Sohn K, Urban CF (2017) Dual transcriptome of the immediate neutrophil and *Candida albicans* interplay. BMC Genomics 18:696

Nosanchuk JD, Gacser A (2008) *Histoplasma capsulatum* at the host–pathogen interface. Microbes Infect 10:973–977

Onomoto K, Onoguchi K, Takahasi K, Fujita T (2010) Type I interferon production induced by Rig-I-like receptors. J Interf Cytokine Res 30:875–881

Panepinto J, Liu L, Ramos J, Zhu X, Valyi-Nagy T, Eksi S, Fu J, Jaffe HA, Wickes B, Williamson PR (2005) The dead-box RNA helicase VAD1 regulates multiple virulence-associated genes in *Cryptococcus neoformans*. J Clin Invest 115:632–641

Park M, Cho YJ, Kim D, Yang CS, Lee SM, Dawson TL Jr, Nakamizo S, Kabashima K, Lee YW, Jung WH (2020) A novel virus alters gene expression and vacuolar morphology in *malassezia* cells and induces a TLR3-mediated inflammatory immune response. mBio 11:e01521–e01520

Pendrak ML, Roberts DD (2011) Ribosomal RNA processing in *Candida albicans*. RNA 17:2235–2248

Peres Da Silva R, Puccia R, Rodrigues ML, Oliveira DL, Joffe LS, Cesar GV, Nimrichter L, Goldenberg S, Alves LR (2015) Extracellular vesicle-mediated export of fungal RNA. Sci Rep 5:7763

Peres Da Silva R, Martins ST, Rizzo J, Dos Reis FCG, Joffe LS, Vainstein M, Kmetzsch L, Oliveira DL, Puccia R, Goldenberg S, Rodrigues ML, Alves LR (2018) Golgi reassembly and stacking protein (GRASP) participates in vesicle-mediated RNA export in *Cryptococcus neoformans*. *Genes (Basel)* 9

Peres Da Silva R, Longo LGV, Cunha J, Sobreira TJP, Rodrigues ML, Faoro H, Goldenberg S, Alves LR, Puccia R (2019) Comparison of the RNA content of extracellular vesicles derived from *Paracoccidioides brasiliensis* and *Paracoccidioides lutzii*. Cells 8

Priest SJ, Yadav V, Roth C, Dahlmann TA, Kuck U, Magwene PM, Heitman J (2022) Uncontrolled transposition following RNAi loss causes hypermutation and antifungal drug resistance in clinical isolates of *Cryptococcus neoformans*. Nat Microbiol 7:1239–1251

Rafiq M, Rivieccio F, Zimmermann AK, Visser C, Bruch A, Kruger T, Gonzalez Rojas K, Kniemeyer O, Blango MG, Brakhage AA (2022) PLB-985 neutrophil-like cells as a model to study *Aspergillus fumigatus* pathogenesis. mSphere 7:e0094021

Rayner S, Bruhn S, Vallhov H, Andersson A, Billmyre RB, Scheynius A (2017) Identification of small RNAs in extracellular vesicles from the commensal yeast *Malassezia sympodialis*. Sci Rep 7:39742

Rehwinkel J, Gack MU (2020) Rig-I-like receptors: their regulation and roles in RNA sensing. Nat Rev Immunol 20:537–551

Reis FCG, Borges BS, Jozefowicz LJ, Sena BAG, Garcia AWA, Medeiros LC, Martins ST, Honorato L, Schrank A, Vainstein MH, Kmetzsch L, Nimrichter L, Alves LR, Staats CC, Rodrigues ML (2019) A novel protocol for the isolation of fungal extracellular vesicles reveals the participation of a putative scramblase in polysaccharide export and capsule construction in *Cryptococcus gattii*. mSphere 4

Ren Z, Tang B, Xing J, Liu C, Cai X, Hendy A, Kamran M, Liu H, Zheng L, Huang J (2022) MTA1-mediated RNA M6A modification regulates autophagy and is required for infection of the rice blast fungus. New Phytol 235:247–262

Riege K, Holzer M, Klassert TE, Barth E, Brauer J, Collatz M, Hufsky F, Mostajo N, Stock M, Vogel B, Slevogt H, Marz M (2017) Massive effect on LncRNAs in human monocytes during fungal and bacterial infections and in response to vitamins A and D. Sci Rep 7:40598

Roche B, Arcangioli B, Martienssen R (2016) RNA interference is essential for cellular quiescence. Science 354:aah5651

Romano N, Macino G (1992) Quelling: transient inactivation of gene expression in *Neurospora crassa* by transformation with homologous sequences. Mol Microbiol 6:3343–3353

Sahr T, Escoll P, Rusniok C, Bui S, Pehau-Arnaudet G, Lavieu G, Buchrieser C (2022) Translocated *Legionella pneumophila* small RNAs mimic eukaryotic micrornas targeting the host immune response. Nat Commun 13:762

Saikh KU (2021) MyD88 and beyond: a perspective on MyD88-targeted therapeutic approach for modulation of host immunity. Immunol Res 69:117–128

Saunte DM, Gaitanis G, Hay RJ (2020) *Malassezia*-associated skin diseases, the use of diagnostics and treatment. Front Cell Infect Microbiol 10:112

Seelbinder B, Wallstabe J, Marischen L, Weiss E, Wurster S, Page L, Loffler C, Bussemer L, Schmitt AL, Wolf T, Linde J, Cicin-Sain L, Becker J, Kalinke U, Vogel J, Panagiotou G, Einsele H, Westermann AJ, Schauble S, Loeffler J (2020) Triple RNA-seq reveals synergy in a human virus-fungus co-infection model. Cell Rep 33:108389

Seth RB, Sun L, Ea C-K, Chen ZJ (2005) Identification and characterization of MAVS, a mitochondrial antiviral signaling protein that activates NF-κB And IRF3. Cell 122:669–682

Shiu PK, Raju NB, Zickler D, Metzenberg RL (2001) Meiotic silencing by unpaired DNA. Cell 107:905–916

Shoham S, Huang C, Chen J-M, Golenbock DT, Levitz SM (2001) Toll-like receptor 4 mediates intracellular signaling without TNF-A release in response to *Cryptococcus neoformans* polysaccharide capsule. J Immunol 166:4620–4626

Shopova IA, Belyaev I, Dasari P, Jahreis S, Stroe MC, Cseresnyes Z, Zimmermann AK, Medyukhina A, Svensson CM, Kruger T, Szeifert V, Nietzsche S, Conrad T, Blango MG, Kniemeyer O, Von Lilienfeld-Toal M, Zipfel PF, Ligeti E, Figge MT, Brakhage AA (2020) Human neutrophils produce antifungal extracellular vesicles against *Aspergillus fumigatus*. mBio 11

Sieber P, Voigt K, Kammer P, Brunke S, Schuster S, Linde J (2018) Comparative study on alternative splicing in human fungal pathogens suggests its involvement during host invasion. Front Microbiol 9:2313

Smith CA, Robertson D, Yates B, Nielsen DM, Brown D, Dean RA, Payne GA (2008) The effect of temperature on natural antisense transcript (Nat) expression in *Aspergillus flavus*. Curr Genet 54:241–269

Subramani A, Griggs P, Frantzen N, Mendez J, Tucker J, Murriel J, Sircy LM, Millican GE, Mcclelland EE, Seipelt-Thiemann RL, Nelson DE (2020) Intracellular *Cryptococcus neoformans* disrupts the transcriptome profile of M1- and M2-polarized host macrophages. PLoS One 15:e0233818

Tabeta K, Hoebe K, Janssen EM, Du X, Georgel P, Crozat K, Mudd S, Mann N, Sovath S, Goode J, Shamel L, Herskovits AA, Portnoy DA, Cooke M, Tarantino LM, Wiltshire T, Steinberg BE, Grinstein S, Beutler B (2006) The Unc93b1 mutation 3d disrupts exogenous antigen presentation and signaling via toll-like receptors 3, 7 and 9. Nat Immunol 7:156–164

Tada H, Nemoto E, Shimauchi H, Watanabe T, Mikami T, Matsumoto T, Ohno N, Tamura H, Shibata KI, Akashi S (2002) *Saccharomyces cerevisiae*-and *Candida albicans*-derived mannan induced production of tumor necrosis factor alpha by human monocytes in a CD14- and toll-like receptor 4-dependent manner. Microbiol Immunol 46:503–512

Takeda K, Kaisho T, Akira S (2003) Toll-like receptors. Annu Rev Immunol 21:335–376

Takeuchi O, Akira S (2010) Pattern recognition receptors and inflammation. Cell 140:805–820

Torres-Martinez S, Ruiz-Vazquez RM (2017) The RNAi universe in fungi: a varied landscape of small RNAs and biological functions. Ann Rev Microbiol 71:371–391

Tosar JP, Segovia M, Castellano M, Gambaro F, Akiyama Y, Fagundez P, Olivera A, Costa B, Possi T, Hill M, Ivanov P, Cayota A (2020) Fragmentation of extracellular ribosomes and tRNAs shapes the extracellular RNAome. Nucleic Acids Res 48:12874–12888

Trieu TA, Calo S, Nicolas FE, Vila A, Moxon S, Dalmay T, Torres-Martinez S, Garre V, Ruiz-Vazquez RM (2015) A non-canonical RNA silencing pathway promotes mRNA degradation in basal fungi. PLoS Genet 11:e1005168

Uehata T, Takeuchi O (2020) RNA recognition and immunity-innate immune sensing and its posttranscriptional regulation mechanisms. Cells 9

Underhill DM, Ozinsky A, Hajjar AM, Stevens A, Wilson CB, Bassetti M, Aderem A (1999) The toll-like receptor 2 is recruited to macrophage phagosomes and discriminates between pathogens. Nature 402:39–43

Van Der Lee R, Buljan M, Lang B, Weatheritt RJ, Daughdrill GW, Dunker AK, Fuxreiter M, Gough J, Gsponer J, Jones DT, Kim PM, Kriwacki RW, Oldfield CJ, Pappu RV, Tompa P, Uversky VN, Wright PE, Babu MM (2014) Classification of intrinsically disordered regions and proteins. Chem Rev 114:6589–6631

Van Prooyen N, Henderson CA, Hocking Murray D, Sil A (2016) CD103+ conventional dendritic cells are critical for TLR7/9-dependent host defense against *Histoplasma capsulatum*, an endemic fungal pathogen of humans. PLoS Pathog 12:e1005749

Väre VY, Eruysal ER, Narendran A, Sarachan KL, Agris PF (2017) Chemical and conformational diversity of modified nucleosides affects tRNA structure and function. Biomol Ther 7:29

Vargas-Junior V, Antunes D, Guimaraes AC, Caffarena E (2022) In silico investigation of riboswitches in fungi: structural and dynamical insights into TPP riboswitches in *Aspergillus oryzae*. RNA Biol 19:90–103

Vermes A, Guchelaar HJ, Dankert J (2000) Flucytosine: a review of its pharmacology, clinical indications, pharmacokinetics, toxicity and drug interactions. J Antimicrob Chemother 46:171–179

Voelz K, Johnston SA, Smith LM, Hall RA, Idnurm A, May RC (2014) 'Division of labour' in response to host oxidative burst drives a fatal *Cryptococcus gattii* outbreak. Nat Commun 5:5194

Wallace EWJ, Maufrais C, Sales-Lee J, Tuck LR, De Oliveira L, Feuerbach F, Moyrand F, Natarajan P, Madhani HD, Janbon G (2020) Quantitative global studies reveal differential translational control by start codon context across the fungal kingdom. Nucleic Acids Res 48:2312–2331

Wang X, Hsueh YP, Li W, Floyd A, Skalsky R, Heitman J (2010) Sex-induced silencing defends the genome of *Cryptococcus neoformans* via RNAi. Genes Dev 24:2566–2582

Wang X, Wang P, Sun S, Darwiche S, Idnurm A, Heitman J (2012) Transgene induced co-suppression during vegetative growth in *Cryptococcus neoformans*. PLoS Genet 8:e1002885
Wang X, Darwiche S, Heitman J (2013) Sex-induced silencing operates during opposite-sex and unisexual reproduction in *Cryptococcus neoformans*. Genetics 193:1163–1174
Wang X, Caffrey-Carr AK, Liu KW, Espinosa V, Croteau W, Dhingra S, Rivera A, Cramer RA, Obar JJ (2020) MDA5 is an essential sensor of a pathogen-associated molecular pattern associated with vitality that is necessary for host resistance against *Aspergillus fumigatus*. J Immunol 205:3058–3070
Wang Y, Cui C, Wang G, Li Y, Wang S (2021) Insects defend against fungal infection by employing microRNAs to silence virulence-related genes. Proc Natl Acad Sci USA 118
Wang X, Cunha C, Grau MS, Robertson SJ, Lacerda JF, Campos A Jr, Lagrou K, Maertens J, Best SM, Carvalho A, Obar JJ (2022) Mavs expression in alveolar macrophages is essential for host resistance against *Aspergillus fumigatus*. J Immunol 209:346–353
Wang S, He B, Wu H, Cai Q, Ramírez-Sánchez O, Abreu-Goodger C, Birch PR, Jin H (2024) Plant mRNAs move into a fungal pathogen via extracellular vesicles to reduce infection. Cell Host Microbe 32:93–105.E6
Warner EF, Bohálová N, Brázda V, Waller ZAE, Bidula S (2021) Analysis of putative quadruplex-forming sequences in fungal genomes: novel antifungal targets? *Microb Genom*:7
Watkins TN, Liu H, Chung M, Hazen TH, Dunning Hotopp JC, Filler SG, Bruno VM (2018) Comparative transcriptomics of *Aspergillus fumigatus* strains upon exposure to human airway epithelial cells. Microb Genom 4
Weiberg A, Wang M, Lin FM, Zhao H, Zhang Z, Kaloshian I, Huang HD, Jin H (2013) Fungal small RNAs suppress plant immunity by hijacking host RNA interference pathways. Science 342:118–123
Wilson RC, Doudna JA (2013) Molecular mechanisms of RNA interference. Annu Rev Biophys 42:217–239
Woods JP (2003) Knocking on the right door and making a comfortable home: *Histoplasma capsulatum* intracellular pathogenesis. Curr Opin Microbiol 6:327–331
Wu B, Peisley A, Richards C, Yao H, Zeng X, Lin C, Chu F, Walz T, Hur S (2013) Structural basis for dsRNA recognition, filament formation, and antiviral signal activation by MDA5. Cell 152:276–289
Wu Y, Xu H, Li Y, Huang D, Chen L, Hu Y, Li L, Zhang D, Huang W (2019) miRNA-344b-1-3p modulates the autophagy of NR8383 cells during *Aspergillus fumigatus* infection via TLR2. Exp Ther Med 18:139–146
Wu-Hsieh BA, Howard DH (1987) Inhibition of the intracellular growth of *Histoplasma capsulatum* by recombinant murine gamma interferon. Infect Immun 55:1014–1016
Xie Y, Yao L, Yu X, Ruan Y, Li Z, Guo J (2020) Action mechanisms and research methods of tRNA-derived small RNAs. Signal Transduct Target Ther 5:109
Xu L-G, Wang Y-Y, Han K-J, Li L-Y, Zhai Z, Shu H-B (2005) Visa is an adapter protein required for virus-triggered IFN-B signaling. Mol Cell 19:727–740
Xu H, Wu Y, Li L, Yuan W, Zhang D, Yan Q, Guo Z, Huang W (2017) MiR-344b-1-3p targets TLR2 and negatively regulates TLR2 signaling pathway. Int J Chron Obstruct Pulmon Dis 12:627–638
Xu B, Luo Q, Gong Y, Li J, Cao J (2021) TLR7 expression aggravates invasive pulmonary aspergillosis by suppressing anti-aspergillus immunity of macrophages. Infect Immun 89
Yakobov N, Fischer F, Mahmoudi N, Saga Y, Grube CD, Roy H, Senger B, Grob G, Tatematsu S, Yokokawa D, Mouyna I, Latge JP, Nakajima H, Kushiro T, Becker HD (2020) RNA-dependent sterol aspartylation in fungi. Proc Natl Acad Sci USA 117:14948–14957
Yang D, Liang Y, Zhao S, Ding Y, Zhuang Q, Shi Q, Ai T, Wu SQ, Han J (2020) ZBP1 mediates interferon-induced necroptosis. Cell Mol Immunol 17:356–368
Yang C, Wu D, Lin H, Ma D, Fu W, Yao Y, Pan X, Wang S, Zhuang Z (2023) Role of RNA modifications, especially m6A, in aflatoxin biosynthesis of *Aspergillus flavus*. J Agric Food Chem 72:726–741
Yoneyama M, Kikuchi M, Natsukawa T, Shinobu N, Imaizumi T, Miyagishi M, Taira K, Akira S, Fujita T (2004) The RNA helicase RIG-I has an essential function in double-stranded RNA-induced innate antiviral responses. Nat Immunol 5:730–737
Yu AM, Gasper PM, Cheng L, Lai LB, Kaur S, Gopalan V, Chen AA, Lucks JB (2021) Computationally reconstructing cotranscriptional RNA folding from experimental data reveals rearrangement of non-native folding intermediates. Mol Cell 81:870–883 e10
Zhang T, Zhao YL, Zhao JH, Wang S, Jin Y, Chen ZQ, Fang YY, Hua CL, Ding SW, Guo HS (2016) Cotton plants export microRNAs to inhibit virulence gene expression in a fungal pathogen. Nat Plants 2:16153
Zhang X, Xu Q, Zi Z, Liu Z, Wan C, Crisman L, Shen J, Liu X (2020) Programmable extracellular vesicles for macromolecule delivery and genome modifications. Dev Cell 55:784–801 e9
Zhang L, Zhang K, Li H, Coelho C, De Souza Goncalves D, Fu MS, Li X, Nakayasu ES, Kim YM, Liao W, Pan W, Casadevall A (2021) *Cryptococcus neoformans*-infected macrophages release proinflammatory extracellular vesicles: insight into their components by multi-omics. mBio 12
Zhang Y, Wang Y, Fan J, Zhu G, Lu L (2022) *Aspergillus fumigatus* elongator complex subunit 3 affects hyphal growth, adhesion and virulence through wobble uridine tRNA modification. PLoS Pathog 18:e1010976
Zhang Y, Jiang J, Ma J, Wei Z, Wang Y, Song B, Meng J, Jia G, De Magalhães JP, Rigden DJ (2023) Directrmdb: a database of post-transcriptional RNA modifications unveiled from direct RNA sequencing technology. Nucleic Acids Res 51:D106–D116
Zou PF, Chang MX, Li Y, Xue NN, Li JH, Chen SN, Nie P (2016) NOD2 In zebrafish functions in antibacterial and also antiviral responses via NF-κB, and also MDA5, RIG-I And MAVS. Fish Shellfish Immunol 55:173–185

The Human Gut Mycobiome and Its Potential as a Regulator of the Host's Metabolic Health

6

Jiarui Chen, Yao Wang, Aimin Xu, and Gianni Panagiotou

Abstract

The human gut ecosystem contains an enormous number of microorganisms, including bacteria, archaea, viruses, and fungi. Compared to bacteria, which account for the majority of the gut microbiome, the role of gut fungal species in relation to the host's metabolic health has received less attention. Nevertheless, with the help of high-throughput sequencing technologies and large-scale human studies, it is now possible to show the influence of gut fungi on several metabolic diseases. In this review, we summarize the current knowledge on the composition of the human gut mycobiome, the methodological developments for characterizing the fungi in the gut, and we provide insights on the possible roles of gut fungi and their

Jiarui Chen and Yao Wang contributed equally with all other contributors.

J. Chen
Department of Microbiome Dynamics, Leibniz Institute for Natural Product Research and Infection Biology – Hans Knöll Institute, Jena, Germany

State Key Laboratory of Pharmaceutical Biotechnology, The University of Hong Kong, Hong Kong, SAR China

Department of Medicine, The University of Hong Kong, Hong Kong, SAR China

Guangdong Provincial Key Laboratory of Food, Nutrition and Health, and Department of Nutrition, School of Public Health, Sun Yat-sen University, Guangzhou, China

Y. Wang
State Key Laboratory of Pharmaceutical Biotechnology, The University of Hong Kong, Hong Kong, SAR China

Department of Medicine, The University of Hong Kong, Hong Kong, SAR China

A. Xu
State Key Laboratory of Pharmaceutical Biotechnology, The University of Hong Kong, Hong Kong, SAR China

Department of Medicine, The University of Hong Kong, Hong Kong, SAR China

Department of Pharmacology and Pharmacy, The University of Hong Kong, Hong Kong, SAR China

G. Panagiotou (✉)
Department of Microbiome Dynamics, Leibniz Institute for Natural Product Research and Infection Biology – Hans Knöll Institute, Jena, Germany

Department of Medicine, The University of Hong Kong, Hong Kong, SAR China

Department of Pharmacology and Pharmacy, The University of Hong Kong, Hong Kong, SAR China

Cluster of Excellence Balance of the Microverse, Friedrich Schiller University, Jena, Germany

Faculty of Biological Sciences, Friedrich Schiller University, Jena, Germany
e-mail: Gianni.Panagiotou@hki-jena.de

A. A. Brakhage et al. (eds.), *Human and Animal Relationships*, The Mycota 6,
https://doi.org/10.1007/978-3-031-64853-3_6

underlying mechanisms on metabolic homeostasis and health.

Keywords

Human gastrointestinal tract · Mycobiome Fungi · ITS sequencing · Metabolic diseases

6.1 Introduction

As a complex organism, the human body contains not only the human genome obtained by inheritance, encoding over 23,000 genes, but also carries a large number of microbial genes from microorganisms using humans as their host. Especially in the gastrointestinal (GI) tract (Turnbaugh et al. 2007), we find over 1000 microorganisms, including bacteria, archaea, viruses, and fungi, encoding more than one million genes (Qin et al. 2010). To date, emerging studies have focused on the bacteria, the predominant microorganisms in the gut microbiome, revealing their important role in influencing the host immune and metabolic functions by secreting either beneficial or detrimental metabolites, such as short-chain fatty acids, vitamins, and endotoxins (Sanna et al. 2019; Waterhouse et al. 2019).

Consequently, the "rare biosphere" in the human gut ecosystem has stimulated interest among scientists, especially regarding the fungal species (Sogin et al. 2006). Despite the relatively low abundance of gut fungal species compared to the bacterial residents, emerging evidence, propelled by the development of high-throughput sequencing techniques and advanced bioinformatics methods, suggests their essential role in maintaining intestinal homeostasis and systemic health (Coker et al. 2019; Jiang et al. 2017). Importantly, the gut mycobiome can affect the host's metabolic health by not only influencing the composition of the gut bacterial microbiome but also through its own metabolic functions. Recent studies provide the first convincing evidence of the tight connection between the host mycobiome and several metabolic diseases, including obesity (Mar Rodríguez et al. 2015), type 1 diabetes, type 2 diabetes (Salamon et al. 2021) and non-alcoholic fatty liver disease (NAFLD) (Demir et al. 2022), further highlighting its crucial role in host homeostasis and metabolic disease development. Hence, we present in this review the current knowledge on the diversity and composition of the human gut mycobiome, the methodological approaches for human gut mycobiome characterization, and the possible connection of gut fungi with metabolic diseases. Eventually, we hope that the evidence presented here summarizing the roles of fungi on human metabolic health will accelerate research in this field to tackle the many methodological challenges associated with mycobiome characterization.

6.2 The Composition of the Human Gut Mycobiome

The human gut mycobiome, also known as the commensal fungi, constitutes a minor component in terms of absolute numbers of the entire gut microbiome. Studies indicate that the gut mycobiome consists of nearly 0.1% of the total microbes in the gut ecosystem (Qin et al. 2010; Arumugam et al. 2011). The overall diversity of the gut mycobiome is commonly regarded as significantly lower than bacterial diversity as seen in multiple studies, and the composition of the gut mycobiome varies greatly among individuals, suggesting that gut fungi may be more personalized (Nash et al. 2017; Scanlan and Marchesi 2008; Shuai et al. 2022a).

When investigating the detailed composition of the human gut mycobiome, multiple studies have identified a plethora of single taxa present in the human gut mycobiome, including 215 species from culture-independent methodologies and 83 species from culture-dependent methodologies, which is nicely summarized in the review of Huseyin et al. (2017). In most studies, it was suggested that *Ascomycota* is the predominant fungi at the phylum level in the gut ecosystem, followed by *Zygomycota* and *Basidiomycota* (Hallen-Adams and Suhr 2017; Hibbett et al.

2007; Hoffmann et al. 2013; Sam et al. 2017). At the genus level, studies performed by Hallen-Adams et al. with fecal samples collected from 45 healthy humans suggest that the majority of the fungal species that inhabit the intestinal niche belong to the *Candida, Cryptococcus, Malassezia, Aspergillus, Saccharomyces, Galactomyces, Trichosporon*, and *Cladosporium* genera (Hallen-Adams et al. 2015). A study performed by Nash et al. using 317 fecal samples from the Human Microbiome Project (HMP) confirmed that the gut mycobiome predominantly includes *Malassezia*, *Candida*, and *Saccharomyces*, while *S. cerevisiae, M. restricta*, and *C. albicans* were identified in 96.8%, 88.3%, and 80.8% of the individuals, respectively (Nash et al. 2017). More recently, Menglei et al. performed the Internal Transcribed Spacer 2 (ITS2) region sequencing in fecal samples from 1244 healthy individuals at two time points and identified 26 core fungal taxa. Eight of them were commonly detected fungi in previous studies, including *Candida, Saccharomyces, Aspergillus, Malassezia,* etc. (Shuai et al. 2022a). Notably, these studies characterized the long-term stability of the human gut mycobiome and identified temporally stable fungi, providing evidence for the colonization of the core mycobiome in part associated with the human life cycle.

To further understand the colonization of fungi in the human gut, the infant gut became an important target of mycobiome studies. In the first year of life, fungi can already be detected in fecal samples but at a relatively low level (Palmer et al. 2007). In contrast to the increasing diversity in the gut bacterial microbiome, the alpha diversity of the gut mycobiome decreases over an infant's first year of life (Laforest-Lapointe and Arrieta 2017). Throughout the development of the gut microbiome and the interactions between gut bacterial and fungal communities, the gut mycobiome reaches a more stable, mature composition with a relatively lower alpha diversity compared to the bacterial counterpart (Nash et al. 2017; Raimondi et al. 2019). Similar to the gut bacterial microbiome, early life events and environmental exposures influence the colonization and shape the composition of the gut mycobiome. These include breastfeeding, the use of antibiotics, or a prolonged hospital stay (Van Belkum et al. 2020). Studies suggest breastfeeding is the key factor in the development of the infant gut mycobiome, with several fungal species belonging to *Malassezia*, *Davidiella*, *Sistotrema,* and *Penicillium* detected in breast milk (Boix-Amorós et al. 2019). Collectively, the early stages of life appear critical for the formation of the gut mycobiome, similar to the gut microbiome (Fig. 6.1).

Moreover, the fungal composition continues to develop and to be modified by environmental stresses, especially the chronic habitual diet and the interactions with gut bacterial communities (Hoffmann et al. 2013) (Fig. 6.1). However, the exact mechanisms and dynamics on the gut fungal composition while facing environmental stresses will need to be explored in future studies.

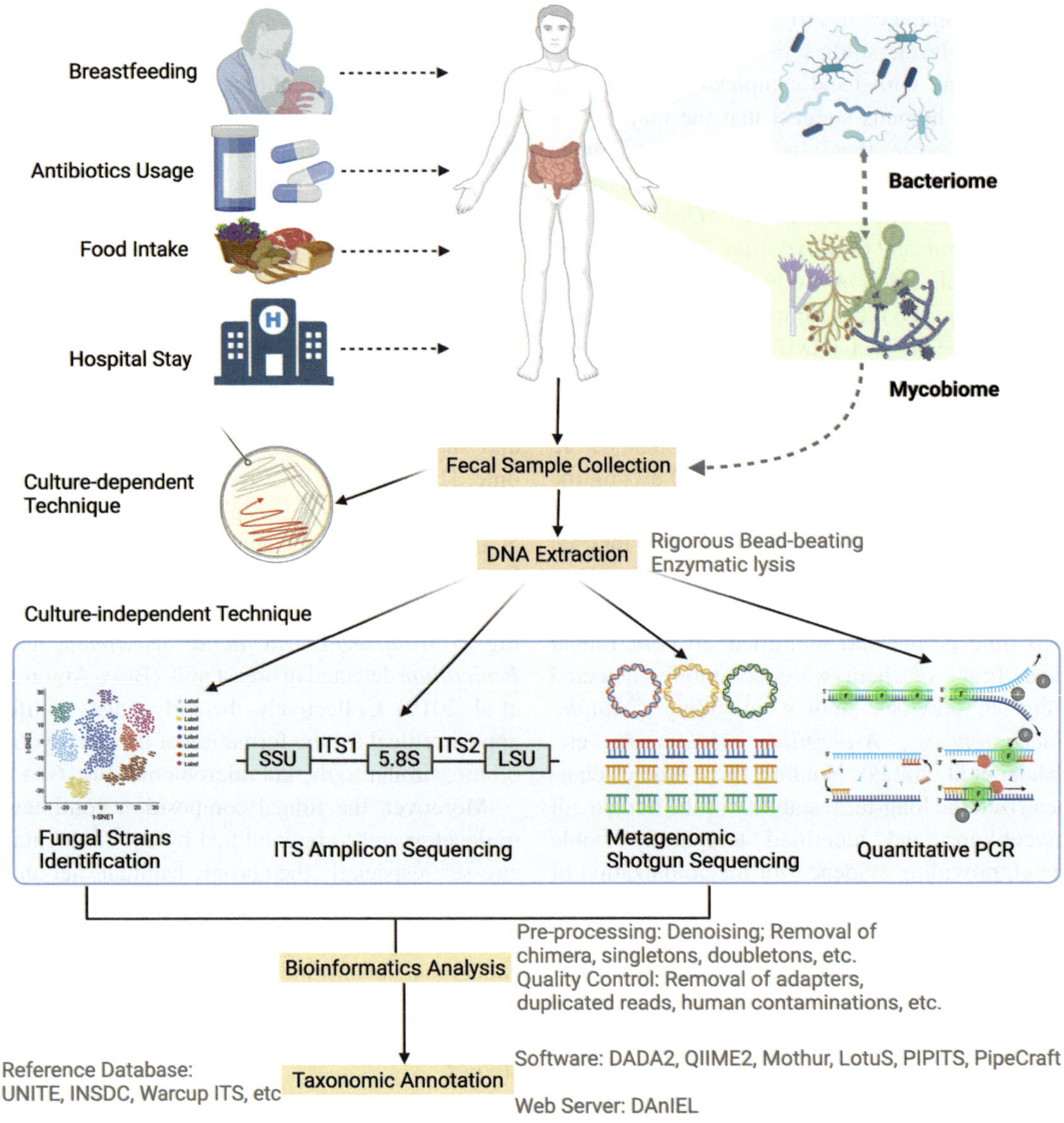

Fig. 6.1 Sources and factors influencing the human gut mycobiome and methodologies for its characterization. The illustrated figure was created using BioRender

6.3 Methodological Developments in Characterizing the Human Gut Mycobiome

6.3.1 Culture-Dependent Methods

The first approaches used to identify and characterize the fungal species in the complex human gut ecosystem included traditional culture-dependent techniques such as microscopy, biochemical assays, and growth on selective media (Finegold et al. 1974; De Hoog et al. 2005; Ouanes et al. 2013). By using complex, specialized media such as Czapek-dox and Dixon's media, more fungal species from the human gut ecosystem could be efficiently cultivated (Gouba et al. 2014). However, the traditional culture-dependent methodologies retain several drawbacks and biases. First of all, culture-dependent techniques are time-consuming since they require lengthy incubation times (Gouba et al. 2013). Furthermore, the growth rates of fast-growing species and non-fastidious species are unbalanced compared to rare fungal species in the culture media since the former may cover the plates more quickly (Griffith et al. 2009). Still, the culture-dependent methods are a basic but important technique that cannot be ignored. This traditional technique can be further combined with other methods, including mass spectrometry (MS), DNA-based analysis of cultures (Gouba et al. 2013), internally transcribed spacer (ITS) sequence analysis (Sauer and Kliem 2010), and the development of differential chromogenic media and semi-diagnostic culturing, to advance the characterization of the human gut mycobiome.

6.3.2 High-Throughput Sequencing Methods

The development of molecular techniques, mainly next-generation sequencing and the expansion of relevant databases, allows for the comprehensive investigation of the human gut mycobiome at a significantly higher resolution (Mardis 2013). Below, we present a brief summary of the major steps and challenges in a typical high-throughput sequencing pipeline for gut mycobiome profiling, including sample processing, DNA extraction, amplicon sequencing, and data analysis (Fig. 6.1).

6.3.2.1 Sample Processing

Sample processing includes both the donor recruitment and fecal sample storage. As the essential first step, the precise inclusion and exclusion criteria of the donor should be defined to align with the requirements of the potential research question. The detailed information of the donor, including lifestyle, overall health state, diet, medical treatments, and especially the use of antibiotics and/or antifungals, should be recorded before and during the study to avoid confounding factors that may introduce bias in the downstream analysis of the gut mycobiome (Seelbinder et al. 2020). After collecting fecal samples from the donors, the samples should be frozen within 12 h in order to prevent the overgrowth of fast-growing fungal species, and multiple freezing cycles should be avoided (Cuthbertson et al. 2015; Nilsson et al. 2019).

6.3.2.2 DNA Extraction

The fungal cell wall is composed of chitin, mannans, glycoproteins, and glucans, which is a different structure compared to the bacterial cell wall (Smits et al. 1999). Therefore, the DNA extraction procedure for the gut mycobiome should be different from the one applied to study the gut bacterial microbiome in order to achieve an efficient lysing of the cell wall and retrieve a qualified DNA yield and quality. Rigorous bead-beating and enzymatic lysis prior to DNA extraction is the commonly used strategy (Angebault et al. 2018; Muñoz-Cadavid et al. 2010). It has been demonstrated that the different DNA extraction methods will significantly affect the fungal DNA yield and quality, but the impact on mycobiome composition and diversity is relatively limited (Rosenbaum et al. 2019). Nevertheless, it is important to choose a consistent DNA extraction strategy that will allow comparative mycobiome analysis from different projects (meta-analysis).

6.3.2.3 Amplicon Sequencing

Amplicon sequencing for the identification of gut fungi depends on targeting the fungal rRNA gene locus. This region has the gene for both the ribosomal small subunit (18S) and the large subunit (26S), and it is separated by the internal transcribed spacer (ITS) regions ITS1 and ITS2, spanning 500–700 bps (Hoggard et al. 2018). To date, human fungi are mainly detected by performing either ITS1 sequencing or ITS2 sequencing. However, ITS1 failed to retrieve sequences for a variety of important fungal species, including *Yarrowia lipolytica* and all three *Malassezia* species, in a mixed community. Moreover, compared to ITS1 sequencing, the ITS2 primers showed a relatively lower bias in identifying several key taxa, suggesting that the ITS2 sequencing is a better option in most cases (Hoggard et al. 2018).

6.3.2.4 Data Analysis

One of the most challenging and nascent topics of mycobiome research is the development of tools and pipelines for analyzing the raw data retrieved from ITS amplicon sequencing (Gabaldón 2019). Starting from the pre-processing of the raw sequencing reads, the steps include filtering of read length, denoising, removal of chimera, singletons, and doubletons, as well as several steps for quality control, including the removal of PCR adaptors, low-quality regions, duplicated reads, and human host contamination (Halwachs et al. 2017). Tools such as Cutadapt and DADA2 were developed for this purpose (Martin 2011; Callahan et al. 2016). Subsequently, the clean reads are clustered into Operational Taxonomic Units (OTUs) by either reference-based or non-reference-based methods. For the reference-based approaches, reads are aligned to fungal reference databases such as UNITE, INSDC, and Warcup ITS (Nilsson et al. 2019). For the non-reference-based strategy, the reads are clustered by *de novo* OTU picking. For the subsequent data analysis, there are several available pipelines, including QIIME2 and mothur, which were designed originally for 16S rRNA analysis (Schloss et al. 2009; Bolyen et al. 2019), and several pipelines generated specifically for ITS sequencing, such as LotuS, PIPITS, and PipeCraft (Gweon et al. 2015; Hildebrand et al. 2014; Anslan et al. 2017). Furthermore, a web server (DAnIEL) was recently developed for integrating all the essential steps of ITS amplicon sequencing analysis as well as the subsequent visualization and statistical analysis (Loos et al. 2021).

6.3.3 Emerging Methods in Mycobiome Analysis

Besides the traditional culture-dependent techniques and ITS amplicon sequencing for studying the human gut mycobiome, emerging methods include single species quantification by qPCR, arrays and microarray assessment, metagenomic shotgun sequencing, and strain level annotation (Fig. 6.1).

6.3.3.1 Quantification

Quantitative PCR (qPCR) can be used to quantify the abundance of specific fungal species based on their DNA markers (Wallander et al. 2013). However, the ability to profile the entire fungal community with qPCR is quite limited compared to the ITS amplicon sequencing because the fungus-specific primers such as *rpb2* primers targeting *Agaricomycetes* for qPCR are often lacking specificity for other groups of fungi and cannot cover all fungal species (Větrovský et al. 2016). Therefore, qPCR is normally used for detecting the abundance of single species with specifically designed primers, e.g., *Arbuscular mycorrhizal* (Voříšková et al. 2017) or *Candida albicans* (Chau et al. 2004). Nevertheless, a qPCR-based abundance estimate is a solid indicator for individual fungal species because the number of rDNA copies remains stable in many species (Amend et al. 2010).

6.3.3.2 Arrays and Microarrays

Furthermore, the hybridization-based array techniques that have been extensively used in the research of clinical microbiology can also be applied for detecting fungi in the human gut. There have been two studies using microarray technologies to assess the abundance of human pathogenic fungi (Leinberger et al. 2005) and *ectomycorrhizal* fungi (Reich et al. 2009). However, despite their ability to retrieve the absolute abundance of fungal species, considering that microarrays can only detect pre-defined fungal species or functions, the use of microarrays is less popular than ITS amplicon sequencing technologies (Zhou et al. 2015).

6.3.3.3 Metagenomic Shotgun Sequencing

The metagenomic shotgun sequencing normally used for detecting the prokaryotic microbiome in the human gut ecosystem can also offer insights into the human gut mycobiome. Compared with ITS amplicon sequencing, the benefit of using metagenomic shotgun sequencing is that it avoids the biases inherent to PCR and primer choice (Logares et al. 2014; Singer et al. 2016). The implementation of metagenomics has become a valuable resource for revisiting the primer sensitivity issue (Klindworth et al. 2013), and it can be used to determine functional profiles by characterizing the prevalence of metabolic genes (Bahram et al. 2018; Sunagawa et al. 2015). However, the shortcoming of performing metagenomics for detecting fungal communities is the limitations of the fungal genome databases compared to bacterial ones, which may lead to a low proportion of sequences reported as fungal. This low abundance of fungal reads is then an additional challenge in the subsequent OTU analysis (Sokol et al. 2017).

6.3.3.4 Fungal Strain Identification

By considering the single nucleotide polymorphisms (SNPs) within genes, the whole-genome re-sequencing alignment to the fungal reference genomes can be used for genotyping fungal species at the strain level (Grünwald et al. 2016; Byrne et al. 2017). The 1000 Fungal Genomes Project provides an important fungal reference genome database for this purpose (Grigoriev et al. 2014). With the help of this database and the implementation of genome-wide association studies, it becomes possible to identify fungal phenotypes in the human gut mycobiome, for example, *Cryptococcus neoformans*, an opportunistic pathogen, as well as the respective virulence factors (Desjardins et al. 2017). Last but not least, combining single-cell technologies and whole-genome amplification can also improve the resolution of fungal species identification to the strain level and even subcellular level (Lin et al. 2014). Based on this, pipelines such as FunOMIC for human mycobiome profiling at strain levels were developed (Xie and Manichanh 2022).

6.4 Impact of the Gut Mycobiome on the Host's Metabolic Health

Despite the fact that the gut mycobiome only comprises less than 0.1% of the totality of microbial cells in the human gut, growing evidence suggests that enteric fungi orchestrate host metabolism homeostasis, playing a fundamental role in the development of obesity and diabetes.

Obesity is a strong risk factor for the onset of type 2 diabetes (T2D). Mar Rodriguez et al. reported that the gut mycobiome of lean individuals was predominated by *Mucor racemosus* and *M. fuscus* compared to obese counterparts (Mar Rodríguez et al. 2015). The first study linking gut fungi and diabetes was reported in 2014, where researchers quantified that *Candida* spp. abundance was significantly higher in fecal samples of patients with diabetes than in non-diabetic individuals by qPCR (Gosiewski et al. 2014). Another study using ITS2 sequencing in the Emirati population identified a shift from *C. albicans* to *C. glabrata* in the patients with T2D (Al Bataineh et al. 2020). Interestingly, a recent study indicates that there is negative feedback between *Candida* spp. and the development of T2D. Diabetic patients are susceptible to *Candida* spp. infection due to the chronic low-grade inflammatory status, which in turn aggravates the inflammation and dysfunction in glycemic control (Rodrigues et al. 2019).

Notably, a number of studies prove that the fungal mycobiome is sensitive to environmental stimuli and could determine the responsiveness of individuals to various interventions. For example, chronic habitual diet is a determinant of the fungal composition. The abundance of *Candida* was positively associated with carbohydrate consumption but inversely associated with protein intake in humans (Hoffmann et al. 2013). Consistently, a recent study also revealed that dairy consumption was negatively correlated with *Candida* abundance in humans (Shuai et al. 2022b). Moreover, intra-kingdom co-abundance of fungal taxa and interkingdom relationships between fungi and bacteria also exhibited diet-specific patterns (Heisel et al. 2017). For instance, when mice were subjected to a standard chow diet, the gut mycobiome was dominated by *Fusarium* and *Alternaria*, whereas *C. albicans* was the most abundant fungi in the high-fat diet (HFD) group (Heisel et al. 2017). Marie et al. found a reduction of fungal populations of *Alternaria*, *Aspergillus,* and *Cladosporium* after a 6-week HFD treatment in C57BL/6 mice, whereas fungal species of *Candida*, *Hanseniaspora,* and *Kurtzmaniella* were increased (van der Merwe et al. 2020). Likewise, low-calorie diet-induced weight loss intervention could effectively restore the abundance of *Mucor* genus in obese patients (Mar Rodriguez et al. 2015). Mice exposed to different diets resulted in persistent differences in fungal communities, which were significantly associated with differential changes in metabolic phenotypes (Mims et al. 2021). Except for dietary, other interventions can significantly impact the human gut mycobiome as well. For instance, Roux-en-Y gastric bypass surgery induced a significant reduction in fungal alpha diversity (Steinert et al. 2020). Therefore, the gut mycobiome is greatly shaped by host states, lifestyle, and interventions, suggesting that in the future it could become a promising target and a therapeutic option for metabolic diseases.

6.5 Gut Mycobiome Metabolic Products as Regulators of Host Metabolism

Like commensal gut bacteria, intestinal fungi are able to produce a wide variety of primary and secondary metabolites, including fatty acids, purines, and amino acids via *de novo* biosynthesis (Gajewski et al. 2017; Chitty and Fraser 2017). Those fungi-derived metabolites can be absorbed and delivered into target organs to affect their function. The absorption of fungal metabolites is mainly tuned by intestinal macrophages, where they are localized in the "balloon-like" protrusions (BLPs) within the epithelium of the distal colon (Chikina et al. 2020). Those BLP-resident macrophages are characterized as CD11c positive and protect epithelial cells by attenuating fungal toxin uptake (Chikina et al. 2020). Of note, fungi, rather than bacteria, specifically stimulate the BLPs formation, suggesting a distinct immunologic mechanism between fungi and bacteria on host metabolism.

Metabolic syndromes such as obesity are characterized by systematic and chronic low-grade inflammation (Ellulu et al. 2017). Prostaglandins, especially prostaglandin E2 (PGE2), are potent immunological regulators that induce inflammation in mammals. Interestingly, studies confirm that *Candida* can directly synthesize PGE2 from arachidonic acid *in vivo* (Noverr et al. 2002). Yun et al. found that overgrowth of fungal *Candida* species is associated with elevated PGE2 levels in humans (Noverr et al. 2002). They further confirmed that Candida-induced PGE2 promotes M2 macrophage polarization, which consequently leads to tissue inflammation. Moreover, other fungi-derived metabolites, such as β-glucans and mannans, also strongly stimulate immune responses in the host. Administration of mannan from *S. cerevisiae* in mice results in the activation of macrophages to trigger IL-17A and TNF-α release, which further exacerbates systematic inflammation (Khmaladze et al. 2014). As for β-glucans, chronic alcohol treatment in mice led to the expansion of the mycobiome population accompanied by an increase in circulating β-glucans concentrations, which further activated hepatic Kupffer cells through C-type lectin-like receptors (CLEC7A) to induce liver inflammation (Yang et al. 2017). In summary, the gut mycobiome could promote the development of metabolic disorders by producing fungi-derived metabolites with immune regulation properties (Fig. 6.2). However, our current understanding of how the gut mycobiome regulates host metabolism is still vague. Further studies are imperative to investigate how the gut microbiome modulates metabolic health at molecular levels.

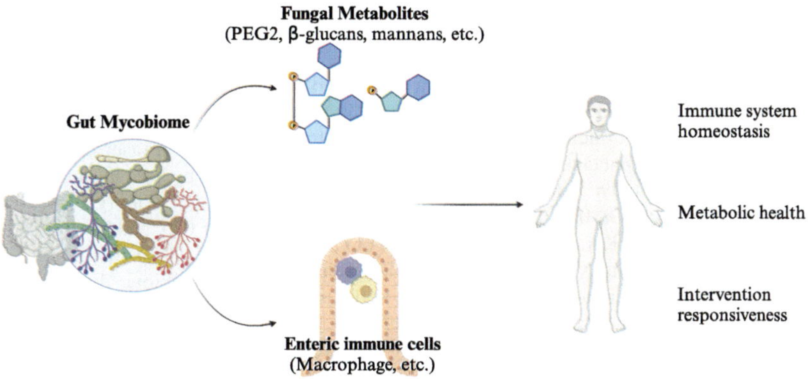

Fig. 6.2 Gut mycobiome regulates host metabolism via functional products. The illustrated figure was created using BioRender

6.6 Future Perspectives

Since growing evidence suggests close associations between intestinal fungi and the development of metabolic diseases, such as obesity, diabetes, and chronic liver disease, there is great interest in the characterization of the gut mycobiome in host health. Unlike bacterial microbiome studies, the specific functions of fungi in host metabolism still need to be validated. Given the fact that fungi only constitute a small part of the whole gut microbiome, it requires single fungal species transplantation to confirm its biological effects. Nevertheless, due to the current limitation in isolating and cultivating gut fungi, only a few studies have attempted to transplant single species into antibiotic-treated mice to investigate their impact *in vivo*. Sun et al. found that anti-fungus treatment effectively suppressed diet-induced obesity, and they identified *Candida parapsilosis* as the most abundant fungal species in obese mice (Sun et al. 2021). They conducted further animal studies to show that intestinal colonization of *C. parapsilosis* was a causal driver in promoting obesity in the fungi-free HFD-fed mice through secretion of lipase, whereas inoculation with lipase-negative *C. parapsilosis* had no significant effect on the development of obesity (Sun et al. 2021). Similar investigations are necessary to provide a cause-and-effect relationship between isolated fungal species and host phenotypes.

Meanwhile, like the bacterial microbiome, there is evidence to demonstrate commensal fungi affect the responsiveness of various interventions. Fecal microbiome transplantation (FMT) is an effective approach for *Clostridioides difficile* infections and inflammatory bowel disease (Quraishi et al. 2017). Leonardi et al. showed that a high abundance of intestinal *Candida* before FMT improves therapeutic responsiveness in ulcerative colitis patients (Leonardi et al. 2020). *Candida* species induced an immune response to produce anti-fungal IgG in CX3CR1^{+} macrophages to protect the host against lethal fungal infections (Doron et al. 2021). Therefore, we reason that the gut mycobiome will also influence the outcomes of other kinds of interventions, such as diet, exercise, surgery, and medication, where the gut microbiome has already been confirmed as a major determinant for therapeutic responsiveness.

Considering the importance of the gut mycobiome in regulating human health, intensifying research in this exciting new frontier is very appealing. In the future, with the development of genomic tools and large-scale omics studies, we could thereby accurately decipher the composition and functions of gut fungi in high resolution, precisely delineate molecular mechanisms of how the gut mycobiome regulates the host immune system and host metabolism, and pave the way for applying the mycobiome as a promising therapeutic approach for health management.

Funding G. P. would like to thank the Federal Ministry for Education and Research (BMBF: https://www.bmbf.de/), Germany, "PerMiCCion" project (Project ID 01KD2101A) for their support.

References

Al Bataineh MT, Dash NR, Bel Lassen P, Banimfreg BH, Nada AM, Belda E, Clement K (2020) Revealing links between gut microbiome and its fungal community in Type 2 Diabetes Mellitus among Emirati subjects: A pilot study. Sci Rep 10(1):9624. https://doi.org/10.1038/s41598-020-66598-2

Amend AS, Seifert KA, Samson R, Bruns TD (2010) Indoor fungal composition is geographically patterned and more diverse in temperate zones than in the tropics. Proc Natl Acad Sci 107(31):13748–13753

Angebault C, Ghozlane A, Volant S, Botterel F, d'Enfert C, Bougnoux M-E (2018) Combined bacterial and fungal intestinal microbiota analyses: Impact of storage conditions and DNA extraction protocols. PLoS One 13(8):e0201174

Anslan S, Bahram M, Hiiesalu I, Tedersoo L (2017) PipeCraft: flexible open-source toolkit for bioinformatics analysis of custom high-throughput amplicon sequencing data. Mol Ecol Resour 17(6):e234–e240

Arumugam M, Raes J, Pelletier E, Le Paslier D, Yamada T, Mende DR, Fernandes GR, Tap J, Bruls T, Batto J-M (2011) Enterotypes of the human gut microbiome. Nature 473(7346):174–180

Bahram M, Hildebrand F, Forslund SK, Anderson JL, Soudzilovskaia NA, Bodegom PM, Bengtsson-Palme J, Anslan S, Coelho LP, Harend H (2018) Structure and function of the global topsoil microbiome. Nature 560(7717):233–237

Boix-Amorós A, Puente-Sánchez F, Du Toit E, Linderborg KM, Zhang Y, Yang B, Salminen S, Isolauri E, Tamames J, Mira A (2019) Mycobiome profiles in breast milk from healthy women depend on mode of delivery, geographic location, and interaction with bacteria. Appl Environ Microbiol 85(9):e02994–e02918

Bolyen E, Rideout JR, Dillon MR, Bokulich NA, Abnet CC, Al-Ghalith GA, Alexander H, Alm EJ, Arumugam M, Asnicar F (2019) Reproducible, interactive, scalable and extensible microbiome data science using QIIME 2. Nat Biotechnol 37(8):852–857

Byrne AQ, Rothstein AP, Poorten TJ, Erens J, Settles ML, Rosenblum EB (2017) Unlocking the story in the swab: a new genotyping assay for the amphibian chytrid fungus Batrachochytrium dendrobatidis. Mol Ecol Resour 17(6):1283–1292

Callahan BJ, McMurdie PJ, Rosen MJ, Han AW, Johnson AJA, Holmes SP (2016) DADA2: high-resolution sample inference from Illumina amplicon data. Nat Methods 13(7):581–583

Chau AS, Mendrick CA, Sabatelli FJ, Loebenberg D, McNicholas PM (2004) Application of real-time quantitative PCR to molecular analysis of Candida albicans strains exhibiting reduced susceptibility to azoles. Antimicrob Agents Chemother 48(6):2124–2131

Chikina AS, Nadalin F, Maurin M, San-Roman M, Thomas-Bonafos T, Li XV, Lameiras S, Baulande S, Henri S, Malissen B, Lacerda Mariano L, Barbazan J, Blander JM, Iliev ID, Matic Vignjevic D, Lennon-Dumenil AM (2020) Macrophages maintain epithelium integrity by limiting fungal product absorption. Cell 183(2):411–428 e416. https://doi.org/10.1016/j.cell.2020.08.048

Chitty JL, Fraser JA (2017) Purine acquisition and synthesis by human fungal pathogens. Microorganisms 5(2). https://doi.org/10.3390/microorganisms5020033

Coker OO, Nakatsu G, Dai RZ, Wu WKK, Wong SH, Ng SC, Chan FKL, Sung JJY, Yu J (2019) Enteric fungal microbiota dysbiosis and ecological alterations in colorectal cancer. Gut 68(4):654–662

Cuthbertson L, Rogers GB, Walker AW, Oliver A, Hoffman LR, Carroll MP, Parkhill J, Bruce KD, van der Gast CJ (2015) Implications of multiple freeze-thawing on respiratory samples for culture-independent analyses. J Cyst Fibros 14(4):464–467

De Hoog G, Matos T, Sudhadham M, Luijsterburg K, Haase G (2005) Intestinal prevalence of the neurotropic black yeast Exophiala (Wangiella) dermatitidis in healthy and impaired individuals. Mycoses 48(2):142–145

Demir M, Lang S, Hartmann P, Duan Y, Martin A, Miyamoto Y, Bondareva M, Zhang X, Wang Y, Kasper P (2022) The fecal mycobiome in non-alcoholic fatty liver disease. J Hepatol 76(4):788–799

Desjardins CA, Giamberardino C, Sykes SM, Yu C-H, Tenor JL, Chen Y, Yang T, Jones AM, Sun S, Haverkamp MR (2017) Population genomics and the evolution of virulence in the fungal pathogen Cryptococcus neoformans. Genome Res 27(7):1207–1219

Doron I, Leonardi I, Li XV, Fiers WD, Semon A, Bialt-DeCelie M, Migaud M, Gao IH, Lin WY, Kusakabe T, Puel A, Iliev ID (2021) Human gut mycobiota tune immunity via CARD9-dependent induction of anti-fungal IgG antibodies. Cell 184(4):1017–1031. e1014. https://doi.org/10.1016/j.cell.2021.01.016

Ellulu MS, Patimah I, Khaza'ai H, Rahmat A, Abed Y (2017) Obesity and inflammation: the linking mechanism and the complications. Arch Med Sci 13(4):851–863

Finegold SM, Attebery HR, Sutter VL (1974) Effect of diet on human fecal flora: comparison of Japanese and American diets. Am J Clin Nutr 27(12):1456–1469

Gabaldón T (2019) Recent trends in molecular diagnostics of yeast infections: from PCR to NGS. FEMS Microbiol Rev 43(5):517–547

Gajewski J, Pavlovic R, Fischer M, Boles E, Grininger M (2017) Engineering fungal de novo fatty acid synthesis for short chain fatty acid production. Nat Commun 8:14650. https://doi.org/10.1038/ncomms14650

Gosiewski T, Salamon D, Szopa M, Sroka A, Malecki MT, Bulanda M (2014) Quantitative evaluation of fungi of the genus Candida in the feces of adult patients with type 1 and 2 diabetes - a pilot study. Gut Pathog 6(1):43. https://doi.org/10.1186/s13099-014-0043-z

Gouba N, Raoult D, Drancourt M (2013) Plant and fungal diversity in gut microbiota as revealed by molecular and culture investigations. PLoS One 8(3):e59474

Gouba N, Raoult D, Drancourt M (2014) Eukaryote culturomics of the gut reveals new species. PLoS One 9(9):e106994

Griffith GW, Ozkose E, Theodorou MK, Davies D (2009) Diversity of anaerobic fungal populations in cattle revealed by selective enrichment culture using different carbon sources. Fungal Ecol 2(2):87–97

Grigoriev IV, Nikitin R, Haridas S, Kuo A, Ohm R, Otillar R, Riley R, Salamov A, Zhao X, Korzeniewski F (2014) MycoCosm portal: gearing up for 1000 fungal genomes. Nucleic Acids Res 42(D1):D699–D704

Grünwald NJ, McDonald BA, Milgroom MG (2016) Population genomics of fungal and oomycete pathogens. Annu Rev Phytopathol 54:323–346

Gweon HS, Oliver A, Taylor J, Booth T, Gibbs M, Read DS, Griffiths RI, Schonrogge K (2015) PIPITS: an automated pipeline for analyses of fungal internal transcribed spacer sequences from the I llumina sequencing platform. Methods Ecol Evol 6(8):973–980

Hallen-Adams HE, Suhr MJ (2017) Fungi in the healthy human gastrointestinal tract. Virulence 8(3):352–358

Hallen-Adams HE, Kachman SD, Kim J, Legge RM, Martínez I (2015) Fungi inhabiting the healthy human gastrointestinal tract: a diverse and dynamic community. Fungal Ecol 15:9–17

Halwachs B, Madhusudhan N, Krause R, Nilsson RH, Moissl-Eichinger C, Högenauer C, Thallinger GG, Gorkiewicz G (2017) Critical issues in mycobiota analysis. Front Microbiol 8:180

Heisel T, Montassier E, Johnson A, Al-Ghalith G, Lin YW, Wei LN, Knights D, Gale CA (2017) High-fat diet changes fungal microbiomes and interkingdom relationships in the murine gut. mSphere 2(5). https://doi.org/10.1128/mSphere.00351-17

Hibbett DS, Binder M, Bischoff JF, Blackwell M, Cannon PF, Eriksson OE, Huhndorf S, James T, Kirk PM, Lücking R (2007) A higher-level phylogenetic classification of the Fungi. Mycol Res 111(5):509–547

Hildebrand F, Tadeo R, Voigt AY, Bork P, Raes J (2014) LotuS: an efficient and user-friendly OTU processing pipeline. Microbiome 2(1):1–7

Hoffmann C, Dollive S, Grunberg S, Chen J, Li H, Wu GD, Lewis JD, Bushman FD (2013) Archaea and fungi of the human gut microbiome: correlations with diet and bacterial residents. PLoS One 8(6):e66019. https://doi.org/10.1371/journal.pone.0066019

Hoggard M, Vesty A, Wong G, Montgomery JM, Fourie C, Douglas RG, Biswas K, Taylor MW (2018) Characterizing the human mycobiota: a comparison of small subunit rRNA, ITS1, ITS2, and large subunit rRNA genomic targets. Front Microbiol 9:2208

Huseyin CE, O'Toole PW, Cotter PD, Scanlan PD (2017) Forgotten fungi—the gut mycobiome in human health and disease. FEMS Microbiol Rev 41(4):479–511

Jiang TT, Shao T-Y, Ang WG, Kinder JM, Turner LH, Pham G, Whitt J, Alenghat T, Way SS (2017) Commensal fungi recapitulate the protective benefits of intestinal bacteria. Cell Host Microbe 22(6):809–816.e804

Khmaladze I, Kelkka T, Guerard S, Wing K, Pizzolla A, Saxena A, Lundqvist K, Holmdahl M, Nandakumar KS, Holmdahl R (2014) Mannan induces ROS-regulated, IL-17A-dependent psoriasis arthritis-like disease in mice. Proc Natl Acad Sci USA 111(35):E3669–E3678. https://doi.org/10.1073/pnas.1405798111

Klindworth A, Pruesse E, Schweer T, Peplies J, Quast C, Horn M, Glöckner FO (2013) Evaluation of general 16S ribosomal RNA gene PCR primers for classical and next-generation sequencing-based diversity studies. Nucleic Acids Res 41(1):e1–e1

Laforest-Lapointe I, Arrieta M-C (2017) Patterns of early-life gut microbial colonization during human immune development: an ecological perspective. Front Immunol 8:788

Leinberger DM, Schumacher U, Autenrieth IB, Bachmann TT (2005) Development of a DNA microarray for detection and identification of fungal pathogens involved in invasive mycoses. J Clin Microbiol 43(10):4943–4953

Leonardi I, Paramsothy S, Doron I, Semon A, Kaakoush NO, Clemente JC, Faith JJ, Borody TJ, Mitchell HM, Colombel JF, Kamm MA, Iliev ID (2020) Fungal trans-kingdom dynamics linked to responsiveness to fecal microbiota transplantation (FMT) therapy in ulcerative colitis. Cell Host Microbe 27(5):823–829 e823. https://doi.org/10.1016/j.chom.2020.03.006

Lin K, Limpens E, Zhang Z, Ivanov S, Saunders DG, Mu D, Pang E, Cao H, Cha H, Lin T (2014) Single nucleus genome sequencing reveals high similarity among nuclei of an endomycorrhizal fungus. PLoS Genet 10(1):e1004078

Logares R, Sunagawa S, Salazar G, Cornejo-Castillo FM, Ferrera I, Sarmento H, Hingamp P, Ogata H, de Vargas C, Lima-Mendez G (2014) Metagenomic 16S rDNA I llumina tags are a powerful alternative to amplicon sequencing to explore diversity and structure of microbial communities. Environ Microbiol 16(9):2659–2671

Loos D, Zhang L, Beemelmanns C, Kurzai O, Panagiotou G (2021) DAnIEL: a user-friendly web server for fungal its amplicon sequencing data. Front Microbiol:2337

Mar Rodríguez M, Pérez D, Javier Chaves F, Esteve E, Marin-Garcia P, Xifra G, Vendrell J, Jové M, Pamplona R, Ricart W (2015) Obesity changes the human gut mycobiome. Sci Rep 5(1):1–15

Mar Rodriguez M, Perez D, Javier Chaves F, Esteve E, Marin-Garcia P, Xifra G, Vendrell J, Jove M, Pamplona R, Ricart W, Portero-Otin M, Chacon MR, Fernandez Real JM (2015) Obesity changes the human gut mycobiome. Sci Rep 5:14600. https://doi.org/10.1038/srep14600

Mardis ER (2013) Next-generation sequencing platforms. Annu Rev Anal Chem 6:287–303

Martin M (2011) Cutadapt removes adapter sequences from high-throughput sequencing reads. EMBnet J 17(1):10–12

Mims TS, Abdallah QA, Stewart JD, Watts SP, White CT, Rousselle TV, Gosain A, Bajwa A, Han JC, Willis KA, Pierre JF (2021) The gut mycobiome of healthy mice is shaped by the environment and correlates with metabolic outcomes in response to diet. Commun Biol 4(1):281. https://doi.org/10.1038/s42003-021-01820-z

Muñoz-Cadavid C, Rudd S, Zaki S, Patel M, Moser S, Brandt M, Gómez B (2010) Improving molecular detection of fungal DNA in formalin-fixed paraffin-embedded tissues: comparison of five tissue DNA extraction methods using panfungal PCR. J Clin Microbiol 48(6):2147–2153

Nash AK, Auchtung TA, Wong MC, Smith DP, Gesell JR, Ross MC, Stewart CJ, Metcalf GA, Muzny DM, Gibbs RA (2017) The gut mycobiome of the Human Microbiome Project healthy cohort. Microbiome 5(1):1–13

Nilsson RH, Anslan S, Bahram M, Wurzbacher C, Baldrian P, Tedersoo L (2019) Mycobiome diversity: high-throughput sequencing and identification of fungi. Nat Rev Microbiol 17(2):95–109

Noverr MC, Toews GB, Huffnagle GB (2002) Production of prostaglandins and leukotrienes by pathogenic fungi. Infect Immun 70(1):400–402. https://doi.org/10.1128/IAI.70.1.400-402.2002

Ouanes A, Kouais A, Marouen S, Sahnoun M, Jemli B, Gargouri S (2013) Apport du milieu chromogène CHROMagar® Candida dans le diagnostic mycologique des levures. J Mycol Médicale 23(4):237–241

Palmer C, Bik EM, DiGiulio DB, Relman DA, Brown PO (2007) Development of the human infant intestinal microbiota. PLoS Biol 5(7):e177

Qin J, Li R, Raes J, Arumugam M, Burgdorf KS, Manichanh C, Nielsen T, Pons N, Levenez F, Yamada T (2010) A human gut microbial gene catalogue established by metagenomic sequencing. Nature 464(7285):59–65

Quraishi MN, Widlak M, Bhala N, Moore D, Price M, Sharma N, Iqbal TH (2017) Systematic review with meta-analysis: the efficacy of faecal microbiota transplantation for the treatment of recurrent and refractory Clostridium difficile infection. Aliment Pharmacol Ther 46(5):479–493. https://doi.org/10.1111/apt.14201

Raimondi S, Amaretti A, Gozzoli C, Simone M, Righini L, Candeliere F, Brun P, Ardizzoni A, Colombari B, Paulone S (2019) Longitudinal survey of fungi in the human gut: ITS profiling, phenotyping, and colonization. Front Microbiol:1575

Reich M, Kohler A, Martin F, Buee M (2009) Development and validation of an oligonucleotide microarray to characterise ectomycorrhizal fungal communities. BMC Microbiol 9(1):1–11

Rodrigues CF, Rodrigues ME, Henriques M (2019) Candida sp. infections in patients with diabetes mellitus. J Clin Med 8(1). https://doi.org/10.3390/jcm8010076

Rosenbaum J, Usyk M, Chen Z, Zolnik CP, Jones HE, Waldron L, Dowd JB, Thorpe LE, Burk RD (2019) Evaluation of oral cavity DNA extraction methods on bacterial and fungal microbiota. Sci Rep 9(1):1–10

Salamon D, Sroka-Oleksiak A, Gurgul A, Arent Z, Szopa M, Bulanda M, Małecki MT, Gosiewski T (2021) Analysis of the gut mycobiome in adult patients with type 1 and type 2 diabetes using next-generation sequencing (ngs) with increased sensitivity—pilot study. Nutrients 13(4):1066

Sam QH, Chang MW, Chai LYA (2017) The fungal mycobiome and its interaction with gut bacteria in the host. Int J Mol Sci 18(2):330

Sanna S, van Zuydam NR, Mahajan A, Kurilshikov A, Vich Vila A, Võsa U, Mujagic Z, Masclee AA, Jonkers DM, Oosting M (2019) Causal relationships among the gut microbiome, short-chain fatty acids and metabolic diseases. Nat Genet 51(4):600–605

Sauer S, Kliem M (2010) Mass spectrometry tools for the classification and identification of bacteria. Nat Rev Microbiol 8(1):74–82

Scanlan PD, Marchesi JR (2008) Micro-eukaryotic diversity of the human distal gut microbiota: qualitative assessment using culture-dependent and-independent analysis of faeces. ISME J 2(12):1183–1193

Schloss PD, Westcott SL, Ryabin T, Hall JR, Hartmann M, Hollister EB, Lesniewski RA, Oakley BB, Parks DH, Robinson CJ (2009) Introducing mothur: open-source, platform-independent, community-supported software for describing and comparing microbial communities. Appl Environ Microbiol 75(23):7537–7541

Seelbinder B, Chen J, Brunke S, Vazquez-Uribe R, Santhaman R, Meyer A-C, de Oliveira Lino FS, Chan K-F, Loos D, Imamovic L (2020) Antibiotics create a shift from mutualism to competition in human gut communities with a longer-lasting impact on fungi than bacteria. Microbiome 8(1):1–20

Shuai M, Fu Y, H-l Z, Gou W, Jiang Z, Liang Y, Miao Z, Xu J-J, Huynh T, Wahlqvist ML (2022a) Mapping the human gut mycobiome in middle-aged and elderly adults: multiomics insights and implications for host metabolic health. Gut

Shuai M, Fu Y, Zhong HL, Gou W, Jiang Z, Liang Y, Miao Z, Xu JJ, Huynh T, Wahlqvist ML, Chen YM, Zheng JS (2022b) Mapping the human gut mycobiome in middle-aged and elderly adults: multiomics insights and implications for host metabolic health. Gut. https://doi.org/10.1136/gutjnl-2021-326298

Singer E, Bushnell B, Coleman-Derr D, Bowman B, Bowers RM, Levy A, Gies EA, Cheng J-F, Copeland A, Klenk H-P (2016) High-resolution phylogenetic microbial community profiling. ISME J 10(8):2020–2032

Smits GJ, Kapteyn JC, van den Ende H, Klis FM (1999) Cell wall dynamics in yeast. Curr Opin Microbiol 2(4):348–352

Sogin ML, Morrison HG, Huber JA, Welch DM, Huse SM, Neal PR, Arrieta JM, Herndl GJ (2006) Microbial diversity in the deep sea and the underexplored “rare biosphere”. Proc Natl Acad Sci U S A 103(32):12115–12120

Sokol H, Leducq V, Aschard H, Pham H-P, Jegou S, Landman C, Cohen D, Liguori G, Bourrier A, Nion-

Larmurier I (2017) Fungal microbiota dysbiosis in IBD. Gut 66(6):1039–1048

Steinert RE, Rehman A, Souto Lima EJ, Agamennone V, Schuren FHJ, Gero D, Schreiner P, Vonlanthen R, Ismaeil A, Tzafos S, Hosa H, Vetter D, Misselwitz B, Bueter M (2020) Roux-en-Y gastric bypass surgery changes fungal and bacterial microbiota in morbidly obese patients-a pilot study. PLoS One 15(7):e0236936. https://doi.org/10.1371/journal.pone.0236936

Sun S, Sun L, Wang K, Qiao S, Zhao X, Hu X, Chen W, Zhang S, Li H, Dai H, Liu H (2021) The gut commensal fungus, Candida parapsilosis, promotes high fat-diet induced obesity in mice. Commun Biol 4(1):1220. https://doi.org/10.1038/s42003-021-02753-3

Sunagawa S, Coelho LP, Chaffron S, Kultima JR, Labadie K, Salazar G, Djahanschiri B, Zeller G, Mende DR, Alberti A (2015) Structure and function of the global ocean microbiome. Science 348(6237):1261359

Turnbaugh PJ, Ley RE, Hamady M, Fraser-Liggett CM, Knight R, Gordon JI (2007) The human microbiome project. Nature 449(7164):804–810

Van Belkum M, Mendoza Alvarez L, Neu J (2020) Preterm neonatal immunology at the intestinal interface. Cell Mol Life Sci 77(7):1209–1227

van der Merwe M, Sharma S, Caldwell JL, Smith NJ, Gomes CK, Bloomer RJ, Buddington RK, Pierre JF (2020) Time of feeding alters obesity-associated parameters and gut bacterial communities, but not fungal populations, in C57BL/6 male mice. Curr Dev Nutr 4(2):nzz145. https://doi.org/10.1093/cdn/nzz145

Větrovský T, Kolařík M, Žifčáková L, Zelenka T, Baldrian P (2016) The rpb2 gene represents a viable alternative molecular marker for the analysis of environmental fungal communities. Mol Ecol Resour 16(2):388–401

Voříšková A, Jansa J, Püschel D, Krüger M, Cajthaml T, Vosátka M, Janoušková M (2017) Real-time PCR quantification of arbuscular mycorrhizal fungi: does the use of nuclear or mitochondrial markers make a difference? Mycorrhiza 27(6):577–585

Wallander H, Ekblad A, Godbold D, Johnson D, Bahr A, Baldrian P, Björk R, Kieliszewska-Rokicka B, Kjøller R, Kraigher H (2013) Evaluation of methods to estimate production, biomass and turnover of ectomycorrhizal mycelium in forests soils–a review. Soil Biol Biochem 57:1034–1047

Waterhouse M, Hope B, Krause L, Morrison M, Protani MM, Zakrzewski M, Neale RE (2019) Vitamin D and the gut microbiome: a systematic review of in vivo studies. Eur J Nutr 58(7):2895–2910

Xie Z, Manichanh C (2022) FunOMIC: pipeline with built-in fungal taxonomic and functional databases for human mycobiome profiling. Comput Struct Biotechnol J

Yang AM, Inamine T, Hochrath K, Chen P, Wang L, Llorente C, Bluemel S, Hartmann P, Xu J, Koyama Y, Kisseleva T, Torralba MG, Moncera K, Beeri K, Chen CS, Freese K, Hellerbrand C, Lee SM, Hoffman HM, Mehal WZ, Garcia-Tsao G, Mutlu EA, Keshavarzian A, Brown GD, Ho SB, Bataller R, Starkel P, Fouts DE, Schnabl B (2017) Intestinal fungi contribute to development of alcoholic liver disease. J Clin Invest 127(7):2829–2841. https://doi.org/10.1172/JCI90562

Zhou J, He Z, Yang Y, Deng Y, Tringe SG, Alvarez-Cohen L (2015) High-throughput metagenomic technologies for complex microbial community analysis: open and closed formats. MBio 6(1):e02288–e02214

The Host's Innate Immune Response to Pathogenic *Candida albicans* and Other Fungal Pathogens

7

Peter F. Zipfel and Christine Skerka

Abstract

Pathogenic fungi such as *Candida albicans* and *Aspergillus fumigatus*, like many other microbial pathogens, evade killing by the host complement system by controlling the action of complement and by exploiting host regulators and inhibitors. The human complement system is the first line of defense against fungal and microbial pathogens. This system can discriminate between self and non-self. Once activated, it is highly efficient in killing and eliminating microbes. Many fungal complement evasion proteins are moonlighting proteins and are involved in basic metabolic reactions such as glycolysis. These fungal moonlighting proteins have metabolic functions in the cytoplasm. However, as cell surfaces are exposed and as secreted proteins, they display complement and immune evasion functions. Here, we review the multiple and often redundant functions of fungal moonlighting proteins from *Candida albicans* and *Aspergillus fumigatus* in acquiring human soluble immune regulators and interacting with cellular receptors. We also describe the role of microRNAs and macrovesicles in the immune battle of fungal pathogens with their human host. The identification and functional characterization of fungal immune evasion proteins is the basis for the identification of new candidates for immune targeting.

Keywords

Candida albicans · Fungal pathogens · Host immune response · Immune complement evasion · Factor H binding · Plasminogen binding

7.1 Introduction

Humans possess an elaborate and highly sophisticated immune system that eliminates infectious microbes, and maintains homeostasis and cellular integrity. However, pathogenic microbes pose a particular challenge. For example, the pathogenic fungus *Candida albicans* controls and blocks virtually every step of an immune attack. To increase our understanding of these processes, it is necessary to identify all of the players (molecules, proteins, fungal virulence factors, and

P. F. Zipfel (✉)
Department of Infection Biology, Leibniz Institute for Natural Product Research and Infection Biology, Hans-Knöll-Institute, Jena, Germany

Institute of Microbiology, Friedrich Schiller University, Jena, Germany
e-mail: peter.zipfel@leibniz-hki.de

C. Skerka
Department of Infection Biology, Leibniz Institute for Natural Product Research and Infection Biology, Hans-Knöll-Institute, Jena, Germany

A. A. Brakhage et al. (eds.), *Human and Animal Relationships*, The Mycota 6,
https://doi.org/10.1007/978-3-031-64853-3_7

immune evasion proteins). The characterization of such molecules from the pathogenic fungus and also from the host allows a deeper understanding of this immune battle and identifies central elements for immune intervention. Such approaches show further that fungi and many, likely all pathogenic microbes, including pathogenic bacteria, viruses, and multicellular parasites, use related and common features for immune evasion. A deep understanding of the mechanisms involved in the "immune battle," as well as identification of molecules and proteins, including microvesicles and RNAs, will provide new therapeutic targets including vaccines, that enable disruption/blockade of fungal propagation in an immunocompetent human host.

7.2 Host Immune Defense with a Focus on Complement

The human host possesses a highly sophisticated immune system to control infectious microbes and to maintain homeostasis and cellular integrity. This immune system comprises two major arms. The innate arm mounts an immediate response that integrates the complement and coagulation cascades with the actions of immune cells such as granulocytes, monocytes, and macrophages. The adaptive arm comprises mainly B lymphocytes (which secrete antigen-specific immunoglobulins) and T lymphocytes (which express antigen-specific T cell receptors) (Janeway 1992; Janeway and Medzhitov 2002).

An invading microbe first encounters the innate immune system, which elicits a toxic response. The adaptive immune system upon first contact takes more time to mount a response, usually 1–2 weeks, but this response is targeted and specific for antigens expressed by the infectious agent or the pathogen.

The complement system, a central component of innate immunity, was discovered and named by the German Nobel laureate Paul Ehrlich om 1901 (Zipfel and Skerka 2009; Zipfel and Skerka 2022). This cascade system, which is activated within seconds of exposure to an infectious agent or a pathogen, progresses explosively to generate toxic compounds that bind to the invader and cause damage. Innate immune cells normally require several hours to migrate to the site of infection, where they become activated, degranulate, secrete cytokines, chemokines, and other immunomodulatory compounds. This innate immune system is evolutionarily older than the adaptive system; thus, the adaptive immune system evolved on top of an already existing innate immune system. Consequently, the two systems interact closely, and there is much cross-talk to foster close cooperation during an immune response.

In this review, we focus on the innate immune response of the human host, particularly the complement system, which is an immediately acting system, that generates highly toxic effector molecules. As mentioned above, exposure to an infectious agent, or pathogenic microbe does activate the complement system within seconds, normally resulting in elimination of the offender within 20 min. However, the human pathogenic fungus *C. albicans* (similar to other pathogenic fungi and microbes) can survive for a long time in an immunocompetent host. This survival is promoted by a battery of fungal evasion proteins, or virulence factors, that appear to control, inactivate, and/or block the complement cascade and subsequent host immune responses (Zipfel et al. 2011; Luo et al. 2013a, b).

The complement system is an efficient and central component of the human immune system; as such, it plays an essential role in immune homeostasis. Its major feature is recognition and elimination of infectious microbes, as well as damaged or modified-self cells. Once these are eliminated, repair mechanisms can be initiated. Complement is activated by one of three pathways: the alternative pathway, the lectin pathway, or the classical pathway. The *alternative pathway* is continuously active, idling at a low level; its level of activation is controlled by inhibitors and regulators. The *lectin pathway* is initiated by mannan components on surfaces, and the *classical pathway* is triggered by human immunoglobulins IgG1, IgG3, and IgM. All three pathways generate C3 convertases, which act on the plasma protein C3. The C3 convertase cleaves C3 and generates C3a and C3b. C3a is an anaphylatoxin that mediates inflammation, cell recruitment, and cell activation, as well as acting as an antifungal and antimicrobial peptide (Coulthard and Woodruff 2015). C3a binds to the C3a receptor (C3aR) expressed by human monocytes, granulocytes, mast cells, dendritic cells, and various T cell subsets. Binding of C3a initiates inflammatory responses, granule release, oxygen radical generation, chemotaxis, histamine release, and smooth muscle cell responses (Valand et al. 2022; Sacks 2010); Cohen et al. 2022.

Newly formed C3b deposits on surfaces. The combined action of circulating, soluble, and membrane-bound regulators directs the cascade, leading to the generation and deposition of cell damaging products onto foreign surfaces and also determines each step of cascade progression. The regulators ensure that the cascade progresses specifically and that the products are directed only to non-self and modified-self surfaces. The alternative pathway triggers a powerful amplification loop that results in massive conversion of C3 into deposition of C3b, a process known as opsonization. Surface-deposited C3b can be processed further to yield inactive C3b, i.e., iC3b, C3dg, and C3d. C3b, along with C3b-derived activation fragments, is recognized by target cells via specific cell receptors, including CR1, CR2, CR3, CR4, and CRIg. The receptors expressed on specific immune cells fulfil targeted biological roles.

Thus, the density and type of C3b product influences, directs, and shapes the immune response. The cascade can also progress further to generate C5 convertases, which cleave the second central complement plasma protein C5 into C5a and C5b. C5a is a powerful anaphylatoxin and binds to two receptors C5aR1 and C5aR2. C5b binds to surfaces and is the focus for assembly of components C6, C7, and C8, along with multiple forms of C9, to form a lytic complex C5b-9 (called the terminal complement complex (TCC) or membrane attack complex [MAC]). The MAC forms a lytic pore that causes target cell lysis. Soluble (s)C5b-9 also has immune activation and inflammatory roles (Dühring et al. 2015; Lukasser-Vogl et al. 2000; Triebel et al. 2003).

7.3 Fungal Complement Inhibition and Immune Evasion

Upon infection of a vertebrate organism, *C. albicans* and other infectious microbial pathogens face a hostile immune system (Luo et al. 2009; Zipfel et al. 2013). An infectious microbe that cannot defend itself against this attack is damaged and killed (Hatinguais et al. 2020; Kong and Jabra-Rizk 2015; Harpf et al. 2020; Rapala-Kozik et al. 2018). However, *C. albicans* and other pathogenic microbes manage to survive and establish an infection. Thus, this pathogen has developed powerful mechanisms that inactivate and/or block the host immune response (Dühring et al. 2015; Erwig and Gow 2016). Given the complexity of the host immune response, pathogenic microbes have developed multiple methods of evasion (Jiménez-López and Lorenz 2013).

Here, we focus on the important human pathogenic fungus *C. albicans*. *C. albicans* is an opportunistic diploid, polymorphic fungal pathogen that can exist as yeast cells, hyphae, or pseudo hyphae (Fig. 7.1). Similar to many other pathogenic microbes, *Candida* can exist in a commensal state; however, the fungus can also cause invasive candidiasis, which has high morbidity and mortality, particularly in immunocompromised and critically ill patients (Netea et al. 2015).

Hyphal candida forms (Fig. 7.1b), which are more invasive, adhere efficiently to host matrices, host tissue surfaces, and host cells. This enables the fungus to cross tissue barriers more easily and to translocate from mucocutaneous sites to the bloodstream. Once in circulation, it can invade other host cells and tissues. Colonization is a prerequisite for invasiveness.

The frequency of opportunistic fungal infections is increasing; indeed, fungal infections have become a serious health problem (WHO report 2022). Clinically, *C. albicans* is the most common fungal pathogen isolated from humans. Infection with this diploid, dimorphic fungus can cause local, as well as systemic, life-threatening infections, particularly in immunocompromised individuals (Zipfel et al. 2008; Harpf et al. 2020). Despite available antifungal drugs, *C. albicans*-associated mortality and morbidity are high (Triebel et al. 2003; Jiménez-López and Lorenz 2013). Over 75% of patients with systemic candidemia die. Simultaneously, drug-resistant *C. albicans* strains are emerging; therefore, the development of new antifungal drugs and/or antifungal vaccines is of great importance.

To better understand why and how *C. albicans* evades the host immune system and spreads within the human host, we must identify the mechanisms

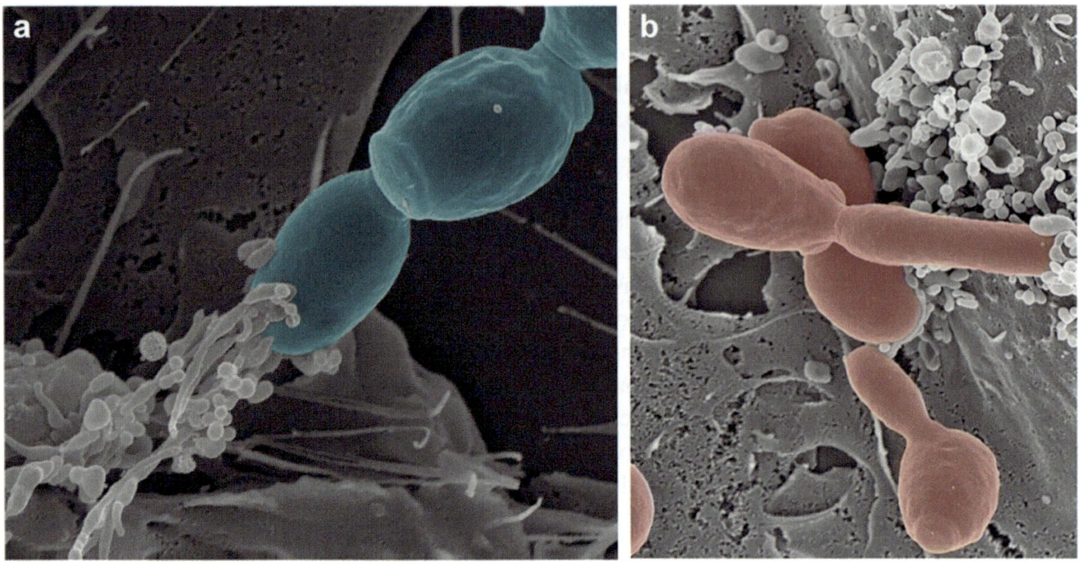

Fig. 7.1 *Candida albicans* a dimorphic human pathogen. (**a**) The human pathogenic yeast *C. albicans* (turquois color) in contact with human endothelial cells. Upon close contact with the pathogenic fungus the human cells form pseudopodia which interact with the pathogenic fungus. (**b**) *C. albicans* (here purple color) can form hyphae which also contact human endothelial and immune cells and are more invasive

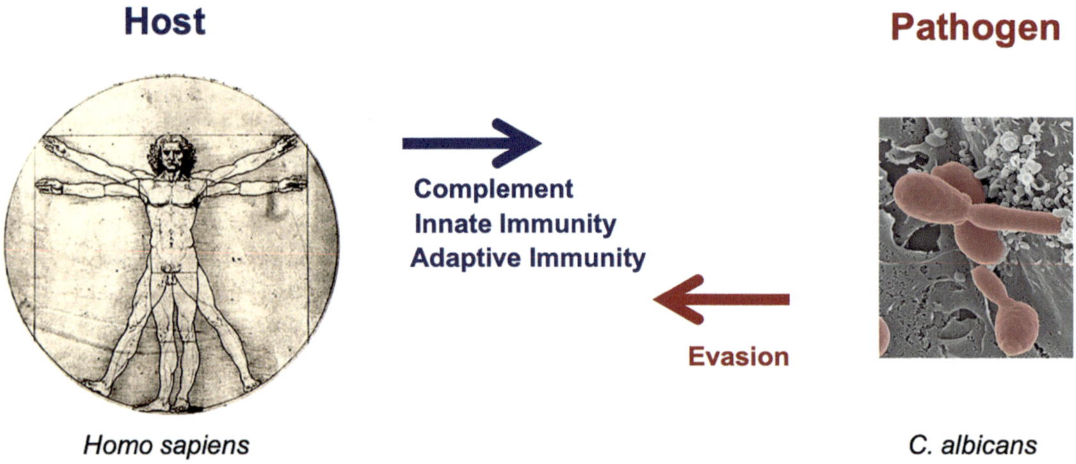

Fig. 7.2 Immune interaction between the human host and *C. albicans,* the pathogenic yeast. The human host uses a sophisticated immune system to recognize and combat invading infectious microbes. This immune response consists of multiple layers, including complement, innate immune responses, and adaptive immune responses. *C. albicans* similar to any other infectious pathogens have developed means to respond to this immune challenge and to block every single level of host immune attack

by which the fungus blocks and interferes with the complement system to avoid an immune attack. We also need to find out how the fungus interacts with human matrices and tissue surfaces. Identification and characterization of immune evasion proteins that facilitate immune escape, and immune cell attack as well as the mechanisms underlying fungus/host cell contact, are necessary. This will increase our understanding of the immunological arms race and enable us to develop new antifungal therapies (Jiménez-López and Lorenz 2013; Erwig and Gow 2016) (Fig. 7.2).

7.4 Fungal Complement and Immune Evasion

7.4.1 Candida and Complement

When *C. albicans* comes in contact with human plasma or serum, it activates all three complement pathways (Davies and Denning 1972; Ray and Wuepper 1976; Ferrante and Thong 1982; Gilmore et al. 1988; Calderone et al. 1988). Many studies show that *C. albicans* binds to many human plasma proteins, as well as to complement receptors exposed on the surface of human innate immune cells (Calderone et al. 1988; Harpf et al. 2020).

7.4.2 Interaction Between *Candida* and Human Plasma Proteins

Candida exploits soluble human regulators for its own benefit to facilitate immune evasion and survival in an immunocompetent host. *Candida* yeast cells and hyphae bind to human plasma proteins, including Factor H, FHL1, C4BP, and plasminogen (Luo et al. 2013a, b; Zipfel et al. 2008; Harpf et al. 2022). When attached to the fungal surface, these human proteins maintain their regulatory activity and inhibit the complement system (Zipfel et al. 2008).

7.4.3 Candida Surface Ligands Gpm1, Pra1, Gpd2, and Hgt1 Bind to Human Factor H

It is important to identify the fungal ligands that bind to human complement regulators. So far **four** *Candida* proteins are known to bind Factor H and FHL1, which are Gpm1, Pra1, Gpd2, and Hgt1 (Fig. 7.3). Indeed, a direct interaction has been confirmed at the protein level. In addition, there are 11 fungal ligands that bind to plasminogen (Crowe et al. 2003; Luo et al. 2013a, b). Thus, *C. albicans*-mediated regulation of complement, and subsequent immune evasion, is dependent on several, not a single, virulence proteins.

Gpm1: Gpm1 (phosphoglycerate mutase) is an isomerase central to the glycolytic pathway. As a glycolytic enzyme, Gpm1 is located in the cell cytosol, where it catalyzes the reversible conversion of 3-phosphoglycerate to 2-phosphoglycerate during glycolysis (Satala et al. 2021). Gpm1, however, is also a moonlighting protein located in the cell wall, where it binds human ligands and mediates immune evasion. Interestingly, similar to other moonlighting proteins, Gpm1 lacks a classical signal peptide; therefore, it is unclear how it is transported to the cell wall. Gpm1 has been identified on the surface of Candida yeast cells, particularly at the tip of the hyphae. Gpm1 binds Factor H and vitronectin (a terminal complement pathway inhibitor and adhesion protein) (Poltermann et al. 2007). Indeed, vitronectin on the fungal surface mediates contact and interaction with human endothelial cells and keratinocytes (Lopez et al. 2014).

Pra1: Initially, Pra1 was identified as a pH-regulated antigen. Cloning of Pra1 from *C. albicans* was reported in 1998. The predicted protein has a mass of 31.4 kDa and carries four N-linked consensus glycosylation sites (Sentandreu et al. 1998; Soloviev et al. 2007). The putative signal peptide includes a potential peptidase cleavage site between amino acids (aa) 15 and 16, which facilitates both surface localization on yeast cells and hyphae, and secretion into the surrounding medium. Secreted Pra1 can also rebind to the fungal surface.

Pra1 expression is upregulated by an increase in pH. The glycoprotein binds Factor H and FHL. Factor H is bound via to domains SCRs 6-7 and SCRs 19-20 H, and to FHL1 via SCRs 6-7, which it shares with shared with Factor H (Luo et al. 2009, 2015, 2018) as well as Factor H and FHL1. *Candida* Pra1 binds other human complement proteins, including C4BP, C3, and the fragments C3a, C3b, iC3b, C3d, and C3d, as well as to complement receptors CR3 (CD11b/CD18), CR4 (CD11c/CD18) and αXβ2 expressed by immune cells (Luo et al. 2010; Soloviev et al. 2011; Losse et al. 2011; vMarcil et al. 2008; Jawhara et al. 2012). In addition, Pra1 expressed on the *Candida* surface regulates the innate immune response

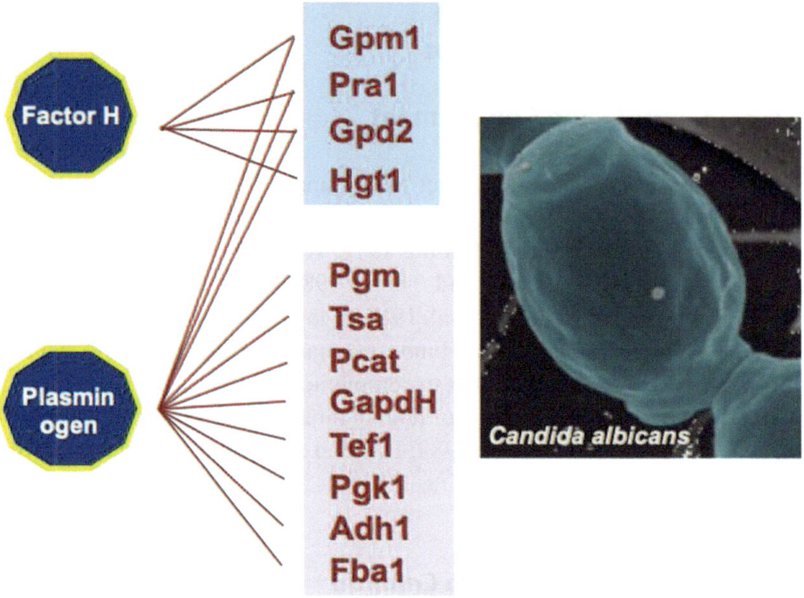

Fig. 7.3 *C. albicans* binds human immune regulators: *C. albicans* has four identified surface proteins that bind Factor H. Factor H is a central human complement and immune regulator. In addition, *C. albicans* uses several surface proteins that bind the zymogen and the inactive proform of the coagulation protease plasminogen. Interestingly three of the fungal Factor H binding proteins bind also plasminogen

(Soloviev et al. 2011) and furthermore inhibits cytokine secretion by T and B cells (Bergfeld et al. 2017).

An additional metabolic role for Pra1 is that of a zinc-binding (zincophore) protein (Citiulo et al. 2012). Interestingly, Pra1 is also a protease that cleaves C3; in this manner, Pra1 reduces the amount of intact C3 in the system, meaning that less is available for cleavage into components that attack the pathogen (see also the section on proteases below) (Luo et al. 2010, 2018).

Secreted Pra1 blocks adhesion, phagocytosis, and opsonophagocytosis, as well as suppressing T cell-mediated immune responses (Luo et al. 2013a, b).

Gpd2: GPd2 (glycerol-3-phospho-3-dehydrogenase 2) is another fungal glycolytic enzyme and moonlighting protein that facilitates fungal immune evasion. It plays a metabolic role in catalyzing the internal transfer of a phosphate group that converts 3-phosphoglycerate (3PG) to 2-phosphoglycerate via an intermediate compound (Marín et al. 2015). Recombinant Gpd2 binds to three human complement regulators: Factor H (via domain SCR 7), FHL1 (via domain SCR 7), and plasminogen. Interestingly, because Factor H binds Gpd2 via SCR 7 and lacks a binding site within the C-terminal domains SCR 19-20, the binding profile is different from that of Gpm1 and Pra1. Again, each of the three human regulators when attached to Gpd2 is functionally active, meaning that the Gpd2 bound regulators inhibit complement. Gpd2-bound plasminogen can be converted by uPA to the active protease plasmin (Luo et al. 2013a, b). Gpd2 is detected on the surface of Candida yeast cells and hyphae. In addition, soluble Gpd2 binds to the surface of hyphae, but not (or at much lower levels) to the surface of *Candida* yeast cells. Gpd2, like many other moonlighting proteins, binds to human cells, suggesting that it also plays a role in fungal adhesion and colonialization.

Hgt1: Hgt1 (high affinity glucose transporter-1) is located in the cell membrane. It belongs to a family of 20 glucose transporters. Hgt1 is a classical membrane protein that harbors 12 membrane spanning domains (UniProt). The protein is present on the cell wall of yeast cells and on hyphae; it binds to Factor H and C4BP. A mutant Candida strain lacking the *hgt1* gene shows reduced binding to both Factor H and C4BP. In addition, the mutant strain lacking Hgt1 shows 25% higher surface deposition of C9 than the wild-type, non-mutant, suggesting that Hgt1 is involved in complement and immune regulation (Lesiak-Markowicz et al. 2011). Based on the reduced ability of the *hgt1*ΔΔ mutant knockout strain to induce rosette formation by complement-coated sheep erythrocytes, it was concluded that Hgt1 also acts as a CR3 analog, and is found on the fungal cell surface (Harpf et al. 2022; Kenno et al. 2019) and on extracellular vesicles (UniProt). So far, the binding sites for the two human immune regulators Factor H or C4BP have not been mapped/localized.

The fungal proteins Gpm1 and Pra1 bind to human Factor H via the same sites, i.e., SCRs 6-7 and SCRs 18-20. Gpd 2 binds to SCRs 6-7 in Factor H, but does not bind to SCRs 19-20. Gpm1, Pra1, and Gpd2 bind FHL1 via SCRs 6-7. Thus, the binding sites in the host proteins are conserved and the fungal virulence proteins target the surface recognition sites of the host regulators. Thereby mimicking self and exploiting the regulatory power, e.g., of the host regulators Factor H and FHL1 for surface complement control (Zipfel et al. 2008). As mentioned above, Factor H and FHL1 bound to the specific ligands at the surface of *Candida* yeast cells maintain their regulatory activities and mediate cofactor activity for the human serine protease Factor I. Thus, the fungus uses human these proteins to degrade and inactivate C3, thereby blocking generation of antifungal C3a, opsonization by C3b complement proteins and opsonophagocytosis by innate immune cells.

7.4.4 *Candida albicans* Surface Ligands that Bind Plasminogen

Plasminogen is a central human, plasma protein and zymogen. When activated and converted by human, but also by microbial activators, plasmin, the active protease is primarily known for its essential role in fibrinolysis. However, plasmin also plays roles in inflammation, complement blockade, and infection; it is also utilized by many pathogens for immune evasion (Barthel et al. 2012; Keragala and Medcalf 2021).

C. albicans similar to multiple pathogenic microbes binds and utilizes human plasminogen for immune evasion (via degradation of complement proteins such as C3 and C4), for tissue and cell destruction, and for matrix degradation. An initial proteomics approach used by Crowe and colleagues identified eight plasminogen-binding *Candida* surface proteins (Crowe et al. 2003). These are Gpm1, thiol-specific antioxidant protein *(Tsa1)*, peroxisomal catalase *(Cta1)*, glyceraldehyde-3-phosphate dehydrogenase *(Tdh3)*, transcription elongation factor *(Tef1)*, phosphoglycerate kinase (*Pgk1*), alcohol dehydrogenase (*Adh1*), and fructose bisphosphate aldolase (*Fba1*). The proteomics approach revealed clear interactions between fungal proteins and human plasminogen, but did not confirm direct binding of any specific fungal ligand to plasminogen. Furthermore, *Candida* has three additional surface ligands for plasminogen, which were not identified in the initial proteome approach. Two of these ligands include the Factor H binding proteins, Pra1 and Gpd2, as well as enolase the central moonlighting protein was identified as a candida plasminogen and also vitronectin, fibronectin ligand (Jong et al. 2003; Satala et al. 2020).

One unexpected result of this approach was the high number of fungal plasminogen ligands identified as surface proteins. Another interesting aspect was that metabolic enzymes with a known and expected cytoplasmic location were identified. In 2003, the surface location of such metabolic enzymes was unexpected; however, this is explained by the modern concept of *moonlighting* functions of by many microbial immune evasion proteins expressed by pathogens (Yadav et al. 2022; Farache et al. 2022). Moonlighting proteins are central for immune escape. In general, moonlighting means that proteins hold down "a second job" by acting at different sites at different times (Singh and Bhalla 2020; Bian et al. 2022). In the microbial world, moonlighting is performed by metabolic enzymes involved in glycolytic pathways; these proteins show novel activity at a site distinct from their "normal" cytoplasmic site of action, i.e., at the cell surface. The role of moonlighting proteins in immune evasion is now well established; indeed, it is carried out by many essential enzymes involved in the tricarboxylic acid cycle and glycolysis (Yu et al. 2021; Satala et al. 2022; Pellon et al. 2022; Arvizu-Rubio et al. 2022; Gancedo et al. 2016; Serrano-Fujarte et al. 2016). An interesting concept is that pathogens in order to maintain a genome of minimal size, must extend biological diversity by generating a functional diversity of existing genes and proteins, such as those encoding central metabolic enzymes.

Interestingly, three fungal Factor H and FHL1 ligands (Gpm1, Pra1, and Gpd2) and enolase also bind plasminogen. In their proteomics approach, Crowe et al. identified Gpm1, but not Pra1 or Gpd2, as ligands for plasminogen. Thus, it can be is speculated that *C. albicans* may use even more than the 12 identified surface proteins to bind plasminogen (Crowe et al. 2003). Furthermore, active plasmin when bound to the fungal, and other microbial virulence proteins acts also as a complement inhibitor (Barthel et al. 2012). Thus, showing the multiple substrate spectra of the central human protease plasmin and highlighting that *C. albicans* exploits multiple ways and mechanisms to block host complement action.

7.5 Candida Proteases Mediating Complement and Immune Escape

Proteases used by candida block complement activation, as well as the many effector functions of the activated cascade.

Candida proteases. *C. albicans* uses fungal proteases, as well as acquired host proteases, to control complement and immune attack, to disintegrate tissue matrices in order to cross tissue barriers, and to damage host endothelial as well as epithelial cells. A combination of fungal and acquired human proteases shape and coordinate complement blockade and immune evasion. For example, multiple proteases target the central complement element C3 and also other complement proteins, and they degrade complement activation products. In addition, proteases degrade the extracellular matrix and damage tissues.

C. albicans expresses ten secreted aspartyl proteases (Saps) (Hube 2006). Three of these degrade complement proteins and act as complement inhibitors, thereby blocking the alternative pathway, the lectin/classical pathway, and opsonophagocytosis. Sap1, Sap2, and Sap3 cleave and inactivate C3 in addition to C3b, C4b, and C5, thereby blocking assembly of the TCC. Thus, the endogenous Sap proteins regulate complement at various levels. Ultimately, inactivation inhibits C3b opsonization, phagocytosis, generation of the proinflammatory antifungal anaphylatoxin C3a, and the TCC (Gropp et al. 2009).

In addition, *C. albicans* uses Pra1 as a protease to facilitate complement and immune escape. Pra1 cleaves C3, thereby reducing the levels of active C3. Pra1 also cleaves and inactivates the antifungal anaphylatoxin C3a which is generated by the host C3 convertases, as well as C3aL formed by Pra1 itself (Luo et al. 2018) (Fig. 7.4). By inactivating intact C3, as well as C3a, C3b, and other C3 effector components, Pra1 weakens and dampens the powerful effects of complement.

Interestingly, Pra1 cleaves human C3 at a site related to, but different from, that used by endogenous human C3 convertase. Human C3 convertases generate two biologically active cleavage fragments, the antifungal anaphylatoxin C3a and the opsonin C3b. The cleavage site used by Pra1 is six aa upstream of that used by human C3 convertases; cleavage generates C3a-like peptides C3aL and C3bL. C3aL harboring a C-terminal six aa truncation is inactivated and degraded directly by fungal Pra1 (Fig. 7.4). C3bL is processed and inactivated by host Factor I, acting together with the cofactor Factor H.

Activated host-derived plasminogen also cleaves and inactivates C3. Thus, the acquired human proenzyme contributes to complement inhibition and inactivation. The active protease plasmin also cleaves matrix components and extracellular matrix proteins, thereby damaging epithelial and endothelial cells (Fig. 7.5).

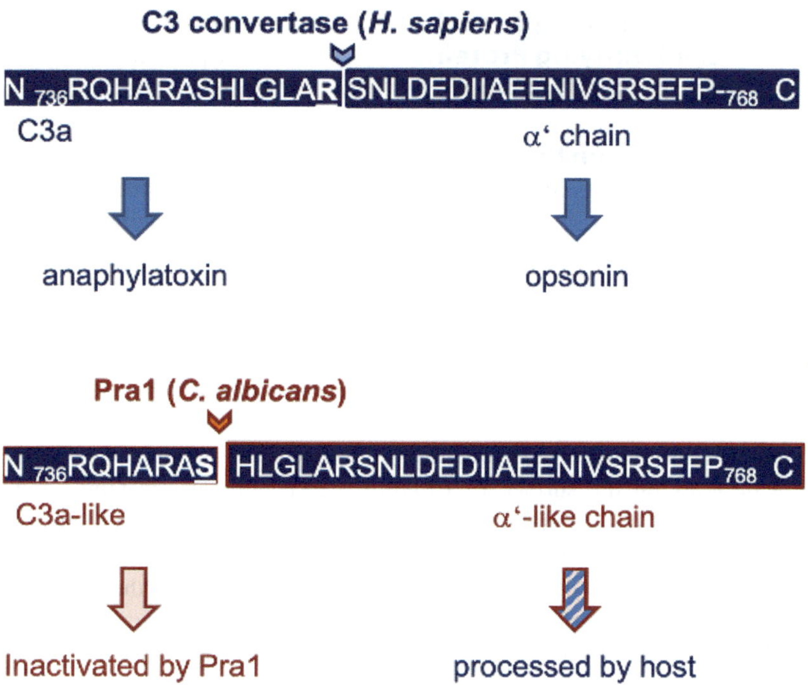

Fig. 7.4 *C. albicans* cleaves C3 at a site related to, but distinct to that of the host convertases: *Upper panel:* human C3 convertases cleave C3 after R_{748}, and generate the antifungal anaphylatoxin C3a and the opsonin C3b. *Lower panel:* Pra from *C. albicans* acts also cleaves as a C3 cleaving proteinase. Pra cleaves C3 after S742, thus six amino acids upstream of the endogenous cleaving site. This generates C3a-like (C3aL) which is further cleaved and inactivated by Pra1 which the C3bL components are processed by host proteins

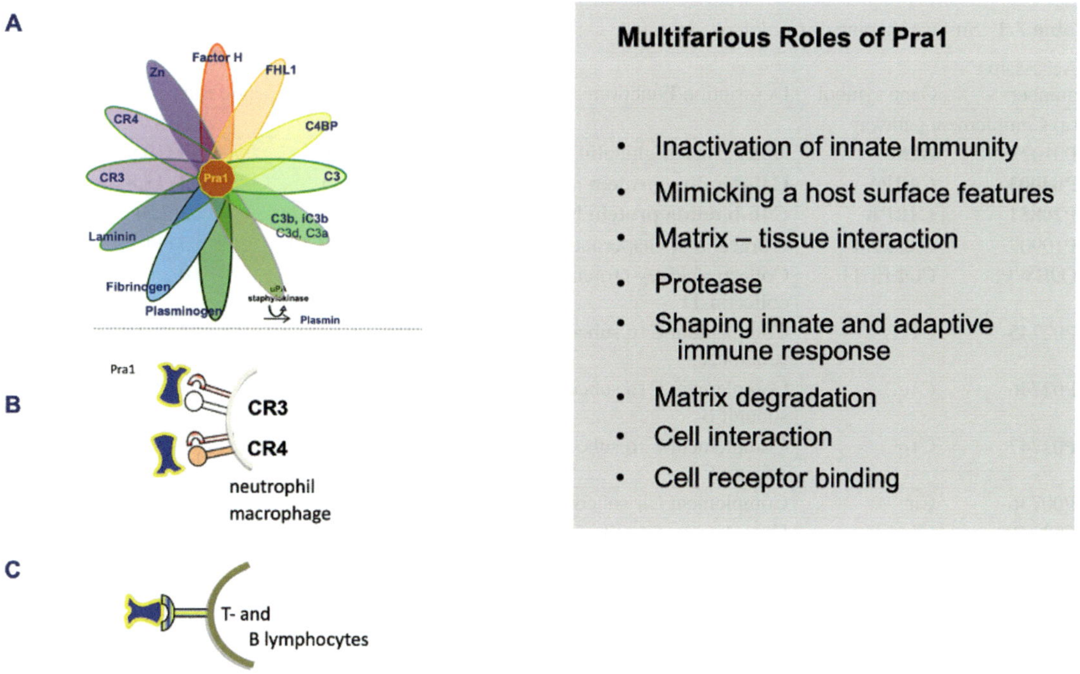

Fig. 7.5 *C. albicans* Pra1 is a multifarious immune evasion proteins A: Pra1 from *C. albicans* binds multiple human plasma proteins and immune modulators, including C3 and the C3 activation fragments, C3b, iC3b, C3d, and C3a, Factor H, FHL1, C4BP, Plasminogen, the surface constituent Laminin, cell surface receptors, CR3, CR4, has metabolic activity and acts as a zincophore. The multiple cellular functions mediated and influenced by Pra1 are depicted in the box on the right

7.6 Proteomics Approaches to Identifying Proteins Expressed by *C. albicans*

7.6.1 Proteins on the Surface of *C. albicans* Yeast Cells and Hyphae

Being both a commensal and a pathogen, *C. albicans* has developed highly efficient strategies for immune evasion, meaning that the pathogen can actively control host complement and immune attack.

Recent ***proteomics approaches*** identified 438 proteins on the surface of *Candida* yeast cells, and 928 on the surface of hyphae (Gil-Bona et al. 2015, 2018; Marín et al. 2015) (Table 7.1). The identified molecules include proteins relevant to cell wall organization, biogenesis, metabolism, and pathogenesis (including stress response).

An analysis of the hyphal **proteasome** identified proteins that play a role in virulence, host immune interactions, and immune evasion.

7.6.2 The Immune Proteasome: Immune "Shaving"

The protein composition of fungal cells and hyphae is complex. However, upon incubation with human plasma, several human proteins attach to the fungal surface. A recent fungal surface proteome-shaping approach based on incubation in normal human serum identified **214** human proteins, including complement and coagulation proteins (Table 7.1). The identified human proteins included known ligands, i.e., Factor H, the α and β chains forming the intact C4 Binding protein (C4BP), as well as human plasminogen. However, neither FHL1 nor FHR1 were identified. Showing that more human pro-

Table 7.1 Surface shaping of *C. albicans* proteins

Accesssion number[a]	Gene symbol	Description/Function	No replicates (peptides on each replicate)[b]	
(a) Complement pathway			NS	HIS
P01023	A2M	Apha-2-macroglobulin	2 (15,11)	3 (10,9,70)
P04003	**C4BPA**	**C4b-binding protein alpha chain**	**4(5,3,4,9)**	**3 (15,8,4)**
P20851	**C4BPB**	**C4b-binding protein beta chain**	**1(2)**	**2(2,3)**
P10909	Clusterin	Extracellular chaperone or ApoJ	4 (11,8,13,19)	3 (11,13,18)
Q9BWP8	COLEC11	Collectin kidney protein 1 or collectin-11	4(5,5,2,4)	0
P02745	C1gA	Complement C1q subcomponent subunit A	0	2 (1,4)
P02746	C1qB	Complement C1q subcomponent subunit B	3 (1,1,1)	3 (5,5,4)
P02747	C1qC	Complement C1q subcomponent subunit C	3 (1,1,2)	3 (7,5,6)
P00736	Cir	Complement Cir subcomponent	2(2,71	3(21,15,14)
Q9NZP8	C1r-like	Complement C1r-like protein	0	3 (6,5,5)
P09871	C1s	Complement C1s subcomponent	2 (1,10)	3 (23,14,11)
P06681	C2	Complement component C2	1(2)	3 (19,15,10)
P01024|	**C3**	**Complement component C3**	**4 (103,106,141,147)**	**3 (141,105,138)**
POCOL4	C4-A	Complement component C4-A	4 (32,32,94,98)	3 (90,61,76)
POCOLS	C4-B	Complement component C4-B	4 (32,33,95,98)	2 (90,62)
P01031	C5	Complement component C5	3 (34,23,21)	3 (68,46,70)
P13671	C6	Complement component C6	3 (20,18,4)	3(21,14, 16)
P10643	C7	Complement component C7	4 (13,9,8,6)	3 (6, 11, 16)
P07357	C8-A	Complement component C8 alpha chain	4 (13,11,6,7)	3 (11,8,8)

(continued)

Table 7.1 (continued)

Accesssion number[a]	Gene symbol	Description/Function	No replicates (peptides on each replicate)[b]	
P07358	C8-B	Complement component C8 beta chain	4 (5,6,16,27)	3 (22,16,13)
P07360	C8-G	C8 gamma chain	4 (6,6,4,1)	3 (6,4,9)
P02748	C9	Complement component C9	4 (17,15,17,22)	3 (24,17,14)
P00751	FB	Complement factor B	4 (8,3,8,10)	3 (47,31,28)
P00746	FD	Complement factor D		3 (2,2,4)
P08603	**FH**	**Complement factor H**	**2 (5,3)**	**3 (18,17,13)**
P05156	FI	Complement factor I	1(1)	3 (3,3,6)
P27918	FP	Complement factor P or Properdin	4 (5,5,2,2)	1(3)
Q15485	Ficolin-2	Collagen/fibrinogen domain-containing protein 2	0	2(1,3)
75,636	Ficolin-3	Collagen/fibrinogen domain-containing protein 3	1(1)	3 (5,2,7)
187	MASP2	Mannan-binding lectin serine protease 2	3 (2,2,5)	1(3)
P80108	GPLD1	Phosphatidylinositol-glycan-specific phospholipase D	0	2 (4,15)
P05155	C1INH	Plasma protease C1 inhibitor (Serpin G1)	0	3 (20,14,24)
P04004	VTN	Vitronectin or protein S	4 (10,12,10,11)	3 (14,11,13)
(b) Coagulation pathway				
P01009	SERPINA1	Alpha-1-antitrypsin (AAT)	4 (4,3,6,12)	3 (37,35,47)
P08697	SERPINF2	Alpha-2-antiplasmin	4 (2.2,9,13)	3 (17,16,18)
P01008	SERPINC1	Antithrombin-III (AT-III)	4 (2,2,5,10)	3 (13,16,25)
P02749	Apoh	Beta-2-glycoprotein 1; APC inhibitor; Apolipoprotein H	2 (1,1)	3 (8,3,4)
Q961Y4	CPB2	Carboxypeptidase B2; Thrombin-activable fibrinolysis inhibitor	4 (2,1,12,10)	3 (5,3,3)
P00740	F9	Coagulation factor IX	4 (3,2,3,5)	3 (5,4,4)
P12259	F5	Coagulation factor V	4 (8,2,11,22)	3 (11,4,8)
P08709	F7	Coagulation factor VII	2(2,3)	2(3,2)
P00742	F10	Coagulation factor X	2(2,2)	3 (6,6,6)
P00748	F12	Coagulation factor XII	1(1)	2 (4,8)
P00488	F13A	Coagulation factor XIII A chain	3 (1,2,1)	2(4,7)
P05160	F13B	Coagulation factor XIII B chain	0	3 (2, 1, 2)
P02671	FGA	Fibrinogen alpha chain	4 (13,12,3,7)	3 (6,6,10)
P02675	FGB	Fibrinogen beta chain	4 (13,10,1,7)	3 (6,11,14)
P02679	FGG	Fibrinogen gamma chain	3 (6,6,4)	3 (2,5,10)
P02751	FN1	Fibronectin; Binds cell surfaces and various compounds (collagen, fibrin, heparin, etc.)	4 (60,62,15, 16)	3 (29,38,24)
P68871	HBB	Hemoglobin subunit beta	4 (7,9,3,10)	3 (9,7,11)
Q9Y251	HPA	Heparanase	2 (1,3)	0
P05546	HCF2	Heparin cofactor 2 (Serpin D1)	4 (8,11,34,28)	3 (29,23,27)
P04196	HRG	Histidine-rich glycoprotein	4 (22,23,26,23)	3 (21,24,17)
Q14520	HABP2	Hyaluronan-binding protein 2	0	3 (4,3,5)
P01042	KNG1	Kininogen-1	4 (6,2,11,15)	3 (11,14,11)

(continued)

Table 7.1 (continued)

Accesssion number[a]	Gene symbol	Description/Function	No replicates (peptides on each replicate)[b]	
P03952	KLKB1	Plasma kallikrein	4 (7,8,2,1)	3 (8,9,10)
P05154	SERPINAS	Plasma serine protease inhibitor	0	2 (3,6)
P00747	**PLG**	**Plasminogen**	**4 (5,6,21,34)**	**3(41,41,15)**
P40197	GP5	Platelet glycoprotein V	2(2,2)	3 (5,2,3)
Q9UK55	SERPINA10	Protein Z-dependent protease inhibitor	4 (1,1,10,9)	3 (6,5,7)
P00734	F2	Prothrombin	4 (16,18,17,25)	3 (21,17,10)
P04070	PROC	Vitamin K-dependent protein C	1 (3)	2 (3,4)
P07225	PROS1	Vitamin K-dependent protein S	3 (1,1,11)	3 (6,3,14)
P04275	WWF	von Willebrand factor	4(18, 16,19,72)	3 (52,36,9)

C. albicans binds numerous human proteins to the surface. The bound proteins included many components of the complement cascade and of the coagulation system. Human proteins that are discussed before and for which ligands are identified are shown in bold characters column 2. Information from (Marín et al. 2015)

[a]Accession number, gene symbol, and function from UniProtKB/Swiss-Prot database (Uniprot 2014)

[b]Proteins were included if they were identified in at least two replicates of one condition with at least 2 pept

teins do bind to *Candida* yeast cells. Attachment of human terminal pathway components is of special interest, as the thick cell wall of *Candida* is considered to efficiently protect the fungus from terminal complement-mediated lyses. However, the surface attachment of all six TCC components (C5, C6, C7, C8, and C9) suggests that the fungus indeed controls TCC action and that the TCC can damage candida. Alternatively, candida may control the additional roles proposed for TCC including inflammation or cytokine response. In addition, multiple proteins of the coagulation pathway, including Plasminogen, several coagulation factors and matrix components as well as matrix-degrading proteases were identified at the surface of yeast cells.

The many human plasma proteins that attach to the fungal surface reveal a complex interplay between the fungal pathogen and the human host. Proteins displayed at the surface of yeast cells and also the hyphal surface form an interface for both fungal and host proteins (Karkowska-Kuleta et al. 2019). Thus, fungal host interaction combines many known and also new proteins on both sides. The interplay of these proteins and their spatiotemporal distribution is of high importance. Their actions are central to understanding immune evasion, attachment to host matrices, and cellular surfaces. Understanding each single molecule, the molecular interaction between fungal and host proteins, as well as integration of these interactions into the immune response activity of human immune cells, is a relevant aspect of infection biology. Ultimately, the fungus acts as a 'wolf in sheep's clothing' to thrive and survive within a human host.

7.7 Vesicles: As a New Fungal Pathogen Interface

In addition to phagocytosis, host immune cells such as neutrophils and monocytes release extracellular traps in response to infection (Urban et al. 2006; Halder et al. 2017). They also release extracellular vesicles (EVs) into the environment (Halder et al. 2020, 2021; Shopova et al. 2020). Extracellular trap formation, referred to as NETosis or METosis depending on whether they originate from neutrophils or monocytes, involves condensation of chromatin, followed by disintegration of the nuclear membrane and the homogenization of nuclear and cytoplasmic components. Subsequently, DNA fibers are expelled from inside the cell, and the released material traps fungal cells to inhibit fungal spread/dissemination and growth. Extracellular traps also contain antifungal components such as myeloperoxidase, neutrophil, or monocytic elastase, and degradative enzymes (Dabrowska et al. 2021; Halder et al. 2021; Shopova et al. 2020), all of which damage the invading microbe. However, NETosis or METosis can damage or kill host immune cells, and damage healthy tissue (Mutua and Gershwin 2021). Therefore, this reaction is restricted to the neutrophil and monocytic cell populations (Irmscher et al. 2019).

EVs are a heterogenous group of membrane vesicles released by prokaryotic and eukaryotic cells, as well as by fungal cells (Raposo and Stoorvogel 2013). These vesicles are important for signaling, cell communication, and coordination of the immune response (Maher et al. 2015; Karkowska-Kuleta et al. 2020). In addition, EVs participate in horizontal signal transfer and can alter the biological function of recipient (host or fungal) cells. The cargo contained in such vesicles includes proteins, lipids, sugars, nucleic acids such as DNA and RNA, and non-coding RNAs such as small RNAs. So called microRNAs (miRNAs) mediate central physiological processes, and can cause pathology or damage to the fungal invader (Raposo and Stoorvogel 2013; Abels and Breakefield 2016). According to their size and content, EVs are grouped into four different categories: apoptotic bodies, microvesicles, exosomes, and exomers.

Apoptotic bodies are the largest vesicles, ranging from 500 to 4000 nm. They are derived from cells undergoing the regulated process of apoptosis, which involves compartmentalization of the dying cell. Apoptotic vesicles contain nuclear fragments and cell organelles, and present so-called "eat me" signals on their outer surface, which trigger uptake by phagocytic cells (Toxicol Pathol 35, 495). A huge number of cells die on a daily basis; therefore, a regulated and efficient clearing process is very important for preventing inflammation.

Microvesicles are smaller, about 200–1000 nm in size, and are derived from outward budding of the plasma membrane of cells. According to the stimulus, these vesicles transport nucleic acids such as plasmid DNA, mRNA, and non-coding RNAs, and proteins, lipids, and metabolites (Sautour et al. 2024; Tricarico et al. 2017).

Exosomes are even smaller, i.e., 30–150 nm, and are of endocytic origin. These vesicles are formed in multi-vesicular bodies (MVBs) within the cytoplasm and are released into the extracellular space by fusion of the MVBs with the extracellular membrane (Yokoi and Ochiya 2021).

Exomers, the smallest vesicles (<50 nm), lack a lipid bilayer membrane (in contrast to EVs, which have a double lipid membrane) (Zhang et al. 2014, 2021; Zhang and Lyden 2019). Exomers are enriched with proteins involved in hypoxia, coagulation, and metabolic pathways, but they also contain DNA, RNA, and miRNAs.

Except for exomers, all EVs are generated via the "endosomal sorting complex responsible for transport" machinery. These protein complexes shuttle the EVs through the cellular cytoplasm to the membrane (Mir and Goettsch 2020).

EVs are important mediators of signaling and cell–cell communication, especially among immune cells within an organism (Szeifert et al. 2021). Microbial infection increases the number of EVs released into the extracellular space. EVs present proteins (e.g., CD9, CD63, CD81, and CD82), called tetraspanins, on their surface, which enable interaction with molecules and integrins expressed on target cells, including MHC (MHC) molecules. This interaction often leads to immune cell stimulation, antigen presentation, and cell maturation and differentiation, thereby enhancing the response to an invading fungus or pathogen (Cai et al. 2018; Zhang et al. 2014).

Depending on their cargo, EVs also have anti-inflammatory and immune suppressive effects. They can carry granular enzymes such as granzyme B, perforin, interferon (IFN), IL-12, and TGF-β, as well as non-coding RNAs. A recent study (Szeifert et al. 2021) reports that EVs released by resting neutrophils exert anti-inflammatory activity, reduce production of reactive oxygen species (ROS), and block cytokine release by neutrophils. However, vesicles collected after exposure of neutrophils to C3b-opsonized *Candida* cells promoted proinflammatory reactions. Such EVs increased ROS production and cytokine secretion from both human neutrophils and activated human endothelial cells. The authors of this report suggested that EVs are "custom made," meaning that they are released to perform specific actions. Thus, release of vesicles depends on environmental factors prevailing at the time of their biogenesis.

Research on EVs, both in general and in the context of fungal infections, is a novel and fast-expanding field of immunobiology. EVs are part of the innate immune response, they are multiple in nature, and their cargo comprises bioactive components. Thus, EVs provide a platform for coordinating the immune response. In general, immune cells release vesicles to communicate with other cells and to coordinate attacks on an invader. Consequently, EVs play important roles in antigen presentation, and in immune cell maturation, activation, and differentiation. Furthermore, EVs shape pro- and anti-inflammatory processes (Keshtkar S et al.). From an evolutionary perspective, EVs have developed to limit the survival of infectious microbes. Accordingly, invading microbes such as *C. albicans* have developed efficient evasion strategies. For example, several pathogenic microbes hide and multiply within cells.

EVs are also formed during infection by *C. albicans* and *Aspergillus fumigatus* (*A. fumigatus*). Notably, vesicles released by *C. albicans* yeast cells induce production of anti-inflammatory EVs by monocytes via binding of soluble ß-glucan to complement receptor 3 (CR3) (expressed by blood monocytes) (Halder et al. 2021). *Candida* cells induce host cells to release vesicles that transport TGF-β on their surfaces, which dampens immune responses by macrophages. Subsequently, the vesicles induce high TGF-β production and release by human endothelial cells (Halder et al. 2017). These actions extend the life of the fungus in the human host. Exosomes released by human monocytes upon interaction with *C. albicans* also transport a number of inhibitory miRNAs (Halder et al. 2022). The miRNA repertoire was recently identified, and the function of miR-15-3p contained in *C. albicans*-induced EVs (miR-15-3p) was examined in detail. This miRNA is transported in exosomes, which are co-stimulated by both the CR3 and TLR4 receptors on monocytes (Halder et al. 2020).

Once released, the exosomes attach to complement C3b/C3d-opsonized fungal surfaces via CR1 receptors expressed on the vesicle surface. Then, the human miRNA transfers into *C. albicans* cells and hyphae. By exploiting the fungal miRNA machinery, the processed miRNA downregulates cell cycle inhibition in *C. albicans*, resulting in increased hyphal growth and filamentation. This cross-species reaction (a human miRNA driving fungal cell growth and differentiation) represents a new sophisticated evasion mechanism used by *C. albicans*.

Polymorphonuclear neutrophilic granulocytes (PMNs) release EVs in response to a fungal infection. However, human PMNs challenged with opsonized *A. fumigatus* conidia release EVs that attack the fungus and create holes in the hyphal surface (Shopova et al. 2020). EVs do this by attaching to and entering the fungal hyphae. Proteomics analysis of EVs revealed the presence of potent antimicrobial peptides such as azurocidin, cathepsin G, and defensin in the vesicles, all of which inhibit fungal growth.

Similarly, C3b-opsonized zymosan, which is a sugar component of the yeast cell wall, induces release of inflammatory EVs by PMNs (Kolonics et al. 2020). In total, immune cells release different types of EVs in response to fungal infections; these vesicles can contain a cocktail of immune-active substances that attack the fungal invader, or components that allow communication with human or fungal cells via cytokines or nucleic acids. The composition of the active cargo is highly regulated and determined by recognition molecules expressed by cells and microbes.

7.8 Other Candida Genes Relevant to Immune Evasion

The *Candida* immune evasion proteins identified so far are multitasking molecules that exert multiple activities at different sites. They reside in the cytoplasm or are exposed on the cell surface, they can be integrated into the fungal cell wall, and they can be secreted as single proteins or contained within vesicles released by *C. albicans*. Most of these protein binds multiple human proteins, but can also show different binding characteristics, interaction partners, functions, and expression patterns in yeast and hyphae. However, in the context of immune evasion, we need to understand the common features of these proteins. Proteins with related and overlapping functions are found in other *Candida* and fungal species, including *C. krusei, C. glabrata, C. tropicalis, C. parapsilosis,* and *A. fumigatus*. They are also found in pathogenic microbes such as Gram-negative and -positive bacteria, helminths, parasites, and viruses.

Thus, complement evasion is a common feature of fungal pathogens as well as other pathogenic microbes that can survive in a human host. The identified ligands for Factor H are expressed on the surface of yeast cells and hyphae. In addition, Pra1 and Gpd2 are enriched at the tips of hyphae. Gpd2, and many other microbial moonlighting proteins, is found in the cytoplasm, consistent with its role as a metabolic enzyme, as well as on the cell surface in the context of immune evasion. In addition to regulating immune responses, moonlight proteins either attach directly to the surface of human cells, and to tissue matrices, or to human adhesion proteins such as vitronectin (Gpd1 is an example of this).

In addition, proteins like Pra1 are surface displayed and are also secreted. These secreted inhibitors act at a distance from the fungal surface and cell wall and thus control immune response distant to the fungal surface. Thus, the fungal topology, and the site of action/inactivation of the host complement machinery, is central and a single fungal evasion protein can control host immune attack at multiple sites.

Knockout mutants of several fungal immune evasion proteins have been generated and tested. As expected, knockout of single genes results in less binding to complement regulators, although the effects are variable. Given that *C. albicans* has at least four binding proteins, e.g., for Factor H and 11 binding plasminogen ligands, the variable effects of these deletion mutants may be due to differential expression, or to alterations of membrane and cell wall integrity.

Candida uses multiple strategies to interfere with complement activation: virulence proteins control and block complement activation and cascade progression, as well as the action of anaphylatoxins C3a, C5a, and opsonophagocytosis, ultimately preventing C5b-9 deposition on the surface.

Although fungal immune evasion has been studied mainly in the context of *C. albicans*, the results are relevant to other *Candida* species such as *C. glabrata, C. parapsillosis,* and the opportunistic fungal pathogen *A. fumigatus*. These species bind human plasma components, including complement and coagulation proteins, particularly Factor H, FHL1, C4BP, and plasminogen. The binding proteins in some of these species have been identified, but it is likely that more exist.

Common characteristics of fungal Immune Escape Immune escape by *C. albicans* shares common features with other *Candida* species and microbial pathogens; the main feature is multitasking moonlighting proteins. In addition to Gpm1, proteome analyses have identified other Candida moonlighting proteins such as enolase, translation elongation factors (Tef, TUF), and GAPDH. Therefore, we can speculate that *C. albicans* also use these surface exposed moonlighting proteins for immune escape. Proteomics approaches also identified other moonlighting proteins with immune evasion functions, including Eno1, PgK1, Fba, Pgi, 1 Adh1, and Tdh1, as well as members of the heat shock protein family (Satala et al. 2020; Pitarch et al. 2006, 2009). Additionally, translation elongation factors not identified by the proteome approach wich are also likely candidates as immune evasion proteins.

Aspf2 in *A. fumigatus* was the first identified factor H-binding protein. Aspf2 also binds FHL1 and plasminogen. Factor H binds to Aspf2 via domains SCRs 6-7 and SCRs 19-20, whereas FHL1 binds to Aspf 2 via SCRs 6-7. In addition, the moonlighting protein enolase also binds to Factor H, FHL1, C4BP, and plasminogen. Mep1 is a complement-cleaving protease expressed by *A. fumigatus*. Mep1 efficiently cleaves and inactivates C3, although its broad substrate specificity means that it also cleaves C4, C5, C3a, C4a, and C5a, and in addition inactivates the pattern-recognition molecules MBL and ficolin-1 (Behnsen et al. 2010; Shende et al. 2018, 2022).

7.9 Summary Immune Escape Proteins as Potential Vaccination Targets

The complement and the coagulation systems are barriers to any invading microbe. Consequently, complement and immune evasion are essential for fungal survival in an immunocompetent human host. Below, we summarize how *C. albicans* and *A. fumigatus* use multiple proteins to control the activity of these host defense systems.

Many fungal proteins that act as surface ligands for host plasma proteins such as Gpm1 and Gpd2 are moonlighting proteins, which are mostly essential metabolic enzymes that function during glycolysis or gluconeogenesis. Given the central metabolic role of these proteins, it is conceptually easy to consider that a pathogenic fungus exploits them and "designs" functions and a new biology. Thus, moonlighting proteins show additional functions at distinct locations. In addition to the usual cytoplasmic location, they are expressed at the cell surface, are secreted, and/or form the cargo of EVs.

Candida Pra1, although not a bona fide moonlighting protein, is a prime example of an immune evasion protein found at multiple sites, and which binds many human ligands and cell receptors. In addition, Pra1 is a protease that targets C3 and other complement proteins.

A single immune evasion protein binds multiple human proteins.

Pra1 binds to cell receptors such as MCP, CR1, and CR3. Fungal Pra1, for example, cleaves C3, C3a, and C5. These proteolytic activities are specific, as Pra1 does not cleave Factor H.

Interactions with multiple host ligands, combined with variable spatial/positioning and multiplicity of function, reveal the power of fungal immune evasion molecules. Thus, potential immune evasion proteins should be evaluated in the context of spatial distribution/location, combined with an in-depth evaluation of their binding profiles and a detailed functional characterization. Expected functions can include metabolic activity, as well as proteolytic activity, matrix contact, receptor interactions, and mediation of cell adhesion and colonization.

The multiple functions of even single immune evasion proteins, show efficiently how the pathogenic fungus can interfere with and block host immune response. In this regard, the immune evasion of *C. albicans* reflects the historic scenario of nine-headed hydras in Greek mythology (Fig. 7.6).

C. albicans The J. Paul Getty Museum

Fig. 7.6 *C. albicans* is the nine or multiheaded hydra in modulating infection. The image on the right shows Heracles, Iolaus, and the Hydra. Image is a Paestan black-figure hydra C6th B.C., The J. Paul Getty Museum

References

Abels ER, Breakefield XO (2016) Introduction to extracellular vesicles: biogenesis, RNA cargo selection, content, release, and uptake. Cell Mol Neurobiol 36(3):301–312. https://doi.org/10.1007/s10571-016-0366-z

Arvizu-Rubio VJ, García-Carnero LC, Mora-Montes HM (2022) Moonlighting proteins in medically relevant fungi. PeerJ 10:e14001. https://doi.org/10.7717/peerj.14001

Barthel D, Schindler S, Zipfel PF (2012) Plasminogen is a complement inhibitor. J Biol Chem 287(22):18831–18842. https://doi.org/10.1074/jbc.M111.323287

Behnsen J, Lessing F, Schindler S, Wartenberg D, Jacobsen ID, Thoen M, Zipfel PF, Brakhage AA (2010) 2020. Secreted Aspergillus fumigatus protease Alp1 degrades human complement proteins C3, C4, and C5. Infect Immun 78(8):3585–3594

Bergfeld A, Dasari P, Werner S, Hughes TR, Song WC, Hortschansky P, Brakhage AA, Hünig T, Zipfel PF, Beyersdorf N (2017) Direct binding of the pH-regulated protein 1 (Pra1) from *Candida albicans* inhibits cytokine secretion by mouse CD4$^+$ T cells. Front Microbiol 8:844. https://doi.org/10.3389/fmicb.2017.00844

Bian X, Jiang H, Meng Y, Li YP, Fang J, Lu Z (2022) Regulation of gene expression by glycolytic and gluconeogenic enzymes. Trends Cell Biol 32(9):786–799. https://doi.org/10.1016/j.tcb.2022.02.003

Cai C, Koch B, Morikawa K, Suda G, Sakamoto N, Rueschenbaum S, Akhras S, Dietz J, Hildt E, Zeuzem S, Welsch C, Lange CM (2018) Macrophage-derived extracellular vesicles induce long-lasting immunity against hepatitis C virus which is blunted by polyunsaturated fatty acids. Front Immunol 9:723. https://doi.org/10.3389/fimmu.2018.00723. PMID: 29706960; PMCID: PMC5906748

Calderone RA, Linehan L, Wadsworth E, Sandberg AL (1988) Identification of C3d receptors on Candida albicans. Infect Immun. 56(1):252–258. https://doi.org/10.1128/iai.56.1.252-258.1988. PMID: 2961702; PMCID: PMC259265

Citiulo F, Jacobsen ID, Miramón P, Schild L, Brunke S, Zipfel P, Brock M, Hube B, Wilson D (2012) Candida albicans scavenges host zinc via Pra1 during endothelial invasion. PLoS Pathog 8(6):e1002777. https://doi.org/10.1371/journal.ppat.1002777

Cohen A, Jeng EE, Voorhies M, Symington J, Ali N, Rodriguez RA, Bassik MC, Sil A (2022) Genome-scale CRISPR screening reveals that C3aR signaling is critical for rapid capture of fungi by macrophages. PLoS Pathog 18(9):e1010237. https://doi.org/10.1371/journal.ppat.1010237

Coulthard LG, Woodruff TM (2015) Is the complement activation product C3a a proinflammatory molecule? Re-evaluating the evidence and the myth. J Immunol 194(8):3542–3548. https://doi.org/10.4049/jimmunol.1403068

Crowe JD, Sievwright IK, Auld GC, Moore NR, Gow NA, Booth NA (2003) Candida albicans binds human plasminogen: identification of eight plasminogen-binding proteins. Mol Microbiol 47(6):1637–1651. https://doi.org/10.1046/j.1365-2958.2003.03390.x

Dabrowska S, Andrzejewska A, Janowski M, Lukomska B (2021) Immunomodulatory and regenerative effects of mesenchymal stem cells and extracellular vesicles: therapeutic outlook for inflammatory and degenerative diseases. Front Immunol 11:591065. https://doi.org/10.3389/fimmu.2020.591065

Davies RR, Denning TJ (1972) Growth and form in Candida albicans. Sabouraudia 10(2):180–188. PMID: 4557879

Dühring S, Germerodt S, Skerka C, Zipfel PF, Dandekar T, Schuster S (2015) Host-pathogen interactions between the human innate immune system and Candida albicans-understanding and modeling defense and evasion strategies. Front Microbiol 6:625. https://doi.org/10.3389/fmicb.2015.00625

Erwig LP, Gow NA (2016) Interactions of fungal pathogens with phagocytes. Nat Rev Microbiol 14(3):163–176. https://doi.org/10.1038/nrmicro.2015.21

Farache D, Antine SP, Lee ASY (2022) Moonlighting translation factors: multifunctionality drives diverse gene regulation. Trends Cell Biol 32(9):762–772. https://doi.org/10.1016/j.tcb.2022.03.006

Ferrante A, Thong YH (1982) Inhibition of serum bactericidal activity by bilirubin. Immunol Lett 4(2):103–105. https://doi.org/10.1016/0165-2478(82)90008-6. PMID: 7037625

Gancedo C, Flores CL, Gancedo JM (2016) The expanding landscape of moonlighting proteins in yeasts. Microbiol Mol Biol Rev 80(3):765–777. https://doi.org/10.1128/MMBR.00012-16

Gil-Bona A, Llama-Palacios A, Parra CM, Vivanco F, Nombela C, Monteoliva L, Gil C (2015) Proteomics unravels extracellular vesicles as carriers of classical cytoplasmic proteins in Candida albicans. J Proteome Res 14(1):142–153. https://doi.org/10.1021/pr5007944

Gil-Bona A, Amador-García A, Gil C, Monteoliva L (2018) The external face of Candida albicans: a proteomic view of the cell surface and the extracellular environment. J Proteome 180:70–79. https://doi.org/10.1016/j.jprot.2017.12.002

Gilmore BJ, Retsinas EM, Lorenz JS, Hostetter MK (1988) An iC3b receptor on Candida albicans: structure, function, and correlates for pathogenicity. J Infect Dis 157(1):38–46. https://doi.org/10.1093/infdis/157.1.38. PMID: 2961817

Gropp K, Schild L, Schindler S, Hube B, Zipfel PF, Skerka C (2009) The yeast Candida albicans evades human complement attack by secretion of aspartic proteases. Mol Immunol 47(2-3):465–475. https://doi.org/10.1016/j.molimm.2009.08.019

Halder LD, Abdelfatah MA, Jo EA, Jacobsen ID, Westermann M, Beyersdorf N, Lorkowski S, Zipfel PF, Skerka C (2017) Factor H binds to extracellular DNA traps released from human blood monocytes in response to Candida albicans. Front Immunol 7:671.

https://doi.org/10.3389/fimmu.2016.00671. PMID: 28133459; PMCID: PMC5233719

Halder LD, Jo EAH, Hasan MZ, Ferreira-Gomes M, Krüger T, Westermann M, Palme DI, Rambach G, Beyersdorf N, Speth C, Jacobsen ID, Kniemeyer O, Jungnickel B, Zipfel PF, Skerka C (2020) Immune modulation by complement receptor 3-dependent human monocyte TGF-β1-transporting vesicles. Nat Commun 11(1):2331. https://doi.org/10.1038/s41467-020-16241-5

Halder LD, Babych S, Palme DI, Mansouri-Ghahnavieh E, Ivanov L, Ashonibare V, Langenhorst D, Prusty B, Rambach G, Wich M, Trinks N, Blango MG, Kornitzer D, Terpitz U, Speth C, Jungnickel B, Beyersdorf N, Zipfel PF, Brakhage AA, Skerka C (2021) Candida albicans induces cross-kingdom miRNA trafficking in human monocytes to promote fungal growth. mBio 13(1):e0356321. https://doi.org/10.1128/mbio.03563-21. Epub 2022 Feb 8. PMID: 35132877; PMCID: PMC8822622

Halder LD, Babych S, Palme DI, Mansouri-Ghahnavieh E, Ivanov L, Ashonibare V, Langenhorst D, Prusty B, Rambach G, Wich M, Trinks N, Blango MG, Kornitzer D, Terpitz U, Speth C, Jungnickel B, Beyersdorf N, Zipfel PF, Brakhage AA, Skerka C (2022) Candida albicans induces cross-kingdom miRNA trafficking in human monocytes to promote fungal growth. mBio 13(1):e0356321. https://doi.org/10.1128/mbio.03563-21

Harpf V, Rambach G, Würzner R, Lass-Flörl C, Speth C (2020) *Candida* and complement: new aspects in an old battle. Front Immunol 11:1471. https://doi.org/10.3389/fimmu.2020.01471

Harpf V, Kenno S, Rambach G, Fleischer V, Parth N, Weichenberger CX, Garred P, Huber S, Lass-Flörl C, Speth C, Würzner R (2022) Influence of glucose on *Candida albicans* and the relevance of the complement FH-binding molecule Hgt1 in a murine model of candidiasis. Antibiotics (Basel) 11(2):257. https://doi.org/10.3390/antibiotics11020257

Hatinguais R, Willment JA, Brown GD (2020) PAMPs of the fungal cell wall and mammalian PRRs. Curr Top Microbiol Immunol 425:187–223. https://doi.org/10.1007/82_2020_201

Hube B (2006) Infection-associated genes of Candida albicans. Future Microbiol 1(2):209–218. https://doi.org/10.2217/17460913.1.2.209. PMID: 17661666

Irmscher S, Brix SR, Zipfel SLH, Halder LD, Mutlutürk S, Wulf S, Girdauskas E, Reichenspurner H, Stahl RAK, Jungnickel B, Wiech T, Zipfel PF, Skerka C (2019) Serum FHR1 binding to necrotic-type cells activates monocytic inflammasome and marks necrotic sites in vasculopathies. Nat Commun 10(1):2961. https://doi.org/10.1038/s41467-019-10766-0

Janeway CA Jr, Medzhitov R (2002) Innate immune recognition. Annu Rev Immunol 20:197–216. https://doi.org/10.1146/annurev.immunol.20.083001.084359

Janeway CA Jr (1992 Jan) The immune system evolved to discriminate infectious nonself from noninfectious self. Immunol Today 13(1):11–16. https://doi.org/10.1016/0167-5699(92)90198-G

Jawhara S, Pluskota E, Verbovetskiy D, Skomorovska-Prokvolit O, Plow EF, Soloviev DA (2012) Integrin $\alpha X\beta_2$ is a leukocyte receptor for Candida albicans and is essential for protection against fungal infections. J Immunol 189(5):2468–2477. https://doi.org/10.4049/jimmunol.1200524

Jiménez-López C, Lorenz MC (2013) Fungal immune evasion in a model host-pathogen interaction: Candida albicans versus macrophages. PLoS Pathog 9(11):e1003741. https://doi.org/10.1371/journal.ppat.1003741

Jong AY, Chen SHM, Stins MF, Kim KS, Tuan TL, Huang SH (2003) Binding of Candida albicans enolase to plasmin(ogen) results in enhanced invasion of human brain microvascular endothelial cells. J Med Microbiol 52(Pt 8):615–622. https://doi.org/10.1099/jmm.0.05060-0

Karkowska-Kuleta J, Satala D, Bochenska O, Rapala-Kozik M, Kozik A (2019) Moonlighting proteins are variably exposed at the cell surfaces of Candida glabrata, Candida parapsilosis and Candida tropicalis under certain growth conditions. BMC Microbiol 19(1):149. https://doi.org/10.1186/s12866-019-1524-5

Karkowska-Kuleta J, Kulig K, Karnas E, Zuba-Surma E, Woznicka O, Pyza E, Kuleta P, Osyczka A, Rapala-Kozik M, Kozik A (2020) Characteristics of extracellular vesicles released by the pathogenic yeast-like fungi *Candida glabrata*, *Candida parapsilosis* and *Candida tropicalis*. Cells 9(7):1722. https://doi.org/10.3390/cells9071722

Kenno S, Speth C, Rambach G, Binder U, Chatterjee S, Caramalho R, Haas H, Lass-Flörl C, Shaughnessy J, Ram S, Gow NAR, Orth-Höller D, Würzner R (2019) *Candida albicans* factor H binding molecule Hgt1p - a low glucose-induced transmembrane protein is trafficked to the cell wall and impairs phagocytosis and killing by human neutrophils. Front Microbiol 9:3319. https://doi.org/10.3389/fmicb.2018.03319

Keragala CB, Medcalf RL (2021) Plasminogen: an enigmatic zymogen. Blood 137(21):2881–2889. https://doi.org/10.1182/blood.2020008951

Kong E, Jabra-Rizk MA (2015) The great escape: pathogen versus host. PLoS Pathog 11(3):e1004661. https://doi.org/10.1371/journal.ppat.1004661

Kolonics F, Szeifert V, Timár CI, Ligeti E, Lőrincz ÁM (2020) The functional heterogeneity of neutrophil-derived extracellular vesicles reflects the status of the parent cell. Cells 9(12):2718. https://doi.org/10.3390/cells9122718. PMID: 33353087; PMCID: PMC7766779

Lesiak-Markowicz I, Vogl G, Schwarzmüller T, Speth C, Lass-Flörl C, Dierich MP, Kuchler K, Würzner R (2011) Candida albicans Hgt1p, a multifunctional evasion molecule: complement inhibitor, CR3 analogue, and human immunodeficiency virus-binding molecule. J Infect Dis 204(5):802–809. https://doi.org/10.1093/infdis/jir455

Lopez CM, Wallich R, Riesbeck K, Skerka C, Zipfel PF (2014) Candida albicans uses the surface protein Gpm1 to attach to human endothelial cells and to keratinocytes via the adhesive protein vitronectin. PLoS One 9(3):e90796. https://doi.org/10.1371/journal.pone.0090796

Losse J, Svobodová E, Heyken A, Hube B, Zipfel PF, Józsi M (2011) Role of pH-regulated antigen 1 of Candida albicans in the fungal recognition and antifungal response of human neutrophils. Mol Immunol 48(15-16):2135–2143. https://doi.org/10.1016/j.molimm.2011.07.007

Lukasser-Vogl E, Gruber A, Lass-Flörl C, Eder A, Høgåsen AK, Morgan BP, Dierich MP, Würzner R (2000 May) Membrane attack complex formation on yeast as trigger of selective release of terminal complement proteins from human polymorphonuclear leukocytes. FEMS Immunol Med Microbiol 28(1):15–23. https://doi.org/10.1111/j.1574-695X.2000.tb01452.x

Luo S, Poltermann S, Kunert A, Rupp S, Zipfel PF (2009) Immune evasion of the human pathogenic yeast Candida albicans: Pra1 is a Factor H, FHL-1 and plasminogen binding surface protein. Mol Immunol 47(2-3):541–550. https://doi.org/10.1016/j.molimm.2009.07.017

Luo S, Hartmann A, Dahse HM, Skerka C, Zipfel PF (2010) Secreted pH-regulated antigen 1 of Candida albicans blocks activation and conversion of complement C3. J Immunol 185(4):2164–2173. https://doi.org/10.4049/jimmunol.1001011

Luo S, Hoffmann R, Skerka C, Zipfel PF (2013a) Glycerol-3-phosphate dehydrogenase 2 is a novel factor H-, factor H-like protein 1-, and plasminogen-binding surface protein of Candida albicans. J Infect Dis 207(4):594–603. https://doi.org/10.1093/infdis/jis718

Luo S, Skerka C, Kurzai O, Zipfel PF (2013b) Complement and innate immune evasion strategies of the human pathogenic fungus Candida albicans. Mol Immunol 56(3):161–169. https://doi.org/10.1016/j.molimm.2013.05.218

Luo S, Hipler UC, Münzberg C, Skerka C, Zipfel PF (2015) Sequence variations and protein expression levels of the two immune evasion proteins Gpm1 and Pra1 influence virulence of clinical Candida albicans isolates. PLoS One 10(2):e0113192. https://doi.org/10.1371/journal.pone.0113192

Luo S, Dasari P, Reiher N, Hartmann A, Jacksch S, Wende E, Barz D, Niemiec MJ, Jacobsen I, Beyersdorf N, Hünig T, Klos A, Skerka C, Zipfel PF (2018) The secreted Candida albicans protein Pra1 disrupts host defense by broadly targeting and blocking complement C3 and C3 activation fragments. Mol Immunol 93:266–277. https://doi.org/10.1016/j.molimm.2017.07.010

Maher CO, Dunne K, Comerford R, O'Dea S, Loy A, Woo J, Rogers TR, Mulcahy F, Dunne PJ, Doherty DG (2015) Candida albicans stimulates IL-23 release by human dendritic cells and downstream IL-17 secretion by Vδ1 T cells. J Immunol 194(12):5953–5960. https://doi.org/10.4049/jimmunol.1403066. Epub 2015 May 11. PMID: 25964489

Marín E, Parra-Giraldo CM, Hernández-Haro C, Hernáez ML, Nombela C, Monteoliva L, Gil C (2015) Candida albicans shaving to profile human serum proteins on hyphal surface. Front Microbiol 6:1343. https://doi.org/10.3389/fmicb.2015.01343

Mir B, Goettsch C (2020) Extracellular vesicles as delivery vehicles of specific cellular cargo. Cells 9(7):1601. https://doi.org/10.3390/cells9071601. PMID: 32630649; PMCID: PMC7407641

Mutua V, Gershwin LJ (2021) A review of neutrophil extracellular traps (NETs) in disease: potential anti-NETs therapeutics. Clin Rev Allergy Immunol 61(2):194–211. https://doi.org/10.1007/s12016-020-08804-7

Netea MG, Joosten LA, van der Meer JW, Kullberg BJ, van de Veerdonk FL (2015) Immune defence against Candida fungal infections. Nat Rev Immunol 15(10):630–642. https://doi.org/10.1038/nri3897

Pitarch A, Jiménez A, Nombela C, Gil C (2006) Decoding serological response to Candida cell wall immunome into novel diagnostic, prognostic, and therapeutic candidates for systemic candidiasis by proteomic and bioinformatic analyses. Mol Cell Proteomics 5(1):79–96. https://doi.org/10.1074/mcp.M500243-MCP200. Epub 2005 Sep 28. PMID: 16195222

Pitarch A, Nombela C, Gil C (2009) Proteomic profiling of serologic response to Candida albicans during host-commensal and host-pathogen interactions. Methods Mol Biol 470:369–411. https://doi.org/10.1007/978-1-59745-204-5_26. PMID: 19089396

Pellon A, Begum N, Sadeghi Nasab SD, Harzandi A, Shoaie S, Moyes DL (2022) Role of cellular metabolism during candida-host interactions. Pathogens 11(2):184. https://doi.org/10.3390/pathogens11020184

Poltermann S, Kunert A, von der Heide M, Eck R, Hartmann A, Zipfel PF (2007) Gpm1p is a factor H-, FHL-1-, and plasminogen-binding surface protein of Candida albicans. J Biol Chem 282(52):37537–37544. https://doi.org/10.1074/jbc.M707280200

Rapala-Kozik M, Bochenska O, Zajac D, Karkowska-Kuleta J, Gogol M, Zawrotniak M, Kozik A (2018) Extracellular proteinases of Candida species pathogenic yeasts. Mol Oral Microbiol 33(2):113–124. https://doi.org/10.1111/omi.12206

Raposo G, Stoorvogel W (2013) Extracellular vesicles: exosomes, microvesicles, and friends. J Cell Biol 200(4):373–383. https://doi.org/10.1083/jcb.201211138

Ray TL, Wuepper KD (1976) Experimental cutaneous candidiasis in rodents. J Invest Dermatol 66(1):29–33. https://doi.org/10.1111/1523-1747.ep12478053. PMID: 1107433

Sacks SH (2010 Mar) Complement fragments C3a and C5a: the salt and pepper of the immune response. Eur J Immunol 40(3):668–670. https://doi.org/10.1002/eji.201040355

Satala D, Satala G, Karkowska-Kuleta J, Bukowski M, Kluza A, Rapala-Kozik M, Kozik A (2020) Structural

insights into the interactions of Candidal enolase with human vitronectin, fibronectin and plasminogen. Int J Mol Sci 21(21):7843. https://doi.org/10.3390/ijms21217843
Satala D, Karkowska-Kuleta J, Zelazna A, Rapala-Kozik M, Kozik A (2020) Moonlighting proteins at the candidal cell surface. Microorganisms 8(7):1046. https://doi.org/10.3390/microorganisms8071046. PMID: 32674422; PMCID: PMC7409194
Satala D, Zelazna A, Satala G, Bukowski M, Zawrotniak M, Rapala-Kozik M, Kozik A (2021) Towards understanding the novel adhesin function of Candida albicans phosphoglycerate mutase at the pathogen cell surface: some structural analysis of the interactions with human host extracellular matrix proteins. Acta Biochim Pol 68(4):515–525. https://doi.org/10.18388/abp.2020_5959
Satala D, Gonzalez-Gonzalez M, Smolarz M, Surowiec M, Kulig K, Wronowska E, Zawrotniak M, Kozik A, Rapala- Kozik M, Karkowska-Kuleta J (2022) The role of Candida albicans virulence factors in the formation of multispecies biofilms with bacterial periodontal pathogens. Front Cell Infect Microbiol 11:765942. https://doi.org/10.3389/fcimb.2021.765942. PMID: 35071033; PMCID: PMC8766842
Sautour M, Théry T, Divoux G, Dupont S, Beney L, Gros CP, Desbois N (2024) Synthesis and characterization of new acid-functionalized porphyrins displaying antimicrobial activity against gram positive bacteria, yeasts and filamentous fungi with or without ultra-high irradiance. Bioorg Med Chem 109:117810. https://doi.org/10.1016/j.bmc.2024.117810. Epub 2024 Jun 13. PMID: 38906069.
Sentandreu M, Elorza MV, Sentandreu R, Fonzi WA (1998) Cloning and characterization of PRA1, a gene encoding a novel pH-regulated antigen of Candida albicans. J Bacteriol 180(2):282–289. https://doi.org/10.1128/JB.180.2.282-289.1998
Serrano-Fujarte I, López-Romero E, Cuéllar-Cruz M (2016) Moonlight-like proteins of the cell wall protect sessile cells of Candida from oxidative stress. Microb Pathog 90:22–33. https://doi.org/10.1016/j.micpath.2015.10.001
Shende R, Wong SSW, Rapole S, Beau R, Ibrahim-Granet O, Monod M, Gührs KH, Pal JK, Latgé JP, Madan T, Aimanianda V, Sahu A (2018) *Aspergillus fumigatus* conidial metalloprotease Mep1p cleaves host complement proteins. J Biol Chem 293(40):15538–15555. https://doi.org/10.1074/jbc.RA117.001476
Shende R, Wong SSW, Meitei HT, Lal G, Madan T, Aimanianda V, Pal JK, Sahu A (2022) Protective role of host complement system in *Aspergillus fumigatus*infection. Front Immunol 13:978152. https://doi.org/10.3389/fimmu.2022.978152
Shopova IA, Belyaev I, Dasari P, Jahreis S, Stroe MC, Cseresnyés Z, Zimmermann AK, Medyukhina A, Svensson CM, Krüger T, Szeifert V, Nietzsche S, Conrad T, Blango MG, Kniemeyer O, von Lilienfeld-Toal M, Zipfel PF, Ligeti E, Figge MT, Brakhage AA (2020) Human neutrophils produce antifungal extracellular vesicles against Aspergillus fumigatus. mBio 11(2):e00596–e00520. https://doi.org/10.1128/mBio.00596-20
Singh N, Bhalla N (2020) Moonlighting proteins. Annu Rev Genet 54:265–285. https://doi.org/10.1146/annurev-genet-030620-102906
Soloviev DA, Fonzi WA, Sentandreu R, Pluskota E, Forsyth CB, Yadav S, Plow EF (2007) Identification of pH-regulated antigen 1 released from Candida albicans as the major ligand for leukocyte integrin alphaMbeta2. J Immunol 178(4):2038–2046. https://doi.org/10.4049/jimmunol.178.4.2038
Soloviev DA, Jawhara S, Fonzi WA (2011) Regulation of innate immune response to Candida albicans infections by αMβ2-Pra1p interaction. Infect Immun 79(4):1546–1558. https://doi.org/10.1128/IAI.00650-10
Szeifert V, Kolonics F, Bartos B, Khamari D, Vági P, Barna L, Ligeti E, Lőrincz ÁM (2021) Mac-1 receptor clustering initiates production of pro-inflammatory, antibacterial extracellular vesicles from neutrophils. Front Immunol 12:671995. https://doi.org/10.3389/fimmu.2021.671995. PMID: 34456905; PMCID: PMC8397541
Tricarico C, Clancy J, D'Souza-Schorey C (2017) Biology and biogenesis of shed microvesicles. Small GTPases 8(4):220–232. https://doi.org/10.1080/21541248.2016.1215283
Triebel T, Grillhösl B, Kacani L, Lell CP, Fuchs A, Speth C, Lass-Flörl C, Steinmann J, Dierich MP, Würzner R (2003) Importance of the terminal complement components for immune defence against Candida. Int J Med Microbiol 292(7-8):527–536. https://doi.org/10.1078/1438-4221-00211
Urban CF, Reichard U, Brinkmann V, Zychlinsky A (2006) Neutrophil extracellular traps capture and kill Candida albicans yeast and hyphal forms. Cell Microbiol. 8(4):668–676. https://doi.org/10.1111/j.1462-5822.2005.00659.x. PMID: 16548892
Valand N, Brunt E, Gazioglu O, Yesilkaya H, Mitchell D, Horley N, Arroo R, Kishore U, Wallis R, Venkatraman GU (2022) Inactivation of the complement lectin pathway by Candida tropicalis secreted aspartyl protease-1. Immunobiology 227(6):152263. https://doi.org/10.1016/j.imbio.2022.152263
vMarcil A, Gadoury C, Ash J, Zhang J, Nantel A, Whiteway M (2008) Analysis of PRA1 and its relationship to Candida albicans- macrophage interactions. Infect Immun 76(9):4345–4358. https://doi.org/10.1128/IAI.00588-07
WHO fungal priority pathogens list to guide research, development and public health action 25 October 2022
Yadav P, Singh R, Sur S, Bansal S, Chaudhry U, Tandon V (2022) Moonlighting proteins: beacon of hope in era of drug resistance in bacteria. Crit Rev Microbiol 26:1–25. https://doi.org/10.1080/1040841X.2022.2036695
Yokoi A, Ochiya T (2021 Sep) Exosomes and extracellular vesicles: Rethinking the essential values in can-

cer biology. Semin Cancer Biol 74:79–91. https://doi.org/10.1016/j.semcancer.2021.03.032

Yu H, Li Y, Li L, Huang J, Wang X, Tang R, Jiang Z, Lv L, Chen F, Yu C, Yuan K (2021) Functional reciprocity of proteins involved in mitosis and endocytosis. FEBS J 288(20):5850–5866. https://doi.org/10.1111/febs.15664

Zhang H, Lyden D (2019) Asymmetric-flow field-flow fractionation technology for exomere and small extracellular vesicle separation and characterization. Nat Protoc. 14(4):1027–1053. https://doi.org/10.1038/s41596-019-0126-x. Epub 2019 Mar 4. PMID: 30833697; PMCID: PMC6733524

Zhang B, Yin Y, Lai RC, Lim SK (2014) Immunotherapeutic potential of extracellular vesicles. Front Immunol 5:518. https://doi.org/10.3389/fimmu.2014.00518. PMID: 25374570; PMCID: PMC4205852

Zhang Z, Denans N, Liu Y, Zhulyn O, Rosenblatt HD, Wernig M, Barna M. Optogenetic manipulation of cellular communication using engineered myosin motors. Nat Cell Biol. 2021;23(2):198–208. doi: https://doi.org/10.1038/s41556-020-00625-2. Erratum in: Nat Cell Biol. 2021;23(3):293. Erratum in: Nat Cell Biol 2021;23(5):565

Zipfel PF, Skerka C (2009) Complement regulators and inhibitory proteins. Nat Rev Immunol 9(10):729–740. https://doi.org/10.1038/nri2620

Zipfel PF, Skerka C (2022) From magic bullets to modern therapeutics: Paul Ehrlich, the German immunobiologist and physician coined the term 'complement'. Mol Immunol 150:90–98. https://doi.org/10.1016/j.molimm.2022.08.002

Zipfel PF, Hallström T, Hammerschmidt S, Skerka C (2008) The complement fitness factor H: role in human diseases and for immune escape of pathogens, like pneumococci. Vaccine 26(Suppl 8):I67–I74. https://doi.org/10.1016/j.vaccine.2008.11.015

Zipfel PF, Skerka C, Kupka D, Luo S (2011) Immune escape of the human facultative pathogenic yeast Candida albicans: the many faces of the Candida Pra1 protein. Int J Med Microbiol 301(5):423–430. https://doi.org/10.1016/j.ijmm.2011.04.010

Zipfel PF, Hallström T, Riesbeck K (2013) Human complement control and complement evasion by pathogenic microbes--tipping the balance. Mol Immunol 56(3):152–160. https://doi.org/10.1016/j.molimm.2013.05.222

8 Mammalian Pattern Recognition Receptors (PRRs) Involved in Recognition of Fungi

Emily Sey, Janet A. Willment, and Gordon D. Brown

Abstract

There are over 100,000 fungal species currently identified, some of which live as commensals in the microbiome of mammalian skin and mucosal surfaces. However, several opportunistic fungi can cause life-threatening invasive infections that lead to over 1.5 million deaths across the globe each year. The four main species that cause invasive infections are *Candida, Aspergillus, Cryptococcus,* and *Pneumocystis*. The immune response to these pathogens is initiated through pattern recognition receptors (PRRs) that recognize pathogen-associated molecular patterns (PAMPs), including components of the fungal cell wall, RNA, and DNA. Following recognition, PRRs initiate phagocytosis, neutrophil extracellular trap (NET) formation, the respiratory burst, inflammasome activation, and production of chemokines and cytokines that drive inflammatory responses and shape the adaptive immune system. This chapter will focus on the PRRs that recognize PAMPs from the four main human fungal pathogens, *Candida albicans*, *Aspergillus fumigatus*, *Cryptococcus neoformans,* and *Pneumocystis jirovecii*. More specifically, we discuss PRRs from the C-type lectin receptor (CLR) family, the Toll-like receptor (TLR) family, and some other receptors that have known fungal ligands and briefly discuss the role of these receptors during antifungal immunity.

Keywords

C-type lectins · Toll-like receptors · Fungal PAMPs · PRRs · Fungal recognition *Candida albicans* · *Aspergillus fumigatus* *Cryptococcus neoformans* · *Pneumocystis*

8.1 Introduction

Fungi can cause *superficial infections* of the skin and mucosa, such as thrush and athlete's foot, and invasive infections of the blood stream and internal organs (Brown et al. 2012). Superficial infections are common, and usually cause mild discomfort but are rarely life threatening. On the other hand, *invasive infections* are rarer but have high mortality rates. Overall, fungal infections kill over 1.5 million people across the globe each year (Bongomin et al. 2017; Houst et al. 2020). Over 600 species are known to cause infection in humans. However, the four main fungal genera that lead to invasive infection are *Aspergillus, Candida, Cryptococcus,* and *Pneumocystis* (Mayer et al. 2013; Fisher et al. 2020). Humans and other mammals regularly encounter species of

E. Sey (✉) · J. A. Willment · G. D. Brown
University of Exeter, Exeter, UK
e-mail: E.Sey@exeter.ac.uk

A. A. Brakhage et al. (eds.), *Human and Animal Relationships*, The Mycota 6,
https://doi.org/10.1007/978-3-031-64853-3_8

these genera. For example, on average, hundreds of Aspergilli spores are inhaled every day and *Candida albicans* is a commensal organism of the gut microbiota, skin, and mucosa in approximately 70% of the population (Hardison and Brown 2012). Despite the high levels of exposure, the incidence of invasive infection is relatively low; due to a highly sophisticated and complex immune system that protects the body from invasion. When this system is dysregulated, the fungus invades and disseminates across the body causing life-threatening conditions. Therefore, individuals most susceptible to fungal infection include those who have a compromised immune system due to disease, such as AIDS, genetic defects, or medical interventions (Pathakumari et al. 2020).

Pattern recognition receptors (PRRs) are germline-encoded receptors that recognize pathogen-associated molecular patterns (PAMPs) that are usually components of the pathogens' cell wall (Hatinguais et al. 2020). The fungal cell wall of most, if not all, fungi is composed of three main components including chitin, glucans, and mannoproteins (Fig. 8.1) (Gow et al. 2017; Garcia-Rubio et al. 2019). The inner cell wall is composed of glucan polymers, which are the most abundant component of the fungal cell wall. The majority of glucans are β-1,3-glucans, but β-1,6; β-1,4; α-1,3; and α-1,4 can also be found. Glucans are anchored, via β-1,4 linkages, to chitin, a linear homopolymer of β-1,4-linked N-acetylglucosamine (GlcNAc). Chitin content in the cell wall varies greatly depending on the fungal species and its morphology. For instance, the cell wall of *Candida* yeasts contains around 2% chitin, while *Candida* hyphae have up to 20% chitin (Chattaway et al. 1968). The outer *Candida* cell wall is composed of highly N- or O-linked mannosylated proteins. Additionally, these mannoproteins can covalently bind to glucans or chitin via their sugar residues or glycosylphosphatidylinositol (GPI) anchored proteins (Free 2013). In addition to chitin, glucans, and mannoproteins, some fungi also contain other cell wall components. For instance, α-glucans are found in the inner cell wall of *Aspergillus* and *Cryptococcus* species (Fig. 8.1). Furthermore, galactomannan is also found in the cell wall of *Aspergillus* species. Melanin is another component found in the outer cell wall of *Aspergillus* conidia, and the inner cell wall of *Cryptococcus*, and is a high molecular weight pigment anchored to chitin and β-glucans (Fig. 8.1). Melanin is negatively charged, hydrophobic and insoluble, which helps protect fungi against stressors (Casadevall et al. 2000; Nosanchuk and Casadevall 2006). *Cryptococcus* contains a unique polysaccharide capsule, mainly composed of glucuronoxylomannan (GXM), and acts as a key virulence factor for this fungus (Casadevall et al. 2019) (Fig. 8.1). Interestingly the mechanisms for chitin synthesis are not found in *Pneumocystis*, however, some studies using probes for chitin have detected its presence in both cyst and troph form (Skalski et al. 2015). The cell wall of *Pneumocystis* cysts contains β-glucans, which are essential for cell wall stability in this morphology, whereas in the troph form, β-glucans are not present (Skalski et al. 2015). The main component of the cell wall in *Pneumocystis* is glycoprotein A, which is carboxylated with mannose residues, and is important for pathogenesis (Skalski et al. 2015).

Recognition of PAMPs expressed by fungi involves several families of PRRs that induce various signalling cascades that ultimately promote pathogen clearance through phagocytosis, production of reactive oxygen species (ROS) and reactive nitrogen species (RNS), and the release of chemokines and cytokines. The responses induced by combinations of PRRs shape the innate and consequently the adaptive immune response and are both critically required to control infection. Depending on the pathogen, specific T helper (Th) cell subsets are required for antifungal immunity (Speakman et al. 2020). Th1 and Th17 are generally considered to be important for the resolution of fungal infections, while the role of Tregs is not completely clear and may be non-protective. Th2 cells are also considered to be non-protective during invasive infections but are often studied in the context of fungal-mediated allergic responses. Historically, B cells showed a largely redundant role in the resolution of fungal disease. However, advances in B cell technology and studies in individuals with genetic defects, B cells have been shown to play an antibody dependent and independent role in fungal clearance (Bromuro et al. 2002;

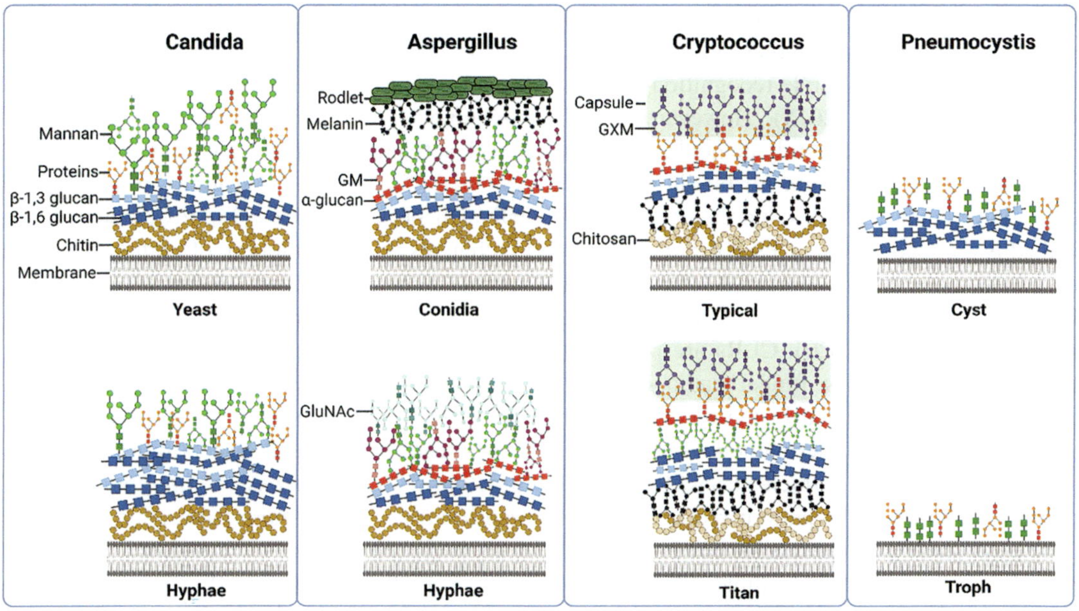

Fig. 8.1 Fungal cell wall with pattern-associated molecular patterns. The cartoon structures depicted on the top row show the main components of the cell wall during normal/commensal state, while on the bottom row are the cell wall structures during pathogenesis. The inner cell wall of most fungi is composed of chitin, β-1,3-glucan and β-1,6-glucan. In *Candida*, the outer cell wall is mainly N-, or O-linked mannans and mannoproteins. During pathogenesis, the β-glucan layer in *Candida* is thicker. In *Aspergillus*, the conidia have an outer hydrophobic rodlet layer and a layer of melanin. The inner cell wall contains typical components but also α-glucan and galactomannan (GM). The hyphal form of *Aspergillus* has no rodlet layer or melanin, and contains reduced glycoproteins while the outer layer contains N-Acetylglucosamine (GluNAc). *Cryptococcus* has an outer capsule layer composed of glucuronoxylomannan (GXM) and contains melanin within its inner cell wall. Compared to typical cells, titan cells have increased mass of all cell wall components and contain mannans in the outer cell wall. *Pneumocystis* lacks the enzymes required for chitin synthesis and the cell wall of the cyst form primarily contains β-glucans, mannoproteins, and glycoprotein A. The troph form of *Pneumocystis* lacks β-glucans, and is therefore not as thick as the cyst form. The main component of the *Pneumocystis* cell wall is glycoprotein A, which is carboxylated with mannose residues

Han et al. 2001; Lanternier et al. 2013). Antibodies have been shown to be involved in the opsonisation of fungal pathogens allowing increased recognition (Opata et al. 2015). Additionally, B cells, via an antibody-independent role, have been shown to enhance T-cell immunity against fungal pathogens (Li et al. 2017). How PRRs contribute to the induction of adaptive and innate responses will be covered below.

The main PRR families involved in fungal recognition are C-type lectin receptors (CLRs) and Toll-like receptors (TLRs), which are the focus of this chapter, but we include a brief description of other receptors that have been implicated in fungal recognition. We will cover the PRRs that have known fungal PAMPs, the downstream signalling from these PRRs and their role during infection. Furthermore, we will focus on the four major fungal pathogens that cause human disease, *Candida*, *Aspergillus*, *Cryptococcus,* and *Pneumocystis*. We will not cover PRRs that have shown to have a physiological function but that have unknown ligands, furthermore, we will not cover the endogenous function of these receptors.

8.2 C-Type Lectin Receptors

The C-type lectin superfamily is characterized by the presence of a C-type lectin domain (CTLD), which has been identified in more than 1000 proteins. The CTLD often confers binding to sugars in a calcium dependent manner and is therefore referred to as a carbohydrate recognition domain (CRD) (Brown et al. 2018). However, several classes within this superfamily do not bind sugars nor do they require calcium for binding. C-type lectins have numerous roles, but their most well-studied is the recognition of PAMPs, or damage-associated molecular patterns (DAMPs) shaping their roles in immunity and homeostasis, respectively (Drouin et al. 2020). CLRs are primarily involved in recognizing fungal pathogens, but have also been implicated in the recognition of bacteria, viruses, and other pathogens (Bermejo-Jambrina et al. 2018; Mnich et al. 2020). The main CLRs involved in pattern recognition are those expressed on myeloid cells, however, members of the selectin and the soluble collectin subfamilies are also involved in regulating antifungal immunity (Hatinguais et al. 2020). In response to infectious pathogens, the CLRs can induce a number of immune responses including, for example, phagocytosis and the respiratory burst (Brown et al. 2012). Additionally, CLRs can induce antigen-presenting cells (APCs) to upregulate co-stimulatory molecules and release T cell polarizing cytokines and chemokines, which attract other immune cells to the site of infection (Speakman et al. 2020). These cell-specific responses can be linked to specific CLRs, which are discussed in detail below for each receptor.

CLRs known to recognize fungal pathogens include the Dectin-1 cluster of receptors, including Dectin-1 and MelLec (Clec-1), the Dectin-2 cluster members, Mincle, Langerin, DC-Specific ICAM-3 grabbing nonintegrin (DC-SIGN), and receptors from other CTLD groups such as Mannose receptor (MR), CD23 and the collectins. Many of these C-type lectins signal directly through, or associate with signalling adaptor molecules (e.g., Fc receptor γ chain (FcRγ) or DAP-12) containing, immunoreceptor tyrosine-based activation motifs (ITAM) or immunoreceptor tyrosine-based inhibitory motifs (ITIM). Following ligand recognition, most CLRs trigger activation of intracellular pathways involving Syk kinase-mediated phosphorylation of tyrosine residues within the ITAM and subsequent induction of Caspase recruitment domain-containing protein 9 (CARD9), a critically important antifungal signalling molecule, or the Raf1 pathway (Geijtenbeek and Gringhuis 2009). Downstream of Syk activation, several signalling pathways are triggered including activation of mitogen-activated protein kinases (MAPK), which leads to changes in transcription of multiple genes, activation of NF-κB, and restructuring of the actin cytoskeleton through Rac and Rho GTPases (Hoving et al. 2014). CARD9 is critical for the regulation of antifungal immunity in both humans and mice. For instance, *in mice*, CARD9 deficiency leads to increased susceptibility following infection with *C. albicans*, *A. fumigatus*, and *C. neoformans*. Furthermore, these mice show decreased release of inflammatory cytokines such as IL-6, TNF-α, and IL-1β and defective neutrophil function such as reduced ROS production and phagocytosis. *In humans*, a loss-of-function mutation in CARD9 leads to profound susceptibility, including infections with *Candida* and *Aspergillus* (Drummond et al. 2018). Signalling from CLRs also includes Syk independent pathways, including Raf-1, NFAT, and the activation of inflammasomes (Gringhuis et al. 2009). The Raf-1 pathway activates transcription factors such as NFAT, IRF1, and IRF5 (Dambuza and Brown 2015). Additionally, ERK and JNK are activated and promote the expression of pro-inflammatory chemokines and cytokines (Ostrop and Lang 2017). Activation of ERK is important for ROS production and has been linked to formation of the NLRP3 inflammasome (Gross et al. 2009; Underhill et al. 2005). The inflammasomes are multi-protein complexes that activate caspase 1, which activates interleukin-1β (IL-1β) and IL-18. In addition, activation of inflammasomes induces a rapid, pro-inflammatory form of cell death called pyroptosis (Guo et al. 2015). In humans signalling through ITIM domains, begins by phosphorylation of tyrosine residues by SRC kinases following ligand recognition. However, unlike the ITAMs Syk is not recruited, instead tyrosine phosphatases are recruited which negatively regulate other signalling pathways (Geijtenbeek and Gringhuis 2009). ITIMs usually associated with

SH2 domain-containing protein tyrosine phosphatase (SHP)-1 and SHP-2, which directly impact ITAM pathways. CLRs with ITIMs such as myeloid inhibitory C-type lectin (MICL) and Dendritic Cell Immunoreceptor (DCIR) do not have known fungal ligands. Some receptors, such as Dendritic cell-specific intracellular adhesion molecule-3-grabbing nonintegrin (DC-SIGN), do not possess typical ITAMs or ITIMs. In fact, DC-SIGN contains a di-leucine motif and a tri-acidic motif, which may modulate activation signals from other PRRs such as the TLRs (Svajger et al. 2010). These motifs also modulate Raf-1 to induce phosphorylation and recruitment of histone acetyltransferases (HATs) to p65, a subunit of NF-κB, leading to increased transcription of inflammatory cytokines including IL-6, IL-10, and IL-12 (Svajger et al. 2010).

8.2.1 Dectin-1

Dectin-1 (Clec-7A), one of the best-characterized CLRs, is expressed on several myeloid cells including monocytes, macrophages, dendritic cells (DCs), and neutrophils (Seo et al. 2013; Willment et al. 2001). Dectin-1 primarily binds β-1,3-glucan, which is commonly found in the cell wall of fungi, unlike other CLRs Dectin-1 does not require calcium for binding (Fig. 8.2) (Thiagarajan et al. 2013; Brown and Gordon 2001). The CRD lacks the conserved classical carbohydrate-binding motifs that other members of the C-type lectin family use to bind carbohydrate ligands. Ligand recognition leads to signalling through the cytoplasmic tail which contains an ITAM-like domain (hem-ITAM) and a tri-acidic residue (Brown and Crocker 2016). Dectin-1 can also be cleaved by proteases such as those produced by *A. fumigatus*, and those produced by neutrophils resulting in reduced antifungal immunity. This is clinically relevant for individuals with cystic fibrosis who have increased neutrophil inflammation and elevated serine proteases. Taken together, this likely contributes to disease pathology (Griffiths et al. 2018).

Cellular responses initiated by Dectin-1 that are dependent on the size of pathogen and the cellular context, include phagocytosis, neutrophil extracellular trap (NET) formation, respiratory burst, autophagy, activation of inflammasomes, and production of lipid mediators, chemokines, and cytokines. The latter impacts multiple cells from both innate and adaptive immune systems (Salazar and Brown 2018). NETs are released by dying neutrophils and are large structures composed of chromatin and antimicrobial proteins designed to capture and immobilize microbes (Hidalgo et al. 2022). Dectin-1 controls the release of NETs based on fungal pathogen size. Large pathogens, such as *C. albicans* hyphae promote release of NETs, while small pathogens that are easily phagocytosed do not. Thus, limiting unnecessary immune pathology associated with the release of NETs (Branzk et al. 2014). Dectin-1 is critical for the control of fungal disease through induction of both innate immunity and Th1 and Th17 adaptive immune responses. For instance, following curdlan stimulation, Dectin-1 can induce several Th17-associated cytokines such as IL-1β, IL-6, and IL-23 both in vitro and in vivo.

The role of Dectin-1 during *C. albicans* infection is well-studied and has led to many insights into the interactions between pathogen and host during infection. For instance, immunity to *C. albicans* via Dectin-1-mediated recognition of β-glucans on *C. albicans* leads to several immune responses including ROS production, and pro-inflammatory cytokine production including IL-12, IL-6, and TNF-α, and anti-inflammatory cytokines, like IL-10, from macrophages. Dectin-1-knockout mice have a significantly reduced survival following *C. albicans* infection (Taylor et al. 2007). Although data opposing these results were published, these discrepancies were subsequently shown to be due to the strain *C. albicans* and not the genetic background of mice used (Saijo et al. 2007; Marakalala et al. 2013). Strain-dependent responses are now commonly observed for various receptors including those in the TLR family (Carmona et al. 2013). Dectin-1 expression on DCs is important for recognition of *C. albicans* at the skin and mucosal surfaces,

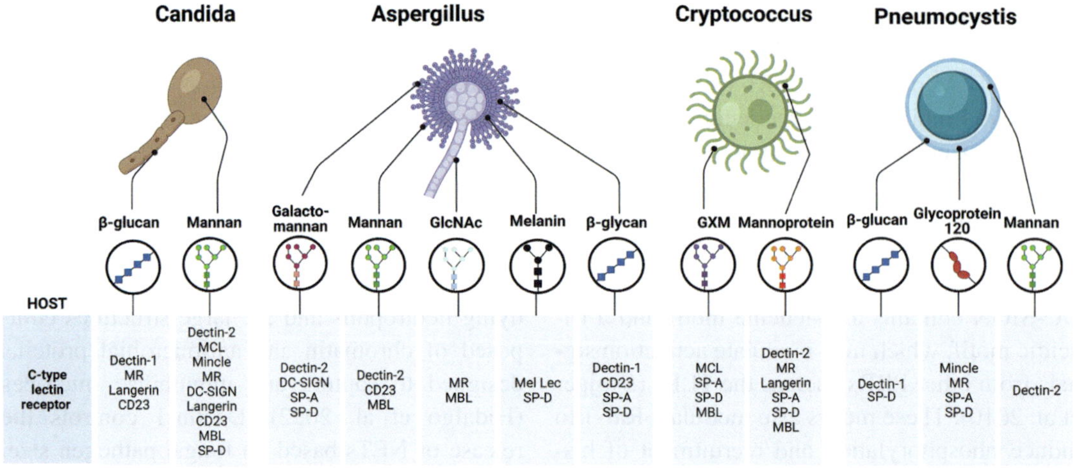

Fig. 8.2 C-type lectin receptors in fungal pathogen recognition. CLRs recognize a variety of fungal pathogen-associated molecular patterns (PAMPs). In *Candida* species, β-glucan is recognized by Dectin-1, Mannose receptor (MR), Langerin, and CD23. While mannans are recognized by Dectin-2, MCL, Mincle, MR, Dendritic cell-specific intracellular adhesion molecule-3-grabbing nonintegrin (DC-SIGN), Langerin, CD23, and the collectins mannose-binding lectin (MBL), and the surfactant protein (SP)-D. In *Aspergillus* species, Galactomannan is recognized by Dectin-2, DC-SIGN, and the collectins SP-A and SP-D. Mannan is recognized by Dectin-2, CD23, and MBL. N-Acetylglucosamine (GlcNAc), present in the outer layer of *Aspergillus* hyphae is recognized by MR and MBL. Melanin, present in the outer layer of conidia is recognized by recently discovered MelLec and the collectin SP-D, but not SP-A. While β-glucan is recognized by Dectin-1, CD23, SP-A, and SP-D. In *Cryptococcus* species, glucuronoxylomannan (GXM) in the capsule is recognized by MCL, and the collectins MBL, SP-A, and SP-D. Cryptococcal mannoproteins are recognized by Dectin-2, MR, Langerin, SP-A, SP-B, and MBL. In *Pneumocystis* species, β-glucans are recognized by Dectin-1 and SP-D, but not SP-A. Glycoprotein 120 is recognized by Mincle, MR, SP-A, and SP-D and mannan is recognized by Dectin-2. Figure created using Biorender.com

such as the oral cavity, gut, and subsequently driving protective adaptive immune responses. This is highlighted in patients with HIV who have low T cell counts, and high prevalence of *oropharyngeal candidiasis* (OPC) (De Repentigny et al. 2004). In the gut, increased susceptibility to *C. albicans* colonization is observed in mice with macrophage-specific Dectin-1 knockout. Furthermore, following systemic *C. albicans* infection, Dectin-1 knockout mice had increased fungal burden in the kidney and intestine. These data suggest an essential role for Dectin-1 in development of adaptive antifungal immunity in the gut (Drummond et al. 2016). In the skin, *C. albicans* yeasts promote Th17 cell polarization via Dectin-1 and produce IL-17A as a key cytokine that drives fungal clearance in the skin (Kashem et al. 2015). *In humans*, polymorphisms in Dectin-1 such as an early-stop-codon mutation Y238X lead to increased risk of recurrent vulvovaginal candidiasis (VVC) (Zahedi et al. 2016). Additionally, these patients have defective neutrophil function and monocytes show reduced production of IL-17 following *C. albicans* stimulation in vitro, however, phagocytosis was unaffected (Ferwerda et al. 2009). Conversely, the I223S mutation in Dectin-1 was shown in HIV-infected patients in Africa to be protective against *oropharyngeal candidiasis* (OPC) (Plantinga et al. 2010). Therefore, Dectin-1 is important for the regulation of innate and adaptive immunity to *C. albicans* in both mice and humans.

Dectin-1 has also been investigated in response to other *Candida* species. For instance, macrophages from Dectin-1 knockout mice show decreased production of pro-inflammatory cytokines including TNF-α IL-6, IL-12, and IL-10 in response to *C. glabrata*. Furthermore, Dectin-1-deficient neutrophils show reduced ROS production (Chen et al. 2017). Additionally, during systemic *C. glabrata* infection Dectin-1 knockout mice also showed increased fungal burden, decreased leukocyte recruitment, and reduced production of IFNγ and IL-17. The latter suggests Th1 and Th17 responses may be regulated by Dectin-1 during *C. glabrata* infection (Ifrim et al. 2013). Thus, Dectin-1 plays a role in immunity to various *Candida* species highlighting the essential role of this receptor during antifungal immunity.

Dectin-1 is the most-studied receptor in the context of *Aspergillus* infection. RAW 264.7 cells, a macrophage-like cell line, overexpressing Dectin-1 show enhanced production of TNF-α, G-CSF, and MIP-2 in vitro in response to *A. fumigatus* compared to parental RAW 264.7 cells (Steele et al. 2005). Furthermore, Dectin-1 recognizes swollen conidia or hyphae more effectively than resting conidia, a morphology associated with lower β-glucan exposure (Steele et al. 2005). *Human DCs* generated from monocytes in vitro, known as moDCs, show macrophage and monocyte-like characteristics. Silencing of Dectin-1 in moDCs using short interfering RNA (siRNA) showed reduced TNF-α and IL-12 in response to *A. fumigatus* germ tubes (Mezger et al. 2008). In mice, Dectin-1 is upregulated on phagocytes following sub-lethal inhalation of *A. fumigatus* conidia (Savers et al. 2016). Dectin-1 knockout mice show increased sensitivity and reduced production of pro-inflammatory cytokines in response to intratracheal administration of *A. fumigatus* compared to wildtype mice (Werner et al. 2009). Thus, Dectin-1 is important for protection during *A. fumigatus* infection. More specifically Dectin-1 has been associated with the adaptive immune response to *A. fumigatus* by promoting Th17 differentiation and inhibiting polarization of Th1 cells (Rivera et al. 2011). In humans, patients with invasive aspergillosis (IA) showed lower levels of Dectin-1-expressing monocytes, which may explain disease susceptibility. Additionally, in humans, polymorphisms in Dectin-1 such as the CLEC7A rs3901533 (T/T) and rs7309123 (G/G) genotypes or the presence of Y238X polymorphism, show increased susceptibility to invasive aspergillosis IA (Sainz et al. 2012; Cunha et al. 2010). Therefore, Dectin-1 plays a protective role against *Aspergillus* in both mice and humans and defects in this receptor lead to increased susceptibility and worse disease outcomes.

Early studies using soluble Dectin-1 Fc-fusion proteins, and antibodies blocking Dectin-1 on alveolar macrophages in vitro, supported recognition of *C. neoformans* spores by Dectin-1 (Giles et al. 2009). However, later studies using CHO cell lines over expressing Dectin-1, and Dectin-1 expression in BWZ reporter cell lines showed limited recognition or phagocytosis of *C. neoformans* spores and yeasts (Walsh et al. 2017). Furthermore, alveolar macrophages and bone marrow (BM) derived DCs (BMDCs) from Dectin-1 knockout mice compared to wildtype mice showed no difference in binding or phagocytosis of *C. neoformans* spores (Walsh et al. 2017). Finally, following either intranasal administration of *C. neoformans* spores Dectin-1 knockout mice showed similar fungal burden and survival as wildtype mice (Walsh et al. 2017; Nakamura et al. 2007). However, CARD9 knockout mice infected with pulmonary *C. neoformans* showed increased mortality and higher CFU burden in the lung than wildtype mice. These mice also showed impaired Th17 polarization and impaired NK cell, and memory T cell recruitment. Thus, the CLR signalling pathway is highly likely to be involved in the resolution of pulmonary cryptococcosis, but not via Dectin-1 (Yamamoto et al. 2014).

Dectin-1 has been shown to bind *P. murina* in vitro using Fc-Dectin-1 fusion proteins. Furthermore, administration of these Fc-proteins in vivo following pulmonary *P. murina* or *P. carinii* infection led to reduced fungal burdens (Ricks et al. 2013; Rapaka et al. 2007). Dectin-1 on macrophages has been shown to recognize *Pneumocystis* species, to induce release of TNF-α and expression of Dectin-1 is upregulated during infection. Additionally, blocking Dectin-1 function led to reduced ROS production in response to *P. carinii* in vitro. Dectin-1 knockout mice are also more susceptible to infection with *P. murina* than wildtype animals. However, cytokine production was equivalent in both mouse groups (Saijo et al. 2007). Therefore, Dectin-1 plays a role in resistance to *Pneumocystis* infection, but the specific mechanisms are yet to be defined. Dectin-1 has also been implicated in driving protective immunity to several other fungal pathogens including, most recently, *Corynespora cassiicola*, a causative agent of phaeohyphomycosis (Drummond et al. 2022).

8.2.2 Dectin-2

Dectin-2 is the best-characterized receptor in the Dectin-2 cluster and is mainly expressed on myeloid cells including tissue resident macrophages, neutrophils, monocytes, and several DC subsets (Taylor et al. 2005). The CTLD of Dectin-2 contains an EPN motif classically associated with recognition of mannose and requires calcium for binding (Kerscher et al. 2013). Dectin-2 recognizes high mannose structures in many fungal pathogens, including *C. albicans, Saccharomyces cerevisiae, Microsporum audounii, Trichophyton rubrum, Pneumocystis, Paracoccidioides brasiliensis, Histoplasma capsulatum, Malassezia*, and capsule-deficient *C. neoformans and A. fumigatus* (Fig. 8.2) (Saijo and Iwakura 2011; Saijo et al. 2010; Tanno et al. 2019; Ishikawa et al. 2013; Chang et al. 2017; Preite et al. 2018; Sato et al. 2006; Kottom et al. 2018; Reedy et al. 2023). Pattern recognition by Dectin-2 leads to coupling with the ITAM-bearing FcRy signalling molecule and activation of NF-κB (Sato et al. 2006). Signalling pathways include the Syk/CARD9 pathways and MAPK pathways. Signalling leads to the secretion of pro-inflammatory cytokines, such as IL-23 and IL-1β, the respiratory burst and inflammasome activation (Kerscher et al. 2013). Mincle, another CLR discussed below, also signals through FcRγ. However, Dectin-2 coupling to FcRγ-Syk promoted an early and strong signal that differed from Mincle signalling through FcRγ, which was delayed. This difference in strength and timing led to different signalling responses for each receptor (Watanabe et al. 2023).

Soluble Dectin-2 Fc-proteins, and RAW macrophages overexpressing Dectin-2 show binding to *C. albicans*, with higher binding to hyphae compared to yeasts (Sato et al. 2006; Vendele et al. 2020). Dectin-2 knockout mice have increased susceptibility to both *C. albicans* and *C. glabrata* infection, and present with higher fungal burdens in the kidneys. Additionally, Dectin-2 knockout mice infected with *C. glabrata* have lower Th1 and Th17 responses, and reductions in IFNγ and IL-17 (Chen et al. 2017). Moreover, in ex vivo Dectin-2 knockout neutrophils and macrophages showed a decreased ability in uptake and killing of *C. glabrata* (Ifrim et al. 2014, 2016). Therefore, Dectin-2 is involved in promoting clearance of both *C. albicans* and *C. glabrata*, and promoting phagocytosis and cytokine release.

Dectin-2 is involved in *A. fumigatus* recognition by pDCs, leading to cytokine release and extracellular trap formation known as pETs, similar to NETs described above (Loures et al. 2015). Dectin-2 binds galactomannan from the cell wall of *A. fumigatus* and in vitro stimulation with galactomannan leads to Syk activation and release of TNF-α (Fig. 8.2) (Reedy et al. 2023). Dectin-2 knockout mice infected with *A. fumigatus* via oropharyngeal aspiration showed increased recruitment of leukocytes (CD45+) to the lung. However, these mice showed similar fungal burdens as wildtype mice (Reedy et al. 2023). Therefore, other receptors may play a compensatory role during *A. fumigatus* in the absence of Dectin-2. In humans, Dectin-2 is upregulated in alveolar macrophages during invasive aspergillosis (IA). Furthermore, a Dectin-2 homozygous deletion mutation (507 del C) resulting in a frameshift (N170I) and early stop codon was identified in a patient with IA. Peripheral blood mononuclear cells (PBMCs) from this patient showed reduced cytokine production in response to *A. fumigatus* in vitro (Griffiths et al. 2021). Thus, Dectin-2 plays a role in recognition of *A. fumigatus* and loss of function in this receptor may cause increased susceptibility to infection in humans.

Dectin-2 recognizes mannoprotein in the cell wall of *C. neoformans* (Fig. 8.2) (Tanno et al. 2019; Nakamura et al. 2015). BMDCs from Dectin-2 knockout mice show reduced phagocytosis of *C. neoformans* (serotype B, B3501) compared to BMDCs from wildtype mice (Kitai et al.

2021). However, Dectin-2 knockout mice, following infection with the same strain of *C. neoformans* showed no significant differences in survival compared to wildtype mice (Nakamura et al. 2015). However, Dectin-2 knockout mice showed a significant increase in IL-4, IL-5, and IL-13, associated with Th2 cells, and increased mucin production, suggesting Dectin-2 may suppress Th2 responses during infection (Nakamura et al. 2015). In humans Dectin-2 rs11045418 genotype, which has an unknown functional change, was found to be associated with susceptibility to pulmonary cryptococcosis but not cryptococcal meningitis (Hu et al. 2015).

Dectin-2 binds *P. carinii* in vitro, demonstrated using Fc-proteins and overexpressing Dectin-2 RAW macrophages (Kottom et al. 2018). Furthermore, in vitro alveolar macrophages from Dectin-2 knockout mice show reduced release of IL-6 and TNF-α in response to *P. murina*. Following pulmonary infection with *P. murina,* Dectin-2 knockout mice did not show an increase in fungal burden, in immunocompetent or immunocompromised mice compared to wildtype mice immunocompetent or immunocompromised mice respectively. Thus, Dectin-2 may be involved in initiating innate immune responses based on in vitro data, but deficiency does not impact overall clearance of infection (Kottom et al. 2018). Thus, the role of Dectin-2 in *Pneumocystis* recognition remains unclear.

8.2.3 MCL (CLECSF8 or Dectin-3)

MCL was originally thought to be specifically expressed on macrophages but is now known to be expressed on neutrophils, monocytes, and several small DC sub-populations. The CTLD of MCL contains calcium-binding residues but lacks the conserved motifs associated with carbohydrate binding. However, MCL has been shown to bind α-mannans and GXM in the fungal cell wall (Fig. 8.2) (Zhao et al. 2014; Huang et al. 2018). The short cytoplasmic domain does not contain any known signalling motifs, and the transmembrane domain lacks charged residues associated with binding to adaptor molecules, such as DAP10 or DAP12. However, MCL has been shown to associate with other CLRs (particularly Mincle and Dectin-2) to signal via Fc receptor gamma (FcγR). Signalling ensues through Syk kinase to induce phagocytosis, the respiratory burst, and release of pro-inflammatory cytokines (Zhu et al. 2013; Zhao et al. 2014).

MCL recognizes α-mannans in the cell wall of *C. albicans* hyphae, this interaction requires the formation of heterodimers with Dectin-2 (Fig. 8.2) (Zhu et al. 2013). During systemic *C. albicans* infection, MCL knockout mice, or wildtype mice receiving MCL blocking antibodies, were more susceptible to infection, and had increased fungal burden in the kidney compared to wildtype mice. Additionally, these mice showed decreased production of pro-inflammatory chemokines and cytokines such as TNF-α, IL-10, and IL-6 (Zhu et al. 2013). MCL knockout mice were also more susceptible to dextran sodium sulfate (DSS)-induced colitis, compared to wildtype mice. These mice showed increased burden of *C. tropicalis* in the gut, which aggravated colitis symptoms (Wang et al. 2016b). Thus, MCL interaction with *Candida* spp. in the gut influences colonic homeostasis.

MCL recognizes GXM of *C. neoformans* serotype AD and *C. gattii* Serotype B, but not common *C. neoformans* serotypes (Fig. 8.2) (Huang et al. 2018). *C. neoformans* H99, a serotype A strain, and the less virulent *C. neoformans* D52 (Serotype D) strain are commonly used in Cryptococcosis infection models. In response to pulmonary infection with *C. neoformans* serotype A, MCL knockout mice showed similar survival and no significant differences in fungal burden (Huang et al. 2018; Campuzano et al. 2017). Additionally, in vitro, DCs and macrophages from MCL knockout mice and wildtype mice showed similar capacity for fungal uptake and killing (Campuzano et al. 2017). However, in mice infected with *C. neoformans* serotype AD or *C. gattii* serotype B, MCL knockout mice showed increased susceptibility to pulmonary infection and more lung injury due to reduced macrophage infiltration and killing activities (Huang et al. 2018). In humans, MCL expression on pDCs was shown to be required for fungistatic activity against *C. neoformans* in vitro, via production of ROS (Hole et al. 2016). Therefore, MCL appears to play a specific role during Cryptococcus infection depending on the serotype. More research is required to define why these differences are observed.

MCL has not been studied in the context of *A. fumigatus* and *Pneumocystis* infections. However, as *A. fumigatus* contains α-mannans in the cell wall this receptor may be involved in recognition of this pathogen. Further research is required to reveal the involvement of this receptor in recognition and immunity to *A. fumigatus* and *Pneumocystis*.

8.2.4 Mincle

Mincle is expressed in myeloid cells, and shows highest expression on macrophages, but has also been found on monocytes, neutrophils, and myeloid DCs. Additionally, it is strongly induced in macrophages following stimulation with LPS, TNF-α, IL-6, or IFN-γ (Patin et al. 2017). The CTLD of Mincle contains an EPN carbohydrate-binding motif and binds fungal α-mannans found in *Malassezia* sp. and *Fonsecaea pedrosoi*. Mincle associates with FcRγ chain via the charged arginine residues in the transmembrane domain (Yamasaki et al. 2008). Surface expression of Mincle is dependent on the presence of the murine MCL stalk region and this interaction facilitates positive expression and enhanced signalling. Downstream signalling is induced through Syk/CARD9 and MAPK pathways (Miyake et al. 2015). Signalling results in increased release of cytokines and chemokines such as TNF-α, CXCL2, CXCL1, and IL-6 (Lee et al. 2017). Additionally, in collaboration with TLRs, or MCL, Mincle induces the respiratory burst and upregulation of Complement Receptor (CR) 3 (Richardson and Williams 2014).

Mincle recognizes *C. albicans* hyphae, in vitro, promoting release of TNF-α by mouse macrophages (Wells et al. 2008). Furthermore, Mincle heterodimerized with Dectin-2 on BMDCs, but not on BMDMs, has been shown to bind *C. albicans*, *C. glabrata*, *C. tropicalis*, and *C. parapsilosis* in vitro. Macrophages from Mincle knockout mice show reduced production of TNF-α in vitro following stimulation with these *Candida* spp. and knockout mice are highly susceptible to systemic *C. albicans* infection (Wells et al. 2008). However, contrary findings have suggested that Mincle knockout is not more susceptible to systemic infection with *C. albicans*, *C. glabrata*, and *C. parapsilosis*, but does show increased susceptibility and fungal burden in the kidney following systemic infection with *C. tropicalis* (Thompson et al. 2021). Therefore, the effect of Mincle during systemic *Candida* infection is still unclear and may be strain and species dependent.

The fungal ligand for Mincle on *A. fumigatus* or *C. neoformans* has not been identified. However, Mincle expression increases in the cornea of rats, mice, and patients with *Aspergillus fumigatus* keratitis (Yu et al. 2018; Zhao et al. 2017a). Additionally, in mice and rats, Mincle expression correlated with increased expression of pro-inflammatory cytokines such as TNF-α and IL-1β in the cornea (Yu et al. 2018, Zhao et al. 2017a). However, the role of Mincle in response to *A. fumigatus* infection at other sites has not been investigated. Following pulmonary infection with *C. neoformans* serotype D in mice, Mincle expression is upregulated in the lung during the early phase of infection. Furthermore, there was a reduction in cytokines associated with the newly classified Th22 cells, including IL-22, TNF-α, and IL-6 (Sato et al. 2020). However, in Mincle knockout mice during pulmonary infection with *C. neoformans* serotype D, there were no differences in fungal burden (Sato et al. 2020). Therefore, other receptors that recognize *C. neoformans* may compensate for the defect in Mincle. Overall, the role of Mincle in response to *A. fumigatus* or *C. neoformans* has not been fully explored and requires further investigation to unpick the role of this receptor in response to these fungi.

Mincle binds the surface glycoprotein (gp)120 in the *Pneumocystis* cell wall (Fig. 8.2) (Kottom et al. 2017). Fc-Mincle fusion proteins showed binding to *P. carinii* and macrophages overexpressing Mincle showed increased binding to *P. carinii* than control macrophages in vitro (Kottom et al. 2017). Mincle knockout mice show increased susceptibility to pulmonary *P. murina* infection and increased fungal burden in the lung. However, Mincle knockout mice did not show any differences in mortality compared to wildtype mice. This lack of effect may be due to increased IL-1R expression, which can promote a variety of innate immune responses, and increased expression of other PRRs such as Dectin-1, Dectin-2, and MCL in the Mincle knockout mice (Kottom et al. 2017).

8.2.5 Mannose Receptor

The MR is mainly expressed in macrophages, immature DCs, and endothelial cells. It contains an N-terminal cysteine-rich domain, a fibronectin (FN) type II domain, and eight C-type lectin domains (CTLDs), which vary in their binding specificity. The cysteine-rich domain binds sulfated sugars, the fibronectin type II domain recognizes collagen, while the CTLDs recognize glycol-structures with terminal mannose, fucose, or N-acetylglucosamine (GlcNAc) (Fig. 8.2) (Taylor et al. 1992; Gazi et al. 2011). The intracellular region of MR does not contain a known signalling domain but has been shown to interact with the ITAM-containing adaptor molecule FcRγ, and other PRRs such as TLR2, to promote intracellular signalling including Rac/Pak/Cdc-42 pathways (Tachado et al. 2007; Rajaram et al. 2017). MR is also known to be proteolytically cleaved by metalloproteases and released into the extracellular space as a soluble form (sMR), a process known as MR shedding. This process is stimulated by binding to fungal particles from *P. carinii, C. albicans, A. fumigatus,* and requires triggering through Dectin-1 (Gazi et al. 2011).

The MR has been shown to bind mannose and glycans from *C. albicans*, using cell lines overexpressing the receptor (Fig. 8.2) (Ezekowitz et al. 1990; Stahl and Ezekowitz 1998). Furthermore, using blocking antibodies in vitro, MR is important for phagocytosis, release of pro-inflammatory cytokines, and ROS production in response to *C. albicans* (Porcaro et al. 2003; Lee et al. 2003; Chieppa et al. 2003). The role of the MR in phagocytosis is mainly involved during later phases of this process as the receptor was not found to be required for fungal uptake (Heinsbroek et al. 2008). Additionally, MR was the main PRR involved in IL-17 production by human PMBCs stimulated in vitro with *C. albicans* (Van De Veerdonk et al. 2009). However, following intraperitoneal infection with *C. albicans*, MR knockout mice show no significant differences in survival, fungal burden, cell recruitment, or antibody production (Lee et al. 2003). Therefore, the role of MR in response to *C. albicans* remains unclear, but is thought to be required for detection of *C. albicans* in the maturing phagosome and modulation of cytokine production.

The MR has been associated with early innate immunity to fungal keratitis. MR expression increased during the initial phase of *A. fumigatus* keratitis infection, along with an increased expression of TNF-α and IL-1β in human and rat cornea. Furthermore, the expression of MR was linked to Dectin-1 in vitro, as addition of curdlan (a Dectin-1 agonist) to human corneal epithelial cells led to increased MR expression, while addition of laminarin (a Dectin-1 antagonist), led to decreased MR expression (Wang et al. 2016a). Additionally, in vitro internalization by pulmonary DCs of *A. fumigatus* conidia but not hyphae was inhibited by addition of MR-reacting mAb, PAM-1 (Bozza et al. 2002). However, *A. fumigatus*-specific ligands for MR have not been identified. Additionally, the role of MR during pulmonary *A. fumigatus* infection in the lung requires further investigation.

Human and murine DCs recognize mannoproteins from *C. neoformans*, and blocking of MR on DCs in vitro, showed decreased fungal uptake (Fig. 8.2) (Mansour et al. 2002). During pulmonary *Cryptococcus* infection, MR knockout mice are more susceptible to infection and have a higher fungal burden in the lungs (Dan et al. 2008). Additionally, removal of carbohydrate residues on the mannoproteins prior to infection in wildtype mice, had the same effect as MR deficiency suggesting these are the ligands on *C. neoformans* for MR (Dan et al. 2008).

The MR on murine macrophages was found to bind gp120 from *P. carinii* (Fig. 8.2) (O'Riordan et al. 1995) and blocking this receptor on human macrophages in vitro, significantly reduces fungal binding and phagocytosis (Zhang et al. 2004). However, following pulmonary infection with *P. carinii*, MR knockout mice showed similar fungal burden as wildtype mice in both immunocompetent and $CD4^{+}$ depleted mice. The MR knockout did show an increased influx of phagocytes into the lung, which was associated with increased pulmonary pathology (Swain et al. 2003). In humans, reduced expression of MR is associated with decreased phagocytosis of *P. carinii* by alveolar macrophages in vitro (Koziel et al. 1998). Therefore, the full role of MR during *Pneumocystis* infection remains unresolved.

8.2.6 DC-SIGN

DC-SIGN is expressed on human DCs and a subset of macrophages and is involved in recognition and phagocytosis of fungi. Although there are eight murine homologues the closest related murine molecule SIGN-R3 is implicated in recognizing fungi and regulating intestinal homeostasis (Eriksson et al. 2013). DC-SIGN recognizes glycans through its EPN motif in its CRD, which binds fructose and mannose (Feinberg et al. 2001). More specifically, DC-SIGN binds N-linked mannan in the cell wall of *C. albicans*, *C. dubliniensis*, *C. glabrata*, and *C. parapsilosis* (Fig. 8.2) (Cambi et al. 2003, 2009), galactomannans in *A. fumigatus* and mannoproteins from *C. neoformans* (Mansour et al. 2006; Li et al. 2018a). Dectin-1 ligation has been shown to promote the recruitment of DC-SIGN to the pathogen contact site and enhance the capture of *C. albicans* (Choraghe and Neumann 2021). However, the effects of the interaction between DC-SIGN and these fungi have not been investigated in vitro or in vivo and so the role of DC-SIGN during antifungal immunity is currently unknown. DC-SIGN, shows allelic variants in humans, with rs4804800 (G), rs11465384 (T), rs7248637 (A), and rs7252229 (C) alleles associated with increased risk of IA (Sainz et al. 2012). Therefore, DC-SIGN may have a role in protective immunity to *Aspergillus* infection.

8.2.7 MelLec

Human MelLec is expressed on endothelial and myeloid cells including monocytes, macrophages, granulocytes, and some populations of DCs, whereas murine MelLec expression is restricted to endothelial cells (Lopez Robles et al. 2017; Thebault et al. 2009; Stappers et al. 2018). MelLec contains a single CTLD, and a cytoplasmic tail with a tyrosine residue within an undefined YSST motif and a tri-acidic motif. Additionally, the cytoplasmic domain contains arginine and tyrosine residues associated with recruitment of ITAM-containing adaptor molecules (Sobanov et al. 2001; Colonna et al. 2000). However, MelLec has not been shown to associate with any of the common adaptor molecules and downstream signalling is unknown.

MelLec was shown to bind 1, 8-dihydroxynaphthalene melanin commonly found in the cell wall of melanized fungi such as *A. fumigatus* conidia (Fig. 8.2) (Stappers et al. 2018). MelLec knockout mice have significantly reduced survival following systemic, but not pulmonary, infection with *A. fumigatus* (Stappers et al. 2018). Using a standard mouse model of allergic sensitization and challenge with live *A. fumigatus* conidia, MelLec knockout mice showed significantly higher fungal burden but reduced pathology, compared to wildtype mice (Tone et al. 2021). In humans, a single nucleotide polymorphism in MelLec is associated with impaired inflammatory response and increased susceptibility of stem cell transplant recipients to disseminated *Aspergillus* infection (Stappers et al. 2018). Therefore, MelLec plays a role in antifungal immunity in both mice and humans.

8.2.8 Langerin

Langerin is expressed on a small subset of DCs known as Langerhans cells, found in the epithelium of mucosal tissues (De Jong et al. 2010).

Langerin binds both mannan and β-glucans, and ligation leads to the formation of a trimer on the cell surface (Fig. 8.2). Intracellular signalling induces the formation of Birbeck granules, intracellular vesicles specific to Langerhans cells (McDermott et al. 2002).

Langerin recognizes both mannan and β-glucan from *C. albicans*, *C. glabrata*, *C. guilliermondii*, *C. krusei*, *C. lusitaniae*, and *C. nivariensis* (Fig. 8.2) (De Jong et al. 2010). During OPC, depletion of Langerin⁺ DCs in mice led to reduced IL-17 production, which is known to be essential for resolution of infection (Sparber et al. 2018). Additionally, in the gastrointestinal tract, Langerin⁺ DCs localized with *C. albicans* and *C. tropicalis* in the Peyer patches, suggesting these cells are responsible for sampling *Candida* species to maintain homeostasis (De Jesus et al. 2015). Therefore, Langerin may be important for immunity to mucosal *Candida* infections. However, more research is required to fully understand the role of this receptor during *Candida* infection.

Langerin shows weak binding to both capsular and acapsular *C. neoformans* and *C. gattii* (De Jong et al. 2010). However, during pulmonary *C. neoformans* infection, Langerin⁺ DCs appear to impair early protection against *C. neoformans* infection in mice (Guasconi et al. 2022). Depletion of Langerin⁺ DCs in mice using a diphtheria toxin treatment model led to lower fungal burdens compared to wildtype mice. Additionally, lung architecture showed less inflammation, suggesting depletion of Langerin⁺ DCs favors a pulmonary protective immune response against *C. neoformans* infection (Guasconi et al. 2022). Therefore, the role of Langerin is non-protective during *Cryptococcus* infection.

8.2.9 CD23

CD23 (FcεRII or Clec4J) is a CLR expressed in mature B cells, eosinophils, monocytes, activated macrophages, follicular dendritic cells, and keratinocytes. CD23 mainly recognizes IgE and is commonly used as a B cell activation marker (Mossalayi et al. 2009; Engeroff and Vogel 2021). Murine CD23 has been shown to bind mannose, GlcNAc, glucose, and fucose, however, the human CD23 does not show binding to these sugars (Fig. 8.2) (Jegouzo et al. 2019).

CD23 has been shown to recognize α-mannan and β-glucan from *C. albicans* and *A. fumigatus* but does not recognize GXM from *Cryptococcus* (Fig. 8.2) (Guo et al. 2018). Following intravenous infection with *C. albicans*, CD23 knockout mice show reduced survival and increased fungal burdens in the kidney and liver. Additionally, these mice had reduced production of IL-6, TNF-α, IL-23, and IL-1β in the kidney (Guo et al. 2018). The role of CD23 in antifungal immunity has further been demonstrated in another model of *C. albicans* infection using JNK-1-deficient mice, which show significantly increased levels of CD23 expression. These mice show a higher survival rate compared to wildtype mice during *C. albicans* infection. Furthermore, blocking the upregulation of CD23 in these mice, increases the fungal burden (Zhao et al. 2017b). Therefore, upregulation of CD23 is important for fungal clearance and survival during *C. albicans* infection. Following intranasal infection with *A. fumigatus* CD23 knockout mice show reduced survival and increased fungal burden in the lung compared to wildtype mice (Guo et al. 2018). Thus, CD23 plays a role in immunity and fungal clearance of both *C. albicans* and *A. fumigatus*. Recognition of *Pneumocystis* by CD23 has not been investigated, however, during *Pneumocystis* infection CD23 is upregulated on alternatively activated macrophages from the lung (Deckman et al. 2017). Suggesting this receptor may play a role during *Pneumocystis* infection.

8.2.10 Collectins

Collectins are soluble CLRs that contain a collagen-like domain and create complexes with up to 27 subunits. Nine collectins have been identified, with three known to be involved in fungal pathogen recognition in mice and humans including, mannose-binding lectin (MBL), and the surfactant proteins SP-A and SP-D (Cummings and Mcever 2009). Collectins recognize components of the fungal cell wall and their main function is to promote microbial aggregation, opsonisation, complement activation, and to modulate the inflammatory response (Cummings and Mcever 2009).

8.2.11 Mannose-Binding Lectin

MBL has been identified as two variants, MBL-A and MBL-C, only one variant, known as MBL is found in humans. MBL is primarily synthesized by the liver, but low mRNA levels have also been found in the small intestine of mice, and in the small intestine and testis tissues of humans (Seyfarth et al. 2006). MBL is released into the blood serum during stress conditions such as infection, and low MBL levels are associated with a predisposition to infection in humans (Neth et al. 2000).

MBL has been shown to bind mannose in the cell wall of *C. albicans* and *C. parapsilosis* leading to activation of complement and aggregation (Fig. 8.2) (Ip and Lau 2004; Neth et al. 2000; Van Asbeck et al. 2008). Additionally, MBL has been shown to promote opsonophagocytosis of *C. albicans* by human neutrophils, by interacting with CR1 independent of complement (Li et al. 2012). Following systemic *C. albicans* infection, mice deficient in MBL-A show similar survival as wildtype mice (Lee et al. 2002). However, mice deficient in both MBL-A and MBL-C show reduced survival during systemic *C. albicans* infection, and phagocytes from these mice showed reduced fungal uptake compared to wildtype mice (Held et al. 2008). Therefore, in MBL-A single knockout mice, MBL-C may compensation for loss of MBL-A. Furthermore, in a DSS model of colitis, MBL double knockout mice showed increased dissemination of *C. albicans* to the kidneys and lungs (Choteau et al. 2016). In humans, MBL protein is increased in the vaginal mucosa during VVC (Pellis et al. 2005). Meta-analysis of polymorphisms in MBL2 codon 54 allele B in humans, resulting in lower MBL production, showed increased susceptibility to recurrent vulvovaginal candidiasis (VVC) (Nedovic et al. 2014). Thus, MBL plays a role in protection against systemic and mucosal *C. albicans* infections.

MBL binds mannose and glucuronoxylomannan (GXM) in the cell wall of *A. fumigatus* (Fig. 8.2) (Neth et al. 2000). MBL binding to *A. fumigatus* conidia promotes fungal aggregation, phagocytosis, and activation of complement (Kaur et al. 2007). Mice with single MBL-A or MBL-C deficiencies have not shown differences in IA compared to wildtype mice, which is likely due to upregulation of the other MBL (Hogaboam et al. 2004). However, in mice, during IA infection, treatment with recombinant MBL increased survival by 80% compared to untreated mice. Additionally, treated mice had decreased fungal burdens and increased TNF-α and IL-1α production (Kaur et al. 2007). Genetic polymorphisms in MBL are associated with worse IA outcomes in patients. Therefore, MBL is important in the resolution of *A. fumigatus* infection, and treatment with recombinant MBL may be a possible therapeutic.

MBL-A and MBL-C were shown to bind mannoproteins and GXM in the cell walls of *C. neoformans* (Fig. 8.2) (Ip and Lau 2004). MBL does not induce aggregation of *C. neoformans* but is involved in activating complement and promoting phagocytosis by neutrophils (Eisen et al. 2008; Van Asbeck et al. 2008). Production of TNF-α by human PBMCs in vitro was enhanced following addition of MBL, compared to *C. neoformans* stimulation alone (Chaka et al. 1997). Patients with cryptococcal meningitis show a defect in MBL function in the CNS but not in the lung, suggesting MBL may be involved following dissemination (Eisen et al. 2008). Thus, MBL also plays a role during immunity to *C. neoformans* infection.

MBL has also been shown to bind *P. carinii* and activate the respiratory burst. Additionally, HIV patients with a genotype associated with low production of MBL were more likely to develop *Pneumocystis* pneumonia (Yanagisawa et al. 2015, 2020). Therefore, although MBL may be used as a potential marker for disease outcome, the ligands for MBL on *Pneumocystis* are not yet known, and no studies have been conducted to investigate the effects of MBL during *Pneumocystis* infection in mice.

8.2.12 SP-D and SP-A

SP-A and SP-D are surfactant proteins mainly found in the alveolar space and at mucosal surfaces in the lung. Extrapulmonary expression of SP-A is restricted, however, SP-D can be found in secretions from several mucosal tissues, such as the intestines.

SP-D and SP-A bind mannose in the cell wall of *C. albicans* to promote aggregation and limit fungal growth (Fig. 8.2) (Van Rozendaal et al. 2000). SP-D strongly binds to *C. albicans* but shows reduced recognition and phagocytosis by macrophages and epithelial cells in vitro (Ordonez et al. 2019). Suggesting SP-D may reduce recognition of *C. albicans* by macrophages. Additionally, in alveolar macrophages and monocytes from humans, SP-A reduced release of pro-inflammatory cytokines induced by *C. albicans* in vitro (Rosseau et al. 1999). However, *C. albicans* treated with SP-D enhanced adhesion and clearance by neutrophils. Therefore, SP-D appears to have different effects depending on the immune cell type involved. The overall effects of these observations during infection have not yet been investigated.

SP-A and SP-D have been shown to bind β-glucan, galactomannan, and galactosaminogalactan in *A. fumigatus* conidia (Fig. 8.2) (Wong et al. 2018; Allen et al. 2001). SP-D has also been shown to bind melanin from *Aspergillus* (Wong et al. 2018). Addition of SP-A and SP-D to human neutrophils and macrophages in vitro improved uptake and killing of *A. fumigatus* conidia. However, addition of SP-D reduced adhesion of *A. fumigatus* to the epithelium and thus promoted clearance by neutrophils (Ordonez et al. 2019). During IA SP-D, but not SP-A, was found to be protective against infection. Furthermore, treatment of immunocompromised mice with intranasal SP-D during IA prolonged their survival and increased production of TNF-α and IFN-γ (Singh et al. 2009; Madan et al. 2010). Immunocompromised SP-D knockout mice with pulmonary IA show similar overall mortality compared to wildtype mice but died earlier and had higher fungal burdens in the lung. These mice also showed reduced production of pro-inflammatory cytokines (Madan et al. 2010; Wong et al. 2018). Conversely, SP-A knockout mice have increased survival following pulmonary *A. fumigatus* infection compared to wildtype mice. This suggests that a lack of SP-A protects mice against infection with this fungal pathogen (Madan et al. 2010). In humans, polymorphisms in SP-A are associated with increased susceptibility to fungal-induced asthma (Saxena et al. 2003). Therefore, SP-D appears to play a protective role to *A. fumigatus* infection, while SP-A is non-protective in mice. However, in humans, SP-A appears to play a protective role. Further investigations are required to fully understand the role of SP-A and SP-D during IA.

SP-A and SP-D have been shown to recognize GXM and mannoproteins in *C. neoformans*, and bind with higher affinity to the acapsular form, compared to the capsular form (Fig. 8.2) (Schelenz et al. 1995; Van De Wetering et al. 2004). SP-D binding to *C. neoformans* in vitro increases recognition and phagocytosis by macrophages (Geunes-Boyer et al. 2009). Furthermore, in SP-D knockout mice, following intranasal *C. neoformans* infection, macrophages show reduced fungal uptake, and there was lower eosinophil infiltration in the lung compared to wildtype mice (Geunes-Boyer et al. 2009; Holmer et al. 2014). SP-A knockout mice show similar susceptibility to intranasal infection with *C. neoformans* as wildtype mice (Giles et al. 2007). This may be due to an increase in the protective SP-D when SP-A is absent, however, further investigation is required to confirm this.

SP-D binds the surface gp120, and β-glucan of *Pneumocystis* (Fig. 8.2) (Madan and Kishore 2020). SP-D and SP-A promote aggregation of *P. carinii*, reducing recognition by alveolar macrophages in vitro (Williams et al. 1996). Following intratracheal infection of *Pneumocystis* in immunocompromised mice, SP-A and SP-D were significantly upregulated, and overexpression of SP-D in these mice led to led to an increased fungal burden (Atochina et al. 2001). This may have been due to increased aggregation by SP-D, preventing recognition by phagocytes (Vuk-Pavlovic et al. 2006). However, during pulmonary *Pneumocystis* infection, SP-A and SP-D deficient mice show increased pathology, reduced production of pro-inflammatory cytokines, and decreased production of ROS and RNS (Atochina et al. 2004). Therefore, SP-A and SP-D may be involved in promoting immunity but may hinder phagocytosis.

8.3 Toll-Like Receptors

Toll-like receptors (TLRs) are crucial for the recognition of infectious as well as pathogenic microorganisms and initiation of immune responses, however, their role in antifungal infections is more nuanced. In humans, 10 functional receptors have been identified named TLR1-10, and in mice 12 receptors have been identified named TLR1-9, and TLR11-13. They have mainly been investigated for their role in recognition of bacteria but are also known to recognize fungi and viruses (Kawai and Akira 2010). TLRs are characterized by the presence of an extracellular domain-containing leucine-rich repeats (LRRs) and an intracellular Toll/IL-1R homology (TIR) domain (Kang and Lee 2011). TLR1, TLR2, TLR4, and TLR6 are located in the plasma membrane, while TLR3, TLR7, TLR8, and TLR9 are found on the membrane of endosomes and lysosomes (Kawasaki and Kawai 2014). TLRs generally exist as homodimers, however, several receptors form heterodimers such as TLR2 with TLR1 or TLR6. TLRs are widely expressed, and found on cells including mast cells, macrophages, DCs, epithelial cells, endothelial cells, and fibroblasts. TLRs are also found in cells of the adaptive immune system such as T cells and B cells (Kawasaki and Kawai 2014).

Members of the TLR family known to recognize fungi include TLR2/1, TLR2/6, TLR2, TLR3, TLR4, TLR7, and TLR9 (El-Zayat et al. 2019). Cell surface TLRs recognize components of the fungal membrane and cell wall including chitin, glycans, and mannans (Fig. 8.3). Intracellularly expressed TLRs recognize pathogen DNA and RNA exposed during the phagocytosis and breakdown of the engulfed pathogen (Blasius and Beutler 2010). Following ligation, TLRs signal via adaptor molecules including Myeloid differentiation primary response 88 (MyD88), MyD88 adapter-like (Mal), TIR-domain-containing adapter-inducing IFN-β (TRIF), and TRIF-related adaptor molecule (TRAM) (El-Zayat et al. 2019). All TLRs, except TLR3, are known to signal via MyD88 which is recruited to the receptor following ligation. MyD88 activates NF-κB and transcription of several pro-inflammatory cytokines (Kawai and Akira 2010). TRIF-dependent pathways are utilized by TLR3 and TLR4, which induce two pathways, one involving the activation of interferon regulatory factor 7 (IRF7), and the other activation of IRF3 which both induce transcription of type I IFN genes (El-Zayat et al. 2019). Mice lacking MyD88 have enhanced susceptibility to fungal infections, however, deficiencies of MyD88 in humans do not lead to increased susceptibility due to redundancy in receptors recognizing commensal fungi. Despite this, humans with TLR4 and TLR2 SNPs have altered responses to both *Candida* and *Aspergillus* infections demonstrating a role for TLRs in the antifungal immune response with a more significant role during co-infections or when immune systems are compromised (Bourgeois and Kuchler 2012). The TLRs are known to contribute to signalling, in some cases synergistically, via other receptors such as Dectin-1 or SIGNR1, modulating the cytokine profile in the case of the former (Dennehy et al. 2008).

8.3.1 TLR2

TLR2 is expressed in immune cells, endothelial cells, and epithelial cells. TLR2 is the only TLR known to pre-form heterodimers with other TLRs, including TLR6, TLR1, and TLR10 (Guan et al. 2010). TLR2 has been shown to recognize fungal β-glucans and phospholipid-mannans in *C. albicans* (Fig. 8.3) (Netea et al. 2006). Additionally, using Fc-TLR2 fusion proteins, TLR2 has also been shown to bind chitin in the

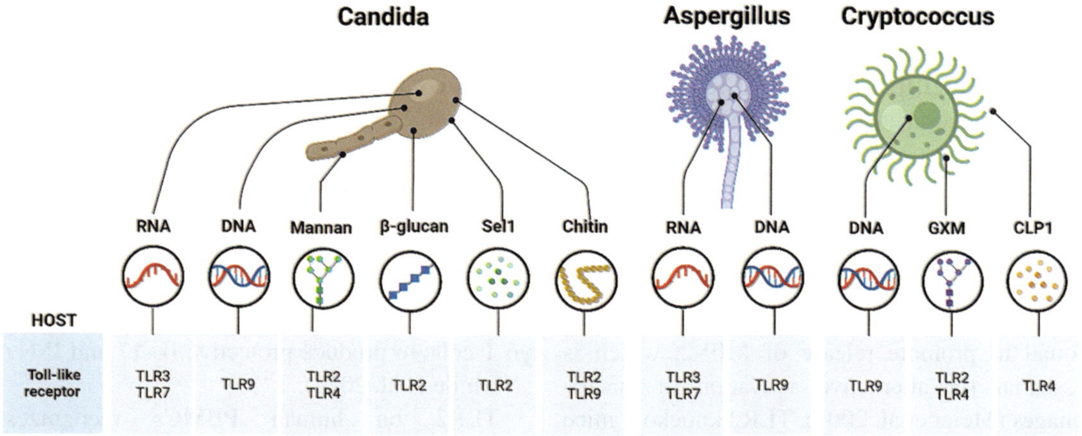

Fig. 8.3 Toll-like receptors in fungal pathogen recognition. TLRs recognize a variety of fungal pathogen-associated molecular patterns (PAMPs). Single-stranded RNA from *Candida* species is recognized by TLR3 and TLR7, while unmethylated CpG DNA is recognized by TLR9. Cell wall PAMPs from *Candida* species including phospholipid-mannans are recognized by TLR2 and O-linked mannans are recognized by TLR4, β-glucans are recognized by TLR2, and chitin is recognized by TLR2 and TLR9. Sel1, which is released by *Candida* species is recognized by TLR2. Single-stranded RNA from *Aspergillus* species is recognized by TLR3 and TLR7, while unmethylated CpG DNA from *Aspergillus* is recognized by TLR9. Unmethylated CpG DNA from *Cryptococcus* is recognized by TLR9. GXM in the cell wall of *Cryptococcus* is recognized by TLR2 and TLR4. CLP1 is secreted by *Cryptococcus* during pathogenesis and is recognized by TLR4. Figure created using Biorender.com

cell wall of *C. albicans* (Fig. 8.3) (Fuchs et al. 2018). In response to *C. albicans* yeasts or hyphae in vitro, macrophages expressing TLR2 show increased production of TNF-α (Tessarolli et al. 2010). Furthermore, in response to *C. albicans* yeast and hyphae in vitro peritoneal macrophages from TLR2 knockout mice show reduced TNF-α production, and splenocytes show reduced IFNγ production compared to cells from wildtype mice (Murciano et al. 2007). Following intraperitoneal challenge with live *C. albicans*, TLR2 knockout mice show lower neutrophil and monocyte recruitment during early stages of infection compared to wildtype animals (Tessarolli et al. 2010). Additionally, these cells showed impaired NOS release, myeloperoxidase activity, phagocytic activity, and chemokine or cytokine production (Tessarolli et al. 2010). Also in the gut of TLR2 knockout mice, DSS treatment (which damages colonic epithelial cells leading to a defective mucosal barrier) followed by *C. albicans* challenge led to increased fungal burden, inflammation, and mortality rates compared to wildtype mice (Choteau et al. 2017). Interestingly, TLR2 knockout mice also show *C. albicans* colonization in the GI tract without the use of antibiotics, which are usually required to allow fungal colonization in mice. Furthermore, in immunosuppressed mice, fungal dissemination from the GI tract was higher in TLR2 knockout mice compared to wildtype mice (Prieto et al. 2016). However, others have not been able to replicate these data and found TLR2 knockout mice had decreased fungal burden and increased production of IL-12 (Bellocchio et al. 2004). These studies used a different strain of *Candida*, which is likely why these differing results were observed, highlighting the importance of fungal strain. Nevertheless, TLR2 seems to be important for barrier protection in the GI tract. TLR2 also plays a role in *C. albicans* invasion at other mucosal barriers. For instance, during VVC TLR2 knockout mice show increased epithelial invasion of *C. albicans* and increased fungal burden in the early stages of infection compared to wildtype mice (Miro et al. 2018). TLR2 has been shown to recognize Sel1, a cysteine-rich protein secreted by *C. albicans* (Fig. 8.3) (Wang et al. 2019). Systemic infection using Sel1-deficient *C. albicans* results in reduced production of pro-inflammatory cytokines such as IL-1β, TNF-α, and IL-6 leading to increased mortality in mice (Wang et al. 2019). Therefore, TLR2 plays a role in both mucosal and systemic immunity to *C. albicans*.

TLR2 recognizes *A. fumigatus* conidia and hyphae, leading to increased release of pro-inflammatory cytokines such as IL-1α and TNF-α by neutrophils and macrophages (Chai et al. 2011; Netea et al. 2003). Furthermore, TLR2 was found to promote release of MIP-2, which is essential for alternative activation of macrophages (Meier et al. 2003). TLR2 knockout mice show reduced survival and increased fungal burden in the lung following intranasal *A. fumigatus* infection compared to wildtype mice. Additionally, the mice that survived the initial *A. fumigatus* infection showed reduced survival following subsequent exposure (Bellocchio et al. 2004). TLR2 is also involved in protection against fungal keratitis. Human corneal epithelial cells upregulate TLR2 following stimulation with *A. fumigatus* in vitro. Furthermore, upregulation of TLR2 in rats with fungal keratitis correlated with and increased IL-1β. Suggesting TLR2 may be involved in protection during this infection (Jie et al. 2009). *Aspergillus flavus* has been shown to regulate the expression of Th17-associated cytokines including IL-17 and IL-10 in human PBMCs. This was proposed to lead to an imbalance between Th17 and Tregs cells in patients with chronic rhinosinusitis with nasal polyposis. The dysregulated Th17 via TLR2 led to resistance to treatment and chronic disease (Rai et al. 2020). Thus, TLR2 recognition of *Aspergillus* plays a role in disease at different sites. Additionally, many of these observations were more pronounced in TLR2 and TLR4 double knockout mice, suggesting both TLR2 and TLR4 are involved in immunity to *Aspergillus* species.

TLR2 heterodimerizes with TLR1 or TLR6 to recognize GXM component of the *C. neoformans* capsule (Fig. 8.3) (Fonseca et al. 2010). Following intraperitoneal challenge with *C. neoformans*, TLR2 knockout mice are more susceptible to infection, show higher fungal burden and have reduced inflammatory cytokines (Yauch et al. 2004; Biondo et al. 2005). However, following intratracheal infection with *C. neoformans*, TLR2 knockout mice show no differences in survival in comparison to wildtype mice (Nakamura et al. 2008). Therefore, route of infection appears to be important for the effect of TLR2 during infection. A *C. neoformans* mutant (Δsgl1) which over-accumulates sterylglucosides has successfully been used to generate a vaccine, dependent on the presence of GXM signalling via TLR2, allowing γδ T cells to produce protective IL-17 and IFNγ (Normile et al. 2022).

TLR2 on human PBMCs recognizes *Pneumocystis* and leads to production of TNF-α and MIP-2. Furthermore, in mice TLR2 is upregulated on murine alveolar macrophages following exposure to *P. murina* in vitro, leading to increased release of TNF-α and MIP-2 (Zhang et al. 2006). During *P. murina* challenge in vivo, TLR2 knockout mice show reduced cytokine release (Zhang et al. 2006). As TLR2 promotes inflammatory macrophages, during *Pneumocystis* pneumonia TLR2 knockout mice showed reduced pulmonary inflammation. However, these mice showed increased fungal burden and increased severity of symptoms (Wang et al. 2008). Thus, the protective role of macrophage cytokine production via TLR2 was required for clearance of *Pneumocystis*.

8.3.2 TLR3

TLR3 recognizes RNA from *C. albicans* and *A. fumigatus* conidia (Fig. 8.3) (Beisswenger et al. 2012). Using human endothelial cell lines (HUVECs) *C. albicans* induced the release of IL-8 via TLR3 signalling. Furthermore, PBMCs from patients with the L412F variant of TLR3, show reduced IFNγ and TNF-α secretion in response to stimulation with *C. albicans* compared to those with the wildtype receptor (Nahum et al. 2012). Additionally, these patients show increased susceptibility to chronic mucocutaneous candidiasis (CMC) (Nahum et al. 2012). Thus, in humans, TLR3 appears to play a role in response to mucosal *C. albicans* infection. In mice, pulmonary infection with *A. fumigatus* conidia in TLR3 knockout mice shows increased susceptibility and higher fungal burden in the lung compared to wildtype mice (Carvalho et al. 2012). Moreover, in stem cell transplant patients, a +95C/A SNP in TLR3 which leads to loss-of-function in DCs, is associated with increased susceptibility to aspergillosis when the SNP was present in donor cells (Beisswenger et al. 2012). Thus, TLR3 may be important for protection against *A. fumigatus* infection. The role of TLR3 during *Cryptococcus* or *Pneumocystis* infection has not been investigated in detail. However, stimulating microglia TLR3 with poly I:C in vitro, enhances phagocytosis of *C. neoformans* (Redlich et al. 2013). Furthermore, stimulating TLR3 with poly I:C in vitro led to increased production of CCL2 by alveolar epithelial cells in response to *Pneumocystis* (Bello-Irizarry et al. 2012). However, further investigation into the role of TLR3 in *Cryptococcus* or *Pneumocystis* has not yet been conducted. Further investigations are required to understand the role of TLR3 in humans and mice during antifungal immunity.

8.3.3 TLR4

TLR4 recognizes O-linked mannans in the cell wall of *C. albicans* and promotes release of pro-inflammatory cytokines such as TNF-α and IL-6 (Fig. 8.3). TLR4 recognition of its ligands and subsequent intracellular signalling is dependent on adaptor proteins MD-2 and CD14 (Ciesielska et al. 2021). In vitro experiments have shown that macrophages from TLR4 knockout mice stimulated with *C. albicans* showed reduced phagocytosis, release of keratinocyte-derived chemokine (KC), macrophage inflammatory protein (MIP-2), and TNF-α compared to wildtype macrophages (Netea et al. 2002). In vivo *C. albicans* infected TLR4 knockout mice show reduced neutrophil recruitment and function and impaired production of chemokines by macrophages (Gasparoto et al. 2010; Meier et al. 2003). However, contradicting results show that TLR4 knockout mice can be more susceptible, less susceptible, or show no change to infection susceptibility compared to wildtype mice, depending on the *C. albicans* or the mouse strain used or the route of infection (Netea and Marodi 2010). *C. albicans* mutants lacking O-glycosylation, are more readily recognized by TLR4, and lead to enhanced activation of macrophages (Lewis et al. 2012). Hence, variability between species in cell wall glycosylation can determine the role of TLR4 during *C. albicans* infection (Glittenberg et al. 2011). In humans, the SNP Thr399Ile in TLR4 was more prevalent in patients with systemic Candidiasis compared to the control group. Additionally, peripheral blood mononuclear cells (PBMCs) from patients carrying this SNP showed enhanced IL-10 release upon *C. albicans* challenge (Van Der Graaf et al. 2005). Therefore, the role of TLR4 in *Candida* infection requires further investigation, where the strain of both mice and *Candida* will be important when deciphering the results.

TLR4 recognizes GXM from *C. neoformans*, but the specific effects of this interaction have not been fully identified (Fig. 8.3). TLR4 also interacts directly with CPL1, which is secreted by *C. neoformans* during pulmonary infection. CPL1 drives alternative activation of macrophages, and in TLR4 knockout mice this process is impaired, suggesting a role for TLR4 in macrophage polarization (Yauch et al. 2004). Additionally, CPL1 induces Th2 cells via TLR4, leading to the creation of a favorable niche for the fungus. Despite these findings following intranasal, intratracheal, intravenous, or intraperitoneal *C. neoformans* infection, TLR4 knockout mice show no differences in overall survival or pathology compared to wildtype mice (Dang et al. 2022). Therefore, other receptors may be able to compensate for the lack of TLR4 in these mice. However, TLR4 knockout mice infected with Δ*cpl1 C. neoformans* showed reduced fungal burden and decreased dissemination to the brain (Dang et al. 2022). Therefore, during *C. neoformans* infection, TLR4 may be detrimental to the resolution of disease. Additionally, these data suggest that TLR4 plays a key role in the recognition of CPL1.

8.3.4 TLR7

TLR7 recognizes single-stranded RNA from *C. albicans* triggering the release of IFN-β and IL-12 p70 (Fig. 8.3) (Biondo et al. 2011; Bourgeois et al. 2011). Following infection with low doses of *C. albicans*, TLR7 knockout mice show increased susceptibility compared to wild-type mice. However, infection of TLR7-deficient mice with higher doses shows no differences in susceptibility compared to wildtype mice (Biondo et al. 2012). Thus, TLR7 may be protective to the resolution of infection during the early stages of infection, and not involved as infection progresses and fungal burden increases. Time point analysis of the effects of infection in TLR7 knockout mice is required to confirm this. TLR7 has also been shown to be important in responses to other *Candida* species. For example, TLR7 is important for IFN-β signalling in DCs in response to *C. glabrata* infection (Bourgeois et al. 2011).

TLR7 also recognizes single-stranded RNA from *A. fumigatus* (Fig. 8.3). During pulmonary *A. fumigatus* infection, TLR7 knockout mice show lower inflammation, tissue injury, fungal burdens, and higher survival compared to wild-type mice (Xu et al. 2021). Furthermore, in wild-type mice TLR7 activation by the R837 ligand led to reduced susceptibility to invasive pulmonary *Aspergillus fumigatus* infection (Xu et al. 2021). Thus, TLR7 appears to be a negative regulator of innate immune responses during *A. fumigatus* infection. However, further investigation is required to establish the cellular effects that lead to these responses.

8.3.5 TLR9

TLR9, an intracellular expressed receptor, recognizes unmethylated CpG DNA in *C. albicans A. fumigatus*, and *C. neoformans* (Fig. 8.3) (Kasperkovitz et al. 2011). TLR9 is also recruited to fungal-containing phagosomes, suggesting a role for this receptor during fungal clearance (Kasperkovitz et al. 2011; Miyazato et al. 2009). Additionally, TLR9 in collaboration with NOD2 and MR, has been shown to recognize chitin in the cell wall of *C. albicans*, leading to IL-10 production and recruitment of eosinophils (Wagener et al. 2014). TLR9 recognizes DNA from *C. albicans* to induce IL-12 production and upregulation of CD40 in BMDCs. However, during *C. albicans* infection, TLR9 knockout mice showed similar susceptibility, fungal burden in the kidney, and survival as wildtype mice (Miyazato et al. 2009; Van De Veerdonk et al. 2008). Therefore, other receptors may be compensating for the loss of TLR9, and the effect on specific cell types will reveal more about the effects of TLR9 during infection.

TLR9 recognizes genomic DNA from *Cryptococcus* and is required for DC recruitment, and production of IL-12, CCL7, and IFN-γ in vivo (Fig. 8.3) (Nakamura et al. 2008). Furthermore, following intranasal or intratracheal infection with *C. neoformans*, TLR9 knockout mice show reduced fungal clearance and lower survival compared to wildtype mice (Wang et al. 2011; Qiu et al. 2012), highlighting a key role for this PRR during *C. neoformans* infection. TLR9 also appears to be important for other Cryptococcal species. For instance, during infection with *C. gattii*, TLR9 knockout mice showed increased mortality and higher incidence of enlarged yeasts known as titan cells. Additionally, TLR9 knockout mice had increased fungal in the spleen and brain compared to wild-type mice (Da Silva-Junior et al. 2021). Titan cells are more resistant to phagocytosis and are linked to increased dissemination and altered immune responses (Dambuza et al. 2018). Therefore, TLR9 may be involved in preventing the transition of *Cryptococcus* into titan cells, thus reducing their dissemination.

8.4 Other Receptors Involved in Fungal Recognition

8.4.1 Scavenger Receptors

Scavenger receptors (SRs) are structurally unrelated and bind multiple ligands including both exogenous and endogenous proteins. There are 12 subclasses of SRs that are mainly known for their role in bacterial pathogen recognition, but relatively few studies have investigated their role in fungal recognition. CD36, SREC-1, and CD5 have been shown to recognize fungal β-glucans and play a role in immune response during fungal infection (Fig. 8.4).

CD36 and SREC-1 recognize β-glucan from *C. albicans* and *C. neoformans* (Fig. 8.4) (Means et al. 2009). However, the effect of CD36 or SREC-1 has not been investigated in the context of *C. albicans* infection. The effects of these receptors have been investigated for *C. neoformans* both in vitro and in vivo. Stimulating macrophages lacking CD36 or SREC-1 in vitro with *C. neoformans* led to reduced production of pro-inflammatory cytokines such as IL-1β, TNF-α, IL-12p40, and MIP-2 compared to wildtype macrophages (Means et al. 2009). Additionally, in vivo following pulmonary *C. neoformans* infection in mice, addition of neutralizing anti-SREC-1 antibody, led to reduced interaction of *C. neoformans* with alveolar macrophages (Means et al. 2009). Furthermore, following intravenous infection with *C. neoformans*, CD36 knockout mice showed increased fungal burden in the spleen and liver compared to wildtype mice (Means et al. 2009). Thus, both CD36 and SREC-1 play a role during *C. neoformans* infection. However, the role of these receptors in response to *C. albicans*, *A. fumigatus*, or *Pneumocystis* has not been investigated.

CD5 binds β-glucans from both *C. albicans*, and *C. neoformans* (Fig. 8.4). Addition of the soluble human ectodomain of CD5 (shCD5) to murine splenocytes in vitro following stimulation with *C. albicans* led to increased release of pro-inflammatory cytokines such as TNF-α and IFN-γ, compared to splenocytes without shCD5 (Velasco-DE Andres et al. 2020). Additionally in vivo, addition of shCD5 during *C. albicans* infection led to increased survival and reduced fungal burdens in the spleen and kidney compared to mice without shCD5 (Velasco-DE Andres et al. 2020). Addition of shCD5 following intranasal *C. neoformans* infection, also led to increased survival and significantly reduced fungal burden in the lung and brain (Velasco-DE Andres et al. 2020). Therefore, CD5 plays an

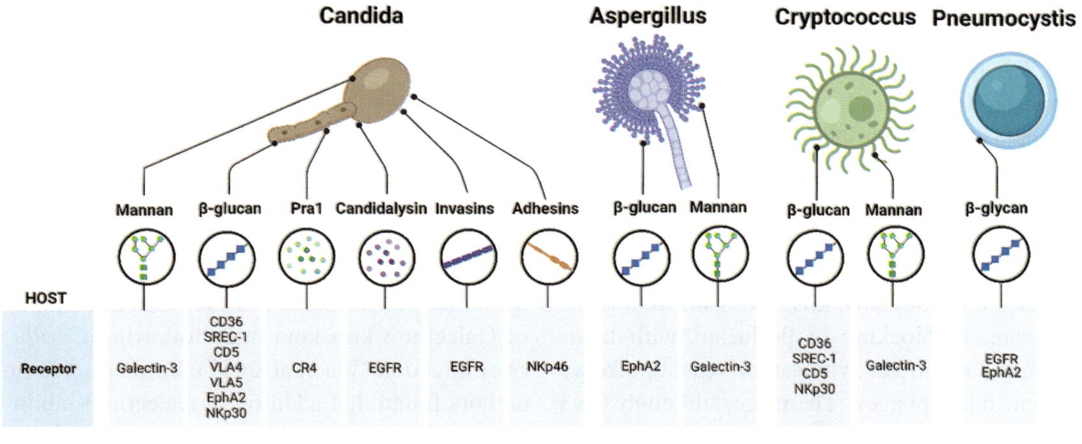

Fig. 8.4 Other receptors in fungal pathogen recognition. Several other receptors that do not belong to the C-type lectin of Toll-like receptor families recognize a variety of fungal pathogen-associated molecular patterns (PAMPs). Galectin-3 recognizes β-1,2-linked oligomannans from *Candida*, and as a heterodimer with Dectin-1 also binds β-glucans. Candida β-glucan is also recognized by scavenger receptors CD36, SREC-1, and CD5, integrins VLA4, VLA5, epithelial cell receptor EphA2, and natural cytotoxicity receptor NKp30. CR4 binds to Pra1 expressed on the cell wall of *C. albicans*. EGRF heterodimerises with EphA2 to bind both Candidalysin, and the invasin Als3. NKp46 binds fungal adhesins Epa1, Epa6, and Epa7 from *Candida galbrata*. EphA2 recognizes β-glucans from *Aspergillus*, while Galectin-3 recognizes *Aspergillus* β-1,2-linked oligomannans. Several receptors recognize β-glucans from *Cryptococcus* including the scavenger receptors CD36, SREC-1, and CD5 and natural cytotoxicity receptor NKp30. β-1,2-linked oligomannans from *Cryptococcus* are recognized by Galectin-3. β-glucans from *Pneumocystis* are recognized by EGFR and EphA2 heterodimers. Figure created using Biorender.com

important role during *C. albicans* and *C. neoformans* infections and could be a possible therapeutic target. The role of CD5 in *Aspergillus* or *Pneumocystis* infection has not been investigated. However, both these fungi contain high levels of β-glucan in their cell wall therefore, CD5 could also be involved in immunity to these fungi.

8.4.2 Integrins

Integrins are a family of heterodimeric receptors that mediate migration and adhesion of immune cells. They are mainly involved in the interaction between the extracellular matrix and the actin cytoskeleton. Integrins can directly recognize fungi and can also act as opsonic receptors that recognize fungi coated with complement factors. Integrins VLA3 (α3β1), VLA5 (α5β1), and complement receptor 3 (CR3) (αmβ2, CD11b/CD18, Mac-1) can bind directly to proteins in the fungal cell wall (Fig. 8.4). While CR3 and CR4 (αxβ2 integrin, CD11c/CD18), were originally thought to recognize *β*-glucan, but are now known to recognize opsonised fungi to trigger phagocytosis and fungal killing by innate immune cells such as macrophages and DCs.

CR3 recognizes unopsonised *C. albicans*, which leads to activation of non-oxidative killing mechanisms in neutrophils such as fungal aggregation and NET formation (Swamydas et al. 2016; Wirnsberger et al. 2014). Additionally, macrophages pre-treated in vitro with CpG-ODN, and an anti-CR3 mAb show reduced uptake of *C. albicans*. CR3 does not recognize β-glucan, as blocking of β-glucans with laminarin, did not affect zymosan or yeast uptake by murine macrophages. Therefore, although CR3 binds fungi, its exact ligand remains unclear. CR3 and CR4 bind pH-regulated Ag 1 (Pra1) a fibrinogen-binding mannoprotein expressed on *C. albicans* cell surface (Soloviev et al. 2007; Jawhara et al. 2012). During systemic *C. albicans* infection, CR4 knockout mice show increased fungal burden in the liver and brain compared to wildtype mice (Jawhara et al. 2012). β1 integrins VLA3 (α3β1) and VLA5 (α5β1) recognize β-glucans. VLA3 (α3β1) induces aggregation by neutrophils, while rVLA5 promotes NETosis (Fig. 8.4). Therefore, integrins have been shown to play a role in the immune response by neutrophils during *C. albicans* infection. However, further research is required to establish the overall effect of integrins during *Candida* infection. Additionally, the role of integrins in response to *Aspergillus*, *Cryptococcus*, or *Pneumocystis* infection has not yet been investigated.

8.4.3 Galectins

The Galectin family contains 15 soluble lectin proteins, of which Galectin-3 has been associated with binding to fungi. Galectin-3 is a β-galactose-binding lectin expressed by macrophages, DCs, and epithelial cells that recognizes β-1,2-linked oligomannans from several fungi including *Candida* spp., *A. fumigatus*, and *C. neoformans* (Fig. 8.4) (Kohatsu et al. 2006; Fradin et al. 2000). Galectin-3 also forms heterodimers with Dectin-1 that show binding to β-glucans (Esteban et al. 2011).

Galectin-3 binding to *C. albicans* leads to fungal cell death in vitro via an unknown mechanism (Kohatsu et al. 2006). In mice, during systemic *Candida* infection with either *C. albicans* or *C. parapsilosis*, Galectin-3 knockout mice showed increased fungal burden in the brain and kidney, respectively, and had significantly reduced survival compared to wildtype mice (Linden et al. 2013). However, contrasting results have shown decreased fungal burden in the brain and kidney of Galectin-3 knockout mice following *C. albicans* infection (Wu et al. 2017). Furthermore, the authors found that addition of Galectin-3 inhibitors led to increased neutrophil ROS production (Wu et al. 2017). These results are likely due to the dose of *C. albicans* used during infection, where Galectin-3 is utilized during clearance of low doses but is detrimental during high-dose infection.

Galectin-3 is also involved in immunity to *C. neoformans* and *A. fumigatus* infection. For instance, the in vitro addition of recombinant Galectin-3 prevents fungal growth and causes lysis of *C. neoformans* extracellular vesicles (EVs) (Almeida et al. 2017). Contradicting results have been observed for the effect of neutrophils expressing Galectin-3 on *A. fumigatus* in vitro, where some research shows increased fungal killing while others showed no effect (Niu et al. 2021; Snarr et al. 2020). These differences are likely due to the tissue which the neutrophils were isolated from, cornea and lung, respectively. In the lung, Galectin-3 expression is upregulated in both humans and mice during pulmonary aspergillosis or cryptococcosis (Snarr et al. 2020; Almeida et al. 2017). Furthermore, Galectin-3 knockout mice with either pulmonary aspergillosis or cryptococcosis show increased fungal burden and higher mortality (Almeida et al. 2017). Additionally, neutrophil recruitment was reduced in *Aspergillus*-infected mice (Snarr et al. 2020). These findings are supported by studies that show that direct binding of Galactin-3 to neutrophils causes increased activation of these cells (Nieminen et al. 2005). During cryptococcosis, Galectin-3 knockout mice showed downregulation of pro-inflammatory cytokine genes associated with Th1 immunity including IL-1β, and IL-12 (Rezende et al. 2021). Therefore, Galectin-3 is involved in modulating the immune response such as adaptive immunity, neutrophil recruitment, and fungal clearance during both *Aspergillus* and *Cryptococcus* infection. The effect of Galaectin-3 during Pneumocystis infection has not been investigated.

8.4.4 EphA2

The non-classical receptor ephrin type-A 2 (EphA2), expressed in mature epithelial cells, has been shown to recognize β-glucans from *C. albicans*, *P. carinii*, and *A. fumigatus* (Fig. 8.4) (Swidergall et al. 2018). During OPC EphA2 knockout mice show reduced production of pro-inflammatory cytokines, such as IL-1β, IFN-γ, and IL-17A, and reduced neutrophil and macrophage recruitment compared to wildtype mice (Swidergall et al. 2018, 2019). Thus, suggesting a key role for this receptor during OPC. EphA2 forms a complex on the cell surface with epidermal growth factor receptor (EGFR). EGFR has been shown to bind Candidalysin, a pore-forming toxin, and Als3 a fungal adhesin. During OPC, EphA2 and EGFR induce the release of inflammatory cytokines such as IL-1α, IL-1β, IL-6, G-CSF, and GM-CSF (Swidergall et al. 2021). EphA2 recognition of *P. carinii* has been linked to IL-6 production by lung epithelial cells (Kottom et al. 2022). However, the overall effects of this interaction during infection have not been investigated. Therefore, EphA2 independently and in collaboration with EGFR is involved in immunity to *Candida* at the epithelial barrier. Further investigations are required to establish the role of this receptor in response to other fungi.

8.4.5 Natural Cytotoxicity Receptors

NKp46 (NCR1) and NKp30 have been shown to interact with fungal ligands. For instance, using Fc-fusion proteins, murine NCR1 and the human ortholog NKp46 were shown to bind fungal adhesins Epa1, Epa6, and Epa7 from *C. glabrata* but did not bind *C. albicans*, *C. parapsilosis*, *C. krusei*, or *C. gattii* (Vitenshtein et al. 2016). Furthermore, NK cells from NCR1 knockout mice showed reduced killing capacity of *C. glabrata* in vitro. Additionally, during *C. glabrata* infection, NKp46 deficient mice showed increased fungal burden in the liver, spleen, and kidney compared to wildtype mice (Vitenshtein et al. 2016). Therefore, NCR1 appears to play a role during *C. glabrata* infection. Further studies are required to identify why this receptor does not bind to other candida species.

NKp30 (NCR3 in mice) was shown to recognize β-1,3 glucan in *C. neoformans* and *C. albicans* (Fig. 8.4) (Li et al. 2013, 2018b). Fungal recognition leads to the formation of NK cell–fungal conjugates, and polarization of cytotoxic granules, which increase the release of perforin and direct killing of *C. neoformans*. In HIV patients, who show a high incidence of fungal infection, NK cells show reduced expression of NKp30 and decreased perforin release compared to healthy controls (Li et al. 2013). Administration of IL-12 and β-1,3 glucan restored NK cell expression of NKp30 and *C. neoformans* killing by NK cells (Li et al. 2013, 2018b). Suggesting NKp30 recognition of β-1,3 glucan and release of IL-12 is important for NK cell-mediated immunity during *C. neoformans* infection. Further research is required to establish a role for NKp30 in response to other fungi.

8.4.6 Pentraxins

Pentraxins are components of the humoral innate immune system, these cyclical disk-like structures are comprised of homomeric domains with a conserved 200 amino acid C terminus with a pentraxin signature. The short pentraxins, such as C-reactive protein (CRP) and serum amyloid P (SAP), and the long pentraxin 3 (Ptx3) bind a variety of microbial ligands (Fig. 8.4). They induce immune effector mechanisms that are facilitated by cross-talk with complement and interactions with Fc receptors, such as CD32 (Porte et al. 2019). SAP proteins are essential for immunity to *A. fumigatus* infection due to their role in phagocytosis and complement activation (Doni et al. 2021). Galactomannans present in *A. fumigatus* conidia are bound by Pentraxin 3 (Ptx3), a multimeric soluble long pentraxin with 10–20 homodomains. Ptx3 is expressed by activated macrophages, DCs, monocytes, and a range of other cell types. Ptx3 is also stored in neutrophilic granules and is associated with NETs following IL-8 stimulation of neutrophils (Razvina et al. 2015). Ptx3 has a non-redundant role in the recognition of *A. fumigatus* by alveolar macrophages and DCs shaping the adaptive immune response in mice (Garlanda et al. 2016). A SNP rs1840680 (AA) in Ptx3 in chronic obstructive pulmonary disease (COPD) patients is associated with lower Ptx3 serum levels and an increased risk of invasive aspergillosis (He et al. 2018). The lectin pathway of complement is modulated by Ptx3, causing increased ficolin-1 deposition on *A. fumigatus* and PTX3's interaction with mannose-binding lectin (MBL) on *C. albicans* enhances the deposition of complement (Ma et al. 2011). However, the effects of Pentraxins during immunity to *C. albicans* infection have not been investigated. Additionally, Pentraxins have not been investigated in the context of *Cryptococcus* of *Pneumocystis* immunity.

8.5 Conclusion

The main fungi that cause invasive infection in humans, including *C. albicans*, *A. fumigatus*, and *C. neoformans,* contain an array of PAMPs that include both cell wall components and secreted proteins. These PAMPs represent ligands for a myriad of immune receptors, of which CLRs and TLRs have been investigated in detail, as discussed. However, recognition of these pathogens is also mediated by several other PRRs, such as the scavenger receptors, integrins, galectins, EphA2, NCRs, and Pentraxins. Many receptors not mentioned in this chapter are also known to be involved in antifungal immunity such as the Nod-like receptors (NLRs) and Rig-like receptors (RLRs), but their specific fungal PAMPs have not yet been identified. Identification of these fungal ligands and better characterization of their roles will help provide an enhanced understanding of how antifungal immunity is induced and regulated during infection. Moreover, we are just beginning to understand the importance of collaboration and synergy between PRRs and how this influences intracellular signalling, cellular responses, and immune outcome. Nevertheless, understanding of the role of PRRs during invasive fungal infections is expanding rapidly. This growing body of knowledge is essential for the development of new therapeutics and vaccines to help reduce the burden of invasive fungal infections on human health across the globe.

Acknowledgments We acknowledge funding from the MRC Centre for Medical Mycology at the University of Exeter (MR/N006364/2 and MR/V033417/1), the NIHR Exeter Biomedical Research Centre, and the Wellcome Trust (97377, 217163, 102705). Additional work may have been undertaken by the University of Exeter Biological Services Unit. The views expressed are those of the author(s) and not necessarily those of the NIHR or the Department of Health and Social Care.

References

Allen MJ, Laederach A, Reilly PJ, Mason RJ (2001) Polysaccharide recognition by surfactant protein D: novel interactions of a C-type lectin with nonterminal glucosyl residues. Biochemistry 40:7789–7798

Almeida F, Wolf JM, Da Silva TA, Deleon-Rodriguez CM, Rezende CP, Pessoni AM, Fernandes FF, Silva-Rocha R, Martinez R, Rodrigues ML, Roque-Barreira MC, Casadevall A (2017) Galectin-3 impacts Cryptococcus neoformans infection through direct antifungal effects. Nat Commun 8:1968

Atochina EN, Beck JM, Scanlon ST, Preston AM, Beers MF (2001) Pneumocystis carinii pneumonia alters expression and distribution of lung collectins SP-A and SP-D. J Lab Clin Med 137:429–439

Atochina EN, Beck JM, Preston AM, Haczku A, Tomer Y, Scanlon ST, Fusaro T, Casey J, Hawgood S, Gow AJ, Beers MF (2004) Enhanced lung injury and delayed clearance of Pneumocystis carinii in surfactant protein A-deficient mice: attenuation of cytokine responses and reactive oxygen-nitrogen species. Infect Immun 72:6002–6011

Beisswenger C, Hess C, Bals R (2012) Aspergillus fumigatus conidia induce interferon-beta signalling in respiratory epithelial cells. Eur Respir J 39:411–418

Bellocchio S, Montagnoli C, Bozza S, Gaziano R, Rossi G, Mambula SS, Vecchi A, Mantovani A, Levitz SM, Romani L (2004) The contribution of the Toll-like/IL-1 receptor superfamily to innate and adaptive immunity to fungal pathogens in vivo. J Immunol 172:3059–3069

Bello-Irizarry SN, Wang J, Olsen K, Gigliotti F, Wright TW (2012) The alveolar epithelial cell chemokine response to pneumocystis requires adaptor molecule MyD88 and interleukin-1 receptor but not toll-like receptor 2 or 4. Infect Immun 80:3912–3920

Bermejo-Jambrina M, Eder J, Helgers LC, Hertoghs N, Nijmeijer BM, Stunnenberg M, Geijtenbeek TBH (2018) C-type lectin receptors in antiviral immunity and viral escape. Front Immunol 9:590

Biondo C, Messina L, Bombaci M, Mancuso G, Midiri A, Beninati C, Cusumano V, Gerace E, Papasergi S, Teti G (2005) Characterization of two novel cryptococcal mannoproteins recognized by immune sera. Infect Immun 73:7348–7355

Biondo C, Signorino G, Costa A, Midiri A, Gerace E, Galbo R, Bellantoni A, Malara A, Beninati C, Teti G, Mancuso G (2011) Recognition of yeast nucleic acids triggers a host-protective type I interferon response. Eur J Immunol 41:1969–1979

Biondo C, Malara A, Costa A, Signorino G, Cardile F, Midiri A, Galbo R, Papasergi S, Domina M, Pugliese M, Teti G, Mancuso G, Beninati C (2012) Recognition of fungal RNA by TLR7 has a nonredundant role in host defense against experimental candidiasis. Eur J Immunol 42:2632–2643

Blasius AL, Beutler B (2010) Intracellular toll-like receptors. Immunity 32:305–315

Bongomin F, Gago S, Oladele RO, Denning DW (2017) Global and multi-national prevalence of fungal diseases-estimate precision. J Fungi (Basel) 3

Bourgeois C, Kuchler K (2012) Fungal pathogens-a sweet and sour treat for toll-like receptors. Front Cell Infect Microbiol 2:142

Bourgeois C, Majer O, Frohner IE, Lesiak-Markowicz I, Hildering KS, Glaser W, Stockinger S, Decker T, Akira S, Müller M, Kuchler K (2011) Conventional dendritic cells mount a type I IFN response against Candida spp. requiring novel phagosomal TLR7-mediated IFN-β signaling. J Immunol 186:3104–3112

Bozza S, Gaziano R, Spreca A, Bacci A, Montagnoli C, Francesco DI, P. & Romani, L. (2002) Dendritic cells transport conidia and hyphae of Aspergillus fumigatus from the airways to the draining lymph nodes and initiate disparate Th responses to the fungus. J Immunol 168:1362–1371

Branzk N, Lubojemska A, Hardison SE, Wang Q, Gutierrez MG, Brown GD, Papayannopoulos V (2014) Neutrophils sense microbe size and selectively release neutrophil extracellular traps in response to large pathogens. Nat Immunol 15:1017–1025

Bromuro C, Torosantucci A, Chiani P, Conti S, Polonelli L, Cassone A (2002) Interplay between protective and inhibitory antibodies dictates the outcome of experimentally disseminated Candidiasis in recipients of a Candida albicans vaccine. Infect Immun 70:5462–5470

Brown GD, Crocker PR (2016) Lectin receptors expressed on myeloid cells. Microbiol Spectr 4

Brown GD, Gordon S (2001) Immune recognition. A new receptor for beta-glucans. Nature 413:36–37

Brown GD, Denning DW, Gow NA, Levitz SM, Netea MG, White TC (2012) Hidden killers: human fungal infections. Sci Transl Med 4:165rv13

Brown GD, Willment JA, Whitehead L (2018) C-type lectins in immunity and homeostasis. Nat Rev Immunol 18:374–389

Cambi A, Gijzen K, De Vries LJ, Torensma R, Joosten B, Adema GJ, Netea MG, Kullberg BJ, Romani L, Figdor CG (2003) The C-type lectin DC-SIGN (CD209) is an antigen-uptake receptor for Candida albicans on dendritic cells. Eur J Immunol 33:532–538

Cambi A, Beeren I, Joosten B, Fransen JA, Figdor CG (2009) The C-type lectin DC-Sign internalizes soluble antigens and HIV-1 virions via a clathrin-dependent mechanism. Eur J Immunol 39:1923–1928

Campuzano A, Castro-Lopez N, Wozniak KL, Leopold Wager CM, Wormley FL, JR. (2017) Dectin-3 is not required for protection against Cryptococcus neoformans infection. PLoS One 12:e0169347

Carmona J, Cruz A, Moreira-Teixeira L, Sousa C, Sousa J, Osorio NS, Saraiva AL, Svenson S, Kallenius G, Pedrosa J, Rodrigues F, Castro AG, Saraiva M (2013) Mycobacterium tuberculosis strains are differentially recognized by TLRs with an impact on the immune response. PLoS One 8:e67277

Carvalho A, De Luca A, Bozza S, Cunha C, D'Angelo C, Moretti S, Perruccio K, Iannitti RG, Fallarino F, Pierini A, Latge JP, Velardi A, Aversa F, Romani L (2012) TLR3 essentially promotes protective class I-restricted memory CD8(+) T-cell responses to Aspergillus fumigatus in hematopoietic transplanted patients. Blood 119:967–977

Casadevall A, Rosas AL, Nosanchuk JD (2000) Melanin and virulence in Cryptococcus neoformans. Curr Opin Microbiol 3:354–358

Casadevall A, Coelho C, Cordero RJB, Dragotakes Q, Jung E, Vij R, Wear MP (2019) The capsule of Cryptococcus neoformans. Virulence 10:822–831

Chai LY, Vonk AG, Kullberg BJ, Verweij PE, Verschueren I, Van Der Meer JW, Joosten LA, Latge JP, Netea MG (2011) Aspergillus fumigatus cell wall components differentially modulate host TLR2 and TLR4 responses. Microbes Infect 13:151–159

Chaka W, Verheul AF, Vaishnav VV, Cherniak R, Scharringa J, Verhoef J, Snippe H, Hoepelman AI (1997) Induction of TNF-alpha in human peripheral blood mononuclear cells by the mannoprotein of Cryptococcus neoformans involves human mannose binding protein. J Immunol 159:2979–2985

Chang TH, Huang JH, Lin HC, Chen WY, Lee YH, Hsu LC, Netea MG, Ting JP, Wu-Hsieh BA (2017) Dectin-2 is a primary receptor for NLRP3 inflammasome activation in dendritic cell response to Histoplasma capsulatum. PLoS Pathog 13:e1006485

Chattaway FW, Holmes MR, Barlow AJ (1968) Cell wall composition of the mycelial and blastospore forms of Candida albicans. J Gen Microbiol 51:367–376

Chen SM, Shen H, Zhang T, Huang X, Liu XQ, Guo SY, Zhao JJ, Wang CF, Yan L, Xu GT, Jiang YY, An MM (2017) Dectin-1 plays an important role in host defense against systemic Candida glabrata infection. Virulence 8:1643–1656

Chieppa M, Bianchi G, Doni A, Del Prete A, Sironi M, Laskarin G, Monti P, Piemonti L, Biondi A, Mantovani A, Introna M, Allavena P (2003) Cross-linking of the mannose receptor on monocyte-derived dendritic cells activates an anti-inflammatory immunosuppressive program. J Immunol 171:4552–4560

Choraghe RP, Neumann AK (2021) Dectin-1-mediated DC-SIGN recruitment to Candida albicans contact sites. Life (Basel) 11

Choteau L, Parny M, Francois N, Bertin B, Fumery M, Dubuquoy L, Takahashi K, Colombel JF, Jouault T, Poulain D, Sendid B, Jawhara S (2016) Role of mannose-binding lectin in intestinal homeostasis and fungal elimination. Mucosal Immunol 9:767–776

Choteau L, Vancraeyneste H, Le Roy D, Dubuquoy L, Romani L, Jouault T, Poulain D, Sendid B, Calandra T, Roger T, Jawhara S (2017) Role of TLR1, TLR2 and TLR6 in the modulation of intestinal inflammation and Candida albicans elimination. Gut Pathog 9:9

Ciesielska A, Matyjek M, Kwiatkowska K (2021) TLR4 and CD14 trafficking and its influence on LPS-induced pro-inflammatory signaling. Cell Mol Life Sci 78:1233–1261

Colonna M, Samaridis J, Angman L (2000) Molecular characterization of two novel C-type lectin-like receptors, one of which is selectively expressed in human dendritic cells. Eur J Immunol 30:697–704

Cummings RD, Mcever RP (2009) C-type lectins. In: Varki A, Cummings RD, Esko JD, Freeze HH, Stanley P, Bertozzi CR, Hart GW, Etzler ME (eds) Essentials of glycobiology, 2nd edn. Cold Spring Harbor Laboratory Press, Cold Spring Harbor, NY

Cunha C, Di Ianni M, Bozza S, Giovannini G, Zagarella S, Zelante T, D'Angelo C, Pierini A, Pitzurra L, Falzetti F, Carotti A, Perruccio K, Latge JP, Rodrigues F, Velardi A, Aversa F, Romani L, Carvalho A (2010) Dectin-1 Y238X polymorphism associates with susceptibility to invasive aspergillosis in hematopoietic transplantation through impairment of both recipient- and donor-dependent mechanisms of antifungal immunity. Blood 116:5394–5402

Dambuza IM, Brown GD (2015) C-type lectins in immunity: recent developments. Curr Opin Immunol 32C:21–27

Dambuza IM, Drake T, Chapuis A, Zhou X, Correia J, Taylor-Smith L, Legrave N, Rasmussen T, Fisher MC, Bicanic T, Harrison TS, Jaspars M, May RC, Brown GD, Yuecel R, Maccallum DM, Ballou ER (2018) The Cryptococcus neoformans Titan cell is an inducible and regulated morphotype underlying pathogenesis. PLoS Pathog 14:e1006978

Dan JM, Kelly RM, Lee CK, Levitz SM (2008) Role of the mannose receptor in a murine model of Cryptococcus neoformans infection. Infect Immun 76:2362–2367

Dang EV, Lei S, Radkov A, Volk RF, Zaro BW, Madhani HD (2022) Secreted fungal virulence effector triggers allergic inflammation via TLR4. Nature 608:161–167

Da Silva-Junior EB, Firmino-Cruz L, Guimaraes-De-Oliveira JC, De-Medeiros JVR, De Oliveira Nascimento D, Freire-De-Lima M, De Brito-Gitirana L, Morrot A, Previato JO, Mendonca-Previato L, Decote-Ricardo D, De Matos Guedes HL, Freire-De-Lima CG (2021) The role of Toll-like receptor 9 in a murine model of Cryptococcus gattii infection. Sci Rep 11:1407

Deckman JM, Kurkjian CJ, Mcgillis JP, Cory TJ, Birket SE, Schutzman LM, Murphy BS, Garvy BA, Feola DJ (2017) Pneumocystis infection alters the activation state of pulmonary macrophages. Immunobiology 222:188–197

De Jesus M, Rodriguez AE, Yagita H, Ostroff GR, Mantis NJ (2015) Sampling of Candida albicans and Candida tropicalis by Langerin-positive dendritic cells in mouse Peyer's patches. Immunol Lett 168:64–72

De Jong MA, Vriend LE, Theelen B, Taylor ME, Fluitsma D, Boekhout T, Geijtenbeek TB (2010) C-type lectin Langerin is a beta-glucan receptor on human Langerhans cells that recognizes opportunistic and pathogenic fungi. Mol Immunol 47:1216–1225

Dennehy KM, Ferwerda G, Faro-Trindade I, Pyz E, Willment JA, Taylor PR, Kerrigan A, Tsoni SV,

Gordon S, Meyer-Wentrup F, Adema GJ, Kullberg BJ, Schweighoffer E, Tybulewicz V, Mora-Montes HM, Gow NA, Williams DL, Netea MG, Brown GD (2008) Syk kinase is required for collaborative cytokine production induced through Dectin-1 and Toll-like receptors. Eur J Immunol 38:500–506

De Repentigny L, Lewandowski D, Jolicoeur P (2004) Immunopathogenesis of oropharyngeal candidiasis in human immunodeficiency virus infection. Clin Microbiol Rev 17:729–759. Table of contents

Doni A, Parente R, Laface I, Magrini E, Cunha C, Colombo FS, Lacerda JF, Campos A Jr, Mapelli SN, Petroni F, Porte R, Schorn T, Inforzato A, Mercier T, Lagrou K, Maertens J, Lambris JD, Bottazzi B, Garlanda C, Botto M, Carvalho A, Mantovani A (2021) Serum amyloid P component is an essential element of resistance against Aspergillus fumigatus. Nat Commun 12:3739

Drouin M, Saenz J, Chiffoleau E (2020) C-type lectin-like receptors: head or tail in cell death immunity. Front Immunol 11:251

Drummond RA, Dambuza IM, Vautier S, Taylor JA, Reid DM, Bain CC, Underhill DM, Masopust D, Kaplan DH, Brown GD (2016) CD4(+) T-cell survival in the GI tract requires dectin-1 during fungal infection. Mucosal Immunol 9:492–502

Drummond RA, Franco LM, Lionakis MS (2018) Human CARD9: a critical molecule of fungal immune surveillance. Front Immunol 9:1836

Drummond RA, Desai JV, Hsu AP, Oikonomou V, Vinh DC, Acklin JA, Abers MS, Walkiewicz MA, Anzick SL, Swamydas M, Vautier S, Natarajan M, Oler AJ, Yamanaka D, Mayer-Barber KD, Iwakura Y, Bianchi D, Driscoll B, Hauck K, Kline A, Viall NS, Zerbe CS, Ferre EM, Schmitt MM, Dimaggio T, Pittaluga S, Butman JA, Zelazny AM, Shea YR, Arias CA, Ashbaugh C, Mahmood M, Temesgen Z, Theofiles AG, Nigo M, Moudgal V, Bloch KC, Kelly SG, Whitworth MS, Rao G, Whitener CJ, Mafi N, Gea-Banacloche J, Kenyon LC, Miller WR, Boggian K, Gilbert A, Sincock M, Freeman AF, Bennett JE, Hasbun R, Mikelis CM, Kwon-Chung KJ, Belkaid Y, Brown GD, Lim JK, Kuhns DB, Holland SM, Lionakis MS (2022) Human Dectin-1 deficiency impairs macrophage-mediated defense against phaeohyphomycosis. J Clin Invest 132

Eisen DP, Dean MM, O'Sullivan MV, Heatley S, Minchinton RM (2008) Mannose-binding lectin deficiency does not appear to predispose to cryptococcosis in non-immunocompromised patients. Med Mycol 46:371–375

El-Zayat SR, Sibaii H, Mannaa FA (2019) Toll-like receptors activation, signaling, and targeting: an overview. Bull Natl Res Centre 43:187

Engeroff P, Vogel M (2021) The role of CD23 in the regulation of allergic responses. Allergy 76:1981–1989

Eriksson M, Johannssen T, Von Smolinski D, Gruber AD, Seeberger PH, Lepenies B (2013) The C-Type Lectin Receptor SIGNR3 Binds to Fungi Present in Commensal Microbiota and Influences Immune Regulation in Experimental Colitis. Front Immunol 4:196

Esteban A, Popp MW, Vyas VK, Strijbis K, Ploegh HL, Fink GR (2011) Fungal recognition is mediated by the association of dectin-1 and galectin-3 in macrophages. Proc Natl Acad Sci U S A 108:14270–14275

Ezekowitz RA, Sastry K, Bailly P, Warner A (1990) Molecular characterization of the human macrophage mannose receptor: demonstration of multiple carbohydrate recognition-like domains and phagocytosis of yeasts in Cos-1 cells. J Exp Med 172:1785–1794

Feinberg H, Mitchell DA, Drickamer K, Weis WI (2001) Structural basis for selective recognition of oligosaccharides by DC-SIGN and DC-SIGNR. Science 294:2163–2166

Ferwerda B, Ferwerda G, Plantinga TS, Willment JA, Van Spriel AB, Venselaar H, Elbers CC, Johnson MD, Cambi A, Huysamen C, Jacobs L, Jansen T, Verheijen K, Masthoff L, Morre SA, Vriend G, Williams DL, Perfect JR, Joosten LA, Wijmenga C, Van Der Meer JW, Adema GJ, Kullberg BJ, Brown GD, Netea MG (2009) Human dectin-1 deficiency and mucocutaneous fungal infections. N Engl J Med 361:1760–1767

Fisher MC, Gurr SJ, Cuomo CA, Blehert DS, Jin H, Stukenbrock EH, Stajich JE, Kahmann R, Boone C, Denning DW, Gow NAR, Klein BS, Kronstad JW, Sheppard DC, Taylor JW, Wright GD, Heitman J, Casadevall A, Cowen LE (2020) Threats posed by the fungal kingdom to humans, wildlife, and agriculture. mBio 11

Fonseca FL, Nohara LL, Cordero RJ, Frases S, Casadevall A, Almeida IC, Nimrichter L, Rodrigues ML (2010) Immunomodulatory effects of serotype B glucuronoxylomannan from Cryptococcus gattii correlate with polysaccharide diameter. Infect Immun 78:3861–3870

Fradin C, Poulain D, Jouault T (2000) beta-1,2-linked oligomannosides from Candida albicans bind to a 32-kilodalton macrophage membrane protein homologous to the mammalian lectin galectin-3. Infect Immun 68:4391–4398

Free SJ (2013) Fungal cell wall organization and biosynthesis. Adv Genet 81:33–82

Fuchs K, Cardona Gloria Y, Wolz OO, Herster F, Sharma L, Dillen CA, Taumer C, Dickhofer S, Bittner Z, Dang TM, Singh A, Haischer D, Schloffel MA, Koymans KJ, Sanmuganantham T, Krach M, Roger T, Le Roy D, Schilling NA, Frauhammer F, Miller LS, Nurnberger T, Leibundgut-Landmann S, Gust AA, Macek B, Frank M, Gouttefangeas C, Dela Cruz CS, Hartl D, Weber AN (2018) The fungal ligand chitin directly binds TLR2 and triggers inflammation dependent on oligomer size. EMBO Rep 19

Garcia-Rubio R, De Oliveira HC, Rivera J, Trevijano-Contador N (2019) The fungal cell wall: Candida, Cryptococcus, and Aspergillus species. Front Microbiol 10:2993

Garlanda C, Jaillon S, Doni A, Bottazzi B, Mantovani A (2016) PTX3, a humoral pattern recognition molecule at the interface between microbe and matrix recognition. Curr Opin Immunol 38:39–44

Gasparoto TH, Tessarolli V, Garlet TP, Torres SA, Garlet GP, Da Silva JS, Campanelli AP (2010) Absence of functional TLR4 impairs response of macrophages after Candida albicans infection. Med Mycol 48:1009–1017

Gazi U, Rosas M, Singh S, Heinsbroek S, Haq I, Johnson S, Brown GD, Williams DL, Taylor PR, Martinez-Pomares L (2011) Fungal recognition enhances mannose receptor shedding through dectin-1 engagement. J Biol Chem 286:7822–7829

Geijtenbeek TB, Gringhuis SI (2009) Signalling through C-type lectin receptors: shaping immune responses. Nat Rev Immunol 9:465–479

Geunes-Boyer S, Oliver TN, Janbon G, Lodge JK, Heitman J, Perfect JR, Wright JR (2009) Surfactant protein D increases phagocytosis of hypocapsular Cryptococcus neoformans by murine macrophages and enhances fungal survival. Infect Immun 77:2783–2794

Giles SS, Zaas AK, Reidy MF, Perfect JR, Wright JR (2007) Cryptococcus neoformans is resistant to surfactant protein A mediated host defense mechanisms. PLoS One 2:e1370

Giles SS, Dagenais TR, Botts MR, Keller NP, Hull CM (2009) Elucidating the pathogenesis of spores from the human fungal pathogen Cryptococcus neoformans. Infect Immun 77:3491–3500

Glittenberg MT, Silas S, Maccallum DM, Gow NA, Ligoxygakis P (2011) Wild-type Drosophila melanogaster as an alternative model system for investigating the pathogenicity of Candida albicans. Dis Model Mech 4:504–514

Gow NAR, Latge JP, Munro CA (2017) The fungal cell wall: structure, biosynthesis, and function. Microbiol Spectr 5

Griffiths JS, Thompson A, Stott M, Benny A, Lewis NA, Taylor PR, Forton J, Herrick S, Orr SJ, Mcgreal EP (2018) Differential susceptibility of Dectin-1 isoforms to functional inactivation by neutrophil and fungal proteases. FASEB J 32:3385–3397

Griffiths JS, White PL, Czubala MA, Simonazzi E, Bruno M, Thompson A, Rizkallah PJ, Gurney M, Da Fonseca DM, Naglik JR, Ingram W, Wilson K, Van De Veerdonk FL, Barnes R, Taylor PR, Orr SJ (2021) A human dectin-2 deficiency associated with invasive aspergillosis. J Infect Dis 224:1219–1224

Gringhuis SI, Den Dunnen J, Litjens M, Van Der Vlist M, Wevers B, Bruijns SC, Geijtenbeek TB (2009) Dectin-1 directs T helper cell differentiation by controlling noncanonical NF-kappaB activation through Raf-1 and Syk. Nat Immunol 10:203–213

Gross O, Poeck H, Bscheider M, Dostert C, Hannesschlager N, Endres S, Hartmann G, Tardivel A, Schweighoffer E, Tybulewicz V, Mocsai A, Tschopp J, Ruland J (2009) Syk kinase signalling couples to the Nlrp3 inflammasome for anti-fungal host defence. Nature 459:433–436

Guan Y, Ranoa DR, Jiang S, Mutha SK, Li X, Baudry J, Tapping RI (2010) Human TLRs 10 and 1 share common mechanisms of innate immune sensing but not signaling. J Immunol 184:5094–5103

Guasconi L, Beccacece I, Volpini X, Burstein VL, Mena CJ, Silvane L, Almeida MA, Musri MM, Cervi L, Chiapello LS (2022) Pulmonary conventional type 1 langerin-expressing dendritic cells play a role in impairing early protective immune response against Cryptococcus neoformans infection in mice. J Fungi (Basel) 8

Guo H, Callaway JB, Ting JP (2015) Inflammasomes: mechanism of action, role in disease, and therapeutics. Nat Med 21:677–687

Guo Y, Chang Q, Cheng L, Xiong S, Jia X, Lin X, Zhao X (2018) C-type lectin receptor CD23 is required for host defense against Candida albicans and Aspergillus fumigatus infection. J Immunol 201:2427–2440

Han Y, Kozel TR, Zhang MX, Macgill RS, Carroll MC, Cutler JE (2001) Complement is essential for protection by an IgM and an IgG3 monoclonal antibody against experimental, hematogenously disseminated candidiasis. J Immunol 167:1550–1557

Hardison SE, Brown GD (2012) C-type lectin receptors orchestrate antifungal immunity. Nat Immunol 13:817–822

Hatinguais R, Willment JA, Brown GD (2020) PAMPs of the fungal cell wall and mammalian PRRs. Curr Top Microbiol Immunol 425:187–223

He Q, Li H, Rui Y, Liu L, He B, Shi Y, Su X (2018) Pentraxin 3 gene polymorphisms and pulmonary aspergillosis in chronic obstructive pulmonary disease patients. Clin Infect Dis 66:261–267

Heinsbroek SE, Taylor PR, Martinez FO, Martinez-Pomares L, Brown GD, Gordon S (2008) Stage-specific sampling by pattern recognition receptors during Candida albicans phagocytosis. PLoS Pathog 4:e1000218

Held K, Thiel S, Loos M, Petry F (2008) Increased susceptibility of complement factor B/C2 double knock-out mice and mannan-binding lectin knockout mice to systemic infection with Candida albicans. Mol Immunol 45:3934–3941

Hidalgo A, Libby P, Soehnlein O, Aramburu IV, Papayannopoulos V, Silvestre-Roig C (2022) Neutrophil extracellular traps: from physiology to pathology. Cardiovasc Res 118:2737–2753

Hogaboam CM, Takahashi K, Ezekowitz RA, Kunkel SL, Schuh JM (2004) Mannose-binding lectin deficiency alters the development of fungal asthma: effects on airway response, inflammation, and cytokine profile. J Leukoc Biol 75:805–814

Hole CR, Leopold Wager CM, Mendiola AS, Wozniak KL, Campuzano A, Lin X, Wormley FL Jr (2016) Antifungal activity of plasmacytoid dendritic cells against cryptococcus neoformans in vitro requires expression of dectin-3 (CLEC4D) and reactive oxygen species. Infect Immun 84:2493–2504

Holmer SM, Evans KS, Asfaw YG, Saini D, Schell WA, Ledford JG, Frothingham R, Wright JR, Sempowski GD, Perfect JR (2014) Impact of surfactant protein D, interleukin-5, and eosinophilia on Cryptococcosis. Infect Immun 82:683–693

Houst J, Spizek J, Havlicek V (2020) Antifungal drugs. Metabolites 10

Hoving JC, Wilson GJ, Brown GD (2014) Signalling C-type lectin receptors, microbial recognition and immunity. Cell Microbiol 16:185–194

Hu XP, Wang RY, Wang X, Cao YH, Chen YQ, Zhao HZ, Wu JQ, Weng XH, Gao XH, Sun RH, Zhu LP (2015) Dectin-2 polymorphism associated with pulmonary cryptococcosis in HIV-uninfected Chinese patients. Med Mycol 53:810–816

Huang HR, Li F, Han H, Xu X, Li N, Wang S, Xu JF, Jia XM (2018) Dectin-3 recognizes glucuronoxylomannan of Cryptococcus neoformans serotype AD and Cryptococcus gattii serotype B to initiate host defense against cryptococcosis. Front Immunol 9:1781

Ifrim DC, Joosten LA, Kullberg BJ, Jacobs L, Jansen T, Williams DL, Gow NA, Van Der Meer JW, Netea MG, Quintin J (2013) Candida albicans primes TLR cytokine responses through a Dectin-1/Raf-1-mediated pathway. J Immunol 190:4129–4135

Ifrim DC, Bain JM, Reid DM, Oosting M, Verschueren I, Gow NA, Van Krieken JH, Brown GD, Kullberg BJ, Joosten LA, Van Der Meer JW, Koentgen F, Erwig LP, Quintin J, Netea MG (2014) Role of Dectin-2 for host defense against systemic infection with Candida glabrata. Infect Immun 82:1064–1073

Ifrim DC, Quintin J, Courjol F, Verschueren I, Van Krieken JH, Koentgen F, Fradin C, Gow NA, Joosten LA, Van Der Meer JW, Van De Veerdonk F, Netea MG (2016) The role of dectin-2 for host defense against disseminated candidiasis. J Interferon Cytokine Res 36:267–276

Ip WK, Lau YL (2004) Role of mannose-binding lectin in the innate defense against Candida albicans: enhancement of complement activation, but lack of opsonic function, in phagocytosis by human dendritic cells. J Infect Dis 190:632–640

Ishikawa T, Itoh F, Yoshida S, Saijo S, Matsuzawa T, Gonoi T, Saito T, Okawa Y, Shibata N, Miyamoto T, Yamasaki S (2013) Identification of distinct ligands for the C-type lectin receptors Mincle and Dectin-2 in the pathogenic fungus Malassezia. Cell Host Microbe 13:477–488

Jawhara S, Pluskota E, Verbovetskiy D, Skomorovska-Prokvolit O, Plow EF, Soloviev DA (2012) Integrin alphaXbeta(2) is a leukocyte receptor for Candida albicans and is essential for protection against fungal infections. J Immunol 189:2468–2477

Jegouzo SAF, Feinberg H, Morrison AG, Holder A, May A, Huang Z, Jiang L, Lasanajak Y, Smith DF, Werling D, Drickamer K, Weis WI, Taylor ME (2019) CD23 is a glycan-binding receptor in some mammalian species. J Biol Chem 294:14845–14859

Jie Z, Wu XY, Yu FS (2009) Activation of Toll-like receptors 2 and 4 in Aspergillus fumigatus keratitis. Innate Immun 15:155–168

Kang JY, Lee JO (2011) Structural biology of the Toll-like receptor family. Annu Rev Biochem 80:917–941

Kashem SW, Riedl MS, Yao C, Honda CN, Vulchanova L, Kaplan DH (2015) Nociceptive sensory fibers drive interleukin-23 production from CD301b+ dermal dendritic cells and drive protective cutaneous immunity. Immunity 43:515–526

Kasperkovitz PV, Khan NS, Tam JM, Mansour MK, Davids PJ, Vyas JM (2011) Toll-like receptor 9 modulates macrophage antifungal effector function during innate recognition of Candida albicans and Saccharomyces cerevisiae. Infect Immun 79:4858–4867

Kaur S, Gupta VK, Thiel S, Sarma PU, Madan T (2007) Protective role of mannan-binding lectin in a murine model of invasive pulmonary aspergillosis. Clin Exp Immunol 148:382–389

Kawai T, Akira S (2010) The role of pattern-recognition receptors in innate immunity: update on Toll-like receptors. Nat Immunol 11:373–384

Kawasaki T, Kawai T (2014) Toll-like receptor signaling pathways. Front Immunol 5:461

Kerscher B, Willment JA, Brown GD (2013) The Dectin-2 family of C-type lectin-like receptors: an update. Int Immunol 25:271–277

Kitai Y, Sato K, Tanno D, Yuan X, Umeki A, Kasamatsu J, Kanno E, Tanno H, Hara H, Yamasaki S, Saijo S, Iwakura Y, Ishii K, Kawakami K (2021) Role of dectin-2 in the phagocytosis of Cryptococcus neoformans by dendritic cells. Infect Immun 89:e0033021

Kohatsu L, Hsu DK, Jegalian AG, Liu FT, Baum LG (2006) Galectin-3 induces death of Candida species expressing specific beta-1,2-linked mannans. J Immunol 177:4718–4726

Kottom TJ, Hebrink DM, Jenson PE, Nandakumar V, Wuthrich M, Wang H, Klein B, Yamasaki S, Lepenies B, Limper AH (2017) The interaction of pneumocystis with the C-type lectin receptor mincle exerts a significant role in host defense against infection. J Immunol 198:3515–3525

Kottom TJ, Hebrink DM, Jenson PE, Marsolek PL, Wuthrich M, Wang H, Klein B, Yamasaki S, Limper AH (2018) Dectin-2 is a C-type lectin receptor that recognizes pneumocystis and participates in innate immune responses. Am J Respir Cell Mol Biol 58:232–240

Kottom TJ, Schaefbauer K, Carmona EM, Limper AH (2022) EphA2 is a lung epithelial cell receptor for pneumocystis beta-glucans. J Infect Dis 225:525–530

Koziel H, Eichbaum Q, Kruskal BA, Pinkston P, Rogers RA, Armstrong MY, Richards FF, Rose RM, Ezekowitz RA (1998) Reduced binding and phagocytosis of Pneumocystis carinii by alveolar macrophages from persons infected with Hiv-1 correlates with mannose receptor downregulation. J Clin Invest 102:1332–1344

Lanternier F, Cypowyj S, Picard C, Bustamante J, Lortholary O, Casanova JL, Puel A (2013) Primary immunodeficiencies underlying fungal infections. Curr Opin Pediatr 25:736–747

Lee SJ, Gonzalez-Aseguinolaza G, Nussenzweig MC (2002) Disseminated candidiasis and hepatic malarial infection in mannose-binding-lectin-A-deficient mice. Mol Cell Biol 22:8199–8203

Lee SJ, Zheng NY, Clavijo M, Nussenzweig MC (2003) Normal host defense during systemic candidiasis in mannose receptor-deficient mice. Infect Immun 71:437–445

Lee WB, Yan JJ, Kang JS, Zhang Q, Choi WY, Kim LK, Kim YJ (2017) Mincle activation enhances neutrophil migration and resistance to polymicrobial septic peritonitis. Sci Rep 7:41106

Lewis SS, Loram LC, Hutchinson MR, Li CM, Zhang Y, Maier SF, Huang Y, Rice KC, Watkins LR (2012) (+)-naloxone, an opioid-inactive toll-like receptor 4 signaling inhibitor, reverses multiple models of chronic neuropathic pain in rats. J Pain 13:498–506

Li D, Dong B, Tong Z, Wang Q, Liu W, Wang Y, Liu W, Chen J, Xu L, Chen L, Duan Y (2012) MBL-mediated opsonophagocytosis of Candida albicans by human neutrophils is coupled with intracellular Dectin-1-triggered ROS production. PLoS One 7:e50589

Li SS, Kyei SK, Timm-Mccann M, Ogbomo H, Jones GJ, Shi M, Xiang RF, Oykhman P, Huston SM, Islam A, Gill MJ, Robbins SM, Mody CH (2013) The NK receptor NKp30 mediates direct fungal recognition and killing and is diminished in NK cells from HIV-infected patients. Cell Host Microbe 14:387–397

Li R, Rezk A, Li H, Gommerman JL, Prat A, Bar-Or A, Canadian B Cells in MS Team (2017) Antibody-independent function of human B cells contributes to antifungal T cell responses. J Immunol 198:3245–3254

Li LY, Zhang HR, Jiang ZL, Chang YZ, Shao CZ (2018a) Overexpression of dendritic cell-specific intercellular adhesion molecule-3-grabbing nonintegrin in dendritic cells protecting against aspergillosis. Chin Med J (Engl) 131:2575–2582

Li SS, Ogbomo H, Mansour MK, Xiang RF, Szabo L, Munro F, Mukherjee P, Mariuzza RA, Amrein M, Vyas JM, Robbins SM, Mody CH (2018b) Identification of the fungal ligand triggering cytotoxic PRR-mediated NK cell killing of Cryptococcus and Candida. Nat Commun 9:751

Linden JR, De Paepe ME, Laforce-Nesbitt SS, Bliss JM (2013) Galectin-3 plays an important role in protection against disseminated candidiasis. Med Mycol 51:641–651

Lopez Robles MD, Pallier A, Huchet V, Le Texier L, Remy S, Braudeau C, Delbos L, Moreau A, Louvet C, Brosseau C, Royer PJ, Magnan A, Halary F, Josien R, Cuturi MC, Anegon I, Chiffoleau E (2017) Cell-surface C-type lectin-like receptor CLEC-1 dampens dendritic cell activation and downstream Th17 responses. Blood Adv 1:557–568

Loures FV, Rohm M, Lee CK, Santos E, Wang JP, Specht CA, Calich VL, Urban CF, Levitz SM (2015) Recognition of Aspergillus fumigatus hyphae by human plasmacytoid dendritic cells is mediated by dectin-2 and results in formation of extracellular traps. PLoS Pathog 11:e1004643

Ma YJ, Doni A, Skjoedt MO, Honore C, Arendrup M, Mantovani A, Garred P (2011) Heterocomplexes of mannose-binding lectin and the pentraxins PTX3 or serum amyloid P component trigger cross-activation of the complement system. J Biol Chem 286:3405–3417

Madan T, Kishore U (2020) Surfactant protein D recognizes multiple fungal ligands: a key step to initiate and intensify the anti-fungal host defense. Front Cell Infect Microbiol 10:229

Madan T, Reid KB, Clark H, Singh M, Nayak A, Sarma PU, Hawgood S, Kishore U (2010) Susceptibility of mice genetically deficient in SP-A or SP-D gene to invasive pulmonary aspergillosis. Mol Immunol 47:1923–1930

Mansour MK, Schlesinger LS, Levitz SM (2002) Optimal T cell responses to Cryptococcus neoformans mannoprotein are dependent on recognition of conjugated carbohydrates by mannose receptors. J Immunol 168:2872–2879

Mansour MK, Latz E, Levitz SM (2006) Cryptococcus neoformans glycoantigens are captured by multiple lectin receptors and presented by dendritic cells. J Immunol 176:3053–3061

Marakalala MJ, Vautier S, Potrykus J, Walker LA, Shepardson KM, Hopke A, Mora-Montes HM, Kerrigan A, Netea MG, Murray GI, Maccallum DM, Wheeler R, Munro CA, Gow NA, Cramer RA, Brown AJ, Brown GD (2013) Differential adaptation of Candida albicans in vivo modulates immune recognition by dectin-1. PLoS Pathog 9:e1003315

Mayer FL, Wilson D, Hube B (2013) Candida albicans pathogenicity mechanisms. Virulence 4:119–128

McDermott R, Ziylan U, Spehner D, Bausinger H, Lipsker D, Mommaas M, Cazenave JP, Raposo G, Goud B, De La Salle H, Salamero J, Hanau D (2002) Birbeck granules are subdomains of endosomal recycling compartment in human epidermal Langerhans cells, which form where Langerin accumulates. Mol Biol Cell 13:317–335

Means TK, Mylonakis E, Tampakakis E, Colvin RA, Seung E, Puckett L, Tai MF, Stewart CR, Pukkila-Worley R, Hickman SE, Moore KJ, Calderwood SB, Hacohen N, Luster AD, Khoury EL, J. (2009) Evolutionarily conserved recognition and innate immunity to fungal pathogens by the scavenger receptors SCARF1 and CD36. J Exp Med 206:637–653

Meier A, Kirschning CJ, Nikolaus T, Wagner H, Heesemann J, Ebel F (2003) Toll-like receptor (TLR) 2 and TLR4 are essential for Aspergillus-induced activation of murine macrophages. Cell Microbiol 5:561–570

Mezger M, Kneitz S, Wozniok I, Kurzai O, Einsele H, Loeffler J (2008) Proinflammatory response of immature human dendritic cells is mediated by dectin-1 after exposure to Aspergillus fumigatus germ tubes. J Infect Dis 197:924–931

Miro MS, Rodriguez E, Vigezzi C, Icely PA, Garcia LN, Peinetti N, Maldonado CA, Riera FO, Caeiro JP, Sotomayor CE (2018) Contribution of TLR2 pathway in the pathogenesis of vulvovaginal candidiasis. Pathog Dis 76

Miyake Y, Masatsugu OH, Yamasaki S (2015) C-type lectin receptor MCL facilitates mincle expression and signaling through complex formation. J Immunol 194:5366–5374

Miyazato A, Nakamura K, Yamamoto N, Mora-Montes HM, Tanaka M, Abe Y, Tanno D, Inden K, Gang X, Ishii K, Takeda K, Akira S, Saijo S, Iwakura Y, Adachi Y, Ohno N, Mitsutake K, Gow NA, Kaku M, Kawakami K (2009) Toll-like receptor 9-dependent activation of myeloid dendritic cells by Deoxynucleic acids from Candida albicans. Infect Immun 77:3056–3064

Mnich ME, Van Dalen R, Van Sorge NM (2020) C-type lectin receptors in host defense against bacterial pathogens. Front Cell Infect Microbiol 10:309

Mossalayi MD, Vouldoukis I, Mamani-Matsuda M, Kauss T, Guillon J, Maugein J, Moynet D, Rambert J, Desplat V, Mazier D, Vincendeau P, Malvy D (2009) CD23 mediates antimycobacterial activity of human macrophages. Infect Immun 77:5537–5542

Murciano C, Yanez A, Gil ML, Gozalbo D (2007) Both viable and killed Candida albicans cells induce in vitro production of TNF-alpha and IFN-gamma in murine cells through a TLR2-dependent signalling. Eur Cytokine Netw 18:38–43

Nahum A, Dadi H, Bates A, Roifman CM (2012) The biological significance of TLR3 variant, L412F, in conferring susceptibility to cutaneous candidiasis, CMV and autoimmunity. Autoimmun Rev 11:341–347

Nakamura K, Kinjo T, Saijo S, Miyazato A, Adachi Y, Ohno N, Fujita J, Kaku M, Iwakura Y, Kawakami K (2007) Dectin-1 is not required for the host defense to Cryptococcus neoformans. Microbiol Immunol 51:1115–1119

Nakamura K, Miyazato A, Xiao G, Hatta M, Inden K, Aoyagi T, Shiratori K, Takeda K, Akira S, Saijo S, Iwakura Y, Adachi Y, Ohno N, Suzuki K, Fujita J, Kaku M, Kawakami K (2008) Deoxynucleic acids from Cryptococcus neoformans activate myeloid dendritic cells via a TLR9-dependent pathway. J Immunol 180:4067–4074

Nakamura Y, Sato K, Yamamoto H, Matsumura K, Matsumoto I, Nomura T, Miyasaka T, Ishii K, Kanno E, Tachi M, Yamasaki S, Saijo S, Iwakura Y, Kawakami K (2015) Dectin-2 deficiency promotes Th2 response and mucin production in the lungs after pulmonary infection with Cryptococcus neoformans. Infect Immun 83:671–681

Nedovic B, Posteraro B, Leoncini E, Ruggeri A, Amore R, Sanguinetti M, Ricciardi W, Boccia S (2014) Mannose-binding lectin codon 54 gene polymorphism and vulvovaginal candidiasis: a systematic review and meta-analysis. Biomed Res Int 2014:738298

Netea MG, Marodi L (2010) Innate immune mechanisms for recognition and uptake of Candida species. Trends Immunol 31:346–353

Netea MG, Van Der Graaf CA, Vonk AG, Verschueren I, Van Der Meer JW, Kullberg BJ (2002) The role of toll-like receptor (TLR) 2 and TLR4 in the host defense against disseminated candidiasis. J Infect Dis 185:1483–1489

Netea MG, Warris A, Van Der Meer JW, Fenton MJ, Verver-Janssen TJ, Jacobs LE, Andresen T, Verweij PE, Kullberg BJ (2003) Aspergillus fumigatus evades immune recognition during germination through loss of toll-like receptor-4-mediated signal transduction. J Infect Dis 188:320–326

Netea MG, Gow NA, Munro CA, Bates S, Collins C, Ferwerda G, Hobson RP, Bertram G, Hughes HB, Jansen T, Jacobs L, Buurman ET, Gijzen K, Williams DL, Torensma R, Mckinnon A, Maccallum DM, Odds FC, Van Der Meer JW, Brown AJ, Kullberg BJ (2006) Immune sensing of Candida albicans requires cooperative recognition of mannans and glucans by lectin and Toll-like receptors. J Clin Invest 116:1642–1650

Neth O, Jack DL, Dodds AW, Holzel H, Klein NJ, Turner MW (2000) Mannose-binding lectin binds to a range of clinically relevant microorganisms and promotes complement deposition. Infect Immun 68:688–693

Nieminen J, St-Pierre C, Sato S (2005) Galectin-3 interacts with naive and primed neutrophils, inducing innate immune responses. J Leukoc Biol 78:1127–1135

Niu Y, Lin J, Li C, Peng X, Jiang N, XU Q, Yin M, Lin H, Gu L, Zhao G (2021) Galectin-3 plays an important pro-inflammatory role in A. fumigatus keratitis by recruiting neutrophils and activating p38 in neutrophils. Int Immunopharmacol 97:107706

Normile TG, Chu TH, Sheridan BS, Del Poeta M (2022) Vaccine protection by Cryptococcus neoformans Deltasgl1 is mediated by gammadelta T cells via TLR2 signaling. Mucosal Immunol 15:1416–1430

Nosanchuk JD, Casadevall A (2006) Impact of melanin on microbial virulence and clinical resistance to antimicrobial compounds. Antimicrob Agents Chemother 50:3519–3528

Opata MM, Hollifield ML, Lund FE, Randall TD, Dunn R, Garvy BA, Feola DJ (2015) B lymphocytes are required during the early priming of CD4+ T cells for clearance of pneumocystis infection in mice. J Immunol 195:611–620

Ordonez SR, Van Eijk M, Escobar Salazar N, De Cock H, Veldhuizen EJA, Haagsman HP (2019) Antifungal activities of surfactant protein D in an environment closely mimicking the lung lining. Mol Immunol 105:260–269

O'Riordan DM, Standing JE, Limper AH (1995) Pneumocystis carinii glycoprotein A binds macrophage mannose receptors. Infect Immun 63:779–784

Ostrop J, Lang R (2017) Contact, collaboration, and conflict: signal integration of Syk-coupled C-type lectin receptors. J Immunol 198:1403–1414

Pathakumari B, Liang G, Liu W (2020) Immune defence to invasive fungal infections: A comprehensive review. Biomed Pharmacother 130:110550

Patin EC, Orr SJ, Schaible UE (2017) Macrophage inducible C-type lectin as a multifunctional player in immunity. Front Immunol 8:861

Pellis V, De Seta F, Crovella S, Bossi F, Bulla R, Guaschino S, Radillo O, Garred P, Tedesco F (2005) Mannose binding lectin and C3 act as recognition molecules for infectious agents in the vagina. Clin Exp Immunol 139:120–126

Plantinga TS, Hamza OJ, Willment JA, Ferwerda B, Van De Geer NM, Verweij PE, Matee MI, Banahan K, O'Neill LA, Kullberg BJ, Brown GD, Van Der Ven AJ,

Netea MG (2010) Genetic variation of innate immune genes in HIV-infected African patients with or without oropharyngeal candidiasis. J Acquir Immune Defic Syndr 55:87–94
Porcaro I, Vidal M, Jouvert S, Stahl PD, Giaimis J (2003) Mannose receptor contribution to Candida albicans phagocytosis by murine E-clone J774 macrophages. J Leukoc Biol 74:206–215
Porte R, Davoudian S, Asgari F, Parente R, Mantovani A, Garlanda C, Bottazzi B (2019) The long pentraxin PTX3 as a humoral innate immunity functional player and biomarker of infections and sepsis. Front Immunol 10:794
Preite NW, Feriotti C, Souza De Lima D, Da Silva BB, Condino-Neto A, Pontillo A, Calich VLG, Loures FV (2018) The Syk-coupled C-type lectin receptors dectin-2 and dectin-3 are involved in Paracoccidioides brasiliensis recognition by human plasmacytoid dendritic cells. Front Immunol 9:464
Prieto D, Carpena N, Maneu V, Gil ML, Pla J, Gozalbo D (2016) TLR2 modulates gut colonization and dissemination of Candida albicans in a murine model. Microbes Infect 18:656–660
Qiu Y, Zeltzer S, Zhang Y, Wang F, Chen GH, Dayrit J, Murdock BJ, Bhan U, Toews GB, Osterholzer JJ, Standiford TJ, Olszewski MA (2012) Early induction of CCL7 downstream of TLR9 signaling promotes the development of robust immunity to cryptococcal infection. J Immunol 188:3940–3948
Rai G, Das S, Ansari MA, Singh PK, Dar SA, Haque S, Gupta N, Sharma S, Ramachandran VG, Sharma S, Jain C, Sharma S (2020) TLR-2 expression and dysregulated human Treg/Th17 phenotype in Aspergillus flavus infected patients of chronic rhinosinusitis with nasal polyposis. Microb Cell Fact 19:215
Rajaram MVS, Arnett E, Azad AK, Guirado E, Ni B, Gerberick AD, He LZ, Keler T, Thomas LJ, Lafuse WP, Schlesinger LS (2017) M. tuberculosis-initiated human mannose receptor signaling regulates macrophage recognition and vesicle trafficking by FcRgamma-chain, Grb2, and SHP-1. Cell Rep 21:126–140
Rapaka RR, Goetzman ES, Zheng M, Vockley J, Mckinley L, Kolls JK, Steele C (2007) Enhanced defense against Pneumocystis carinii mediated by a novel dectin-1 receptor Fc fusion protein. J Immunol 178:3702–3712
Razvina O, Jiang S, Matsubara K, Ohashi R, Hasegawa G, Aoyama T, Daigo K, Kodama T, Hamakubo T, Naito M (2015) Differential expression of pentraxin 3 in neutrophils. Exp Mol Pathol 98:33–40
Redlich S, Ribes S, Schutze S, Eiffert H, Nau R (2013) Toll-like receptor stimulation increases phagocytosis of Cryptococcus neoformans by microglial cells. J Neuroinflammation 10:71
Reedy JL, Crossen AJ, Negoro PE, Harding HB, Ward RA, Vargas-Blanco DA, Timmer KD, Reardon CM, Basham KJ, Mansour MK, Wuthrich M, Fontaine T, Latge JP, Vyas JM (2023) The C-type lectin receptor dectin-2 is a receptor for Aspergillus fumigatus galactomannan. mBio 14(1):e0318422
Rezende CP, Brito P, Da Silva TA, Pessoni AM, Ramalho LNZ, Almeida F (2021) Influence of galectin-3 on the innate immune response during experimental Cryptococcosis. J Fungi (Basel) 7
Richardson MB, Williams SJ (2014) MCL and Mincle: C-type lectin receptors that sense damaged self and pathogen-associated molecular patterns. Front Immunol 5:288
Ricks DM, Chen K, Zheng M, Steele C, Kolls JK (2013) Dectin immunoadhesins and pneumocystis pneumonia. Infect Immun 81:3451–3462
Rivera A, Hohl TM, Collins N, Leiner I, Gallegos A, Saijo S, Coward JW, Iwakura Y, Pamer EG (2011) Dectin-1 diversifies Aspergillus fumigatus-specific T cell responses by inhibiting T helper type 1 CD4 T cell differentiation. J Exp Med 208:369–381
Rosseau S, Hammerl P, Maus U, Gunther A, Seeger W, Grimminger F, Lohmeyer J (1999) Surfactant protein A down-regulates proinflammatory cytokine production evoked by Candida albicans in human alveolar macrophages and monocytes. J Immunol 163:4495–4502
Saijo S, Iwakura Y (2011) Dectin-1 and Dectin-2 in innate immunity against fungi. Int Immunol 23:467–472
Saijo S, Fujikado N, Furuta T, Chung SH, Kotaki H, Seki K, Sudo K, Akira S, Adachi Y, Ohno N, Kinjo T, Nakamura K, Kawakami K, Iwakura Y (2007) Dectin-1 is required for host defense against Pneumocystis carinii but not against Candida albicans. Nat Immunol 8:39–46
Saijo S, Ikeda S, Yamabe K, Kakuta S, Ishigame H, Akitsu A, Fujikado N, Kusaka T, Kubo S, Chung SH, Komatsu R, Miura N, Adachi Y, Ohno N, Shibuya K, Yamamoto N, Kawakami K, Yamasaki S, Saito T, Akira S, Iwakura Y (2010) Dectin-2 recognition of alpha-mannans and induction of Th17 cell differentiation is essential for host defense against Candida albicans. Immunity 32:681–691
Sainz J, Lupianez CB, Segura-Catena J, Vazquez L, Rios R, Oyonarte S, Hemminki K, Forsti A, Jurado M (2012) Dectin-1 and DC-SIGN polymorphisms associated with invasive pulmonary Aspergillosis infection. PLoS One 7:e32273
Salazar F, Brown GD (2018) Antifungal Innate Immunity: A Perspective from the Last 10 Years. J Innate Immun 10:373–397
Sato K, Yang XL, Yudate T, Chung JS, Wu J, Luby-Phelps K, Kimberly RP, Underhill D, Cruz PD Jr, Ariizumi K (2006) Dectin-2 is a pattern recognition receptor for fungi that couples with the Fc receptor gamma chain to induce innate immune responses. J Biol Chem 281:38854–38866
Sato Y, Sato K, Yamamoto H, Kasamatsu J, Miyasaka T, Tanno D, Miyahara A, Kagesawa T, Oniyama A, Kawamura K, Yokoyama R, Kitai Y, Umeki A, Ishizuka S, Takano K, Shiroma R, Nakahata N, Kawakami K, Kanno E, Tanno H, Yamasaki S, Hara H, Ishii K, Kawakami K (2020) Limited role of mincle in the host defense against infection with Cryptococcus deneoformans. Infect Immun 88
Savers A, Rasid O, Parlato M, Brock M, Jouvion G, Ryffel B, Cavaillon JM, Eberl G, Ibrahim-Granet O (2016) Infection-mediated priming of phagocytes pro-

tects against lethal secondary Aspergillus fumigatus challenge. PLoS One 11:e0153829

Saxena S, Madan T, Shah A, Muralidhar K, Sarma PU (2003) Association of polymorphisms in the collagen region of SP-A2 with increased levels of total IgE antibodies and eosinophilia in patients with allergic bronchopulmonary aspergillosis. J Allergy Clin Immunol 111:1001–1007

Schelenz S, Malhotra R, Sim RB, Holmskov U, Bancroft GJ (1995) Binding of host collectins to the pathogenic yeast Cryptococcus neoformans: human surfactant protein D acts as an agglutinin for acapsular yeast cells. Infect Immun 63:3360–3366

Seo BS, Lee SH, Lee JE, Yoo YC, Lee J, Park SR (2013) Dectin-1 stimulation selectively reinforces Lps-driven IgG1 production by mouse B cells. Immune Netw 13:205–212

Seyfarth J, Garred P, Madsen HO (2006) Extra-hepatic transcription of the human mannose-binding lectin gene (mbl2) and the Mbl-associated serine protease 1-3 genes. Mol Immunol 43:962–971

Singh M, Madan T, Waters P, Sonar S, Singh SK, Kamran MF, Bernal AL, Sarma PU, Singh VK, Crouch EC, Kishore U (2009) Therapeutic effects of recombinant forms of full-length and truncated human surfactant protein D in a murine model of invasive pulmonary aspergillosis. Mol Immunol 46:2363–2369

Skalski JH, Kottom TJ, Limper AH (2015) Pathobiology of Pneumocystis pneumonia: life cycle, cell wall and cell signal transduction. FEMS Yeast Res 15

Snarr BD, St-Pierre G, Ralph B, Lehoux M, Sato Y, Rancourt A, Takazono T, Baistrocchi SR, Corsini R, Cheng MP, Sugrue M, Baden LR, Izumikawa K, Mukae H, Wingard JR, King IL, Divangahi M, Satoh MS, Yipp BG, Sato S, Sheppard DC (2020) Galectin-3 enhances neutrophil motility and extravasation into the airways during Aspergillus fumigatus infection. PLoS Pathog 16:e1008741

Sobanov Y, Bernreiter A, Derdak S, Mechtcheriakova D, Schweighofer B, Duchler M, Kalthoff F, Hofer E (2001) A novel cluster of lectin-like receptor genes expressed in monocytic, dendritic and endothelial cells maps close to the NK receptor genes in the human NK gene complex. Eur J Immunol 31:3493–3503

Soloviev DA, Fonzi WA, Sentandreu R, Pluskota E, Forsyth CB, Yadav S, Plow EF (2007) Identification of pH-regulated antigen 1 released from Candida albicans as the major ligand for leukocyte integrin alphaMbeta2. J Immunol 178:2038–2046

Sparber F, Dolowschiak T, Mertens S, Lauener L, Clausen BE, Joller N, Stoitzner P, Tussiwand R, Leibundgut-Landmann S (2018) Langerin+ DCs regulate innate IL-17 production in the oral mucosa during Candida albicans-mediated infection. PLoS Pathog 14:e1007069

Speakman EA, Dambuza IM, Salazar F, Brown GD (2020) T Cell Antifungal Immunity and the Role of C-Type Lectin Receptors. Trends Immunol 41:61–76

Stahl PD, Ezekowitz RA (1998) The mannose receptor is a pattern recognition receptor involved in host defense. Curr Opin Immunol 10:50–55

Stappers MHT, Clark AE, Aimanianda V, Bidula S, Reid DM, Asamaphan P, Hardison SE, Dambuza IM, Valsecchi I, Kerscher B, Plato A, Wallace CA, Yuecel R, Hebecker B, Da Gloria Teixeira Sousa M, Cunha C, Liu Y, Feizi T, Brakhage AA, Kwon-Chung KJ, Gow NAR, Zanda M, Piras M, Zanato C, Jaeger M, Netea MG, Van De Veerdonk FL, Lacerda JF, Campos A, Carvalho A, Willment JA, Latge JP, Brown GD (2018) Recognition of DHN-melanin by a C-type lectin receptor is required for immunity to Aspergillus. Nature 555:382–386

Steele C, Rapaka RR, Metz A, Pop SM, Williams DL, Gordon S, Kolls JK, Brown GD (2005) The beta-glucan receptor dectin-1 recognizes specific morphologies of *Aspergillus fumigatus*. PLoS Pathog 1:e42

Svajger U, Anderluh M, Jeras M, Obermajer N (2010) C-type lectin DC-SIGN: an adhesion, signalling and antigen-uptake molecule that guides dendritic cells in immunity. Cell Signal 22:1397–1405

Swain SD, Lee SJ, Nussenzweig MC, Harmsen AG (2003) Absence of the macrophage mannose receptor in mice does not increase susceptibility to Pneumocystis carinii infection in vivo. Infect Immun 71:6213–6221

Swamydas M, Gao JL, Break TJ, Johnson MD, Jaeger M, Rodriguez CA, Lim JK, Green NM, Collar AL, Fischer BG, Lee CC, Perfect JR, Alexander BD, Kullberg BJ, Netea MG, Murphy PM, Lionakis MS (2016) CXCR1-mediated neutrophil degranulation and fungal killing promote Candida clearance and host survival. Sci Transl Med 8:322ra10

Swidergall M, Solis NV, Lionakis MS, Filler SG (2018) EphA2 is an epithelial cell pattern recognition receptor for fungal beta-glucans. Nat Microbiol 3:53–61

Swidergall M, Solis NV, Wang Z, Phan QT, Marshall ME, Lionakis MS, Pearlman E, Filler SG (2019) EphA2 Is a neutrophil receptor for Candida albicans that stimulates antifungal activity during oropharyngeal infection. Cell Rep 28:423–433 e5

Swidergall M, Solis NV, Millet N, Huang MY, Lin J, Phan QT, Lazarus MD, Wang Z, Yeaman MR, Mitchell AP, Filler SG (2021) Activation of EphA2-EGFR signaling in oral epithelial cells by Candida albicans virulence factors. PLoS Pathog 17:e1009221

Tachado SD, Zhang J, Zhu J, Patel N, Cushion M, Koziel H (2007) Pneumocystis-mediated IL-8 release by macrophages requires coexpression of mannose receptors and TLR2. J Leukoc Biol 81:205–211

Tanno D, Yokoyama R, Kawamura K, Kitai Y, Yuan X, Ishii K, De Jesus M, Yamamoto H, Sato K, Miyasaka T, Shimura H, Shibata N, Adachi Y, Ohno N, Yamasaki S, Kawakami K (2019) Dectin-2-mediated signaling triggered by the cell wall polysaccharides of Cryptococcus neoformans. Microbiol Immunol 63:500–512

Taylor ME, Bezouska K, Drickamer K (1992) Contribution to ligand binding by multiple carbohydrate-recognition domains in the macrophage mannose receptor. J Biol Chem 267:1719–1726

Taylor PR, Reid DM, Heinsbroek SE, Brown GD, Gordon S, Wong SY (2005) Dectin-2 is predominantly myeloid restricted and exhibits unique activation-dependent

expression on maturing inflammatory monocytes elicited in vivo. Eur J Immunol 35:2163–2174
Taylor PR, Tsoni SV, Willment JA, Dennehy KM, Rosas M, Findon H, Haynes K, Steele C, Botto M, Gordon S, Brown GD (2007) Dectin-1 is required for beta-glucan recognition and control of fungal infection. Nat Immunol 8:31–38
Tessarolli V, Gasparoto TH, Lima HR, Figueira EA, Garlet TP, Torres SA, Garlet GP, Da Silva JS, Campanelli AP (2010) Absence of TLR2 influences survival of neutrophils after infection with Candida albicans. Med Mycol 48:129–140
Thebault P, Lhermite N, Tilly G, Le Texier L, Quillard T, Heslan M, Anegon I, Soulillou JP, Brouard S, Charreau B, Cuturi MC, Chiffoleau E (2009) The C-type lectin-like receptor CLEC-1, expressed by myeloid cells and endothelial cells, is up-regulated by immunoregulatory mediators and moderates T cell activation. J Immunol 183:3099–3108
Thiagarajan PS, Yakubenko VP, Elsori DH, Yadav SP, Willard B, Tan CD, Rodriguez ER, Febbraio M, Cathcart MK (2013) Vimentin is an endogenous ligand for the pattern recognition receptor Dectin-1. Cardiovasc Res 99:494–504
Thompson A, Da Fonseca DM, Walker L, Griffiths JS, Taylor PR, Gow NAR, Orr SJ (2021) Dependence on mincle and dectin-2 varies with multiple Candida species during systemic infection. Front Microbiol 12:633229
Tone K, Stappers MHT, Hatinguais R, Dambuza IM, Salazar F, Wallace C, Yuecel R, Morvay PL, Kuwano K, Willment JA, Brown GD (2021) MelLec exacerbates the pathogenesis of Aspergillus fumigatus-induced allergic inflammation in mice. Front Immunol 12:675702
Underhill DM, Rossnagle E, Lowell CA, Simmons RM (2005) Dectin-1 activates Syk tyrosine kinase in a dynamic subset of macrophages for reactive oxygen production. Blood 106:2543–2550
Van Asbeck EC, Hoepelman AI, Scharringa J, Herpers BL, Verhoef J (2008) Mannose binding lectin plays a crucial role in innate immunity against yeast by enhanced complement activation and enhanced uptake of polymorphonuclear cells. BMC Microbiol 8:229
Van Der Graaf CA, Netea MG, Verschueren I, Van Der Meer JW, Kullberg BJ (2005) Differential cytokine production and Toll-like receptor signaling pathways by Candida albicans blastoconidia and hyphae. Infect Immun 73:7458–7464
Van De Veerdonk FL, Netea MG, Jansen TJ, Jacobs L, Verschueren I, Van Der Meer JW, Kullberg BJ (2008) Redundant role of TLR9 for anti-Candida host defense. Immunobiology 213:613–620
Van De Veerdonk FL, Marijnissen RJ, Kullberg BJ, Koenen HJ, Cheng SC, Joosten I, Van Den Berg WB, Williams DL, Van Der Meer JW, Joosten LA, Netea MG (2009) The macrophage mannose receptor induces IL-17 in response to Candida albicans. Cell Host Microbe 5:329–340
Van De Wetering JK, Coenjaerts FE, Vaandrager AB, Van Golde LM, Batenburg JJ (2004) Aggregation of Cryptococcus neoformans by surfactant protein D is inhibited by its capsular component glucuronoxylomannan. Infect Immun 72:145–153
Van Rozendaal BA, Van Spriel AB, Van De Winkel JG, Haagsman HP (2000) Role of pulmonary surfactant protein D in innate defense against Candida albicans. J Infect Dis 182:917–922
Velasco-DE Andres M, Catala C, Casado-Llombart S, Martinez-Florensa M, Simoes I, Garcia-Luna J, Mourglia-Ettlin G, Zaragoza O, Carreras E, Lozano F (2020) The lymphocytic scavenger receptor CD5 shows therapeutic potential in mouse models of fungal infection. Antimicrob Agents Chemother 65
Vendele I, Willment JA, Silva LM, Palma AS, Chai W, Liu Y, Feizi T, Spyrou M, Stappers MHT, Brown GD, Gow NAR (2020) Mannan detecting C-type lectin receptor probes recognise immune epitopes with diverse chemical, spatial and phylogenetic heterogeneity in fungal cell walls. PLoS Pathog 16:e1007927
Vitenshtein A, Charpak-Amikam Y, Yamin R, Bauman Y, Isaacson B, Stein N, Berhani O, Dassa L, Gamliel M, Gur C, Glasner A, Gomez C, Ben-Ami R, Osherov N, Cormack BP, Mandelboim O (2016) NK cell recognition of Candida glabrata through binding of NKp46 and NCR1 to fungal ligands Epa1, Epa6, and Epa7. Cell Host Microbe 20:527–534
Vuk-Pavlovic Z, Mo EK, Icenhour CR, Standing JE, Fisher JH, Limper AH (2006) Surfactant protein D enhances Pneumocystis infection in immune-suppressed mice. Am J Physiol Lung Cell Mol Physiol 290:L442–L449
Wagener J, Malireddi RK, Lenardon MD, Köberle M, Vautier S, Maccallum DM, Biedermann T, Schaller M, Netea MG, Kanneganti TD, Brown GD, Brown AJ, Gow NA (2014) Fungal chitin dampens inflammation through IL-10 induction mediated by NOD2 and TLR9 activation. PLoS Pathog 10:e1004050
Walsh NM, Wuthrich M, Wang H, Klein B, Hull CM (2017) Characterization of C-type lectins reveals an unexpectedly limited interaction between Cryptococcus neoformans spores and Dectin-1. PLoS One 12:e0173866
Wang SH, Zhang C, Lasbury ME, Liao CP, Durant PJ, Tschang D, Lee CH (2008) Decreased inflammatory response in Toll-like receptor 2 knockout mice is associated with exacerbated Pneumocystis pneumonia. Microbes Infect 10:334–341
Wang JP, Lee CK, Akalin A, Finberg RW, Levitz SM (2011) Contributions of the MyD88-dependent receptors IL-18R, IL-1R, and TLR9 to host defenses following pulmonary challenge with Cryptococcus neoformans. PLoS One 6:e26232
Wang Q, Zhao G, Lin J, Li C, Jiang N, Xu Q, Wang Q, Zhang J (2016a) Role of the mannose receptor during Aspergillus fumigatus infection and interaction with dectin-1 in corneal epithelial cells. Cornea 35:267–273
Wang T, Pan D, Zhou Z, You Y, Jiang C, Zhao X, Lin X (2016b) Dectin-3 deficiency promotes colitis development due to impaired antifungal innate immune responses in the gut. PLoS Pathog 12:e1005662
Wang W, Deng Z, Wu H, Zhao Q, Li T, Zhu W, Wang X, Tang L, Wang C, Cui SZ, Xiao H, Chen J (2019) A small secreted protein triggers a TLR2/4-dependent

inflammatory response during invasive Candida albicans infection. Nat Commun 10:1015
Watanabe M, Motooka D, Yamasaki S (2023) The kinetics of signaling through the common FcRgamma chain determine cytokine profiles in dendritic cells. Sci Signal 16:eabn9909
Wells CA, Salvage-Jones JA, Li X, Hitchens K, Butcher S, Murray RZ, Beckhouse AG, Lo YL, Manzanero S, Cobbold C, Schroder K, Ma B, Orr S, Stewart L, Lebus D, Sobieszczuk P, Hume DA, Stow J, Blanchard H, Ashman RB (2008) The macrophage-inducible C-type lectin, mincle, is an essential component of the innate immune response to Candida albicans. J Immunol 180:7404–7413
Werner JL, Metz AE, Horn D, Schoeb TR, Hewitt MM, Schwiebert LM, Faro-Trindade I, Brown GD, Steele C (2009) Requisite role for the dectin-1 beta-glucan receptor in pulmonary defense against Aspergillus fumigatus. J Immunol 182:4938–4946
Williams MD, Wright JR, March KL, Martin WJ 2nd (1996) Human surfactant protein A enhances attachment of Pneumocystis carinii to rat alveolar macrophages. Am J Respir Cell Mol Biol 14:232–238
Willment JA, Gordon S, Brown GD (2001) Characterization of the human beta-glucan receptor and its alternatively spliced isoforms. J Biol Chem 276:43818–43823
Wirnsberger G, Zwolanek F, Stadlmann J, Tortola L, Liu SW, Perlot T, Jarvinen P, Durnberger G, Kozieradzki I, Sarao R, De Martino A, Boztug K, Mechtler K, Kuchler K, Klein C, Elling U, Penninger JM (2014) Jagunal homolog 1 is a critical regulator of neutrophil function in fungal host defense. Nat Genet 46:1028–1033
Wong SSW, Rani M, Dodagatta-Marri E, Ibrahim-Granet O, Kishore U, Bayry J, Latge JP, Sahu A, Madan T, Aimanianda V (2018) Fungal melanin stimulates surfactant protein D-mediated opsonization of and host immune response to Aspergillus fumigatus spores. J Biol Chem 293:4901–4912
Wu SY, Huang JH, Chen WY, Chan YC, Lin CH, Chen YC, Liu FT, Wu-Hsieh BA (2017) Cell intrinsic galectin-3 attenuates neutrophil Ros-dependent killing of Candida by modulating CR3 downstream Syk activation. Front Immunol 8:48
Xu B, Luo Q, Gong Y, Li J, Cao J (2021) TLR7 expression aggravates invasive pulmonary aspergillosis by suppressing anti-Aspergillus immunity of macrophages. Infect Immun 89
Yamamoto H, Nakamura Y, Sato K, Takahashi Y, Nomura T, Miyasaka T, Ishii K, Hara H, Yamamoto N, Kanno E, Iwakura Y, Kawakami K (2014) Defect of CARD9 leads to impaired accumulation of gamma interferon-producing memory phenotype T cells in lungs and increased susceptibility to pulmonary infection with Cryptococcus neoformans. Infect Immun 82:1606–1615
Yamasaki S, Ishikawa E, Sakuma M, Hara H, Ogata K, Saito T (2008) Mincle is an Itam-coupled activating receptor that senses damaged cells. Nat Immunol 9:1179–1188
Yanagisawa K, Ogawa Y, Uchiumi H, Gohda F, Mawatari M, Ishizaki T, Mitsui T, Yokohama A, Handa H, Tsukamoto N, Nojima Y (2015) Gene polymorphisms of mannose-binding lectin confer susceptibility to Pneumocystis pneumonia in Hiv-infected patients. J Infect Chemother 21:769–775
Yanagisawa K, Wichukchinda N, Tsuchiya N, Yasunami M, Rojanawiwat A, Tanaka H, Saji H, Ogawa Y, Handa H, Pathipvanich P, Ariyoshi K, Sawanpanyalert P (2020) Deficiency of mannose-binding lectin is a risk of Pneumocystis jirovecii pneumonia in a natural history cohort of people living with HIV/AIDS in Northern Thailand. PLoS One 15:e0242438
Yauch LE, Mansour MK, Shoham S, Rottman JB, Levitz SM (2004) Involvement of CD14, toll-like receptors 2 and 4, and MyD88 in the host response to the fungal pathogen Cryptococcus neoformans in vivo. Infect Immun 72:5373–5382
Yu GR, Lin J, Zhang J, Che CY, Peng XD, Li C, Hu LT, Zhu GQ, He K, Zhao GQ (2018) Mincle in the innate immune response of mice fungal keratitis. Int J Ophthalmol 11:539–547
Zahedi N, Abedian Kenari S, Mohseni S, Aslani N, Ansari S, Badali H (2016) Is human Dectin-1 Y238X gene polymorphism related to susceptibility to recurrent vulvovaginal candidiasis? Curr Med Mycol 2:15–19
Zhang J, Zhu J, Imrich A, Cushion M, Kinane TB, Koziel H (2004) Pneumocystis activates human alveolar macrophage NF-kappaB signaling through mannose receptors. Infect Immun 72:3147–3160
Zhang C, Wang SH, Lasbury ME, Tschang D, Liao CP, Durant PJ, Lee CH (2006) Toll-like receptor 2 mediates alveolar macrophage response to Pneumocystis murina. Infect Immun 74:1857–1864
Zhao XQ, Zhu LL, Chang Q, Jiang C, You Y, Luo T, Jia XM, Lin X (2014) C-type lectin receptor dectin-3 mediates trehalose 6,6′-dimycolate (TDM)-induced Mincle expression through CARD9/Bcl10/MALT1-dependent nuclear factor (NF)-kappaB activation. J Biol Chem 289:30052–30062
Zhao G, Xu Q, Lin J, Chen W, Cui T, Hu L, Jiang N (2017a) The role of Mincle in innate immune to fungal keratitis. J Infect Dev Ctries 11:89–97
Zhao X, Guo Y, Jiang C, Chang Q, Zhang S, Luo T, Zhang B, Jia X, Hung MC, Dong C, Lin X (2017b) JNK1 negatively controls antifungal innate immunity by suppressing CD23 expression. Nat Med 23:337–346
Zhu LL, Zhao XQ, Jiang C, You Y, Chen XP, Jiang YY, Jia XM, Lin X (2013) C-type lectin receptors Dectin-3 and Dectin-2 form a heterodimeric pattern-recognition receptor for host defense against fungal infection. Immunity 39:324–334

Infection Models for Human Pathogenic Fungi

9

Ilse D. Jacobsen

Abstract

Infection models are necessary tools to dissect the processes leading to pathogenesis, to identify fungal and host factors that contribute to disease or resistance, and to test the efficacy of antifungal compounds and adjunctive therapies. This chapter aims to provide an overview of the most commonly used infection models, their respective advantages and disadvantages, and the context in which they have been employed in the research of human fungal pathogens.

Keyword

Infection models · Pathogenic fungi
Alternative models · Mammalian models
Invertebrate models · Cell culture

9.1 Introduction

Infection models date back to the onset of modern microbiology when Robert Koch included them as an essential component in the "Koch's Postulates," designed to judge evidence of microbial disease causation (Opal 2009). Since then, they have not only been employed to identify microbes that cause disease, but also to identify virulence factors, mechanisms underlying pathogenesis, and host factors that contribute to susceptibility or resistance. While Koch and his contemporaries used various warm-blooded host species, laboratory rodents, especially mice, soon became the most commonly used species for pragmatic reasons (Szabo and MacCallum 2011). In parallel, cell culture was increasingly used to study the interaction of microbes with distinct types of host cells. Over the last decades, invertebrate models for fungal infections have been developed, and so-called mini-hosts are increasingly used to study human pathogenic fungi (Madende et al. 2020; Singulani et al. 2018; Smith and Casadevall 2021). Thereby, large varieties of different infection models are currently available to study human pathogenic fungi. The challenge is to select the model best suitable to answer the research question while balancing ethical considerations, practical restrictions, and translatability of findings to humans or other host species of interest. The aim of this chapter is to provide an overview of (1) points to be considered when selecting a model using examples from the field, (2) the advantages and disadvantages of the different infection model types, and (3) to illustrate the type of scientific questions the different models have been applied to success-

I. D. Jacobsen (✉)
Research Group Microbial Immunology, Leibniz Institute for Natural Product Research and Infection Biology – Hans Knoell Institute, Jena, Germany

Institute of Microbiology, Friedrich Schiller University Jena, Jena, Germany
e-mail: Ilse.Jacobsen@leibniz-hki.de

A. A. Brakhage et al. (eds.), *Human and Animal Relationships*, The Mycota 6,
https://doi.org/10.1007/978-3-031-64853-3_9

Table 9.1 Overview of the advantages and disadvantages of different types of infection models

Type of infection model	Advantages	Disadvantages
Cell culture	No (cell lines) or limited (primary cells) ethical concerns/legal limitations	Limited time window for observations
	Relatively inexpensive, easy to standardize	Recapitulates only specific aspects of infection
	Dissection of different steps of the infection process	No or limited cross-talk between different cell types/organ systems
	Easy visualization of cellular interactions	
Invertebrate models	No ethical concerns/legal limitations	*C. elegans, D. melanogaster*: no experiments at 37 °C
	Inexpensive, no specialized facility required (*C. elegans, G. mellonella*)	*G. mellonella*: limited molecular tools
	High-throughput possible	*D. melanogaster*: special facility required
	C. elegans: direct visualization	No adaptive immunity
		Limitations in translation (tissue-specific interactions, routes of infection)
Mammalian models	Anatomy, physiology, and immunity similar to humans	Ethical concerns, legal restrictions
	Pathogenesis, host/organ-specific environment similar to humans	Specialized facilities and experienced personnel required
	Distinct models for different types/routes of infection and risk factors	High complexity can complicate analysis of distinct cellular/molecular mechanisms
	Interplay between different types of cells, soluble factors, organ systems	
	Mice: genetically tractable, molecular tools available	

fully. Table 9.1 provides a summary of the main advantages and disadvantages of the different infection models.

9.2 Model Complexity

Infections in humans involve different cell types and organ systems: During pulmonary aspergillosis, for example, the initial steps include germination of spores on respiratory epithelium, followed by interaction with alveolar macrophages and recruited immune cells, leading to local and systemic immune responses. Fungal invasion of blood vessels can lead to bleeding, thrombosis, and in some cases to fungal dissemination to other organs and progression to cerebral aspergillosis. Each step involves different types of host cells, and is shaped by cytokines and chemokines produced by both epithelial and immune cells. This complex interplay also occurs in murine infection models, and lung anatomy as well as epithelial and immune responses are largely conserved between men and mice. This makes it likely that the overall infection outcome, for example, reduced virulence of an *A. fumigatus* strain lacking a specific gene, can be translated to human hosts. However, identifying the exact type of interaction in which the mutant fungus is at a disadvantage can be challenging because the multitude of interactions cannot be monitored simultaneously in vivo. Thus, while murine models in this example accurately reflect the level of complexity observed in human infections, the complexity itself can make it difficult to identify distinct cellular and molecular mechanisms that lead to reduced virulence.

One way to identify the mechanisms that lead to differences in virulence, is to use models of reduced complexity, for example, host cells of a specific type, that are challenged with the fungus in vitro. Host–fungal interactions in cell culture can be directly visualized, and the system is easier to analyze. This has been proven to be a highly successful approach in dissecting the function of distinct molecules and their subsequent impact on pathogenesis: After the discovery that a melanin-deficient *A. fumigatus* mutant is less virulent, studies using macrophages allowed to identify the role of conidial 1,8-dihydroxynaphtalene-melanin for inhibition of phagolysosome maturation (Chamilos and Carvalho 2020). Some in vitro models can be expanded by step-wise addition of immune cells to an epithelial model, thereby addressing not only the direct protective or detrimental effect of the immune cells but also providing a system in which potential crosstalk between different cell types can be investigated. A recent example is the development of a lung-on-chip model that allows visualization of conidia germination on epithelial cells and macrophage activity (Hoang et al. 2022). However, due to the reduced complexity of in vitro models, predictions of in vivo infections can be difficult, especially if pathogens adapt to varying host niches over time. One example is β-glucan unmasking in *C. albicans*, which is triggered by the attack of immune cells in vivo, resulting in altered exposure of an important PAMP, and thereby affecting subsequent interactions with the host immune system (Hopke et al. 2016; Wheeler and Fink 2006).

Thus, both complex in vivo and less complex in vitro models have specific advantages and disadvantages; in many cases, a combination of models is necessary to identify both the mechanisms of host–pathogen interactions and their consequences for pathogenesis and infection outcome.

9.3 Ethical Considerations

Infection experiments that lead to clinical disease will ultimately cause some level of suffering to experimental animals. In acknowledgment of the ability of vertebrates to experience suffering, and the responsibility of humans for the wellbeing of animals in their care, animal welfare laws regulate the use of vertebrates in research. Permission by specific regulatory bodies is usually required, and in many countries the regulatory process includes submission of experimental details and attention to the 3R principle (replacement, reduction, refinement) (MacArthur Clark 2018). Mortality was traditionally used as the main endpoint in animal experiments, but within the concept of the 3Rs, regulations in most countries require that (vertebrate) animals are euthanized at pre-defined humane endpoints. While mammalian models have the advantage to share most physiological, anatomical, and immunological properties with humans, ethical (and practical) considerations limit their application for large-scale screening purposes and strain comparisons.

The use of invertebrate alternative model hosts, such as larvae of the greater wax moth *Galleria mellonella*, the fruit fly *Drosophila melanogaster*, or the worm *Caenorhabditis elegans*, is less constrained by ethical considerations. Due to the considerably lower developmental stage such invertebrates are unlikely to process physical impairment in the same way as mammals or other vertebrates, and it is assumed that their subjective perception does not facilitate suffering comparable to mammals. Consequently, the use of such invertebrates in infection experiments is not legally restricted, and they are accepted as alternatives for replacement of mouse experiments to comply with the 3R framework.

For cell culture, infection experiments per se do not raise ethical concerns, but the origin of cells must be taken into consideration. Immortalized cell lines can be maintained and expanded long-term, but the preparation of primary cells requires a donor. Harvesting of the necessary cells from mice often require the animal to be killed; while humane killing imposes less suffering on a mouse than an infection would, the killing itself requires ethical justification, and is legally regulated in many countries. In contrast, primary cells from humans can be obtained with minimal harm from healthy donors if they are present in blood or tissue that is removed for other primary reasons during surgical procedures. This, however, requires informed consent of the volunteer or patient, and often an ethical review by an ethics board. Guidelines on consent might limit the use of material to explicitly mentioned research only. Therefore, primary cells might not be as readily available as immortalized cell lines or invertebrate alternative hosts.

9.4 Available Molecular and Genetic Tools, and Other Practical Considerations

The amount of information that can be extracted from an infection experiment depends to a considerable extend on the availability of different read-out methods and tools to manipulate host or pathogen. On the pathogen side, methods developed in the context of one infection model can be easily transferred to another: Methods for the quantification of fungal burden work with mouse organs, invertebrate hosts, and in tissue culture once the protocol has been adapted to specific circumstances. Tools to analyze the host response, however, are often host-specific. For example, antibodies targeting signaling pathways in or molecules produced by immune cells, such as cytokines, are widely available for human and murine systems. In comparison, the selection for rats is more limited, and such tools have often not yet been developed for other mammals or avian species. In consequence, in-depth analysis of host responses might only be possible in mice or humans. To test hypotheses, targeted modulation of the host by genetic modification or antibody-mediated depletion of selected immune cells is often employed. One reason why mice are the predominant mammalian species in infection research is the relative ease by which targeted genetic modifications can be achieved. Similarly, a large number of mutants have been generated in the model invertebrates *Drosophila melanogaster* and *Caenorhabditis elegans*, whereas genetic modification of *Galleria mellonella* was achieved only recently (Dinh et al. 2021). Thus, the choice of a specific infection model might be influenced by the availability of tools to analyze or manipulate the host.

Recent years have seen considerable progress in methods for genetic manipulation of cell lines, at least partially driven by the advancement of CRISPR/Cas technology (Shalem et al. 2014). This allows a move from murine cells derived from transgenic animals to human cells with specific modifications in the future, which will also allow us to better assess the extent to which results can be translated to humans.

Besides ethics and availability of tools to analyze infection models, other practical considerations might affect the choice of a specific model: Working with live animals requires adequate facilities and specialized staff, especially if the species of choice has to be bred on site. Thus, experiments in mice can only be conducted if an appropriate vivarium is available, and the use of *Drosophila* usually requires a special room to breed enough flies for experiments. In contrast, cell culture experiments and breeding of *Caenorhabditis* require only appropriate incubators, a safety bench, and simple microscopes, all of which are often available in microbiology labs. One of the advantages of *Galleria* is that larvae can be purchased and easily maintained without the need for specialized equipment. Furthermore, even research-quality *Galleria* larvae are substantially cheaper to purchase than laboratory mice, which makes them an attractive option if funding is limited or for high-throughput approaches that require high numbers of animals.

In summary, choosing an infection model is a complex process influenced by the suitability of the research question to be answered, ethical considerations, the availability of read-out methods, and practical considerations. The following part will outline aspects of host–fungal interactions that are typically addressed by the different types of infection models, and discuss advantages and disadvantages to be considered.

9.5 Cell Culture Models

Cell culture models can be roughly divided by (1) the cell type and number of cell types used within the model; (2) the use of immortal cell lines versus primary cells; and (3) the use of monolayers versus multilayered systems resembling tissue.

9.5.1 Cell Lines

An example of a simple cell culture model is the use of epithelial cell lines derived from carcinomas, such as TR146 cells. This cell line is derived from human buccal epithelium and grows as well-differentiating keratinizing squamous cell carcinoma. The advantage of this model is that the cell line is commercially available and can be cultured over several passages, making it a logistically convenient model. TR146 cells have been widely employed as a mucosa model for oral candidiasis with methods established to analyze adhesion, invasion and host cell damage as the three key steps in pathogenesis (Wächtler et al. 2011; Mogavero and Hube 2021). Using this model, several *C. albicans* adhesins were identified (de Groot et al. 2013). It facilitated the discovery that *C. albicans* invades host cells by two distinct mechanisms, active penetration mediated by fungal activity, and induced endocytosis relying on host activity and interaction of fungal factors such as Als3 and Ssa1 with host molecules such as cadherins and the epidermal growth factor receptor (Wächtler et al. 2012; Sun et al. 2010; Phan et al. 2005, 2007; Solis et al. 2017). Furthermore, this model played a central role in the discovery of the *C. albicans* peptide toxin candidalysin (Moyes et al. 2016) and other virulence factors (Martin et al. 2011; Richardson et al. 2018). In addition, the model can be used to evaluate the impact of drug treatment on host–fungal interactions (Puri et al. 2019). Similar models were established and used successfully to study the interaction of *C. albicans* with intestinal and vaginal epithelial cells and endothelial cells (Sheppard and Filler 2014; Grubb et al. 2008; Mikamo et al. 2018; Dalle et al. 2010).

9.5.2 Primary Cells

Cell lines, especially macrophages, have also been useful to study the interaction between human pathogenic fungi and immune cells. Examples include a study leading to the discovery that *A. fumigatus* melanin inhibits phagolysosome maturation (Thywissen et al. 2011), and the use of genetically modified cell lines to elucidate the contribution of TLR9 to the response to fungi (Khan et al. 2016). However, primary immune cells are used more frequently than cell lines to investigate interactions between fungi and immune cells. One reason is the lack of cell lines for some types of immune cells, such as neutrophils or dendritic cells. While neutrophil-like cells have been used successfully (Rafiq et al. 2022), their response to pathogens might differ from primary cells, as it has been observed for macrophage cell lines (Andreu et al. 2017; Bourgeois et al. 2009; Kortmann et al. 2015). Another reason is the relative ease with which primary immune cells can be isolated from peripheral blood, making them readily available. The use of human cells furthermore facilitates the identification of host factors predisposing to fungal infection and the underlying mechanisms by comparing cells from patients and healthy volunteers (Cunha et al. 2010, 2014; Stappers et al. 2018). Murine immune cells, usually isolated from either spleen or bone marrow, are also commonly used; in combination with genetically modified mice, such approaches allow evaluation of the impact of immune receptors, signaling pathways, and other molecular and cellular mechanisms (Khan et al. 2016; Taylor et al. 2007; Saijo et al. 2007; Ding et al. 2021; Wellington et al. 2014). While primary cells have the advantages over cell lines described above, the recruitment of human donors and the limited life span of primary cells, requiring immediate processing of samples, can make work with primary cells more challenging compared to the use of cell lines. If primary murine cells are used, sampling often requires the animal to be sacrificed, which should be considered from an ethical perspective. For both cell lines and primary cells, protocols for the quantification of phagocytosis, fungal killing, and immune cell damage have been firmly established and allow to dissect the interactions at the cellular and molecular level (recently reviewed in Last et al. 2021).

9.5.3 Complex Cell Culture Models

Models consisting of a single cell type are technically easy to work with but represent a simplified version of the in vivo situation. If the possible crosstalk between different types of cells is of interest, more complex models are necessary. One example is the supplementation of a multilayered oral epithelial cell culture model with neutrophils in a transwell system, which was used to demonstrate how epithelial cells and immune cells interact during *Candida* infection (Wagener et al. 2012). A technically more refined approach is organ-on-chip models (Last et al. 2021). These systems consist of epithelial cells separated by a membrane from endothelial cells, with the option to perfuse either side separately, and to introduce immune cells into the circulatory side (Last et al. 2021). Due to the higher technical demands, such models have not yet been used widely to study the interaction of fungi with host cells, but recent studies show promising potential: Introduction of macrophages to lung epithelium-on-chip was used recently to determine the extent to which the immune cells prevent epithelial infection by *A. fumigatus* in an organ-like setting (Hoang et al. 2022). Gut-on-a-chip can be used to study the interaction of *Candida* with commensal bacteria in a system that shares more features with the gut as an organ compared to single-layer cell culture models (Graf et al. 2019). A further step in increasing complexity and mimicking the in vivo situation in vitro is the use of ex vivo models and organoids. One example is anticoagulated whole blood as an ex vivo infection model. In combination with bioinformatical modelling, this experimental approach was shown to be useful to determine the relative contribution of different types of immune cells to fungal killing in a setting mimicking bloodstream infection (Hünniger et al. 2014; Machata et al. 2020). Another example is the use of organ explants, such as skin. Human skin explants are commercially available and have been used to establish a model to study the infection of burn wounds by *C. albicans* (von Müller et al. 2020). Finally, 3D culture systems utilizing stem cells to generate organoids are increasingly used in infection research (Kim et al. 2020; Aguilar et al. 2021), and might extend the spectrum of available in vitro methods in the near future.

This short overview shows that cell culture infection models have been essential for our understanding of host–fungal interactions on the molecular, cellular, and to some extent the tissue level. Cell lines are convenient models due to the relative ease of use and the suitability for high-throughput screening. Primary cells have been especially important for immunological studies. The complexity of cell culture models can be increased by combining different cell types or by using ex vivo models. With increasing complexity, however, the technical demands increase and the suitability for high-throughput analyses decreases.

9.6 Invertebrate Models

Invertebrate models have emerged as important model systems to study fungal infections in recent decades (Fusco-Almeida et al. 2023). Their advantage over cell culture models is that they represent complete animals, with different organ systems that can interact during and be affected by infection. The use of invertebrates is subject to fewer ethical and regulatory issues, making them an attractive alternative to mammalian models. Furthermore, the small size and short generation time, or commercial availability in a ready-to-use developmental stage, makes them cheaper than traditional animal models and allows high throughput. Many aspects of innate immunity are conserved between mammals and invertebrates (Trevijano-Contador and Zaragoza 2018; Hoffmann et al. 1999), and as this presents the first line of defense against fungal pathogens (Salazar and Brown 2018), important aspects of pathogenesis are likely conserved, and similar fungal virulence strategies apply to invertebrate and mammalian hosts alike. However, there are substantial differences in the body plan, and not all organs of humans have an equivalent in invertebrates. Therefore, some tissue-specific infections and distinct routes of human infection, such as inhalation of spores, cannot be mimicked in these hosts. Adaptive immunity only evolved in vertebrates, and thus cannot be studied in invertebrates. As recently pointed out in a review on *Galleria mellonella* as a model system, the strength and relevancy of invertebrate models depend on the extent to which pathogenesis, host response, and relevance of fungal virulence factors are comparable to human hosts (Smith and Casadevall 2021). In this area, more in-depth research is needed to better understand how fungal infection leads to host disease in invertebrates, and to which extent findings can be translated to humans. In the following, the three main model systems, *Caenorhabditis elegans*, *Drosophila melanogaster*, and *Galleria mellonella* will be briefly described with their specific advantages and disadvantages.

9.6.1 *Caenorhabditis elegans*

The nematode *C. elegans* has been used as a model organism for genetics and neurobiology in the 1960s. Since then, *C. elegans* has also been established as an in vivo model for immunology and infectious diseases. Advantages are its rapid life cycle, physiological simplicity, a fully sequenced genome, and the possibility of genetic manipulation. *C. elegans* can be maintained and bred easily in the lab using *Escherichia coli* OP50 on agar plates as a food source.

The small size of the worms (adults are approx. 1 mm in length) makes direct injection of microbes impossible; thus, only oral infections are feasible. These can be easily achieved by placing the worms on pathogen lawns on agar plates (Powell and Ausubel 2008). Ingestion of microbes leads to intestinal infection, which due to the transparency of its cuticle can be directly observed microscopically if fluorescent fungal strains are used. *C. elegans* has been used for a variety of human fungal pathogens to determine virulence as well as drug efficacy testing, including *Candida* spp., *A. fumigatus*, and *Penicillium marneffei* (Madende et al. 2020; Ahamefule et al. 2020, 2021; Okoli et al. 2009; Huang et al. 2014; Lionakis and Kontoyiannis 2012; Panayidou et al. 2014). An interesting aspect is the use of *C. elegans* to investigate the impact of bacterial–fungal interactions on virulence (Poupet et al. 2019). One important limitation is that *C. elegans* cannot be maintained at 37 °C.

9.6.2 *Drosophila melanogaster*

Similar to *C. elegans*, the fruit fly *D. melanogaster* is a model organism for genetics and development in biology that has been intensely studied for over a century. In recent decades, it has also become a model for human diseases based on genetic similarities (Ugur et al. 2016). The thorough characterization and the well-developed tools for genetic manipulation make it a useful model in infectious disease research, and it has been used to study a wide variety of bacteria, fungi, and viruses (Harnish et al. 2021). The *Drosophila* defense mechanisms are well studied (Buchon et al. 2014), and include toll as a pattern recognition receptor alongside conserved pathways such as NF-κB and JNK (a MAPK cascade) (Harnish et al. 2021; Buchon et al. 2014). As wild-type flies are resistant to most types of fungal infections, toll-deficient mutants are commonly used.

Drosophila can be infected by different routes: Orally by providing the pathogen in food, cutaneously by rolling anesthetized flies in spores, and by injection, with the latter two requiring anesthesia (Lionakis 2011). The virulence of different fungal pathogens including *C. albicans*, *C. glabrata*, *C. neoformans*, *Aspergillus*, and Mucorales has been studied in *D. melanogaster*, and the fly can also be used to test antimicrobial compounds (Lionakis et al. 2005; Chamilos et al. 2006; Apidianakis et al. 2004; Brunke et al. 2015). However, exact dosing of compounds in *Drosophila* is difficult, as oral uptake cannot be accurately measured and repeated injection leads to injury (Lionakis 2011). Although *Drosophila* is a model organism and has been shown to be suitable as a fungal infection model, it has not been used widely. One reason is probably its small size (approx. 3 mm) which requires technical expertise in handling and special rooms to prevent escape of genetically modified and/or infected flies. Similar to *C. elegans*, another important limitation is that *Drosophila* cannot be maintained at 37 °C.

9.6.3 *Galleria mellonella*

Larvae of the greater wax moth *G. mellonella* have gained significant importance as a model host to study fungal infections (Binder et al. 2016). As *Galleria* larvae are used as fishing bait, they are usually readily available from various sources and inexpensive. Compared to *C. elegans* and *D. melanogaster*, they have the distinct advantage of larger size (2–3 cm at fifth instar, the developmental stage used for infection experiments), making them easier to handle and allowing injection of defined volumes into the hind pro legs (Fallon et al. 2012). Systemic infection induced by this route is the most commonly used for fungal infection, although oral administration has also been described for studies on bacteria (Lange et al. 2019). Furthermore, purchased larvae can be maintained within a Petri dish, without the need of feeding during experiments. Another advantage is the possibility to maintain *Galleria* at 37 °C, the core body temperature of humans. These advantages made *Galleria* the most widely used alternative animal host in fungal research over the last two decades. It has been used for *Candida*, *Aspergillus*, *Cryptococcus*, Histoplasma, Paracoccidioides, and Mucorales (Singulani et al. 2018; Pereira et al. 2018; Durieux et al. 2021; Stempinski et al. 2022; Jacobsen 2019; Thomaz et al. 2013), as well as in studies investigating antifungal compounds (Jemel et al. 2020).

As *G. mellonella* has been used only for a comparatively short period of time as an infection model, it is overall not as well characterized as other invertebrates such as *C. elegans* and *D. melanogaster*. The genome sequence of *G. mellonella* was only recently published in 2018 (Lange et al. 2018), and genetic manipulation techniques for *G. mellonella* are not yet available. The lack of standardized housing and breeding conditions can affect the insects' immune response and thereby limit reproducibility (Firacative et al. 2020). Although current research is aimed at a better understanding of how *Galleria* responds to pathogens—a necessary prerequisite to understand to which extent results obtained in this model can be transferred to mammals, including humans—the processes during infection leading to larval death remain a "black box" to some extent to date (Smith and Casadevall 2021).

9.7 Mammalian Models

Animal infection models using vertebrates, especially mammals, are essential in infectious disease research, and their use is mandatory in the approval of new antifungal drugs for demonstrating efficacy and investigating pharmacodynamic and pharmacokinetic properties. Various vertebrate model hosts have been used in fungal research, and besides mammals, zebrafish larvae, embryonated chicken eggs, and adult birds have been used (Jacobsen et al. 2012; Rosowski et al. 2018; Clemons and Stevens 2005). For translation to human patients, however, mammalian models are considered the gold standard due to similarities in anatomy and physiology. Another prerequisite for translation is the ability of the pathogen to infect the species of choice and the development of a disease that closely resembles the pathological alterations in human patients. Most fungal pathogens infecting humans are not host-specific (*Pneumocystis* being a notable exception), and many human pathogenic fungi are also observed as infectious agents in veterinary practice in a variety of mammals (Seyedmousavi et al. 2018). Various mammalian species, such as macaques, piglets, rabbits, guinea pigs, and rats (Clemons and Stevens 2005; Cambier et al. 2017; Conti et al. 2014; Desoubeaux and Cray 2018; Dei-Cas et al. 1998; Jensen 2022; Hohl 2014; Normile et al. 2020; MacCallum 2012; Chesnay et al. 2022), can be used in fungal infection models. Economic factors and ease of handling, however, make mice the most commonly used animal model for most human pathogenic fungi. Furthermore, the availability of genetically modified strains allows to investigate the contribution of immune receptors and signaling pathways in mice. To model underlying risk factors, medication can be used to achieve immunosuppression. Most of our current knowledge on the pathogenesis of fungal infections and the host immune response comes from studies using mouse models.

To reflect the different types of disease caused by fungi depending on not only the fungal species but also route of infection, anatomical region involved, and underlying host risk factors, different mouse infection models have been developed. For example, distinct models exist for *C. albicans* skin infection, oropharyngeal candidiasis, vulvovaginal candidiasis, peritonitis, catheter-associated infection, and disseminated (systemic) infection (MacCallum 2012). Similarly, varieties of infection models have been developed for mucormycoses (Jacobsen 2019) and other fungal pathogens. The following provides brief introductions to the main types of infection models.

9.7.1 Systemic (Disseminated) Infection

Disseminated or systemic fungal infections affecting various internal organs are life-threatening conditions that can be caused by a variety of fungal pathogens including *Candida*, dimorphic fungi, and molds. They usually originate from colonized mucosal surfaces (*Candida*), as sequelae of primary pulmonary infection (most dimorphic fungi and molds), or after infection of indwelling medical devices (e.g., catheters) or inoculation of fungi by trauma. The easiest way to model disseminated infections is the direct injection of pathogens into the bloodstream of experimental animals, usually the lateral tail vein in mice. As this approach circumvents steps of colonization or establishment of infection in the lung prior to dissemination, the contribution of fungal factors or host defense mechanisms essential in these early processes might be missed. However, this approach is technically easy, can be standardized, and used for a variety of fungal pathogens, with or without immunosuppression, which makes it a common approach in testing antifungal compounds (Jacobsen 2019; Clemons and Stevens 2005; Clemons et al. 2007; Thomson et al. 2017). Furthermore, since dissemination from other sites of infection, such as the lung, occurs only infrequently in patients with fungal infections like pulmonary aspergillosis, and is rarely observed in animal models, investigations on the disseminated form of infection can only be reliably done using intravenous infection. For systemic candidiasis, the tail vein infection model is very well characterized and frequently used (MacCallum 2013). As an alternative, dissemination of *Candida* can be induced following gastrointestinal colonization by a combination of intestinal barrier disturbance (DSS-induced colitis) and neutropenia (Koh et al. 2008). This differs from tail vein injection regarding organ targeting and likely reflects dissemination in human patients more accurately. However, the additional experimental steps required and the applicability to neutropenic patients only limits the use of this model.

9.7.2 Pulmonary Infection

With the exception of candidiasis and dermatomycoses, inhalation of infectious particles is a common route of infection with fungal pathogens, and the lung is the primary target organ. Respiratory infection in mice can be achieved using inhalation chambers (Sheppard et al. 2004), intranasal application, or intratracheal instillation (Desoubeaux and Cray 2018). The latter two are most commonly used, for various fungi that cause pulmonary infections. The technical details can differ between studies, but generally short-term anesthesia is required for intranasal and intratracheal application. Although a controlled dose is applied intranasally and intratracheally, not all spores reach the lower airways and lung parenchyma. This should be considered if studies using different application techniques are compared. Pilot experiments, in which mice are sacrificed shortly after infection, can be useful to determine the effective infectious dose reaching the lungs. Furthermore, intranasal infection might lead to upper respiratory tract infection, which should be considered in cryptococcosis models (Coelho et al. 2019).

Most types of fungal pneumonia require some level of immunosuppression; corticosteroids, induction of leucopenia by cyclophosphamide, or depletion of neutrophils using monoclonal antibodies are most commonly used (Jacobsen 2019). In immunocompetent individuals, some fungi like *Aspergillus* can cause allergic reactions, leading to allergic bronchopulmonary aspergillosis (ABPA). In contrast to models for invasive pulmonary infections, in which infection is achieved by a single application of spores, ABPA models require additional sensitization which can be achieved by different means (reviewed in Templeton et al. 2010).

9.7.3 Mucosal and Skin Infection

Gastrointestinal colonization with fungi is common, and the gut serves as an important reservoir for fungi and origin of disseminated fungal infections. Fungal infections of the gastrointestinal tract, however, are rare. In contrast, infections of the oral and vaginal tract by fungi, predominately *Candida*, are common (Telles et al. 2017; Goncalves et al. 2016). To study these types of infection, distinct mouse models have been developed. Oral infection can be achieved by sublingual placement of swabs saturated with a fungal suspension in anesthetized mice (Conti et al. 2014; Solis and Filler 2012). Immunocompetent animals clear the infection within a couple of days, but progressive infection can be achieved by using cortisone acetate for immunosuppression (Conti et al. 2014; Solis and Filler 2012). Thus, this model mimics oropharyngeal candidiasis in immunocompromised patients, but not immunocompetent individuals suffering from fungal denture stomatitis. Denture stomatitis can be induced in mice and rats, but this requires manufacturing of devices fitted to the oral cavity of rodents (Moraes et al. 2022).

High levels of circulating estrogen levels have been linked to vulvovaginal candidiasis in women (He et al. 2022), and are also a prerequisite in rodent models of this disease (Cassone and Sobel 2016). Vaginal infection is achieved by vaginal application of yeast suspensions after pretreatment with subcutaneously injected estradiol (Yano and Fidel Jr. 2011).

Skin infection by *Candida* can be induced by intradermal injection (Conti et al. 2014), or prolonged exposure of saturated patches to shaved skin at the back of the mouse (Iwasawa et al. 2022). For *Malassezia*, an important pathogen causing dermatomycosis, a similar approach was used with two important modifications: Infection was established on the ears instead of the skin at the back, facilitating easy quantification of inflammation by measuring ear swelling, and the epidermal barrier was disrupted by mild tape stripping prior to infection (Sparber and LeibundGut-Landmann 2019). With these modifications, a single topical application of fungal cells in olive oil is sufficient to establish infection.

References

Aguilar C, Alves da Silva M, Saraiva M, Neyazi M, Olsson IAS, Bartfeld S (2021) Organoids as host models for infection biology—a review of methods. Exp Mol Med 53(10):1471–1482. https://doi.org/10.1038/s12276-021-00629-4

Ahamefule CS, Qin Q, Odiba AS, Li S, Moneke AN, Ogbonna JC et al (2020) *Caenorhabditis elegans*-based *Aspergillus fumigatus* infection model for evaluating pathogenicity and drug efficacy. Front Cell Infect Microbiol 10:320. https://doi.org/10.3389/fcimb.2020.00320

Ahamefule CS, Ezeuduji BC, Ogbonna JC, Moneke AN, Ike AC, Jin C et al (2021) *Caenorhabditis elegans* as an infection model for pathogenic mold and dimorphic fungi: applications and challenges. Front Cell Infect Microbiol 11:751947. https://doi.org/10.3389/fcimb.2021.751947

Andreu N, Phelan J, de Sessions PF, Cliff JM, Clark TG, Hibberd ML (2017) Primary macrophages and J774 cells respond differently to infection with Mycobacterium tuberculosis. Sci Rep 7:42225. https://doi.org/10.1038/srep42225

Apidianakis Y, Rahme LG, Heitman J, Ausubel FM, Calderwood SB, Mylonakis E (2004) Challenge of *Drosophila melanogaster* with *Cryptococcus neoformans* and role of the innate immune response. Eukaryot cell 3(2):413–419. https://doi.org/10.1128/ec.3.2.413-419.2004

Binder U, Maurer E, Lass-Florl C (2016) *Galleria mellonella*: An invertebrate model to study pathogenicity in correctly defined fungal species. Fungal Biol 120(2):288–295. https://doi.org/10.1016/j.funbio.2015.06.002

Bourgeois C, Majer O, Frohner I, Kuchler K (2009) *In vitro* systems for studying the interaction of fungal pathogens with primary cells from the mammalian innate immune system. Methods Mol Biol 470:125–139. https://doi.org/10.1007/978-1-59745-204-5_11

Brunke S, Quintin J, Kasper L, Jacobsen ID, Richter ME, Hiller E et al (2015) Of mice, flies—and men? Comparing fungal infection models for large-scale screening efforts. Dis Model Mech 8(5):473–486. https://doi.org/10.1242/dmm.019901

Buchon N, Silverman N, Cherry S (2014) Immunity in *Drosophila melanogaster*—from microbial recognition to whole-organism physiology. Nat Rev Immunol 14(12):796–810. https://doi.org/10.1038/nri3763

Cambier L, Heinen MP, Mignon B (2017) Relevant animal models in dermatophyte research. Mycopathologia 182(1-2):229–240. https://doi.org/10.1007/s11046-016-0079-3

Cassone A, Sobel JD (2016) Experimental models of vaginal candidiasis and their relevance to human candidiasis. Infect Immun 84(5):1255–1261. https://doi.org/10.1128/iai.01544-15

Chamilos G, Carvalho A (2020) *Aspergillus fumigatus* DHN-melanin. Curr Top Microbiol Immunol 425:17–28. https://doi.org/10.1007/82_2020_205

Chamilos G, Lionakis MS, Lewis RE, Lopez-Ribot JL, Saville SP, Albert ND et al (2006) *Drosophila melanogaster* as a facile model for large-scale studies of virulence mechanisms and antifungal drug efficacy in *Candida* species. J Infect Dis 193(7):1014–1022. https://doi.org/10.1086/500950. JID35316 [pii]

Chesnay A, Paget C, Heuzé-Vourc'h N, Baranek T, Desoubeaux G (2022) *Pneumocystis* pneumonia: pitfalls and hindrances to establishing a reliable animal model. J Fungi 8(2):129

Clemons KV, Stevens DA (2005) The contribution of animal models of aspergillosis to understanding pathogenesis, therapy and virulence. Med Mycol 43(Suppl 1):S101–S110

Clemons KV, Capilla J, Stevens DA (2007) Experimental animal models of coccidioidomycosis. Ann N Y Acad Sci 1111:208–224. https://doi.org/10.1196/annals.1406.029

Coelho C, Camacho E, Salas A, Alanio A, Casadevall A (2019) Intranasal inoculation of *Cryptococcus neoformans* in mice produces nasal infection with rapid brain dissemination. mSphere 4(4). https://doi.org/10.1128/mSphere.00483-19

Conti HR, Huppler AR, Whibley N, Gaffen SL (2014) Animal models for candidiasis. Curr Protoc Immunol 105:19.6.1–19.6.7. https://doi.org/10.1002/0471142735.im1906s105

Cunha C, Di Ianni M, Bozza S, Giovannini G, Zagarella S, Zelante T et al (2010) Dectin-1 Y238X polymorphism associates with susceptibility to invasive aspergillosis in hematopoietic transplantation through impairment of both recipient- and donor-dependent mechanisms of antifungal immunity. Blood 116(24):5394–5402. https://doi.org/10.1182/blood-2010-04-279307

Cunha C, Aversa F, Lacerda JF, Busca A, Kurzai O, Grube M et al (2014) Genetic PTX3 deficiency and aspergillosis in stem-cell transplantation. N Engl J Med 370(5):421–432. https://doi.org/10.1056/NEJMoa1211161

Dalle F, Wachtler B, L'Ollivier C, Holland G, Bannert N, Wilson D et al (2010) Cellular interactions of *Candida albicans* with human oral epithelial cells and enterocytes. Cell Microbiol 12(2):248–271. https://doi.org/10.1111/j.1462-5822.2009.01394.x. CMI1394 [pii]

de Groot PW, Bader O, de Boer AD, Weig M, Chauhan N (2013) Adhesins in human fungal pathogens: glue with plenty of stick. Eukaryot Cell 12(4):470–481. https://doi.org/10.1128/ec.00364-12

Dei-Cas E, Brun-Pascaud M, Bille-Hansen V, Allaert A, Aliouat EM (1998) Animal models of pneumocystosis. FEMS Immunol Med Microbiol 22(1–2):163–168. https://doi.org/10.1111/j.1574-695X.1998.tb01201.x

Desoubeaux G, Cray C (2018) Animal models of aspergillosis. Comp Med 68(2):109–123

Ding X, Kambara H, Guo R, Kanneganti A, Acosta-Zaldívar M, Li J et al (2021) Inflammasome-mediated GSDMD activation facilitates escape of *Candida albicans* from macrophages. Nat Commun 12(1):6699. https://doi.org/10.1038/s41467-021-27034-9

Dinh H, Semenec L, Kumar SS, Short FL, Cain AK (2021) Microbiology's next top model: *Galleria* in the molecular age. Pathog Dis 79(2). https://doi.org/10.1093/femspd/ftab006

Durieux MF, Melloul É, Jemel S, Roisin L, Dardé ML, Guillot J et al (2021) *Galleria mellonella* as a screening tool to study virulence factors of *Aspergillus fumigatus*. Virulence 12(1):818–834. https://doi.org/10.1080/21505594.2021.1893945

Fallon J, Kelly J, Kavanagh K (2012) *Galleria mellonella* as a model for fungal pathogenicity testing. Methods Mol Biol 845:469–485. https://doi.org/10.1007/978-1-61779-539-8_33

Firacative C, Khan A, Duan S, Ferreira-Paim K, Leemon D, Meyer W (2020) Rearing and maintenance of *Galleria mellonella* and its application to study fungal virulence. J Fungi (Basel) 6(3). https://doi.org/10.3390/jof6030130

Fusco-Almeida AM, de Matos Silva S, Dos Santos KS, de Lima Gualque MW, Vaso CO, Carvalho AR et al (2023) Alternative non-mammalian animal and cellular methods for the study of host-fungal interactions. J Fungi (Basel) 9(9). https://doi.org/10.3390/jof9090943

Goncalves B, Ferreira C, Alves CT, Henriques M, Azeredo J, Silva S (2016) Vulvovaginal candidiasis: Epidemiology, microbiology and risk factors. Crit Rev Microbiol 42(6):905–927. https://doi.org/10.3109/1040841x.2015.1091805

Graf K, Last A, Gratz R, Allert S, Linde S, Westermann M et al (2019) Keeping *Candida* commensal: how lactobacilli antagonize pathogenicity of *Candida albicans* in an in vitro gut model. Dis Model Mech 12(9). https://doi.org/10.1242/dmm.039719

Grubb SE, Murdoch C, Sudbury P, Saville SP, Lopez-Ribot JL, Thornhill MH (2008) *Candida albicans*—endothelial cell interactions: a key step in the pathogenesis of systemic candidiasis. Infect Immun 76(10):4370–4377

Harnish JM, Link N, Yamamoto S (2021) *Drosophila* as a model for infectious diseases. Int J Mol Sci 22(5). https://doi.org/10.3390/ijms22052724

He Y, Tang R, Deng J, Cai T, He P, Wu J et al (2022) Effects of oestrogen on vulvovaginal candidosis. Mycoses 65(1):4–12. https://doi.org/10.1111/myc.13385

Hoang TNM, Cseresnyés Z, Hartung S, Blickensdorf M, Saffer C, Rennert K et al (2022) Invasive aspergillosis-on-chip: A quantitative treatment study of human *Aspergillus fumigatus* infection. Biomaterials 283:121420. https://doi.org/10.1016/j.biomaterials.2022.121420

Hoffmann JA, Kafatos FC, Janeway CA, Ezekowitz RA (1999) Phylogenetic perspectives in innate immunity. Science 284(5418):1313–1318. https://doi.org/10.1126/science.284.5418.1313

Hohl TM (2014) Overview of vertebrate animal models of fungal infection. J Immunol Methods 410:100–112. https://doi.org/10.1016/j.jim.2014.03.022

Hopke A, Nicke N, Hidu EE, Degani G, Popolo L, Wheeler RT (2016) Neutrophil attack triggers extracellular trap-dependent *Candida* cell wall remodeling and altered immune recognition. PLoS Pathog 12(5):e1005644. https://doi.org/10.1371/journal.ppat.1005644

Huang X, Li D, Xi L, Mylonakis E (2014) *Caenorhabditis elegans*: a simple nematode infection model for *Penicillium marneffei*. PLoS One 9(9):e108764. https://doi.org/10.1371/journal.pone.0108764

Hünniger K, Lehnert T, Bieber K, Martin R, Figge MT, Kurzai O (2014) A virtual infection model quantifies innate effector mechanisms and *Candida albicans* immune escape in human blood. PLoS Comput Biol 10(2):e1003479. https://doi.org/10.1371/journal.pcbi.1003479

Iwasawa MT, Miyachi H, Wakabayashi S, Sugihira T, Aoyama R, Nakagawa S et al (2022) Epidermal clearance of *Candida albicans* is mediated by IL-17 but independent of fungal innate immune receptors. Int Immunol 34(8):409–420. https://doi.org/10.1093/intimm/dxac019

Jacobsen ID (2019) Animal models to study mucormycosis. J Fungi (Basel) 5(2). https://doi.org/10.3390/jof5020027

Jacobsen ID, Grosse K, Hube B (2012) Embryonated chicken eggs as alternative infection model for pathogenic fungi. Methods Mol Biol 845:487–496. https://doi.org/10.1007/978-1-61779-539-8_34

Jemel S, Guillot J, Kallel K, Botterel F, Dannaoui E (2020) *Galleria mellonella* for the evaluation of antifungal efficacy against medically important fungi, a narrative review. Microorganisms 8(3). https://doi.org/10.3390/microorganisms8030390

Jensen HE (2022) Animal models of invasive mycoses. APMIS 130(7):427–435. https://doi.org/10.1111/apm.13110

Khan NS, Kasperkovitz PV, Timmons AK, Mansour MK, Tam JM, Seward MW et al (2016) Dectin-1 controls TLR9 trafficking to phagosomes containing β-1,3 glucan. J Immunol 196(5):2249–2261. https://doi.org/10.4049/jimmunol.1401545

Kim J, Koo BK, Knoblich JA (2020) Human organoids: model systems for human biology and medicine. Nat Rev Mol Cell Biol 21(10):571–584. https://doi.org/10.1038/s41580-020-0259-3

Koh AY, Kohler JR, Coggshall KT, Van Rooijen N, Pier GB (2008) Mucosal damage and neutropenia are required for *Candida albicans* dissemination. PLoS Pathog 4(2):e35. https://doi.org/10.1371/journal.ppat.0040035

Kortmann J, Brubaker SW, Monack DM (2015) Cutting edge: inflammasome activation in primary human macrophages is dependent on flagellin. J Immunol 195(3):815–819. https://doi.org/10.4049/jimmunol.1403100

Lange A, Beier S, Huson DH, Parusel R, Iglauer F, Frick JS (2018) Genome sequence of *Galleria mellonella* (greater wax moth). Genome Announc 6(2). https://doi.org/10.1128/genomeA.01220-17

Lange A, Schäfer A, Frick JS (2019) A *Galleria mellonella* oral administration model to study commensal-induced innate immune responses. J Vis Exp 145. https://doi.org/10.3791/59270

Last A, Maurer M, Mosig AS, Gresnigt MS, Hube B (2021) *In vitro* infection models to study fungal-host interactions. FEMS Microbiol Rev 45(5). https://doi.org/10.1093/femsre/fuab005

Lionakis MS (2011) *Drosophila* and *Galleria* insect model hosts: new tools for the study of fungal virulence, pharmacology and immunology. Virulence 2(6):521–527. https://doi.org/10.4161/viru.2.6.18520.18520 [pii]

Lionakis MS, Kontoyiannis DP (2012) *Drosophila melanogaster* as a model organism for invasive aspergillosis. Methods Mol Biol 845:455–468. https://doi.org/10.1007/978-1-61779-539-8_32

Lionakis MS, Lewis RE, May GS, Wiederhold NP, Albert ND, Halder G et al (2005) Toll-deficient *Drosophila* flies as a fast, high-throughput model for the study of antifungal drug efficacy against invasive aspergillosis and *Aspergillus* virulence. J Infect Dis 191(7):1188–1195

MacArthur Clark J (2018) The 3Rs in research: a contemporary approach to replacement, reduction and refinement. Br J Nutr 120(s1):S1–S7. https://doi.org/10.1017/s0007114517002227

MacCallum DM (2012) Hosting infection: experimental models to assay *Candida* virulence. Int J Microbiol 2012:363764. https://doi.org/10.1155/2012/363764

MacCallum DM (2013) Mouse model of invasive fungal infection. Methods Mol Biol 1031:145–153. https://doi.org/10.1007/978-1-62703-481-4_17

Machata S, Sreekantapuram S, Hünniger K, Kurzai O, Dunker C, Schubert K et al (2020) Significant differences in host-pathogen interactions between murine and human whole blood. Front Immunol 11:565869. https://doi.org/10.3389/fimmu.2020.565869

Madende M, Albertyn J, Sebolai O, Pohl CH (2020) *Caenorhabditis elegans* as a model animal for investigating fungal pathogenesis. Med Microbiol Immunol 209(1):1–13. https://doi.org/10.1007/s00430-019-00635-4

Martin R, Wachtler B, Schaller M, Wilson D, Hube B (2011) Host-pathogen interactions and virulence-associated genes during *Candida albicans* oral infections. Int J Med Microbiol 301(5):417–422. https://doi.org/10.1016/j.ijmm.2011.04.009. S1438-4221(11)00026-9 [pii]

Mikamo H, Yamagishi Y, Sugiyama H, Sadakata H, Miyazaki S, Sano T et al (2018) High glucose-mediated overexpression of ICAM-1 in human vaginal epithelial cells increases adhesion of *Candida albicans*. J Obstet Gynaecol 38(2):226–230. https://doi.org/10.1080/01443615.2017.1343810

Mogavero S, Hube B (2021) *Candida albicans* interaction with oral epithelial cells: adhesion , invasion, and damage assays. Methods Mol Biol 2260:133–143. https://doi.org/10.1007/978-1-0716-1182-1_9

Moraes GS, Albach T, Sugio CY, de Oliveira FB, Neppelenbroek KH, Urban VM (2022) Experimental animal models for denture stomatitis: A methodological review. Lab Anim 56(4):331–343. https://doi.org/10.1177/00236772211069249

Moyes DL, Wilson D, Richardson JP, Mogavero S, Tang SX, Wernecke J et al (2016) Candidalysin is a fungal peptide toxin critical for mucosal infection. Nature 532(7597):64–68. https://doi.org/10.1038/nature17625

Normile TG, Bryan AM, Del Poeta M (2020) Animal models of *Cryptococcus neoformans* in identifying immune parameters associated with primary infection and reactivation of latent infection. Front Immunol 11:581750. https://doi.org/10.3389/fimmu.2020.581750

Okoli I, Coleman JJ, Tampakakis E, An WF, Holson E, Wagner F et al (2009) Identification of antifungal compounds active against *Candida albicans* using an improved high-throughput *Caenorhabditis elegans* assay. PLoS One 4(9):e7025. https://doi.org/10.1371/journal.pone.0007025

Opal SM (2009) A brief history of microbiology and immunology. In: Artenstein A (ed) Vaccines: A biography. Springer, New York

Panayidou S, Ioannidou E, Apidianakis Y (2014) Human pathogenic bacteria, fungi, and viruses in *Drosophila*: disease modeling, lessons, and shortcomings. Virulence 5(2):253–269. https://doi.org/10.4161/viru.27524

Pereira TC, de Barros PP, Fugisaki LRO, Rossoni RD, Ribeiro FC, de Menezes RT et al (2018) Recent advances in the use of *Galleria mellonella* model to study immune responses against human pathogens. J Fungi (Basel) 4(4). https://doi.org/10.3390/jof4040128

Phan QT, Fratti RA, Prasadarao NV, Edwards JE Jr, Filler SG (2005) N-cadherin mediates endocytosis of Candida albicans by endothelial cells. J Biol Chem 280(11):10455–10461

Phan QT, Myers CL, Fu Y, Sheppard DC, Yeaman MR, Welch WH et al (2007) Als3 is a *Candida albicans* invasin that binds to cadherins and induces endocytosis by host cells. PLoS Biol 5(3):e64

Poupet C, Saraoui T, Veisseire P, Bonnet M, Dausset C, Gachinat M et al (2019) *Lactobacillus rhamnosus* Lcr35 as an effective treatment for preventing *Candida albicans* infection in the invertebrate model *Caenorhabditis elegans*: First mechanistic insights. PLoS One 14(11):e0216184. https://doi.org/10.1371/journal.pone.0216184

Powell JR, Ausubel FM (2008) Models of *Caenorhabditis elegans* infection by bacterial and fungal pathogens. Methods Mol Biol 415:403–427. https://doi.org/10.1007/978-1-59745-570-1_24

Puri S, Kumar R, Rojas IG, Salvatori O, Edgerton M (2019) Iron chelator deferasirox reduces *Candida albicans* invasion of oral epithelial cells and infection levels in murine oropharyngeal candidiasis. Antimicrob Agents Chemother 63(4). https://doi.org/10.1128/aac.02152-18

Rafiq M, Rivieccio F, Zimmermann AK, Visser C, Bruch A, Krüger T et al (2022) PLB-985 neutrophil-like cells as a model to study *Aspergillus fumigatus* pathogenesis. mSphere 7(1):e0094021. https://doi.org/10.1128/msphere.00940-21

Richardson JP, Ho J, Naglik JR (2018) *Candida*-epithelial interactions. J Fungi (Basel) 4(1). https://doi.org/10.3390/jof4010022

Rosowski EE, Knox BP, Archambault LS, Huttenlocher A, Keller NP, Wheeler RT et al (2018) The zebrafish as a model host for invasive fungal infections. J Fungi (Basel) 4(4):136. https://doi.org/10.3390/jof4040136

Saijo S, Fujikado N, Furuta T, Chung SH, Kotaki H, Seki K et al (2007) Dectin-1 is required for host defense against Pneumocystis carinii but not against Candida albicans. Nat Immunol 8(1):39–46. https://doi.org/10.1038/ni1425

Salazar F, Brown GD (2018) Antifungal innate immunity: a perspective from the last 10 years. J Innate Immun 10(5–6):373–397. https://doi.org/10.1159/000488539

Seyedmousavi S, Bosco SMG, de Hoog S, Ebel F, Elad D, Gomes RR et al (2018) Fungal infections in animals: a patchwork of different situations. Med Mycol 56(suppl_1):165–187. https://doi.org/10.1093/mmy/myx104

Shalem O, Sanjana NE, Hartenian E, Shi X, Scott DA, Mikkelson T et al (2014) Genome-scale CRISPR-Cas9 knockout screening in human cells.

Science 343(6166):84–87. https://doi.org/10.1126/science.1247005
Sheppard DC, Filler SG (2014) Host cell invasion by medically important fungi. Cold Spring Harb Perspect Med 5(1):a019687. https://doi.org/10.1101/cshperspect.a019687
Sheppard DC, Rieg G, Chiang LY, Filler SG, Edwards JE Jr, Ibrahim AS (2004) Novel inhalational murine model of invasive pulmonary aspergillosis. Antimicrob Agents Chemother 48(5):1908–1911
Singulani JL, Scorzoni L, de Oliveira HC, Marcos CM, Assato PA, Fusco-Almeida AM et al (2018) Applications of invertebrate animal models to dimorphic fungal infections. J Fungi (Basel) 4(4). https://doi.org/10.3390/jof4040118
Smith DFQ, Casadevall A (2021) Fungal immunity and pathogenesis in mammals versus the invertebrate model organism *Galleria mellonella*. Pathog Dis 79(3). https://doi.org/10.1093/femspd/ftab013
Solis NV, Filler SG (2012) Mouse model of oropharyngeal candidiasis. Nat Protoc 7(4):637–642. https://doi.org/10.1038/nprot.2012.011. nprot.2012.011 [pii]
Solis NV, Swidergall M, Bruno VM, Gaffen SL, Filler SG (2017) The aryl hydrocarbon receptor governs epithelial cell invasion during oropharyngeal candidiasis. mBio 8(2). https://doi.org/10.1128/mBio.00025-17
Sparber F, LeibundGut-Landmann S (2019) Infecting mice with *Malassezia* spp. to study the fungus-host interaction. J Vis Exp 153. https://doi.org/10.3791/60175
Stappers MHT, Clark AE, Aimanianda V, Bidula S, Reid DM, Asamaphan P et al (2018) Recognition of DHN-melanin by a C-type lectin receptor is required for immunity to *Aspergillus*. Nature 555(7696):382–386. https://doi.org/10.1038/nature25974
Stempinski PR, Smith DFQ, Casadevall A (2022) *Cryptococcus neoformans* virulence assay using a *Galleria mellonella* larvae model system. Bio Protoc 12(15). https://doi.org/10.21769/BioProtoc.4480
Sun JN, Solis NV, Phan QT, Bajwa JS, Kashleva H, Thompson A et al (2010) Host cell invasion and virulence mediated by *Candida albicans* Ssa1. PLoS Pathog 6(11):e1001181. https://doi.org/10.1371/journal.ppat.1001181
Szabo EK, MacCallum DM (2011) The contribution of mouse models to our understanding of systemic candidiasis. FEMS Microbiol Lett 320(1):1–8. https://doi.org/10.1111/j.1574-6968.2011.02262.x
Taylor PR, Tsoni SV, Willment JA, Dennehy KM, Rosas M, Findon H et al (2007) Dectin-1 is required for beta-glucan recognition and control of fungal infection. Nat Immunol 8(1):31–38. https://doi.org/10.1038/ni1408
Telles DR, Karki N, Marshall MW (2017) Oral fungal infections: diagnosis and management. Dent Clin North Am 61(2):319–349. https://doi.org/10.1016/j.cden.2016.12.004
Templeton SP, Buskirk AD, Green BJ, Beezhold DH, Schmechel D (2010) Murine models of airway fungal exposure and allergic sensitization. Med Mycol 48(2):217–228. https://doi.org/10.3109/13693780903420658
Thomaz L, García-Rodas R, Guimarães AJ, Taborda CP, Zaragoza O, Nosanchuk JD (2013) *Galleria mellonella* as a model host to study *Paracoccidioides lutzii* and *Histoplasma capsulatum*. Virulence 4(2):139–146. https://doi.org/10.4161/viru.23047
Thomson P, Mayayo E, López-Fernández L, Guarro J, Capilla J (2017) Combined antifungal therapy against systemic murine infections by rare *Cryptococcus* species. Mycoses 60(2):112–117. https://doi.org/10.1111/myc.12569
Thywissen A, Heinekamp T, Dahse HM, Schmaler-Ripcke J, Nietzsche S, Zipfel PF et al (2011) Conidial dihydroxynaphthalene melanin of the human pathogenic fungus *Aspergillus fumigatus* interferes with the host endocytosis pathway. Front Microbiol 2:96. https://doi.org/10.3389/fmicb.2011.00096
Trevijano-Contador N, Zaragoza O (2018) Immune Response of *Galleria mellonella* against human fungal pathogens. J Fungi (Basel) 5(1). https://doi.org/10.3390/jof5010003
Ugur B, Chen K, Bellen HJ (2016) *Drosophila* tools and assays for the study of human diseases. Dis Model Mech 9(3):235–244. https://doi.org/10.1242/dmm.023762
von Müller C, Bulman F, Wagner L, Rosenberger D, Marolda A, Kurzai O et al (2020) Active neutrophil responses counteract *Candida albicans* burn wound infection of *ex vivo* human skin explants. Sci Rep 10(1):21818. https://doi.org/10.1038/s41598-020-78387-y
Wächtler B, Wilson D, Haedicke K, Dalle F, Hube B (2011) From attachment to damage: defined genes of *Candida albicans* mediate adhesion, invasion and damage during interaction with oral epithelial cells. PLoS One 6(2):e17046. https://doi.org/10.1371/journal.pone.0017046
Wächtler B, Citiulo F, Jablonowski N, Forster S, Dalle F, Schaller M et al (2012) *Candida albicans*-epithelial interactions: dissecting the roles of active penetration, induced endocytosis and host factors on the infection process. PLoS One. 7(5):e36952. https://doi.org/10.1371/journal.pone.0036952. PONE-D-11-12698 [pii]
Wagener J, Mailänder-Sanchez D, Schaller M (2012) Immune responses to *Candida albicans* in models of *in vitro* reconstituted human oral epithelium. Methods Mol Biol 845:333–344. https://doi.org/10.1007/978-1-61779-539-8_22
Wellington M, Koselny K, Sutterwala FS, Krysan DJ (2014) *Candida albicans* triggers NLRP3-mediated pyroptosis in macrophages. Eukaryot Cell 13(2):329–340. https://doi.org/10.1128/EC.00336-13. EC.00336-13 [pii]
Wheeler RT, Fink GR (2006) A drug-sensitive genetic network masks fungi from the immune system. PLoS Pathog 2(4):e35
Yano J, Fidel PL Jr (2011) Protocols for vaginal inoculation and sample collection in the experimental mouse model of *Candida* vaginitis. J Vis Exp 58. https://doi.org/10.3791/3382

Part III

Techniques

Transcriptomic Analyses of Host Colonisation in Fungal Pathogens of Humans

10

Elaine Bignell

Abstract

The mammalian host environment presents a uniquely complex challenge to fungi, most of which have fine-tuned physiologies selected for and honed, in wider ecological niches. Survival, persistence, and proliferation in mammalian hosts require agile perception and adaptation to the multitude of stimuli, metabolic demands, and insults encountered. Therefore, pathogenicity requires not only the possession of inventories of virulence traits but also the capacity to regulate their expression with an exquisite degree of temporal precision. More than any other methodology applied to understand this process, transcriptomic approaches have illuminated the critical events that initiate and substantiate successful host encounters, revealing that the sequence and timing of multiple biotic and abiotic influences, coupled with host immune status and pathogen responses determine the pathogenic identity of fungi in their hosts as well as disease outcomes. Given the plethora of studies that have focused upon responses to individual stresses and host immune cell types, the emphasis of this chapter will be upon the outputs and new insights gained from fungal transcriptomes derived from mammalian host environments. Here, the power (but not the technical details that are excellently reviewed elsewhere (Hovhannisyan and Gabaldón. Transcriptome sequencing approaches to elucidate host–microbe interactions in opportunistic human fungal pathogens. In: Rodrigues (ed) Fungal physiology and immunopathogenesis. Springer, Cham, 2019)), of state-of-the-art methodologies are highlighted, with an emphasis upon the research questions and the different approaches that have led to new insights, and a look towards emerging technologies that will enable us to address important unanswered questions.

Keywords

Transcriptomics · Host response · *Aspergillus fumigatus* · *Candida* · *Cryptococcus* RNA-seq · Microarray · Infection

10.1 The Host-Adapting *Aspergillus fumigatus* Transcriptome

Studies analysing the host-adapting *A. fumigatus* transcriptome (Fig. 10.1 and Table 10.1) have captured pathogen adaptation to the host environment in germlings harvested from

E. Bignell (✉)
Medical Research Council Centre for Medical Mycology, University of Exeter, Exeter, UK
e-mail: E.Bignell@exeter.ac.uk

A. A. Brakhage et al. (eds.), *Human and Animal Relationships*, The Mycota 6,
https://doi.org/10.1007/978-3-031-64853-3_10

bronchoalveolar lavage (BAL) or homogenised tissues of the leukopenic or corticosteroid treated murine lung, in naturally occurring infection of the canine sinus, or in response to exposure to whole human blood. Methodologies have included RNA amplification coupled to microarray technology, RNA-Seq, and/or Nanostring technologies.

Early studies, using microarray approaches (McDonagh et al. 2008; Bertuzzi et al. 2014), sought to define the differences in gene expression profiles of lab (YPD) and in-host cultured fungal cells and the temporal basis of host adaptation relative to ungerminated spores. McDonagh et al. (2008) produced the first snapshot of the host-adapting *A. fumigatus* transcriptome, harvesting pre-invasive fungal germlings from BAL fluid of leukopenic mice at 12–14 h post-infection, and having validated the fidelity of transcript abundances in amplified and non-amplified control RNA samples, gene expression data were captured for 9075 *A. fumigatus* genes. The in vivo transcriptome resembled that of in vitro responses to iron limitation, alkaline challenge, and nutrient starvation. Locational bias of differentially regulated genes mapped those upregulated during host colonisation to the subtelomeric regions of the *A. fumigatus* genome, including multiple clusters of physically colocalised genes.

Gene Ontology (GO) terms amongst differentially expressed genes revealed transport of metal ions, cations, carbohydrates, siderophore iron, and amino acids. Genes that were upregulated in vivo included 11 siderophore biosynthesis/transport-encoding genes, 2 ferric-chelate reductase genes, 13 amino acid permease genes, and 3 proline permease genes, as well as the general amino acid permease, Gap1 and 9 genes annotated as maltose permeases or transporters. Genes encoding antigens causing inflammatory irregularity during allergic *A. fumigatus* disease were also upregulated, including elastinolytic metalloprotease (Mep), an aorsin-like serine protease and 3 dipeptidyl peptidases.

Functional categories of ergosterol biosynthesis, heme biosynthesis, aerobic respiration, ribosome biogenesis, protein biosynthesis, and processing were downregulated relative to in vitro growth. This might have reflected the poor nutritional value of murine lung relative to YPD and/or reduced growth during host adaptation compared to broth culture. However, despite frequent reports of metabolic dampening in microbial systems under stress, the in vivo *A.*

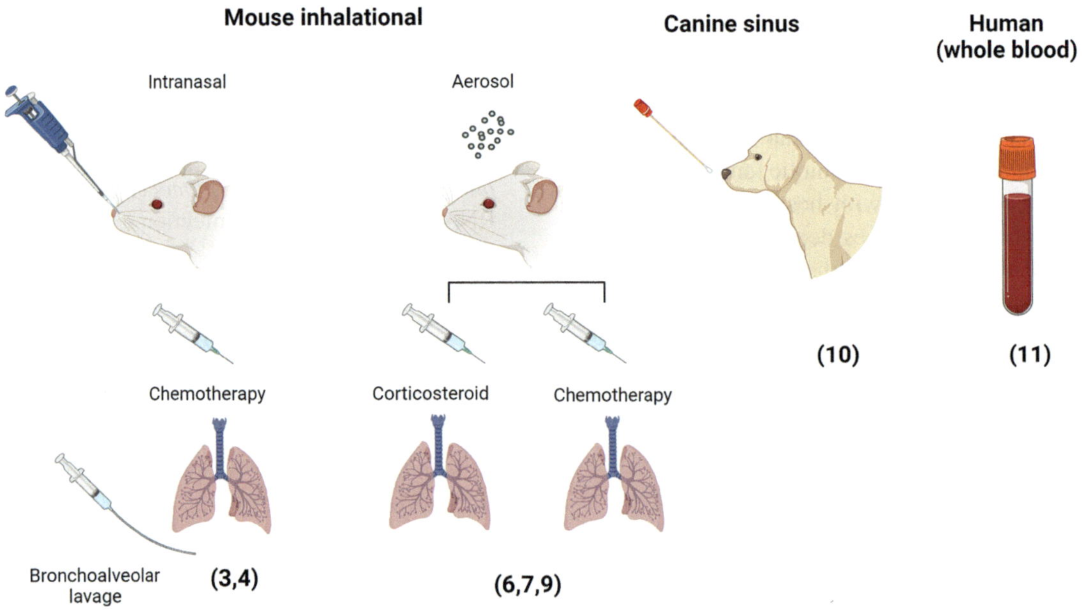

Fig. 10.1 Studies analysing the host-adapting *A. fumigatus* transcriptome. Numbers in parentheses indicate reference numbers. Created with BioRender.com

Table 10.1 Studies analysing the host-adapting fungal transcriptome

Pathogen	Strain	Host	Sample size	Infection route	Inoculum	Time points	Sample type	Technology	Total genes analysed	Comparison	Ref.
Aspergillus fumigatus	Af293	Mouse, CD1, male immunocompromised (cyclophosphamide + cortisone acetate)	*n* = 24 per group, 5 biological replicates	Intranasal	1 × 10^8 spores/ mouse	12–14 h	B.A.L.	Oligo microarrays	9075	YPD liquid culture	McDonagh et al. (2008)
Aspergillus fumigatus	ATCC46645 and isogenic *ΔpacC*	Mouse, CD1, male immunocompromised (cyclophosphamide + cortisone acetate)	*n* = 8 per time-point	Intranasal	1 × 10^8 spores/ mouse	4, 8, 12, 16 h	B.A.L.	Oligo microarrays	9075	Ungerminated spores	Bertuzzi et al. (2014)
Aspergillus fumigatus	CBS144.89 and isogenic *ΔsrbA*	Mouse, CD1, male and female, immunocompromised (Kenalog)	Not specified	Aerosol	40 ml of 1 × 10^9 conidia	2, 3, and 4 days	Lung	nCounter	12	In vitro, LGMM, 16 h +/– hypoxia chamber 30–120 m	Chung et al. (2014)
Aspergillus fumigatus	Af293	Mouse, BALB/c, male immunocompromised (cortisone acetate)	*n* = 3	Aerosol	12 ml of 1 × 10^9 spores/ml	2, 4, and 5 days	Lung	nCounter	467	AMM 24 h 37 °C or 20 h + 4 h—nitrate or 1.5^8 conidia in—zinc media for 20 h, and additional depletion for 4 h of both nitrogen and zinc	Liu et al. (2021)
Aspergillus fumigatus	ATCC46645	Human blood	10 ml blood, *n* = 2 biological reps per condition	Ex vivo whole-blood infection model	2 ml mycelium from overnight (16 h) culture in minimal medium	30 and 180 min at 37 °C in vitro	Ex vivo human blood	RNA-Seq	Not specified	In vitro minimal medium	Irmer et al. (2015)
Aspergillus fumigatus	CBS144.89	Mouse, CD1, female immunocompromised (Kenalog) or immunocompromised (cyclophosphamide + Kenalog)	*n* = 3 per time point	Intranasal	12 ml of 1 × 10^9 spores/ml	2, 3 days post-infection	Lung	RNA-Seq	3345	Inter-time point and inter-model comparisons	Kale et al. (2017)
Aspergillus fumigatus	Veterinary isolate from canine nasal passage	Dogs with chronic non-invasive sino-nasal aspergillosis (SNA)	Four sino-nasal plaques from three different dogs	Natural infection		RNA-Seq		nCounter	9213	N/A	Valdes et al. (2020)
Aspergillus fumigatus	Af293, Δ*acuK*, ΔacuM	Mouse, BALB/c, male, immunocompromised (cortisone acetate)	*n* = 3 per group	Aerosol	12 ml of 1 × 10^9 spores/ml	4 days	Lung	nCounter	71		Pongpom et al. (2015)

(continued)

Table 10.1 (continued)

Pathogen	Strain	Host	Sample size	Infection route	Inoculum	Time points	Sample type	Technology	Total genes analysed	Comparison	Ref.
Candida albicans	K1 (clinical isolate)	Mouse, ICR, immunocompromised (cyclophosphamide)	*n* = 10 mice per time-point	Lateral tail vein injection	Inoculum 5.5 log_{10} cells/ml. Inoculum size not specified	Mid-log-phase growth	Kidney	Microarray	6737	YPD liquid culture at 37 °C	Andes et al. (2005)
Candida albicans	SC5314 and NGY152	New Zealand white rabbits	*n* = 2 rabbits per fungal isolate	Maginal ear vein	1–4 × 10^7 yeast cells per kg body weight	3 days	Micro-abscesses shaved from kidney tissue	Microarray	6333	Microabcesses versus in vitro culture (RPMI)	Walker et al. (2009)
Candida albicans	SC5314	Mouse, BALB/c, female or *Galleria mellonella*	3–4 mice per time-point, *n* = 2 biological reps	Lateral tail vein injection	250 µl of a 10^5 cells/ml suspension	16 and 48 h	Kidney	RNA enrichment (Sure Select) + RNA-Seq	5365	In vitro reference spiked with non-infected host material	Amorim-Vaz et al. (2015)
Candida species	*C. albicans* SC5314 *C. glabrata* ATCC2001 *C. tropicalis* DSM4959 *C. parapsilosis* GA1 *C. haemulonii* CBS 5149	Human blood	*n* = 3 per condition from 3 independent donors	Ex vivo whole-blood infection model	1 × 10^6 cells per ml	15, 30, 60, 120, 240 min	Ex vivo human blood	RNA-Seq	6815 (*C. albicans*) 5281 (*C. glabrata*) 6258 (*C. tropicalis*) 5837 (*C. parapsilosis*)	Inter time-point and species comparisons	Kammer et al. (2020)
Candida species	*C. auris* B8441 *C. albicans* SC5314 *C. glabrata* ATCC2001 *C. tropicalis* DSM4959 *C. parapsilosis* GA1 *C. haemulonii* CBS 5149	Human blood	*n* = 3 per condition from 3 independent donors	Ex vivo whole-blood infection model	1 × 10^6 cells per ml	15, 30, 60, 120, 240 min	Ex vivo human blood	RNA-Seq	5421 *C. auris* 6815 (*C. albicans*) 5281 (*C. glabrata*) 6258 (*C. tropicalis*) 5837 (*C. parapsilosis*)	Inter time-point and species comparisons	Allert et al. (2022)

Candida albicans	SC5314	Mouse, BALB/c, female	$n = 3$ mice per condition and time-point	Lateral tail vein injection	2.5×10^4 to 6.25×10^6 cfu/g body weight	0, 8, 12, 24, 72 h the liver 8 h p.i., the kidneys 12 h p.i., and the kidneys 24 h p.i.	Spleen, liver, and kidney	Microarray	6569	YPD liquid culture at 37 °C	Hebecker et al. (2016)
Candida albicans	CAI-4, *Δefg1*	Mouse, BALB/c, gender not specified	$n = 2$ (cecum) or $n = 4$ (ileum), experiment performed twice	Oral gavage	0.1 ml *C. albicans* cells at 5×10^8 cells/ml	3 days	Cecum and ileum	Microarray	6394	YPD liquid culture at 37 °C	Pierce et al. (2013)
Candida albicans	SC5314 and isogenic null mutants: *Δrim101*, *Δrob1*, *Δrpn4*, *Δsut1*, *Δzap1*	Mouse, BALB/c, male	$n = 3$ per strain and time-point	Lateral tail vein injection	1×10^6 yeast phase cells	12, 24, 48 h	Kidney	nCounter	248 environment-responsive genes and 231 transcription factor genes		
Candida albicans	SC5314	Mouse, C57BL/6, female, oestrogen treated	$n = 3$ per group, 2 independent experiments	Intravaginal	5×10^6 yeast cells	3 days	Vaginal tissue	RNA-Seq	52 high-expressed genes (having ≥100 reads in 2 of 3 mice)	N/A	Bruno et al. (2015)
C. neoformans var. grubii of molecular type VNI and VNII	G0 and HC1	Human, AIDS, and cryptococcal infection	2 human patients, 1 strain from each patient	Natural infection	N/A	N/A	CSF	RNA-Seq	6809 (for G0) and 6778 (for HC1)	Ex vivo CSF, YPD	Chen et al. (2014)
C. neoformans	31 clinical isolates typed as *C. neoformans*	Human, AIDS, and cryptococcal infection	31 human patients, 1 strain from each patient	Natural infection	N/A	N/A	CSF	RNA-Seq	6962	Inter-strain comparisons	Yu et al. (2021)

(continued)

Table 10.1 (continued)

Pathogen	Strain	Host	Sample size	Infection route	Inoculum	Time points	Sample type	Technology	Total genes analysed	Comparison	Ref.
C. neoformans	H99	Mouse, C57BL/6, female, immunocompetent; cynomolgus monkey *Macaca fascicularis*		Intranasal	10^5 fungal cells in 50 µl PBS	7 days	Lung and brain	RNA-Seq	5671	Inter-species comparison	Li et al. (2019)
C. neoformans	H99	Mouse, C57/BL6 male, immunocompetent	*n* = 6 mice per strain (pooled)	Intratracheal	1 × 105 CFU/30 µl	6 and 240 h	BAL	RNA-Seq		YPD liquid culture at 37°C	Freitas et al. (2022)
C. neoformans	H99, CHC193, A5-35-17, A1-35-8, C23, Bt85, Tu401-1	New Zealand white rabbits	*n* = 3 per fungal isolate(pooled)	Intracisternal	0.3 ml of a 3.3 × 109 c.f.u./ml suspension of yeast cells	24 h	CSF	RNA-Seq	6821	YPD, SLIM, tert-butylhydroperoxide, PG, MP	Yu et al. (2020)
C. neoformans	KN99-mCherry	Mouse, female, C57BL/6NJ	*n* = 6 mice pooled	Intranasal	2.5 × 104 *Cryptococcus* cells in 50 µl of 1XPBS	14 days	Lung (ex vivo sorted cells)	RNA-Seq	8056	In vitro/ex vivo/in vivo	Denham et al. (2022)

fumigatus transcriptome resembled that of rapamycin-mediated TOR kinase inhibition, perhaps signifying fungal starvation in the lung relative to YPD laboratory culture.

A subsequent study (Bertuzzi et al. 2014) from the same group, adopted a similar methodology to examine the temporal landscape of gene expression during colonisation of the leukopenic murine lung. Relative to ungerminated *A. fumigatus* spores a total of 3733 genes were found to be differentially regulated at a minimum of one time point over a time series of 4, 8, 12, and 16 h post-infection. This permitted the capture of stage-specific gene expression during invasive colonisation of the leukopenic murine lung, the first reported longitudinal study of gene expression during mammalian pulmonary infection. The study revealed respiration, metabolism, and amino acid biosynthesis as being prioritised during early infection of the leukopenic host, whilst cation transport, secondary metabolism, and iron metabolism were subsequently emphasised during the commencement of invasive growth. Notably, upregulated expression of secreted gene products was sustained throughout the time series. Comparative analysis of host adaptation in a null mutant of the pH-responsive transcription factor PacC, revealed a multitude of secreted gene products under PacC regulatory control during host colonisation. In keeping with the critical role of PacC in mediation of epithelial invasion and cytotoxicity, PacC is the only transcription factor to date that has been shown to govern both early- and late-induced damage of respiratory epithelia during invasive growth of *A. fumigatus* (Okaa et al. 2023).

These studies demonstrated that comparative analysis of mutant and wild-type transcriptomes can help to pinpoint the in vivo regulons of transcription factors that drive fungal pathogenesis, however, it is not yet possible to decipher whether observed differences of gene expression result directly or indirectly from deletion of the transcriptional regulator. Such insights would require ChIP-Seq methodology, but this requires far more starting material than a standard tissue harvest will yield. Chung et al. (Liu et al. 2021) circumnavigated this problem by establishing correlations between gene promoters bound and/or activated by SrebP during in vitro culture, and those expressed in vivo using Nanostring nCounter technology. In this way, it was possible to demonstrate that SrbA-governs ergosterol biosynthetic and iron uptake gene expression under hypoxic growth conditions in vitro as well as playing a role in nitrate assimilation and heme biosynthesis.

ChIP-Seq analysis using an SrbA-specific antibody was applied to fungal mycelia after 4 and 12 h of hypoxic in vitro culture revealing a subset of 30 genes associated with SrbA binding that were located within 1 kb upstream of translational start sites. Transcript abundances of 12 mRNAs whose encoding genes had been identified as being SrbA-dependent from in vitro analyses, were assessed in an in vivo murine model of IPA, using Nanostring nCounter technology. This revealed SrbA-dependent hypoxic induction of ergosterol biosynthesis genes that were also upregulated in vivo consistent with their SrbA-dependent induction, and inferring exposure to hypoxia in infected tissues. A putative siderophore-iron transporter critical for response to low iron also showed strong SrbA dependency in vitro and high transcript levels in vivo, establishing a direct link between *A. fumigatus* SrbA, hypoxia adaptation, and iron homeostasis during infection. The study also identified and functionally characterised a second SREBP family member, designated SrbB, which is a direct transcriptional target of SrbA in hypoxia.

Liu et al (Chung et al. 2014) expanded Nanostring nCounter technology to assess transcript levels of 467 *A. fumigatus* transcription factor-encoding genes, most of the transcription factors encoded by the *A. fumigatus* genome. By applying highly sensitive nCounter technology it was possible to circumnavigate mRNA amplification and achieve direct analysis of tissue-based pathogen gene expression over timescales (up to 5 days) of host invasion not captured from analyses of BAL samples (Bertuzzi et al. 2014; McDonagh et al. 2008). The approach revealed the phenomenon of steady state invasive growth, a gene expression 'plateau' where fungal gene expression profile at days 4 and 5 was highly similar. Versus growth in a minimal medium 125 and 85 of 467 transcription factor genes (27 and 18%, respectively, of the entire genomic cohort) were, respectively, up and downregulated. In keeping with previous studies induction of a significant number of iron acquisition genes occurred during infectious growth, and genes encoding proteins that act to repress iron uptake and siderophore biosynthesis (such as *sreA*) were downregulated. Sparsity of zinc and nitrogen was also inferred from the transcriptional profiles of relevant transcription factors. Three transcription factors featured amongst the 30 most highly abundant transcripts of the in vivo transcriptome, namely PacC, SrbA, and HacA, although notably the fold-change induction of these genes relative to in vitro culture did not reach significance. Secondary metabolism genes, particularly of the gliotoxin biosynthetic gene cluster were exceptionally highly induced (up to 10,000-fold relative to in vitro culture). Infectious growth likely prompts selective biosynthesis of specific mycotoxins as some secondary metabolites such as fumagillin and pseurotin were repressed. From a database of 129 transcriptomes, it was possible to conclude that the in vivo transcriptome is unique, being tailored to both nutritional limitation and phagocyte attack that could be replicated in vitro via depletion of nitrogen, iron, and/or zinc.

Analysis of *A. fumigatus* gene expression in blood (Irmer et al. 2015) revealed an early phase of sensing the environment followed by shut down of most uptake and energy consumption mechanisms via a resting mycelial stage, prompting the conclusion that *A. fumigatus* is hardly able to propagate at all in blood. Interestingly given the likelihood of haemorrhage during invasion of highly vascularised pulmonary tissue this might in part explain the plateau affect reported by Lui et al. (2021). In blood, the functional categories most highly upregulated are transcription, protein synthesis, and transport. Those most significantly downregulated are cell cycle and DNA processing. Arginine and histidine, pyrimidine, purine, and carbohydrate metabolism are favoured over that of aspartate and pyruvate. Like growth in the lung (McDonagh et al. 2008), the general amino acid permease was found to be upregulated in blood. The in-blood transcriptome was found to exhibit little similarity to that of exposure to alkaline stress, oxidative stress, neutrophils, dendritic cells, or starvation.

Attempts to capture both host and pathogen gene expression have been few for aspergillosis. Kale et al. (2017) applied dual RNA-seq to the comparative study of chemotherapy and steroid-induced aspergillosis, achieving good resolution of fungal gene expression at early but not late stages of infection. At day 2 of infection fungal genes upregulated in the steroid-treated host were enriched for those associated with translation, cation, and glutamate transport, extending to include secondary metabolism, siderophore-iron transport, and metal ion transport by day 3. Early stages of IPA in a chemotherapy model of infection involved an emphasis on genes involved in fatty acid and triterpene metabolism. Transcripts of the genes encoding gliotoxin biosynthetic functions were more abundant in the steroid model relative to the chemotherapy model. Global analysis of predicted secreted gene products revealed context-specific gene expression suggesting that the pathogen tailors the repertoire of secreted products according to the niche in which it grows.

Dual RNA-Seq analysis of naturally occurring canine sinus infection has been documented (Valdes et al. 2020). Sino-nasal aspergillosis (SNA) is the most common canine *Aspergillus* infection. The fungus can colonise the nasal passage for months eventually leading to symptomatic disease involving sneezing, nasal discharge, and tissue destruction. Analysis of sino-nasal plaques revealed that 9213 genes (92.8% of all genes of the Af293 genome) were found to be expressed. Fungal plaques were found to exhibit a pH of 8 at the time of isolation suggesting that alkalization of the sino-nasal area is part of the fungal strategy to colonise and counter the host defences. Accordingly, the expression of a fungal zinc chelator (AspF2) and alkaline zinc transporter (ZrfC) are maintained at stable medium to high levels. Further evidence of a fungal adaptation response includes upregulation of genes promoting synthesis of immune-modulatory metabolites, such as gliotoxin, adaptation to nutritional stress, and phenotypic and genetic plasticity. The finding that IL-10 was expressed in SNA is suggestive of a dampened Th1 response that would prevent excessive tissue damage whilst accommodating survival of the host-adapting pathogen, leading to the suggestion that the host relies upon nutritional immunity and inflammatory responses, whilst the adaptive responses of *A. fumigatus* enable growth of the pathogen and contribute to a positive feedforward loop that suppresses the host immune response. The established 'warzone' promotes chronic, symptomatic colonisation and genetic selection of fungal colony variants that are highly adapted to host survival.

10.2 The Host-Adapting *Candida* Transcriptome

Candida species adopt cosmopolitan lifestyles existing as commensal organisms of healthy individuals but frequently causing a range of oral or vaginal mucosal infections. Bloodstream infection, occurring in settings of immunocompromise or post-surgical infection can progress to become acute, life-threatening, invasive infections of deep-seated organs such as the kidney, liver, and brain. Although gastrointestinal colonisation by *Candida* spp. is widespread amongst healthy humans, severe systemic infections are thought to arise due to the escape of commensal *Candida* spp. from the gastrointestinal (GI) tract. Dissemination via the bloodstream presents a harsh environment where access to nutrients is limited and the threat to host immune cells is immediate. *Candida* spp. also causes chronic mucosal infections, significantly vulvovaginal candidiasis (VVC) that is caused most frequently by *Candida albicans*.

Studies analysing the host-adapting *Candida* spp. transcriptomes (Fig. 10.2 and Table 10.1) have captured pathogen adaptation to the host environment in fungal cells harvested from kidneys of infected rabbits, the spleen, liver, kidneys, vaginal tissue, and GI tracts of infected mice, or in response to exposure to whole human blood. Methodologies have included RNA enrichment coupled to RNA-Seq, RNA-Seq, microarray, and Nanostring technologies.

Early studies of the host-adapting *Candida albicans* transcriptome tackled problems of low pathogen versus host biomass by using sequential lysis of mammalian and fungal cells (Andes et al. 2005) or larger mammal infection models with a focus on acquiring and pooling tissues from multiple fungal lesions (Walker et al. 2009).

After infection of mouse (Andes et al. 2005) or rabbit (Walker et al. 2009) kidney, fungal genes involved in adhesion, stress adaptation, and carbon assimilation sources were upregulated in kidney lesions versus growth in RPMI or YPD whilst genes involved in glucose utilisation were downregulated. This suggests that *C. albicans* cells that colonise the kidney are less glycolytically active than cells growing in RPMI or YPD and alternative carbon assimilation pathways, such as fatty acid β-oxidation, the glyoxylate cycle, and the TCA cycle are favoured.

Using biotinylated bait probes to capture *C. albicans* sequences, Amorim-Vaz et al (2015) analysed *C. albicans* transcriptomes directly from early and late stages of systemic infection in mice or the invertebrate *Galleria mellonella*, achieving enrichment for fungal versus host transcripts of up to 1600-fold allowing detection over 86% of *C. albicans* genes via RNA-Seq at early and late stages of infection. Genes that were highly induced included those involved in stress response, adhesion, iron acquisition, and biofilm formation, with surprising conservation of the pathogen transcriptome regardless of host model, and involving about 1200 genes that are commonly regulated in the two infection models at early and late stages of infection, relative to in vitro grown cells. 1169 genes of the host infection signature were analysed via gene set enrichment analysis (GSEA) using custom lists of *C. albicans* genes differentially expressed in response to host cells, or responsive to conditions mimicking host environments, or expressed in host tissues, revealing two major clusters of genes with positive and negative values. Positively regulated genes included around 300 that are also upregulated in biofilms as well as

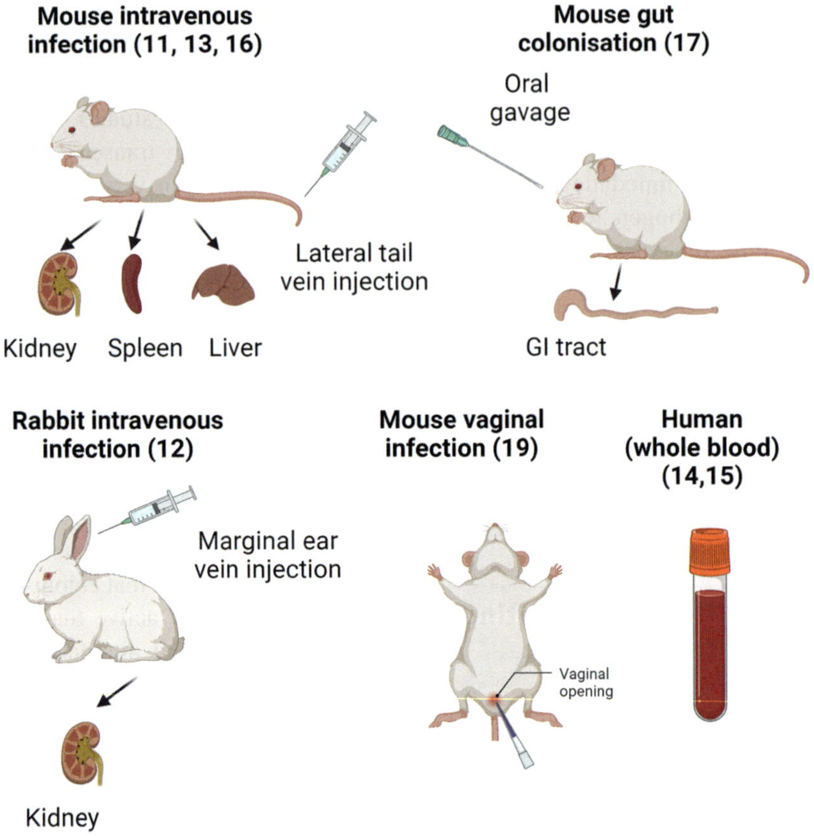

Fig. 10.2 Studies analysing the host-adapting *Candida* spp. transcriptomes. Numbers in parentheses indicate reference numbers. Created with BioRender.com

iron-regulated genes that increased their expression over time in both hosts. Downregulated genes were enriched for processes such as amino acid biosynthesis, glycolysis, and electron transport, likely reflecting less active metabolism in the host due to the less favourable growth environment.

Four *Candida* species, *C. albicans*, *C. glabrata*, *C. parapsilosis*, and *C. tropicalis*, predominate as fungal pathogens causing bloodstream infections but are phylogenetically diverse and differentially virulent. Recently a fifth species, *C. auris*, has emerged and quickly spread giving cause for clinical concern due to frequent multidrug resistance. *C. auris* is most often found on the skin and in the hospital suggesting that this species might be differentially adapted for survival outside the human body. Comparative dual species RNA-Seq analysis of the five species in a human whole-blood infection model allowed an opportunity to probe common and divergent pathogenic strategies (Kammer et al. 2020; Allert et al. 2022). The host response was found to comprise a conserved species-independent response involving strong upregulation of proinflammatory processes, however, beyond a core fungal response, pathogen transcriptomes differed significantly.

Shutdown of protein biosynthesis, rRNA processing, translation, and purine biosynthesis was a common response for all fungal species. Glycolytic and fatty acid synthesis genes were also downregulated. General stress response genes such as those encoding heat shock proteins, and genes encoding hydrolytic enzymes were commonly upregulated. Although the five genomes share approximately 3700 orthologues, only 87 were differentially regulated in all species at any time point during infection. *C. albicans*, *C. tropicalis*, and *C. parapsilosis* were found to mount immediate and sustained upregulation of glyoxylate and fermentative genes indicative of glucose restriction. Neither *C. glabrata* or *C. auris* reacted this way. In the case of *C. glabrata*, transporters for carbohydrates, amino acids, and ammonium were downregulated; and in the case of *C. auris* upregulation of genes involved in translation was observed instead of no upregulation of the glyoxylate cycle or fermentative genes, a signature unique to *C. auris*. Whilst *C. albicans* upregulated adhesins within a rapid timeframe, orthologous genes of *C. tropicalis* or *C. parapsilosis* were not upregulated and in *C. glabrata* only three of 67 predicted adhesins in *C. glabrata* were immediately upregulated. The relatively inert transcriptional status of *C. glabrata* compared to the other *Candida* species extended to multiple other gene families and functional categories, including the alternative glyoxylate cycle and ROS detoxifying enzymes. Indeed only 1.3–8.2% of the *C. glabrata* transcriptome was differentially regulated over the time course compared to about 15% of the transcriptome of *C. auris*. Neither morphogenesis nor related virulence factors such as Ece1 peptide (candidalysin) are universally conserved features of all *Candida* pathogenic landscapes, but secreted proteases and lipases are encoded by the *C. auris* genome some of which were significantly upregulated during blood infection. Over half of 659 *C. auris*-specific genes were differentially expressed during at least one time point of infection, but included genes having putative roles in iron acquisition during blood infection.

C. auris amino acid and ammonium transporter genes are upregulated during blood infection as well as genes of the expanded MFS class, known for their role in antifungal resistance, a transcriptomic signature unique to *C. auris* and likely significant for the transport of lipids, ions, and metabolites during host adaptation. It was noted that the upregulation of MFS transporter genes, including *MDR1* in *C. auris* may contribute to *C. auris* azole resistance by preparing *C. auris* for antifungal exposure. The expanded family of siderophore transporter genes is also expanded and upregulated during blood infection. Notably, *C. albicans*, *C. tropicalis*, and *C. parapsilosis* did not exhibit such transcriptional signatures.

Dual species transcriptome profiling has been applied to decipher organ-specific *C. albicans* adaptations, comparing gene expression in murine kidney, liver, and spleen (Hebecker et al. 2016). Integration of host and pathogen transcriptional data into an inter-species regulatory network connected *C. albicans* cell wall remodelling and metabolism to organ-specific immune responses. Compared to YPD culture, 917 *C. albicans* genes were differentially expressed in liver, and 908 and 875 in kidney at 12 and 24 h, respectively. Cell wall and cell surface modulating gene functions were significantly upregulated in all samples, including hypha-associated genes involved in adhesion. A starvation response was evident from the fungal transcriptome in kidney at 24 h versus emphasis upon catabolic processes and carbohydrate transport in the liver, potentially reflecting organ-specific nutrient supply.

Network analysis revealed three disconnected sub-networks each composed of mouse and *Candida* clusters, the first involving downregulated host genes involved in tissue homeostasis, and remodelling linked to upregulated fungal genes affecting iron homeostasis and biofilm formation. The second sub-network comprised fungal genes upregulated in the liver but downregulated in the kidney (predicted as negatively regulating the expression of IL6, CXCL10, and Interferon- and TNF-receptors in the liver) comprising cell surface-associated fungal gene products that induce filamentation and affect cell wall composition, in turn affecting immune cell interaction. This theme of organ-specific regulatory interactions involving host immune response, fungal metabolism, and morphology was also maintained in the third sub-network. Filamentation-associated genes featured as the most highly upregulated fungal genes in the kidney and intriguingly despite the known absence of filamentation in murine liver after intravenous infection, both hypha-associated genes and those normally associated with filamentation were found to be upregulated in the liver. It is therefore possible that *C. albicans* induces hypha-associated gene expression in the liver environment that is counteracted by the activity of phagocytes.

Combined analyses of host and pathogen signatures in the different organ niches revealed that innate defence mechanisms are slower to activate in the kidney, possibly due to lower abundance of resident immune cells and slow recruitment of additional immune cells, thereby promoting fungal growth, iron acquisition, and metabolic adaptation. A swifter immune response in the liver may prevent fungal filamentation, hence the observed differential expression of cell wall and cell surface modifying enzymes.

During invasive disease, gene expression dynamics of *C. albicans* environmentally responsive genes and transcription factor genes revealed early and late phases of infection, respectively characterised by induction of zinc and iron limitation, Rim101-dependent and hyphal genes, or genes related to phagocytosis by macrophages. The transcriptomes of mutants lacking each of six virulence-essential transcription factors (Rim101, Efg1, Zap1, Rob1, Rpn4, Sut1) revealed distinct gene expression profiles during infection versus in vitro growth. Therefore, pathogen gene expression is exquisitely tailored to the host environment via transcriptomic signatures that are not expressed in laboratory culture. Concordant with the findings of Hebecker et al. (2016) kidney colonisation initiates slowly, with an eventual 100-fold increase in fungal burden by 24 h, when leukocyte infiltration begins.

A probe set including 248 environmentally responsive genes chosen from previously described genome-wide datasets was developed and expression changes over the time course of invasive infection were compared to 166 other published gene expression studies, revealing significant correlations.

Early up genes involved genes expressed during the yeast-to-hyphal transition, and were reflected in transcriptomes of mutants defective in the regulation of hyphal formation. Hyphal formation, iron limitation, zinc limitation, and Rim101 activation were identified as driving forces that govern early gene expression responses. The late up genes involved those induced by growth on acetate as a carbon source. It was postulated that the late gene expression responses may be driven by tissue damage and an inflammatory cell influx and indeed two phases of host gene expression were also discernible.

Early up transcription factor genes included regulators of hyphal formation (Ume6, Tec1), zinc acquisition (Zap1), and iron acquisition (Hap43, Sef1) whilst later up transcription factors were mostly uncharacterised genes. Amongst the 30 most highly expressed transcription factor genes at 48 h post-infection, 18 of 24 that had been previously studied had been found to be required for proliferation or virulence.

Rim101 was selected to pursue regulatory relationships during infection because *RIM101* is the most highly expressed transcription factor gene at 48 h post-infection. Assessing the expression of 148 environmentally responsive genes in the *rim101Δ/Δ* mutant 33 RNAs were found to be downregulated versus a complemented strain and expression of 14 RNAs was increased. As expected of a mutant exhibiting deficiency of hyphal growth, genes downregulated in the *rim101Δ/Δ* mutant included many hypha-induced genes.

Similar to gene expression in the liver (Hebecker et al. 2016) *C. albicans* cells colonising the GI tract express genes associated with hyphal growth, despite the fact that most cells are in the yeast form (Pierce et al. 2013). Noting that genes encoding proteins regulated by Efg1p during hyphal growth, such as *ECE1*, were highly expressed during GI tract colonisation, Pierce et al. aimed to understand the role of the *C. albicans* transcription factor Efg1p, a major regulator of hyphal formation, during GI tract colonisation by undertaking a comparative analysis of *C. albicans* gene expression during colonisation of different regions of the GI tract. In keeping with different environments of the ileum and cecum, gene expression in these organs differed significantly. The ileum is an alkaline, oxygen-rich niche whilst the cecum is rich in short-chain fatty acids due to greater numbers of commensal bacteria.

RNA was extracted from *C. albicans*-colonised GI tract contents at day 3 post-colonisation. Consistent with low availability of glucose in the cecum, genes involved in galactose metabolism and gluconeogenesis were upregulated in fungal cells colonising the cecum, whilst genes involved in alcohol and phosphate metabolism, haemoglobin utilisation, various heat shock genes, and oxidative stress response, were upregulated in fungal cells colonising both the cecum and the ileum. Relative to laboratory culture significant enrichment was detected for genes upregulated in response to hyphal induction, temperature, osmotic shock, carbon and nitrogen starvation, growth in galactose, and alkaline growth. Differentially expressed genes of fungal cells isolated from the cecum were not enriched for any GO process categories and might represent genes that are unexpressed during laboratory growth being colonisation specific. Over half of the genes differentially expressed in the ileum versus laboratory growth were not similarly regulated in the cecum. Genes supporting acetyl-CoA metabolism and the TCA cycle were active during colonisation of the ileum as were alkaline-induced genes. The ileum is characterised by a higher pH than that of the cecum, however, a *rim101*$^{-}$ null mutant inoculated into mice by oral gavage was able to colonise as well as the WT, thus although the pH-responsive transcription factor Rim101p is required for systemic virulence, it is not required for GI colonisation.

efg1 null mutant cells were able to grow to higher levels in the GI tract than WT cells but were hypersusceptible to host immune responses. Expression of adhesin and hypha-regulated genes, secreted aspartyl proteases, and a superoxide dismutase were found to be egf1 dependent. Other genes were expressed independently of Egf1 integrity, such as the hyphal wall protein gene *HWP1*, and the pH-responsive glycosidase gene *PHR1*. Higher expression of carnitine metabolism genes was demonstrated as likely contributing to the enhanced growth capabilities of the efg1 null mutant.

Consistent with organ-specific findings of Hebeacker et al, analyses of gut colonisation suggested that coregulation of metabolism and host interactions by the efg1 transcription factor allows *C. albicans* to adopt two different survival strategies: increased growth or protection from the host immune response. Growth limitation in a healthy host promotes expression of *C. albicans* genes that will protect it against the host immune system, whilst overgrowth of *C. albicans* may shift towards increased pathogenesis.

Similar to the hyper-colonising phenotype of efg1 null cells in the GI tract, a null mutant of the paralogous *EFH1*-encoded transcription factor, which is highly expressed during growth of *C. albicans* in the intestinal tract, exhibits increased colonisation of the murine intestinal tract (White et al. 2007). Strains engineered to overexpress efh1 exhibited reduced levels of intestinal and oral colonisation in athymic mice, therefore, efh1 is important for determining the population size of *C. albicans* during colonisation. However, efh1 was not required for invasive infection indicating that genes contributing to commensal colonisation can be distinct from genes that are required for virulence. Microarray analysis of fungal genes expressed during colonisation of the newborn piglet oral and oesophageal cavities revealed *EFH1* and *ECE1* (candidalysin) gene expression.

Using RNA-Seq, Bruno et al. (2015)characterised the early host vaginal and fungal transcriptomes of mice during vulvovaginal candidiasis (VVC). Network analysis of host genes that were differentially expressed between infected and naive mice predicted NLRP3 inflammasome activation. Accordingly, intravaginal challenge of Nlrp3$^{-/-}$ mice with *C. albicans* demonstrated severely reduced levels of polymorphonuclear leukocytes (PMN) and inflammatory cytokines in vaginal lavage fluid. RNA-Seq analysis of *C. albicans* genes highlighted upregulation of hypha-associated secreted aspartyl proteinases 4, 5, and 6 (SAP4–6), which are known inflammasome activators. Concordantly SAP null mutants, including a Δ*SAP4–6* triple null, induced significantly less PMN influx and inflammatory cytokine during intravaginal challenge, despite being able to colonise as efficiently as a wild-type isolate. A comprehensive analysis of the *C. albicans* transcriptome during VVC was precluded by inability to obtain sufficient reads.

10.3 The Host-Adapting *Cryptococcus* Transcriptome

Cryptococcus neoformans is the leading cause of fungal meningitis. Individuals suffering from $CD4^+$ lymphocyte deficiency (a primary cause being HIV/AIDS) are particularly susceptible. Candidate gene studies, undertaken in laboratory isolates, had established the early knowledge base, delineating several important virulence functions from whole animal studies, such as the capsular polysaccharide, the ability to grow at 37 °C, and the production of, e.g. melanin, urease, and phospholipase.

Upon availability of genome sequences, additional impetus for undertaking in vivo transcriptome studies derived from the need to understand several key (Chen et al. 2014) stages of disease: disease initiation in the lung; survival and proliferation in the lung (namely in-tissue dormancy; reactivation of latent infection with fungal proliferation; yeast dissemination and traversal of the blood–brain barrier; proliferation in brain tissue and the subarachnoid space) and the recognition that new treatments might target one, several or all of these processes (Yu et al. 2021; Chen et al. 2014).

Studies analysing the host-adapting *Cryptococcus* transcriptome (Fig. 10.3 and Table 10.1) have captured pathogen adaptation to the host environment in fungal cells harvested from cerebrospinal fluid (CSF) of infected humans, mice, or rabbits, and in the lungs and brains of infected mice and monkeys. Recently, RNA-Seq has been the methodology of choice for such studies.

The cells of *Cryptococcus* spp. switch morphology in response to environmental conditions, a type of phenotypic plasticity that may enhance survival. Several morphotypes have been described, including spores, hyphae and pseudohyphae, seed cells, microcells, yeast cells, and titan cells. Desiccated yeast cells or spores are thought to be infectious propagules and ecological niches include trees, soil, bird excreta, and amoeboid, avian, mammalian, and other host niches.

To address relative pathogenicity of clinical and environmental *C. neoformans* isolates, Yu et al. (2020) studied the transcriptomes of seven *C. neoformans* isolates grown in five different in vitro and in vivo conditions. The four major lineages of *C. neoformans* var. *grubii* differ in origin and genotype whereby some lineages are frequently isolated from pigeon guano, whilst the VNB lineage is found in association with certain trees. The selected isolates were of varying origins and virulence profiles and included clinical isolates, as well as strains isolated from pigeon guano or trees. Transcriptomes were studied during growth on rich medium or nutrient-limited medium, or media containing pigeon guano. Gene expression during macrophage co-culture and during CNS infection of the rabbit were also studied. RNA-Seq analysis identified lineage-specific gene expression, particularly of genes involved in oxidative stress resistance, iron detoxification, and capsule production, located with a biased distribution towards subtelomeric regions of the genome and often organised into gene clusters. In addition to transcriptome architectures clustering by condition tested, they also could be segregated according to lineage. Lineage-specific genes that were differentially expressed across all tested conditions were mostly associated with membrane transport, perhaps arising because of differential evolution of lineages in association with different environmental niches. Accordingly, lineage-specific genes linked to nitrogen source usage were upregulated in lineages isolated from pigeon guano, particularly genes involved in urea metabolism.

Genes upregulated in in vivo conditions but downregulated in vitro conditions, nearly half of which were known to be virulence-associated including those related to lysine biosynthesis, inositol phosphate and inositol lipid metabolism, nitrogen metabolism, DNA recombination and repair, cell cycle, chromosome organisation, and stress responses. Sodium efflux, oligopeptide transport, quorum sensing, calcium homeostasis, oxygen sensing via *SRE1*, thermal sensing and response via the Bck1-Mkk2-Mpk1 signalling

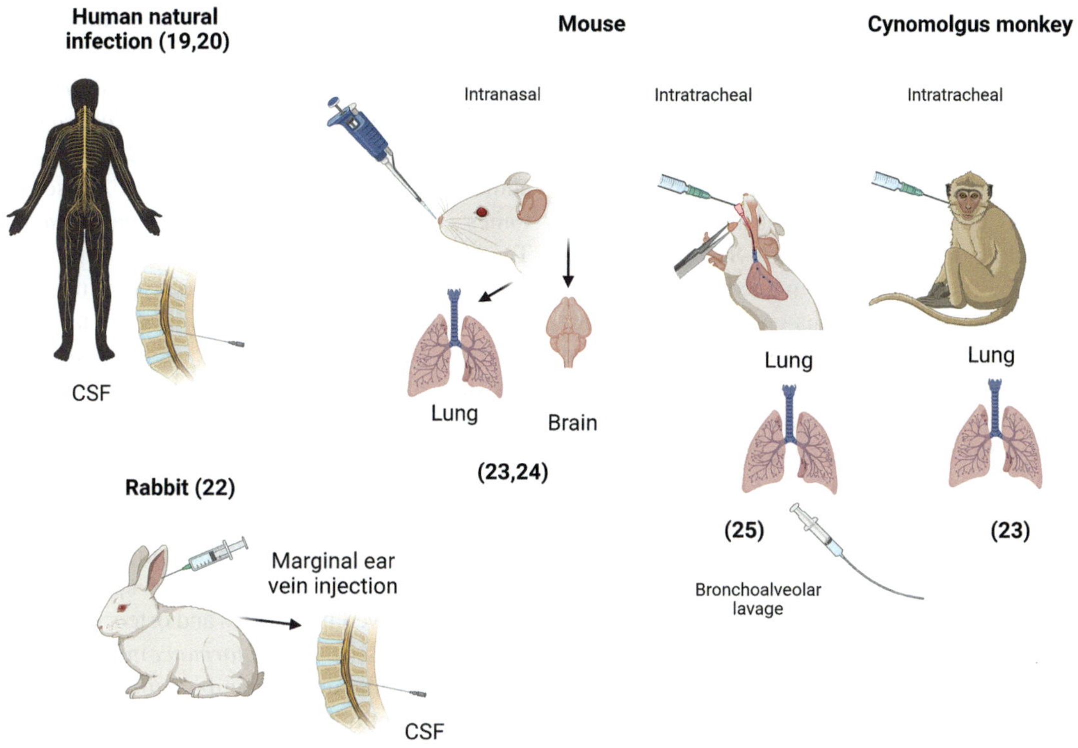

Fig. 10.3 Studies analysing the host-adapting *Cryptococcus* transcriptomes. Numbers in parentheses indicate reference numbers. Created with BioRender.com

pathway, and pH-sensing via *RIM101* were all upregulated in vivo.

To identify differences that may contribute to the higher virulence of clinical isolates, gene expression was analysed in pairs of closely related clinical and environmental isolates. Relative to environmental isolates certain clinical isolates differentially regulated genes associated with cell wall biosynthesis and zinc uptake.

Founding studies of in-patient adaptation were conducted by Chen et al (2014) who extracted RNA from yeast cells directly harvested from the cerebrospinal fluid (CSF) of two AIDS patients with cryptococcal meningitis prior to antifungal therapy. Using RNA-Seq the transcriptomes of the two isolates were comparatively analysed under three conditions: in vivo CSF, ex vivo CSF, and YPD, intended to replicate CSF exposure and nutritionally replete culture. Despite the genetic diversity of the two clinical isolates, remarkable similarity of transcriptional responses was observed whereby transcriptomes of the two clinical isolates under each test condition were more similar than that of individual isolates under different stresses. In-patient growth was associated with higher metabolic activity than *ex vivo* culture in human CSF, evidenced by upregulation in CSF of genes related to cell metabolism, gene expression, and ribosome biogenesis, consistent with fungal proliferation in the host environment. Divergent gene expression between strains was mainly associated with expression of transporters. Twenty genes were found to be CSF-specific (versus YPD) regardless of in vivo or ex vivo CSF exposure, including *RIM101*, *ENA1*, and *CFO1*, whilst six genes lacked functional annotations. In vivo-specific genes included sulfiredoxin (*SRX1*), reported as having a critical role in oxidative stress resistance and plausibly important for pathogen survival in the presence of monocytes/macrophages in CSF or microglial cells in the brain. Upregulation of *SIT1* expression, required for expression of virulence factors under low-iron conditions as well as *CFO1* indicates pathogen reliance upon iron acquisition functions in CSF. Differential regulation of classical virulence factors, such as capsule, was not observed, perhaps because such virulence factors are expressed during the early stages of infection.

Understanding of the human-adapting *C. neoformans* transcriptome was further deepened by expanding the scale of RNA-Seq analyses to encompass yeast transcriptomes from 31 *C. neoformans* isolates of various genotypes, and comparing in-host transcriptomes to those captured from multiple other experimental conditions (nutrient replete in vitro growth; artificial CSF; rabbit CSF from experimental cryptococcosis at two time points) thus revealing in vivo induction of carbon metabolism, fatty acid synthesis, transport, cell wall biosynthesis and stress responses (Yu et al. 2021; Chen et al. 2014). This pan-isolate RNA-Seq approach provided a powerful means to probe how the *Cryptococcus* genetic background affects gene expression in the CNS. A major output of the study was a list of the 50 most highly abundant *C. neoformans* transcripts common to all of the isolates tested. Seven such genes had been previously characterised (*CMP1*, *CIG1*, *ENA1*, *RIM101*, *CDA1*, *CQS1*, and *RCK2*) and relevance for virulence demonstrated. Two genes were known to be essential (*FKS1* and *EF3*). Multiple cell wall-associated functions were identified, along with another eight genes (governing ionic homeostasis and iron uptake) that are potentially regulated in a *RIM101*-dependent manner. Genes involved in fatty acid and glucose metabolism were also represented amongst the conserved upregulated functions.

Comparison of transcriptomes drawn from human and rabbit CSF revealed significant differences between these conditions, with heterogeneity of pathogen response perhaps the result of extended periods of time in the host prior to harvest of the human CSF. Indeed, in a longitudinal study of rabbit infection there are marked differences over time in expression of genes involved in regulation of gene expression, protein processing, and ribosomal biogenesis (day 4 compared to the initial infection day). This might reflect yeast growth arrest during adaptation to new microniches within the host.

Seeking more physiologically relevant models of human infection, Li et al. (2019) developed a nonhuman primate model of *C. neoformans* infection in the cynomolgus monkey *Macaca fascicularis*, using dual RNA-Seq and undertook additional systematic comparison to transcriptomes of mouse-colonising *C. neoformans*. Twenty percent of the differentially expressed host genes were shared between two models, indicating distinct species-specific responses. In keeping with other studies *C. neoformans* was shown to activate metal sequestration and dampen sugar metabolism, during infection. Interestingly, at 7 days post-infection, *C. neoformans* cells were detected in the brain of mice, but not of monkeys thus gene expression in lung tissue served as the basis of comparative analysis. Conserved host responses included upregulation of genes involved in diabetes and osteoclastogenesis in response to *C. neoformans* infection, the genetic or chemical, manipulation of which altered susceptibility to fungal infection.

Adaptation to diverse tissue niches is an essential prerequisite for *C. neoformans* pathogenicity. Morphogenetic shifts coupled with host cues can facilitate this. Denham et al (2022) described a small morphotype called "seed" cells as having a high propensity for dissemination that is critical for extrapulmonary organ entry. Freitas et al (2022) also demonstrated that during murine pulmonary infection cell size reduction (to ≤ 3 µm) is important for initial fungal adaptation during infection.

Dissecting the temporal, transcriptional basis of such adaptation revealed multiphasic upregulation of transcripts, initially those related to ribosome biogenesis at the beginning (6 h) of infection and latterly (10 days) involving inositol biosynthesis, energy production, and the proteasome (Freitas et al. 2022). Transcriptome data suggest that initial pathogenic strategy revolves around host adaptation and replication, followed by later shift towards host resistance and virulence. Considering the bronchoalveolar space as the initial relevant compartment of the host-pathogen interaction, differentially expressed genes (relative to YPD and between in vivo time points) were sought. Within the first 6 h of infection, 75 of 532 upregulated transcripts were involved in ribosomal regulation. Later gene expression favours the upregulation of genes involved in inositol metabolism, 16 of which were upregulated after 10 days. Genes consistently upregulated in vivo included those supporting respiration, electron transport, pentose phosphate metabolism, and electron transport.

In an independent study, having observed that small ex vivo morphologies derived exhibit similar organ entry profiles upon reinfection into mice, Denham et al. (2022) hypothesised that transcriptome analysis might reveal the morphogenetic stimulus, going on to test this via comparative transcriptomic analysis of *C. neoformans* cells grown in various media as well as for small, intermediate and large-sized *ex vivo C. neoformans* cells. Ex vivo cell populations shared conserved cohorts of 699 and 291 up- and downregulated genes, respectively. Gene ontology term (GO-term) enrichment analysis revealed that phosphate ion transport genes were induced in conditions that promote seed cell formation and by iterative elimination of media components it was discovered that phosphate drives *C. neoformans* cell populations towards smaller morphotypes. Phosphate exposure was sufficient to stimulate cell size reduction, thus phosphate limitation imposed by alkaline environments would likely offset such morphogenesis. *C. neoformans* infection was associated with a progressive reduction in pH over a 3–17 day time course, therefore a model is proposed by which a feed-forward loop of seed cell formation and dissemination is established via acidification of the infectious niche coupled with phosphate release caused by tissue damage that drives size variation and dissemination of *C. neoformans* cells.

10.4 Take Home Points and Perspectives

The primary infectious niches of human fungal pathogens are predominantly reliant upon innate and nutritional immunity to offset pathogen attack whilst fungal growth requires metabolic, adaptive programs of gene expression that promote tissue invasion and nutrient release. Transcriptomic analyses have revealed that fungal virulence is promoted through induction of genes that heighten the propensity for invasive growth and dissemination, and phenotypic and genetic plasticity are common features of prolonged infectious growth. Although in vitro culture can recapitulate in-host gene expression it is not universally reflective of host colonisation, in particular, the opportunity to define the effects of feed-forward mechanisms underpinning pathogenicity is lost in such experimental settings. Technological advances will also be important for deciphering such mechanisms since current approaches for in vivo transcriptional profiling of human fungal pathogens have relied exclusively upon average mRNA abundance measures drawn from mass sampling of heterogeneous micro-niches of infected tissues.

Individual fungal isolates undergo genetic and phenotypic switching thus temporal analyses of the in vivo host–pathogen interaction are essential for defining host adaptations. Between strain differences are commonplace and laboratory isolates are often not representative of natural populations. In-host transcriptomes routinely reveal virulence-associated pathogen functions and are a powerful means of understanding pathobiology, however, roughly half of all host-specific gene expression lies with genes of unknown function. Additional strategies will be needed to determine the functions of such genes, computational analysis of the enormous wealth of available transcriptome data will be an important approach.

References

Allert S, Schulz D, Kammer P, Grossmann P, Wolf T, Schauble S, Panagiotou G, Brunke S, Hube B (2022) From environmental adaptation to host survival: Attributes that mediate pathogenicity of Candida auris. Virulence 13:191–214

Amorim-Vaz S, Tran Vdu T, Pradervand S, Pagni M, Coste AT, Sanglard D (2015) RNA enrichment method for quantitative transcriptional analysis of pathogens in vivo applied to the fungus Candida albicans. mBio 6:e00942–e00915

Andes D, Lepak A, Pitula A, Marchillo K, Clark J (2005) A simple approach for estimating gene expression in Candida albicans directly from a systemic infection site. J Infect Dis 192:893–900

Bertuzzi M, Schrettl M, Alcazar-Fuoli L, Cairns TC, Munoz A, Walker LA, Herbst S, Safari M, Cheverton AM, Chen D, Liu H, Saijo S, Fedorova ND,

Armstrong-James D, Munro CA, Read ND, Filler SG, Espeso EA, Nierman WC, Haas H, Bignell EM (2014) The pH-responsive PacC transcription factor of Aspergillus fumigatus governs epithelial entry and tissue invasion during pulmonary aspergillosis. PLoS Pathog 10:e1004413

Bruno VM, Shetty AC, Yano J, Fidel PL Jr, Noverr MC, Peters BM (2015) Transcriptomic analysis of vulvovaginal candidiasis identifies a role for the NLRP3 inflammasome. mBio 6

Chen Y, Toffaletti DL, Tenor JL, Litvintseva AP, Fang C, Mitchell TG, McDonald TR, Nielsen K, Boulware DR, Bicanic T, Perfect JR (2014) The Cryptococcus neoformans transcriptome at the site of human meningitis. mBio 5:e01087–e01013

Chung D, Barker BM, Carey CC, Merriman B, Werner ER, Lechner BE, Dhingra S, Cheng C, Xu W, Blosser SJ, Morohashi K, Mazurie A, Mitchell TK, Haas H, Mitchell AP, Cramer RA (2014) ChIP-seq and in vivo transcriptome analyses of the Aspergillus fumigatus SREBP SrbA reveals a new regulator of the fungal hypoxia response and virulence. PLoS Pathog 10:e1004487

Denham ST, Brammer B, Chung KY, Wambaugh MA, Bednarek JM, Guo L, Moreau CT, Brown JCS (2022) A dissemination-prone morphotype enhances extrapulmonary organ entry by Cryptococcus neoformans. Cell Host Microbe 30:1382–400 e8

Freitas GJC, Gouveia-Eufrasio L, Emidio ECP, Carneiro HCS, de Matos Baltazar L, Costa MC, Frases S, de Sousa Araujo GR, Paixao TA, Sossai BG, Caza M, Kronstad JW, Peres NTA, Santos DA (2022) The dynamics of Cryptococcus neoformans cell and transcriptional remodeling during infection. Cells 11

Hebecker B, Vlaic S, Conrad T, Bauer M, Brunke S, Kapitan M, Linde J, Hube B, Jacobsen ID (2016) Dual-species transcriptional profiling during systemic candidiasis reveals organ-specific host-pathogen interactions. Sci Rep 6:36055

Irmer H, Tarazona S, Sasse C, Olbermann P, Loeffler J, Krappmann S, Conesa A, Braus GH (2015) RNAseq analysis of Aspergillus fumigatus in blood reveals a just wait and see resting stage behavior. BMC Genomics 16:640

Kale SD, Ayubi T, Chung D, Tubau-Juni N, Leber A, Dang HX, Karyala S, Hontecillas R, Lawrence CB, Cramer RA, Bassaganya-Riera J (2017) Modulation of immune signaling and metabolism highlights host and fungal transcriptional responses in mouse models of invasive pulmonary aspergillosis. Sci Rep 7:17096

Kammer P, McNamara S, Wolf T, Conrad T, Allert S, Gerwien F, Hunniger K, Kurzai O, Guthke R, Hube B, Linde J, Brunke S (2020) Survival strategies of pathogenic Candida species in human blood show independent and specific adaptations. mBio 11

Li H, Li Y, Sun T, Du W, Li C, Suo C, Meng Y, Liang Q, Lan T, Zhong M, Yang S, Niu C, Li D, Ding C (2019) Unveil the transcriptional landscape at the Cryptococcus-host axis in mice and nonhuman primates. PLoS Negl Trop Dis 13:e0007566

Liu H, Xu W, Bruno VM, Phan QT, Solis NV, Woolford CA, Ehrlich RL, Shetty AC, McCraken C, Lin J, Bromley MJ, Mitchell AP, Filler SG (2021) Determining Aspergillus fumigatus transcription factor expression and function during invasion of the mammalian lung. PLoS Pathog 17:e1009235

McDonagh A, Fedorova ND, Crabtree J, Yu Y, Kim S, Chen D, Loss O, Cairns T, Goldman G, Armstrong-James D, Haynes K, Haas H, Schrettl M, May G, Nierman WC, Bignell E (2008) Sub-telomere directed gene expression during initiation of invasive aspergillosis. PLoS Pathog 4:e1000154

Okaa UJ, Bertuzzi M, Fortune-Grant R, Thomson DD, Moyes DL, Naglik JR, Bignell E (2023) Aspergillus fumigatus drives tissue damage via iterative assaults upon mucosal integrity and immune homeostasis. Infect Immun 91:e0033322

Pierce JV, Dignard D, Whiteway M, Kumamoto CA (2013) Normal adaptation of Candida albicans to the murine gastrointestinal tract requires Efg1p-dependent regulation of metabolic and host defense genes. Eukaryot Cell 12:37–49

Pongpom M, Liu H, Xu W, Snarr BD, Sheppard DC, Mitchell AP, Filler SG (2015) Divergent targets of Aspergillus fumigatus AcuK and AcuM transcription factors during growth in vitro versus invasive disease. Infect Immun 83:923–933

Valdes ID, Hart de Ruijter ABP, Torres CJ, Breuker JCA, Wosten HAB, de Cock H (2020) The sino-nasal warzone: transcriptomic and genomic studies on sino-nasal aspergillosis in dogs. NPJ Biofilms Microbiomes 6:51

Walker LA, Maccallum DM, Bertram G, Gow NA, Odds FC, Brown AJ (2009) Genome-wide analysis of Candida albicans gene expression patterns during infection of the mammalian kidney. Fungal Genet Biol 46:210–219

White SJ, Rosenbach A, Lephart P, Nguyen D, Benjamin A, Tzipori S, Whiteway M, Mecsas J, Kumamoto CA (2007) Self-regulation of Candida albicans population size during GI colonization. PLoS Pathog 3:e184

Yu CH, Chen Y, Desjardins CA, Tenor JL, Toffaletti DL, Giamberardino C, Litvintseva A, Perfect JR, Cuomo CA (2020) Landscape of gene expression variation of natural isolates of Cryptococcus neoformans in response to biologically relevant stresses. Microb Genom 6

Yu CH, Sephton-Clark P, Tenor JL, Toffaletti DL, Giamberardino C, Haverkamp M, Cuomo CA, Perfect JR (2021) Gene expression of diverse Cryptococcus isolates during infection of the human central nervous system. mBio 12:e0231321

Proteomics and Its Application to the Human Pathogenic Fungus *Aspergillus fumigatus*

11

Thomas Krüger and Olaf Kniemeyer

Abstract

The human pathogenic fungus *Aspergillus fumigatus* is a ubiquitously occurring saprophytic mold, which can cause diseases ranging from allergic reactions to life-threatening invasive infections. The release of the *A. fumigatus* genome in 2005 heralded the genomic era for *A. fumigatus*. Since then, omics-technologies including proteomics have provided new insights into the biology and pathogenicity of this fungus. The proteomics field is driven primarily by the technological developments in mass spectrometry and bioinformatics, but also sample preparation including the enrichment of peptides with posttranslational modifications have been substantially improved. Proteomics created new knowledge about the localization of *A. fumigatus* proteins, the primary and secondary metabolisms of this fungus, its development, signal transduction, stress response, and virulence.

Keywords

Mass spectrometry · Posttranslational modifications · Human pathogenic fungi Proteins · Proteomics · Genomics · Secretome Cell wall · *Aspergillus fumigatus*

11.1 Introduction

Human pathogenic fungi have a high impact on human's health, but their medical importance is still underappreciated. It is estimated that more than 150 million people suffer from serious fungal diseases (Bongomin et al. 2017; Brown et al. 2012). They particularly occur in individuals with immunosuppressed immune functions, such as AIDS and cancer patients, after organ transplantation and corticosteroid therapy. The most predominant and clinically relevant pathogens are the yeasts *Candida albicans, Candida auris,* and *Cryptococcus neoformans*, and the mould *Aspergillus fumigatus*. All four have been ranked

T. Krüger · O. Kniemeyer (✉)
Department of Molecular and Applied Microbiology, Leibniz Institute for Natural Product Research and Infection Biology – Hans Knöll Institute, Jena, Germany
e-mail: olaf.kniemeyer@leibniz-hki.de

A. A. Brakhage et al. (eds.), *Human and Animal Relationships*, The Mycota 6,
https://doi.org/10.1007/978-3-031-64853-3_11

and categorized into the highest priority group of health-threatening fungi in 2022 by the WHO (Antimicrobial Resistance Division 2022).

In contrast to the previous version of this chapter in the second addition of this book, the *Candida albicans* proteome is not covered due to the massively increased number of studies on fungal proteomics. Accordingly, we will focus on the *A. fumigatus* proteome. For the *C. albicans* proteome, the reader is referred to a number of excellent reviews on this topic (Wijayawardene et al. 2023; Shibasaki et al. 2018; Culibrk et al. 2016; Krüger et al. 2015; Kniemeyer et al. 2011).

Aspergillus fumigatus belongs to the large group of filamentous ascomycetes fungi. It is a ubiquitously distributed saprophyte, which degrades organic matter in the soil. *A. fumigatus* propagates via conidia, the asexual reproductive spores of fungi. They are spread via the air and are relatively small in size (2–3 μM diameter). For this reason, *A. fumigatus* conidia can be easily inhaled and reach the lung alveoli. Especially in immunocompromised patients, conidia germinate and can produce invasive infections in the lung (pulmonary invasive aspergillosis, IPA), which can disseminate to other organs (Latge and Chamilos 2019). Recent global estimates have found that around 2.1 million cases of IPA occur annually with an average fatality rate of more than 30% (https://gaffi.org).

In numerical terms, other *A. fumigatus* infections are even more prevalent: Around 3 million patients suffer from chronic pulmonary aspergillosis (CPA), more than 4 million people are affected by allergic bronchopulmonary aspergillosis (ABPA) and 6.5 million by severe asthma with fungal sensitization(SAFS) (Bongomin et al. 2017). All diseases are associated with underlying medical conditions. CPA occurs in patients with structural lung disease, such as sarcoidosis, chronic obstructive pulmonary disease (COPD) or sequela of tuberculosis. It is characterized by destructive fungal growth in lung cavities (Izumikawa et al. 2014). ABPA is a complex allergic inflammatory reaction, which can be mainly found in asthma and cystic fibrosis patients (Roboubi et al. 2023). SAFS is a broad term for severe asthma and raised *Aspergillus*-specific IgE (Agarwal et al. 2023).

Despite the increasing awareness of fungal infections and ongoing research, *Aspergillus* mycoses are still associated with high fatal outcomes. One of the reasons for the high mortality of invasive aspergillosis is the difficult diagnosis, since most clinical signs are not specific to this infection. The diagnostic criteria include host factors (e.g., immunosuppressive therapy), radiological signs, and microbiological detection (Donnelly et al. 2020). A culture-based identification or microscopic detection of fungal structures is often not possible. For this reason, molecular-based test assays play an important role in diagnosis. The detection of the fungal cell wall components β-1,3 glucan or galactomannan in patient samples like serum or BAL is often used for the identification of *Aspergillus* infections (Scharmann et al. 2023; Lass-Florl 2019). In addition, PCR testing of serum, plasma, or BAL is used (Keri and Chandrasekar 2023). Besides difficult diagnosis, the rapid emergence of antifungal-drug resistance poses a serious clinical problem (Fisher et al. 2022), especially since only three different antifungal drug classes are in clinical use (Ullmann et al. 2018), namely azoles, echinocandins, and the polyene amphotericin B.

Given these challenges, there is a strong demand in medical mycology for new methodological approaches to gain detailed insights into host–fungal pathogen interactions and to improve the medical practice of diagnosis and therapeutic interventions. Current omics-technologies are capable of contributing to the advancement in the field of medical mycology, due to the rapid progress in DNA sequencing, MS technologies, and machine learning algorithms (Wekesa and Kimwele 2023).

The release of the genome of *A. fumigatus* in 2005 herald the era of *Aspergillus fumigatus* genomics. Nowadays, 337 different *A. fumigatus* genomes are accessible on the NCBI taxonomy browser webpage (Schoch et al. 2020) and many more genomes were sequenced for population genomics studies (Rhodes et al. 2022; Barber et al. 2021; Horta et al. 2022). This forms the basis for transcriptome, proteome, metabolome, and integrative systems biology studies on *A. fumigatus* biology and pathogenicity. In the era of "fungiomics" (Ball et al. 2020), proteomics has and will have a high impact on the development of this field. Proteins are the active agents within the cell and proteomic techniques offer a global and integrated perspective of the entire proteins encoded by a genome including posttranslational modifications and isoform formations.

In the following, we will provide a general overview about the methods and technologies used in proteomics. Subsequently, recent advances in the proteomic analysis of *A. fumigatus* are summarized.

11.2 Methods of Proteome Analysis

11.2.1 Emergence and Challenges

The word proteomics was invented in the mid-1990s and is defined as the study of the "protein complement of the genome" (Wilkins et al. 1996). Proteomic studies were initially just represented by quantitative 2D-polyacrylamide gel electrophoresis (2D-GE) with a subsequent identification of the proteins by Edman degradation (Edman 1950) or mass spectrometry analysis. Since then, gel-free methods have been established based on coupling of liquid chromatography with tandem mass spectrometry (LC-MS/MS). The term proteomics now covers all kinds of studies on global changes of the protein pattern within a cell or tissue (expression proteomics) as well as surveys of protein complexes and the protein structures and functions (functional proteomics). Approaches like 2D-GE, protein and antibody arrays, yeast two-hybrid approaches, co-immunoprecipitation assays, GFP-tagged gene libraries, NMR spectroscopy as well as X-ray crystallography are still valuable to study protein components of a biological system, but certainly, the overwhelming majority of proteome analyses today are based on LC-MS/MS techniques, which are covered here.

All scientists working in the field of proteomics are confronted with the high diversity of proteins due to posttranslational modifications, splice variations, and the high dynamic range of protein concentrations in biological systems. The dynamic range for eukaryotic cells is estimated to be in the range 10^5–10^6 for eukaryotic cells and 10^{10} for serum (Patterson and Aebersold 2003). Many proteins are of low abundance, e.g., 50% of the protein content of yeast is the product of 100 genes and the rest of the proteins are just present at a low abundance ranging over 50–10,000 copies (Futcher et al. 1999). When studying host–pathogen interactions the sample complexity and dynamic range can be even higher depending on the multiplicity of infection (MOI) (Krüger et al. 2015).

Unlike DNA-based methods, the amount of starting material cannot be amplified for proteomic studies. The detection of all protein components of a biological system is still challenging, but with recent developments in sample preparation, LC-MS/MS techniques and analytical instrumentation, and especially in silico approaches to annotate and match tandem mass spectra to peptides and proteins, we are getting closer toward a complete coverage of all protein components of a biological system (Bekker-Jensen et al. 2017; Gao et al. 2021; Müller et al. 2020; de Godoy et al. 2008).

11.2.2 Sample Preparation

Proteomics sample preparation is an essential foundation for a successful proteome analysis. Correct implementation and standardization of protocols have a significant impact on the analysis depth and the data quality. The choice of adequate methods is largely determined by the scientific questions that precede the analysis and further determine the conclusions that can be drawn from the chosen analytical approach. Proteome analysis can be classified based on two main strategies. Top-down analyses are applied for the analysis of intact proteins, mainly to characterize different proteoforms of posttranslationally or chemically modified proteins for translational research or in the pharmaceutical industry (Toby et al. 2019). Bottom-up analysis is the method of choice for the identification and relative quantification of thousands of proteins in complex proteome samples. This most common approach involves an indirect protein identification based on a site-specific proteolytic digestion of proteins to peptides. The physicochemical properties of shorter proteolytic peptides facilitate chemical analysis compared to the challenges associated with the determination of native proteins. In general, proteolytic peptides are chemically more stable, of lower hydrophobicity and therefore easier to solubilize, and the distribution of different charge states remains manageable without the need for sophisticated deconvolution of precursor ions in mass spectrometry. Posttranslational modifications can also be determined by specific enrichment methods and even quantified in a site-specific manner, although this approach does not allow proteoform characterization. Proteolytic digestion of proteins is also associated with an increasing analyte complexity. Where there were initially thousands of proteins, there are ultimately millions of peptides carrying various types of posttranslational modifications. Bottom-up proteomics therefore places special demands on the sample preparation procedure.

Depending on the scientific question and the study design, different strategies are available to study cellular and extracellular proteins. Cellular protein isolation is carried out either from whole cell lysates or from subcellular compartments and organelles including cell membranes, nuclei, mitochondria, and peroxisomes. Most recently, more and more studies try to tackle the challenges of proteomics analysis of single cells by downscaling and automation of the sample preparation procedure. In addition, proteome studies can also focus on extracellular proteins on the surface of the cells (surfome), secreted into the surrounding medium (secretome), or associated with extracellular vesicles.

An integral part of any protein isolation strategy is the interplay of disruption, solubilization, and precipitation techniques. Due to the higher physicochemical stability of fungal cells and especially dormant spores compared to mammalian cells, effective lysis methods for disruption and permeabilization of the intact cell structure are of great importance in order to isolate and purify fungal proteins. Mechanical methods include friction forces by cell grinding with mortar and pestle under liquid nitrogen, shear forces by bead beating homogenization or freeze-thaw cycles as well as cavitation forces by ultrasonic treatment or application of pressure cells (French press). Chemical and biochemical lysis methods involve membrane solubilization techniques and the enzymatic degradation of nucleic acids (nucleases) as well as cell wall components such as polysaccharides and peptidoglycans (e.g., with Lysozyme). When the ordered structure of the cell is disrupted by the lysis process, which includes the permeabilization and destruction of membranes, then proteins that do not come into contact with each other under native conditions due to their different localizations in the cell can theoretically interact with each other. It is, therefore, advisable to add protease inhibitors and establish denaturing conditions during cell lysis. Isolation of proteins from certain subcellular compartments is even more demanding. It requires strategies for fractionated cell lysis, which involves lysis of the fungal cell wall and cell membrane without destroying the membrane of the organelle or compartment of interest such as mitochondria, nuclei, or vesicles, that needs to be isolated and separated from the other cell components. Besides the conventional method of subcellular fractionation after cell lysis, organelle-localizable reactive molecules (ORMs) have been developed that are composed of both an organelle-localizing and a chemically reactive moiety. ORMs can penetrate membranes and spontaneously accumulate in target organelles of living cells, where the protein labeling is facilitated to selectively tag mitochondrial or nuclear proteins, for instance (Zhu et al. 2019).

Different strategies for the solubilization of proteins and membranes exist, such as solubilization with acid-cleavable detergents such as Rapigest, ProteaseMax, or Progenta (Wu et al. 2011; Lo et al. 2013). These acid-labile surfactants are fully compatible with mass spectrometry analysis. However, it has been reported that acid-labile surfactants can introduce artifacts during sample preparation by hydrophobic modification of cysteine residues and are therefore not recommended in studies of lipid modifications of proteins (Ji et al. 2015). It is also known that some cleavable surfactants generate amine products after acid-induced degradation that interfere with quantitative labeling techniques based on amine-reactive NHS-esters such as tandem mass tags (TMT) and isobaric tags for relative and absolute quantitation (iTRAQ) (Zhang 2015). Recently, an Azo-based photocleavable surfactant has been developed, which avoids the disadvantages of most acid-labile surfactants and is also advisable when studying acid-labile post-translational modifications (Brown et al. 2020). More commonly, easily removable detergents (e.g., *n*-octyl-β-D-glucopyranoside [OGP], sodium dodecyl sulfate [SDS], and sodium deoxycholate [SDC]), chaotropic agents (e.g., urea, thiourea, and guanidine) or organic solvents (e.g., methanol, acetonitrile, and 2,2,2-trifluoroethanol) are used to solubilize a broad range of proteins (Wu et al. 2011; Lo et al. 2013; Russell et al. 2001; Schoof et al. 2021). Protein isolation and clean-up is usually achieved by protein precipitation with TCA (Link and LaBaer 2011), acetone (Nickerson and Doucette 2020), TCA/acetone (Damerval et al. 1986), ethanol (van Oss 1989), or methanol-chloroform-water (Wessel and Flügge 1984).

Gel-based methods such as SDS-PAGE (sodium dodecyl sulfate-polyacrylamide gel electrophoresis) and 2D-GE were once dominant to reduce sample complexity and to consider the high dynamic range of proteomic samples. In the process, protein digestion is also performed in-gel followed by a subsequent extraction of the proteolytic peptides (Shevchenko et al. 2006). Although these gel-based methods are still relevant for certain applications, they are becoming less important for proteome analysis in general as gel-free methods are more versatile and less labor intensive. In-solution protein digestion protocols include protein solubilization and denaturation as well as the reduction and alkylation of cysteine thiol residues. Furthermore, several sophisticated one-pot reaction workflows for processing and clean-up of protein samples have been developed over the years to improve and simplify the sample preparation process and to minimize sample loss. The filter-aided sample preparation (FASP) utilizes a molecular weight cut-off (MWCO) membrane and needs several centrifugation steps for the exchange of buffers and reagents (Wiśniewski et al. 2009). Another filter-aided approach employs suspension trapping filters (S-Trap) with a three-dimensional porous material that traps protein suspensions while depleting interfering substances (HaileMariam et al. 2018). The SP3 protocol is based on ethanol-driven solvation capture and paramagnetic hydrophilic beads for the exchange and removal of components used for lysis, protein solubilization, and enzymatic digestion (Hughes et al. 2019). Another method optimized to have minimal sample loss utilizes an in-StageTip (iST, stop-and-go-extraction tip) approach based on solid-phase retention and final elution of proteolytic peptides (Kulak et al. 2014). Besides the aggregation or trapping of proteins on filters, beads, or resins, the sample preparation by easy extraction and digestion (SPEED) takes a completely different and particularly time-saving approach by using acid lysis with neat trifluoroacetic acid (TFA) and subsequent neutralization with Tris buffer (Doellinger et al. 2020).

Without reduction and alkylation (e.g., carbamidomethylation) of thiol residues, cysteine-containing peptides are underrepresented in bottom-up analysis. Hence, the derivatization of cysteine thiols is a well-established approach in proteomics workflows. The reduction can be performed by dithiothreitol (DTT), β-mercaptoethanol, or tris-2(-carboxyethyl)-phosphine (TCEP), for instance. The latter has the advantage of being significantly more stable and not a thiol itself, making it possible to carry out the reduction and alkylation in one step without getting quenched by the alkylating reagent (Müller and Winter 2017; Getz et al. 1999).

Iodoacetamide is the most common reagent to perform carbamidomethylation in order to prevent oxidation (formation of disulfide bridges or sulfoxides) and block the free thiol groups of proteins. It has been shown that two iodoacetamide molecules can produce a 2-acetamidoacetamide covalent adduct to lysine residues ($C_4H_6N_2O_2$; 114.0429 Da) with the exact same atomic composition as the diglycyl group, known from the ubiquitination of the ε-amino group of lysine residues after tryptic digestion (Nielsen et al. 2008). To prevent misleading artifact formation, it is necessary to avoid iodoacetamide at least when studying posttranslational modifications such as the ubiquitination or SUMOylation of proteins. However, iodine-containing reagents such as iodoacetic acid and iodoacetamide are significantly prone to unspecific offsite alkylation reactions (single and double adducts) at other amino acids such as methionine, lysine, aspartic acid, and the peptide/protein N-terminus and consequently their usage was generally questioned and the application of either chloroacetamide or acrylamide has been recommended (Müller and Winter 2017; Nielsen et al. 2008; Suttapitugsakul et al. 2017).

For the proteolytic digestion of proteins, trypsin is the gold standard due to its widespread availability and its highly specific cleavage specificity at the C-terminal side after the basic amino acids arginine and lysine (Olsen et al. 2004). As a consequence, tryptic peptides can be determined by LC-MS/MS as double or multiple-charged precursor ions since at least both the N-terminus and the C-terminal amino acid residue can accept protons. Typical are overnight digestion protocols at 37 °C, but especially with optimization of the buffer, complete proteome digestion can be accomplished in less than 1-hour incubations at 70 °C (Gutierrez et al. 2018). Sequence coverage of certain proteins highly depends on the amino acid composition. Lysine- and arginine-poor protein domains such as transmembrane domains, certain posttranslational modifications, and proline in proximity to the theoretical cleavage site can result in low sequence coverage. On the other hand, highly lysine- and arginine-rich domains can result in short peptide lengths below six amino acids with a lower probability of peptide uniqueness necessary for unambiguous protein assignments of peptide spectrum matches. Several alternative proteases exist to overcome the restrictions of trypsin specificity, in particular, chymotrypsin, LysC, LysN, AspN, GluC, and ArgC (Giansanti et al. 2016), as well as newer proteases such as LysargiNase (Huesgen et al. 2015), and ProAlanase (Samodova et al. 2020). A schematic overview of the fungal proteomics workflow is shown in Fig. 11.1.

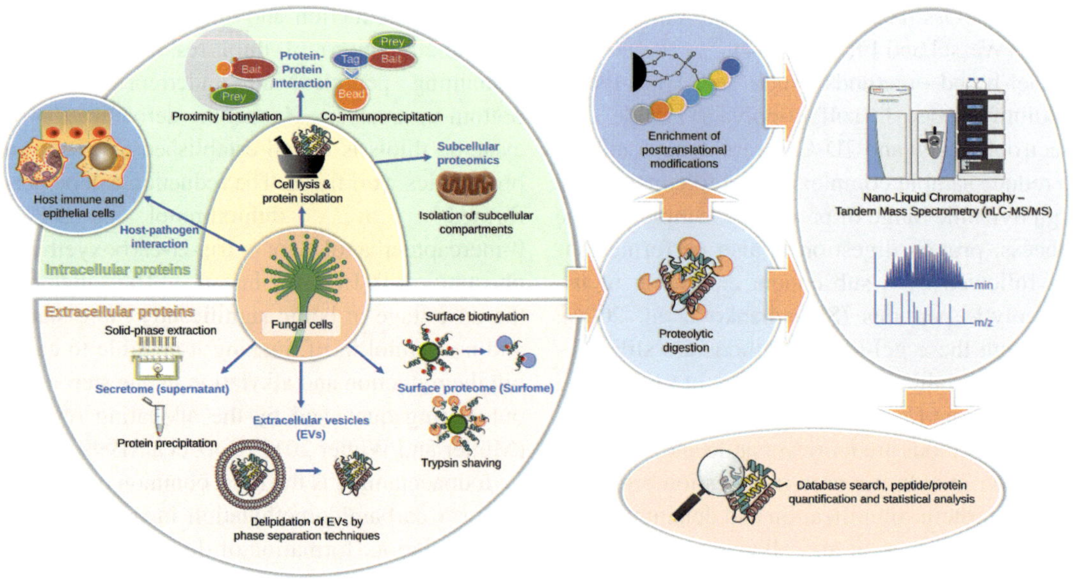

Fig. 11.1 Overview of the proteomic workflow from sample processing to LC-MS/MS analysis for fungal proteomics

11.2.3 Liquid Chromatography-Tandem Mass Spectrometry (LC-MS/MS)

The liquid chromatographic separation of complex mixtures of millions of peptides carrying various modifications is a challenging task. Separation performance, sensitivity, reproducibility, and robustness of the chromatographic system are important factors with a substantial impact on the success of an analysis. In addition, the chromatography has a significant influence on the analysis time and sample throughput. Specific for proteomics applications is the use of capillary columns with the trend toward ever smaller inner diameters and cross-sectional areas of the packed capillaries to reduce the eluent volume flowing through the column into the nanoscale range. Typical inner diameters are 75 µm at a flow rate of 200–400 nl/min. This has a strong concentration effect leading to superior sensitivities in mass spectrometric peptide detection. The downside of LC systems with extremely low nanoflow rates is the lower robustness, quantitative reproducibility, throughput, and user-friendliness as well as the higher maintenance effort in comparison to microflow systems. As a consequence, an increasing number of proteomics scientists prefer higher flow rates in the microscale range, whenever sample material is not the limiting factor and especially when relative and absolute quantification of proteins is performed. Comparison of cap(illary)LC (150 µm i.d., 1 µL/min) and µLC (1 mm i.d., 50 µL/min) setups showed 10 times better sensitivity of the capLC system, which is preferred for limited sample amounts, while the µLC separation achieved higher analysis throughput and separation efficiency and is preferred whenever the protein amount is not limited (Sui et al. 2022). Nowadays, it is possible to identify >7000 proteins by single-shot µLC-MS/MS analysis of 50 µg total protein from mammalian cells based on a typical UHPLC setup with a flowrate of 50 µL/min (Bian et al. 2021b; Bian et al. 2020). High-throughput microflow LC-MS/MS was proven to be extremely robust allowing >50 analyses per day and >14,000 samples could be analyzed on a single column without loss of chromatographic performance (Bian et al. 2021a). To improve the sensitivity of microflow applications, postcolumn solvent addition of DMSO/ACN and DMSO/MeOH mixtures was shown to be very efficient especially for more hydrophilic peptides, which increases the peptide/protein identification rate by up to 30% (Distler et al. 2019).

Nevertheless, nanoflow applications are still the most common choice in proteomics research. In order to increase the separation performance by enhancing the peak capacity (maximum number of resolvable peaks) and the number of theoretical plates (equilibrium stages between mobile and stationary phase as a measure of separation efficiency), capillary columns are packed under high pressure with very small hydrophobic silica particles (C18, C8, C4), typically in the scale of 1.5–2.0 µm particle sizes, which is common for UHPLC applications. Separation performance is further enhanced by the application of relatively long columns, usually 15–75 cm, depending on the selected gradient duration, which enables relatively sharp peaks even when using ultralong gradients over several hours of separation (Köcher et al. 2012; Anagnostopoulos et al. 2017). For a long period of time, there were few developments in the field of chromatographic separation of peptides and proteins. However, recent developments show that there is still a need and a potential for optimization:

The pillar array column (PAC) technology is the first chromatography based on micro-chip columns (PharmaFluidics, Thermo Fisher Scientific). The backbone of the separation bed is formed by a perfectly ordered array of narrow pillars. It eliminates heterogeneous flow paths of conventional packed columns and minimizes dispersion effects, which allows sharper peaks and increased sensitivities (Stejskal et al. 2021; Stadlmann et al. 2019). Micropillar chromatography minimizes pressure fluctuations and increases the robustness and reproducibility. The relatively low backpressure allows separation channels of exceptional length (up to 2 m) and separation power (Stejskal et al. 2021; Stadlmann et al. 2019).

Flow rates in the nanoscale range usually require peptide loading on a trapping column as flushing the sample loop with several microliters of volume results in massively delayed sample transfer into the chromatographic system. Peptide trapping allows sample loading at higher flow rates (usually 5–30 μL/min) and online desalting of the sample. However, it is a common approach to clean up the peptide sample prior to the LC-MS/MS analysis by solid-phase extraction (SPE) with centrifugal or pipette tips packed with C18-silica, especially the STAGE (stop and go extraction) tip approach is widely used (Rappsilber et al. 2007). It was therefore obvious to directly combine offline SPE sample loading onto packed tips with the online chromatographic separation. In the EvoSep system, disposable SPE tips, which are loaded and desalted offline are placed in an autosampler. From here, each individual tip can be picked up and placed in the injection port to elute the peptides onto the analytical column (Bache et al. 2018). The particular chromatographic system leads to significantly less sample loss, carryover, variation, and analysis time. The EvoSep chromatographic system connected to a state-of-the-art mass spectrometry instrument enables quantification of >5000 proteins within 24 min gradients (Bekker-Jensen et al. 2020).

Until the year 2000, tandem MS systems were able to acquire a maximum of only one MS/MS scan per second (Link et al. 1999; Shishkova et al. 2016), which theoretically allowed up to 3600 peptide identifications per hour excluding lag and equilibration times. To increase the analysis depth, excessive sample pre-fractionation was required so that each fraction could be measured separately to overcome the mass spectrometry limitations in scan speed and cycle time. However, this was accompanied by significantly longer analysis times. Fractionation on the protein level could be achieved with conventional gel-based methods such as SDS-PAGE or 2D-GE. Separated protein bands or spots were excised, destained, and in-gel digested followed by the extraction of the proteolytic peptides. As the high-throughput MALDI-TOF analysis is usually more efficient for the mass spectrometric readout of 2D-GE spots, gel-free multidimensional separation techniques were developed for LC-MS/MS applications. Similar to the 2-dimensional separation according to charge (isoelectric focusing) and size (SDS-PAGE) in gel-based applications, proteins were fractionated by size exclusion and ion exchange chromatography in gel-free applications often referred to as MudPIT (Multi-dimensional Protein Identification Technology) approach.

The tremendous technological progress in the last 20 years of mass spectrometry makes multidimensional protein offline fractionation increasingly obsolete. It is now possible to identify thousands of proteins in a single-shot LC-MS/MS run (Hebert et al. 2018; Meier et al. 2018; Kawashima et al. 2022). An important step forward was the invention of the Orbitrap technology by Alexander Makarov (Hu et al. 2005; Makarov 2019). The Orbitrap features a decent space charge capacity, a broad dynamic range and superior mass resolution, and mass accuracy. Nowadays, the patented proprietary orbitrap technology (Thermo Fisher) is the leading mass analyzer in mass spectrometry-based proteomics as it usually outperforms quadrupole time-of-flight (QTOF) instruments. The latest development of the Asymmetric Track Lossless (Astral) high-resolution mass analyzer used in conjunction with the Orbitrap technology can accomplish an ultrafast MS2 scan speed of up to 200 Hz with exceptional proteome coverage and sample throughput (Heil et al. 2023). Ion and mass analyzers used in mass spectrometry-based proteomics are illustrated in Fig. 11.2. To close the technological gap, competing vendors successfully coupled the ion mobility spectrometry (IMS) with their QTOF instruments to increase speed, selectivity, and sensitivity by inclusion of another dimension of peptide separation (Meier et al. 2015; Iacob et al. 2008; Shvartsburg and Smith 2008). While mass spectrometry is basically the movement of ions in an electric field under vacuum based on their *m/z* ratio as a function of their moment of inertia, ion mobility is based on the movement of ions in an electric field

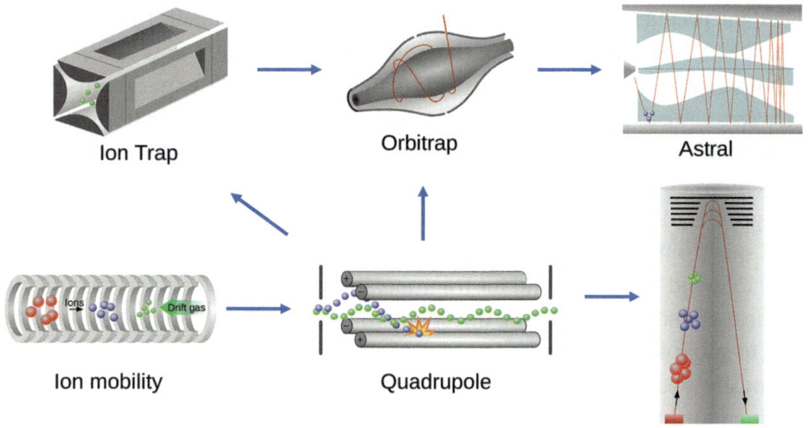

Fig. 11.2 Overview of the different mass analyzers available for mass spectrometry-based proteomics. The quadrupole mass analyzer is often used ass mass filter in combination with mass analyzers, e.g., Orbitrap or TOF

under drift gas flow, based on their three-dimensional conformation as a function of the collision cross section (CCS). CSS is the measure of the effective area of an ion for interaction with neutral gas particles. The most prominent IMS techniques coupled to QTOF instruments are the drift tube ion mobility spectrometry (DTIMS, Agilent) (Ma et al. 2017), traveling wave ion mobility spectrometry (TWIMS, Waters) (Shvartsburg and Smith 2008), and trapped ion mobility spectrometry (TIMS, Bruker) (Meier et al. 2021), which all together allow sensitive determination of the CSS value of each analyte in order to distinguish isobaric ions based on their ion mobility. Meanwhile, there are also ion mobility interfaces available used to filter, fractionate, or separate ions before entering the mass spectrometer. IMS interfaces, namely differential mobility separation (DMS, Sciex) (Walker et al. 2018) and high-field asymmetric waveform ion mobility spectrometry (FAIMS, Thermo Fisher) (Bonneil et al. 2015) can be installed as an add-on between the ion source and the inlet of the mass spectrometer.

Different technologies exist to generate fragment ions from a selected precursor ion. The most common approach is the collision-induced dissociation (CID), in which an electric field accelerates the precursor ions into an inert neutral gas (N_2, He, Ar). Covalent bonds are broken due to the high kinetic energy and repeated collisions with the gas molecules. An orbitrap-specific CID technique is the higher energy C-trap dissociation (HCD), in which the precursor ions pass through a C-trap before and after fragmentation in a multipole collision cell (Olsen et al. 2007). In comparison to the CID resonant-excitation, which is based on a dynamic radio frequency (RF) dependent on the *m/z* value of the precursor ion, HCD fragmentation can be performed at a fixed RF amplitude. It is therefore favorable for reporter ion quantification since it does not suffer from low mass cutoff. CID and HCD mainly generate b- and y-ions resulting from a cleavage of the peptide bonds starting from the N- and C-terminus of the peptide, respectively. An alternative fragmentation method favorable for top-down proteomics or the bottom-up analysis of posttranslational modifications (PTMs) is the electron transfer dissociation (ETD). The method mainly generates c- and z-ions resulting from a cleavage of the peptide backbone between the amine and the alpha C-atom. Polycyclic aromatic hydrocarbons such as fluoranthene are used to generate reactive radical anions. In contact with the peptide cations unstable radical cations are formed through electron transfer leading to the peptide dissociation while PTMs remain intact (Mao et al. 2011).

The most prominent type of tandem mass spectrometric acquisition in untargeted proteomics is the so-called data-dependent acquisition (DDA) mode. Each scan cycle consists of a full scan event for MS^1 ion screening and sequential data-dependent MS^2 scans of collision-induced fragment ions derived from the most abundant precursor ions (topN). The higher the scan speed of the mass analyzer the more precursor ions can be theoretically selected by the mass filter (e.g., quadrupole) for isolation and further fragmentation in a collision cell. The number of selected precursor ions also depends on the elution profile of the peptide peaks that enter the mass spectrometer at a certain retention time. In order to avoid overly redundant fragmentation of the most abundant precursors, different strategies exist to provoke MS2 fragmentation of peptides of lower abundance, such as deep sample fractionation or separation techniques and triggering data-dependent MS2 scans primarily at the apex of the peak. However, the most effective approach is the so-called dynamic exclusion, in which the *m/z* value of a certain precursor ion that already underwent MS2 fragmentation is blacklisted for a defined period of time, depending on the elution profile of the chromatographic separation. This approach increases the dynamic range of the analysis and facilitates a broader proteome coverage. Even though for the analysis of affinity-purified samples that are used to study protein-protein interactions based on the spectral counting approach, it was shown that high-confident interactors can be identified with better sensitivity and more representative spectral counts by eliminating the dynamic exclusion step (Zhang et al. 2022).

In general, standard DDA methods leave much to be desired in terms of the dynamic range and the identification of low-abundance peptides including those posttranslationally modified. As an alternative approach, data-independent acquisition (DIA) has been developed (Venable et al. 2004), which allows a less biased peptide identification and does not depend so much on the abundance of a peptide or protein. DIA is most often based on overlapping mass-to-charge ranges that are isolated and fragmented sequentially either of 10–25 amu or in a dynamic fashion between 5 and 500 amu based on the precursor ion density. As a consequence, the resulting MS^2 spectra are of chimeric nature based on a multitude of precursor ions. In order to deconvolute the multiplexed spectra and to connect related fragment ions to a specific precursor ion, the use of spectral libraries collected from DDA experiments was initially mandatory. Dependence on DDA experiments impairs the analytical throughput and restricts the identification of peptides on those that were previously identified by DDA. Various conditions, extensive prefractionation, and repeated measurements are necessary to gain comprehensive spectral libraries. To circumvent the limiting factors, deep learning models have been developed for in silico prediction of spectral libraries (Lou et al. 2021; Sinitcyn et al. 2021; Tsou et al. 2015; Tiwary et al. 2019; Yang et al. 2020; Searle et al. 2020). Such direct DIA approaches can either be used alone or supplemented with experimental libraries. This is especially important for proteome analysis of non-canonical organisms, where the availability of experimental DDA data is limited and for smaller projects where the effort to create an extensive experimental spectral library would not be justified. In the SWATH MS approach, DIA was combined with a targeted data extraction strategy for fragment ion-based peptide quantification based on spectral library information (Gillet et al. 2012).

Spectral library predictions of fragment ion intensities and retention times by the machine learning framework INFERYS (Zolg et al. 2021) have been successfully applied for the database search of DDA experiments in order to boost peptide identifications. The combination of INFERYS 2.0 with the recently developed CHIMERYS algorithm (MSAID GmbH, Munich, Germany) allows the deconvolution of rather complex DDA-based chimera spectra by comparing experimental spectra with MS/MS spectra predicted by INFERYS. Chimera spectra are heterogeneous MS/MS spectra derived from the fragmentation of several co-isolated precursor ions with nearly identical *m/z* values and retention times (Dorfer et al. 2018). The artificial intelligence-powered cloud-native search algorithm allows the identification of several peptides per single MS/MS scan. This has to be mentioned in the context of mass spectrometry instrument settings, as the full potential of the CHIMERYS algorithm is only reached by increasing the isolation width for precursor ions. Depending on the performance and selectivity of the quadrupole, which is still the most efficient mass analyzer for precursor ion isolation, standard search algorithms require narrow isolation widths (m/z 0.3–2.0 amu) to prevent the ambiguity of chimera spectra. This is even more important for MS2 reporter ion quantification (TMT, iTRAQ) due to the ratio distortion effect of co-isolated precursors. However, the CHIMERYS algorithm leads to a protein identification boost of 30–40% (preliminary data presented by Thermo Fisher Scientific at the Proteomic Forum 2022 in Leipzig, Germany) when the occurrence of chimera spectra is fostered by wide window acquisitions based on broader isolation widths such as m/z 4.0 amu or even beyond. Such an improvement of protein identifications solely through improved software is comparable to the progress that can be expected from at least two new mass spectrometry instrument generations.

11.2.4 Quantification Techniques

The vast majority of proteome studies aim to explore modulations in biosynthesis, turnover, and posttranslational modifications of proteins by comparison of growth and stress conditions or any treatment of interest against standardized control conditions. In microbiology research, the function and underlying molecular mechanisms of proteins or even site-specific protein modifications, can be further elucidated by comparison of genetically modified strains with wild-type strains. Various methods have been established to relatively quantify differences in protein abundances or in the frequency of site-specific posttranslational modifications. In general, proteomics methods can be divided into label-free quantification (LFQ) and stable isotope labeling techniques. LFQ approaches are either based on spectral counting, which quantifies the protein abundance based on the number of peptide spectrum matches assigned to a protein by MS^2 identification or on the sum of MS^1 precursor ion signal intensities or peak areas (area under the curve, AUC) (Neilson et al. 2011).

Stable isotope labeling techniques can be distinguished in both metabolic and chemical labeling methods. Metabolic labeling has the significant advantage that it allows multiplexing of different samples before sample preparation. The label is incorporated during cultivation at the organism level. As a result, technical variation can be kept to an absolute minimum and is only limited to the cultivation. Label incorporation requires predominant quantitative metabolic utilization of the labeled source by the organism of interest. SILAC (Stable isotope-labeling by amino acids in cell culture) is the most common metabolic labeling approach. Traditionally, arginine labeled with six ^{13}C atoms is used as heavy amino acid (Arg-6) that can be easily compared against the light version (Arg-0) by the defined mass shift of 6 Da on the MS1 precursor ion level (Ong et al. 2002). It should be considered that arginine undergoes a widespread metabolic conversion to proline depending on growth conditions such as the buffer concentration. This can have undesired side effects such as double-labeled peptides and thus additional unwanted peaks (Ong and Mann 2007). In order to quantitatively compare all tryptic peptides in terms of their abundance, it is preferred to label them with both arginine and lysine. Besides ^{13}C, labeling with ^{15}N is also possible, while the use of deuterated amino acids is not recommended due to the possible $^{2}H/^{1}H$ exchange and the fact that deuterated peptides show a retention time shift and elute earlier than the light version of the same peptide (Zhang et al. 2001). This makes precise quantification difficult. Labeling of other amino acids such as tyrosine has been used to study tyrosine kinase signaling and posttranslational phosphorylation (Ibarrola et al. 2004). Several variations of the SILAC method have been established over the years. For the pulsed SILAC (pSILAC) method, labeled amino acids are supplemented temporarily to monitor the de novo protein synthesis and to identify translationally regulated proteins (Schwanhäusser et al. 2009). In general, SILAC requires almost complete incorporation of the labeled amino acids for precise quantification. Therefore, auxotrophy for the labeled amino acid in the studied organism is usually a prerequisite to ensure effective metabolic labeling and precise quantification. The incorporation yield of the heavy amino acid into newly synthesized proteins and the overall turnover needs to be evaluated for any organism that should be examined with SILAC. In order to eliminate the need for almost complete incorporation of labeled amino acids and to increase the level of multiplexity, which limits the applicability of the conventional SILAC approach, another variation, called NeuCode SILAC has been established (Merrill et al. 2014). The NeuCode (neutron encoding) technique only utilizes heavy amino acids with a different stable isotope coding. This eliminates the need for a complete incorporation and allows high orders of multiplicity, when high-resolution mass spectrometry is applied (Merrill et al. 2014). Alternatively, it is also possible to perform metabolic labeling with ^{13}C- or ^{15}N-containing salts or other metabolized additives other than amino acids. Here, it is also not essential to achieve complete labeling. However, the delta mass between labeled and unlabeled peptides is initially unknown in the MS analysis and must be identified using suitable software (Gouw et al. 2010).

Besides the in vivo metabolic labeling, chemical labeling techniques are performed in vitro and can be differentiated into quantification techniques for precursor ions (MS1 level) and reporter ions (MS2 level). Label-based precursor ion quantification techniques include techniques that have been applied for protein-level labeling such as ICPL (isotope-coded protein labeling) (Kellermann and Lottspeich 2012) and ICAT (isotope-coded affinity tags) (Gygi et al. 1999). ICPL is an amino-reactive labeling approach used to tag stable isotope derivatives at the free amino groups of intact proteins (Kellermann and Lottspeich 2012). In contrast, ICAT is a cysteine thiol-reactive labeling approach containing a stable isotope-coded group, a linker group, and a biotin for affinity purification of biotinylated peptides after proteolytic cleavage (Gygi et al. 1999). Drawbacks of ICAT are the bulky affinity group that increases the complexity of MS/MS spectra interpretation and the limitation to Cys-containing proteins. On the other hand, the cysteine thiol reactivity predestines the method for use in quantitative redox proteomics analysis (Lindemann and Leichert 2012).

On the peptide level, dimethyl labeling can be used to quantify proteins. With this universal and inexpensive labeling technique both the N-terminus and the ε-amino group of Lys residues are derivatized via reductive amination with formaldehyde and cyanoborohydride (Hsu et al. 2003). Multiplexing is facilitated by labeling with light (+28 Da), intermediate (+32 Da), and heavy versions (+36 Da) of the dimethyl derivatization based on the incorporation of ^{2}H (D) and ^{13}C isotopes (Kovanich et al. 2012). Further non-isobaric labeling techniques for labeling of primary amines include the mTRAQ triplex reagents that have been initially designed for MRM quantification of peptides and proteins (Kang et al. 2010). Analogously, mTMT has been developed for duplex precursor-based quantification with a mass difference of 11 Da between light and heavy-labeled peptides (Paulo and Gygi 2019). Both mTRAQ and mTMT are modified versions of the isobaric mass tags used for highly multiplexed reporter ion quantification on the MS2 level, known as iTRAQ (up to 8-plex) and TMT (up to 16-plex) (Ross et al. 2004; Thompson et al. 2003). The isobaric reagents consist of an amino-reactive *N*-hydroxysuccinimidyl (NHS) ester group, a mass normalizer and a mass reporter group, which have a different distribution of stable isotopes. Labeled and multiplexed peptides cannot be distinguished on the precursor ion level (MS1). The MS1 ion signal is the sum of the peptide proportion from all samples, whereby an increase in sensitivity is achieved. Only after MS2 fragmentation, observation of the peak intensities of the set of specific reporter ions allow peptide quantification for each uniquely labeled sample. Reporter ion quantification minimizes the technical variation included by the analytical system and increases the sensitivity by multiplexing. However, an important drawback of the iTRAQ or TMT quantification is the ratio compression phenomenon induced by isotopic impurities and especially the quadrupole-based co-isolation of nearly isobaric peptides resulting in chimera spectra that may distort the reporter ion ratios (Li et al. 2014). To overcome the ratio distortion issue by co-isolated precursor ions, reduction of the sample complexity by peptide pre-fractionation (Sandberg et al. 2014; Lau et al. 2011) and MS3 detection of reporter ions have been established as the reporter ion fragmentation requires higher collision energies compared to the energy necessary to generate *b*- and *y*-ion series by peptide bond breakage (McAlister et al. 2014; Cao et al. 2015).

11.2.5 Single Cell Proteomics (SCP)

Most single cell studies are focused on gene expression in specific cells. However, changes in protein abundances and their posttranslational modifications are much closer to the cellular phenotype than those of genomic or even transcriptomic origin as proteins are the main driver of cellular functionality. Single cell proteomics provides new insights into the complexity of cellular systems, its heterogenicity, and signaling networks. SCP techniques are also very well suited to detect rare cells and to analyze limited sample material. From the isolation of the single cell over the cell lysis, the proteolytic digestion, optional labeling strategies, the chromatography and mass spectrometry acquisition to the bioinformatic analysis, every step in SCP needs to be optimized in order to get most out of the very restricted sample material. Single mammalian cells contain 200–400 pg of protein, much less than that used in standard proteomics analysis (e.g., 100 μg). The miniaturized sample preparation requires a higher degree of automation to increase reproducibility, robustness, throughput, and sensitivity.

In order to establish a single cell proteomics platform, well-characterized cell lines such as HeLa or K562 can be used for a proof-of-concept demonstration of the applied methodology. However, biological samples will require robust methods to isolate cells of interest (Couvillion et al. 2019). Single cells can be isolated by manual or automated picking under a microscope. Especially the manual approach is labor-intensive and allows only low throughput. For higher throughput, selective separation of cells can be implemented by affinity-based methods, flow cytometry, and fluorescence-activated cell sorting (FACS) (Spitzer and Nolan 2016). Affinity-based techniques are inherently antibody centered. This can limit its applicability as foreknowledge about the target is needed (Couvillion et al. 2019). Other methods to harvest specific cells of interest from tissues include laser capture microdissection (Espina et al. 2006) and image-based single cell isolation. The latter is a very effective method, which has been utilized by a commercial SCP solution (CellenONE proteoCHIP, Cellenion, Lyon, France) that can handle the complete sample preparation workflow from the isolation of the cell to the digested sample measured by LC-MS/MS analysis. Cell suspensions can be obtained by careful cell scraping or trypsin treatment in order to disaggregate the cells and disrupt the cell-to-cell contact. In the cellenONE device, 1–50 μL cell suspensions can be used as starting material to produce droplets by a glass capillary. The capillary tip consists of an ejection zone and a sedimentation zone and is surrounded by a piezoelectric ceramic, which induces acoustic waves, when voltage is applied. This acoustic-based precision dispensing generates a droplet of a defined size depending on the voltage, pulse, and nozzle size and usually has a volume of 100–500 pL per droplet. Whenever a single cell occurs in the ejection zone of the tip, the cell is dispensed into the target. In order to image and count the cells, bright field transmission or fluorescence can be utilized. The throughput currently allows isolation of up to 40 single cells per minute.

The isolated cells are then treated with a reagent mixture containing all necessary components to lyse the single cell and to proteolytically digest its proteins. Sample preparation for SCP requires different features: Miniaturization is needed to handle reduced reaction volumes. Method simplification makes it possible to dispense any vial exchange in order to minimize sample loss. A microfluidic nanodroplet approach has been developed for miniaturized sample preparation in nanoPOTS (nanodroplet

Processing in One-pot for Trace Sample) (Zhu et al. 2018a, b). This allows 200 nl digests instead of 100 μL reaction volumes commonly used in standard proteomics applications. Automation reduces the susceptibility to errors that manual sample handling is prone to and allows simultaneous processing and high throughput. Evaporation of small sample volumes during preparation and storage is an important issue. Different strategies are used to overcome this problem: Addition of 5% DMSO has the welcome side effect of also keeping hydrophobic peptides solubilized. Furthermore, active sample hydration with water or technical maintenance of high humidity (~85%) in single cell devices can be utilized to keep proteins and peptides in solution. During storage in the autosampler, multiwell plates can be covered by special foils or liquids to prevent sample evaporation.

Peptide quantification by TMT multiplexing can significantly increase sensitivity and thus overall peptide and protein identification rates in SCP applications as all quantification channels are finally unified into one multiplexed sample. A higher degree of multiplexing by the development of more and more different isobaric tandem mass tags with different reporter ions allows the use of extra channels for background evaluation (empty quan channel) or simply to amplify the MS2 signal (booster or carrier channel, not for quantification) by labeling of a sample pool providing more material (Schoof et al. 2021). The method was initially termed Single Cell ProtEomics by Mass Spectrometry (SCoPE-MS). SCoPE-MS augments each isobaric labeled single cell set with a carrier sample of about 200 cells that provide significantly more ions for peptide identification and reduces the effect of surface adhesion issues of low concentrated samples (Budnik et al. 2018; Specht et al. 2021). As Boosting to Amplify Signal with Isobaric Labeling (BASIL) strategy, it has been also applied for phosphoproteomics analysis with limited sample material (Yi et al. 2019; Chua et al. 2020). Although the signal boost in multiplexed SCP by the carrier channel provides considerable advantages over label-free SCP strategies, it also suffers from ratio distortion and signal spillover effects from isotopic impurities of the isobaric labels (Furtwängler et al. 2022). Long injection times are needed for single cell channels in order to allow robust reporter ion quantification. To overcome these restrictions, newly developed real-time searches (RTS) implemented for Tribrid mass spectrometry instruments based on a quadrupole mass analyzer, a linear ion trap (LIT) and an orbitrap (OT) have been applied to optimize and increase both proteome coverage and quantitative accuracy. The so-called RETICLE (RTS enhanced quantification of single cell spectra) method utilizes fast RTS linear ion trap scans to preselect MS1 precursor ions for MS2 reporter ion quantification in the orbitrap (Furtwängler et al. 2022).

Further improvements in single cell proteome coverage can be achieved by optimization of the chromatographic separation such as the usage of micropillar columns or the application of capillary columns with smaller column diameters (10–30 μm instead of 75 μm), which allow extremely low flowrates (down to 10–50 nl/min instead of 250–300 nl/min) with a massive gain in sensitivity by the strong concentration effect and the improved ionization efficiency. It was shown that columns with 30 μm inner diameter (i.d.) increase signal intensity by more than threefold relative to standard proteomics columns with 75 μm i.d., which allows an increase in peptide identifications of more than 30% (Zhu et al. 2018c). Further miniaturization leads to significant performance gains, especially when the sample material is limited. By using a 200-pg HeLa digest, peptide separation with 20 μm i.d. columns was increased compared to 30 μm i.d. columns by 75% and 32% at the peptide and protein level, respectively (Cong et al. 2020). To reduce dead volumes and to minimize dispersion effects the analytical columns are usually manufactured with an integrated emitter for direct nanospray ionization.

In general, applications with limited sample material and low protein concentrations benefit from developments in single cell proteomics techniques.

11.2.6 Database Search and Data Curation

Before the availability of genomic data and database searches, Edman degradation was used to characterize the sequence of peptides (Edman 1950). With the improvement of mass spectrometric technologies and the possibility to perform de novo peptide sequencing by assignment of peptide fragment ions from tandem mass spectra, several software tools have been developed to identify unknown or novel peptides (Dancík et al. 1999; Taylor and Johnson 2001; Ma et al. 2003; Frank and Pevzner 2005; Chi et al. 2013). Nevertheless, proteomics is basically an analytical tool of the post-genomic era. The better the quality of the sequenced genome and its annotation, including alternative splicing variants, the better the proteome analysis, which is based on it. The data evaluation also benefits from a comprehensive functional annotation of the proteins in the database in order to extract biological meaning from the extensive data by functional enrichment analysis. Besides the most common search algorithms, Sequest (Eng et al. 1994) and Mascot (Perkins et al. 1999), numerous other protein database search engines have been developed over the years to match tandem mass spectrometry data with protein databases, in particular Protein Prospector (Clauser et al. 1999), X!Tandem (Craig and Beavis 2004), OMSSA (Geer et al. 2004), MS-GF+ (Kim et al. 2010), Andromeda (Cox et al. 2011), Byonic (Bern et al. 2012), Comet (Eng et al. 2013), MS Amanda (Dorfer et al. 2014), and MS Fragger (Kong et al. 2017).

Further techniques to improve peptide identifications involve the match-between-run approach implemented in the freely available proteomics software MaxQuant (Tyanova et al. 2016). The matching-between-runs feature allows peptides not identified in one sample (e.g., due to insufficient MS/MS spectra information) to be identified by matching peptides in other samples that contain the same or similar sequence. Successfully identified spectra from other samples are used as a library in order to match peptide precursors within tight mass-to-charge and retention time windows. The feature can also reduce the occurrence of false positives by increasing the likelihood that actually identified peptides are consistent across multiple samples by comparing multiple data (Tyanova et al. 2016).

The field of proteomics has grown at a considerable rate and a diverse set of new technologies has been applied to investigate the change of protein expression on a global scale. Attention has to be paid to a proper experimental design of proteomics studies, which allows producing statistically significant data. The required level of stringency in the generation and analysis of proteomic data should not be underestimated (Wilkins et al. 2006). The impact of technical and biological variations on the error propagation and statistical significance of the results, the right criteria for the correct identification and quantification of proteins, and the correct use of bioinformatics tools has to be considered (Biron et al. 2006; Urfer et al. 2006; Krüger et al. 2013). As a consequence, proteomic journals have initiated the development of a set of minimum guidelines for the publication of proteomics research (Wilkins et al. 2006). Furthermore, a proteomics standards initiative (PSI) was founded within the Human Proteome Organisation (HUPO) in 2002. The PSI work group aims at defining community standards for data representation in proteomics to facilitate systematic data capture, comparison, exchange, and verification (Deutsch et al. 2017). Following the "minimum information about a microarray experiment" (MIAME) guidelines (Brazma et al. 2001), the corresponding "minimum information about a proteomics experiment" (MIAPE) requirements have been established (Taylor et al. 2006). Besides the development of MIAPE reporting requirements, the PSI is working on platform-independent data exchange formats, guidelines for quantitative mass spectrometry experiments (Martínez-Bartolomé et al. 2013), controlled vocabularies (Mayer et al. 2013), and the promotion of publicly available proteomics databases (Hermjakob 2006; Deutsch et al. 2015). Guidelines have been also established for reporting ion mobility mass spectrometry measurements (Gabelica et al. 2019). The ProteomeXchange consortium provides globally coordinated data submission and dissemination (Vizcaíno et al. 2014) for proteomics data repositories such as PRIDE (Perez-Riverol et al. 2022), PeptideAtlas (Desiere et al. 2006), MassIVE (Wang et al. 2018), jPOST (Okuda et al. 2017), Panorama (Sharma et al. 2018), and iProX (Chen et al. 2022).

All the recent advancements in the proteomics field, such as MS instrumentation, improvements in search algorithms, and protocols, has had also a great impact on the study of the *A. fumigatus* proteome, which will be discussed in the following sections.

11.3 Proteomics Studies of the Human Pathogenic Fungus *A. fumigatus*

11.3.1 Secondary Metabolism

The saprophytic fungus *A. fumigatus* is characterized by its versatile catabolic metabolism and its ability to produce a diverse set of secondary metabolites (Tekaia and Latge 2005; Bignell et al. 2016). These secondary metabolites are often formed at certain developmental stages and are associated with the virulence of this fungus (Steenwyk et al. 2020; Moon et al. 2022). The redox-active nonribosomal peptide gliotoxin produced by *A. fumigatus* has been well characterized (Scharf et al. 2016) and the connection between gliotoxin production and development was proven by a comparative proteomic study on *A. fumigatus* mutants lacking the central regulator of asexual sporulation, AbaA and BrlA. Both strains exhibited reduced levels of enzymes involved in gliotoxin biosynthesis including GliT, which is discussed below (Shin et al. 2015).

Several studies addressed the question of how this cytotoxic metabolite impacts *A. fumigatus* physiology itself. Owens et al. (2014) compared the H_2O_2 and gliotoxin-induced stress response in *A. fumigatus*. Both compounds are expected to lead to ROS formation. The proteomic data revealed that gliotoxin led to a diminution of de novo purine biosynthesis, while purine salvage pathways were induced. The level of the gliotoxin-neutralizing oxidoreductase GliT increased only upon gliotoxin exposure, which excluded an ROS-induced induction of this enzyme.

Besides GliT, *A. fumigatus* avoids self-poisoning by detoxifying gliotoxin with the help of the S-methyltransferase GtmA. A label-free quantitative proteomic analysis of a mutant strain lacking gtmA (Δ*gtmA*) in comparison to the wild type revealed a significant remodeling of secondary metabolism in the knock-out strain: Gliotoxin production increased, while many secondary metabolite biosynthetic enzymes in the subtelomeric region of chromosome VIII decreased in abundance. The authors concluded that GtmA facilitates the biosynthesis of other secondary metabolites encoded by biosynthetic gene clusters (Doyle et al. 2018), whereas the exact mechanism of this effect remains unclear, but a redox-regulatory mechanism could be involved. A similar picture was found in an *A. fumigatus* mutant lacking a biosynthetic gene involved in the biosynthesis of the histidine derivative ergothioneine, which exhibits antioxidant properties. A comparative proteome analysis of the Δ*egtA* mutant with the wild type revealed a substantial remodeling of the mutant's proteome (Sheridan et al. 2016) including an attenuated gliotoxin production, which was observed at both the proteome and the metabolome levels, and an increase in glutathione biosynthesis. This suggests that these two sulfur-containing, redox-active metabolites are interlinked.

The reduced form of gliotoxin with two reactive thiol groups, designated as dithiol gliotoxin, chelates Zn^{2+}, and inhibits Zn^{2+}-dependent metalloenzymes. Consequently, externally added zinc can relieve the autotoxicity of dithiol gliotoxin in *A. fumigatus*. A comparative study on the proteomic response of a mutant lacking the aforementioned gliotoxin detoxifying enzyme GliT to gliotoxin versus Zn^{2+}/gliotoxin exposure confirmed these activities. The presence of gliotoxin alone led to decreased abundance of zinc-ion binding proteins, but increased abundance of zinc homeostasis regulating proteins such as ZrfB and AspF2 (Saleh et al. 2018). These data demonstrate that GliT and GtmA protects *A. fumigatus* from the zinc-chelating properties of dithiol gliotoxin.

In particular external stimuli affect secondary metabolite production in fungi. A proteomic study based on label-free quantitative proteomics using an orbitrap mass spectrometer demonstrated on the proteome and metabolome level that gliotoxin production is stimulated by exposure to bacterial culture filtrates of the human pathogenic bacterium *Klebsiella pneumoniae* (Curtis et al. 2023). Such an interplay between these two microorganisms in the patient's lung could exacerbate infection due to increased gliotoxin production. In a similar way, Margalit et al. (2022) investigated the effect of the *Pseudomonas aeruginosa* secretome on the *A. fumigatus* proteome. In contrast, proteins involved in gliotoxin biosynthesis were decreased in *A. fumigatus* cultures exposed to *P. aeruginosa* culture supernatants, while the level of enzymes involved in the biosynthesis of other secondary metabolites increased in abundance, such as pseurotin A, verruculogen and fumitremorgin. It can be speculated that the proteomic response of *A. fumigatus* during interaction with other pathogenic bacteria is species dependent. However, more data are needed for firm conclusions.

11.3.2 Proteomic Changes During Development

The proteome of *A. fumigatus* alters significantly during asexual development, which starts with a dormant conidium that produces finally a mass of intertwined hyphae. The first attempts to get an overview of these changes were still based on 2D-gel electrophoresis. The first proteome reference map was published in 2009 and included the mycelial and mitochondrial fraction of this fungus (Vodisch et al. 2009), while a year later a conidial proteome was made available (Teutschbein et al. 2010). Out of 449 different proteins, 57 were over-represented in dormant conidia in comparison to mycelium. These included enzymes involved in ROS detoxification, dihydroxynaphthalene (DHN)-melanin biosynthesis, and conidial rodlet layer formation, all of which confer resistance to environmental stressors. Enzymes and proteins necessary for a rapid reactivation of protein biosynthesis upon germination, e.g., ribosomal proteins, elongation factors, aminoacyl-tRNA synthetases, and amino acid biosynthesis were quite abundant in dormant conidia. In addition, the enzymes pyruvate decarboxylase and alcohol dehydrogenase, both involved in ethanol fermentation, were detectable in dormant conidia and led the authors to assume that alcoholic fermentation occurs during dormancy or early germination. Later, quantitative shotgun proteomic approaches provided deeper insights into the proteome dynamics during *A. fumigatus* development. The most thorough descriptions of the conidial proteome of *A. fumigatus* to date detected more than 5000 proteins, few of which were also present in 48-h mycelium (Kelani et al. 2023). An LC-MS/MS proteomic approach based on isobaric iTRAQ-tagging identified 461 different proteins during early development over a period of 16 h (Cagas et al. 2011b). The data confirmed that conidial germination is initiated by building the translation machinery and generating ATP via respiration. A similar study postulated developmental stage candidate biomarkers and highlighted the finding that the conidial proteome is dominated by small, lineage-specific proteins with so far uncharacterized function (Suh et al. 2012). A morpho-type specific, CipC-like mycelial protein was also described by Bauer et al. (2010) in a 2D-gel-based study. Proteomic information about the stage-specific abundance of *A. fumigatus* proteins could serve as a basis for the discovery of biomarkers of active infection or as possible therapeutic targets.

11.3.3 Signaling

A large array of upstream regulators affects developmental processes and secondary metabolite production in *Aspergillus* species. Signaling occurs at the protein level and proteomics represents a valuable tool for gaining deeper insights into the detailed mechanisms.

One of the earliest proteome studies on *A. fumigatus* signaling characterized its cyclic AMP (cAMP) transduction pathway. The protein kinase A (PkA), which is also described as cAMP-dependent protein kinase, is a central regulator of fungal metabolism, virulence, and morphology (Zhao et al. 2006). Here, the proteome of a pkaC1-overexpressing strain, encoding the catalytic subunit of PKA, was characterized by 2D-gel electrophoresis in comparison to the corresponding wild-type strain (Grosse et al. 2008). More than two-third of the proteins with altered abundance showed decreased levels in the mutant strain. In particular metabolic enzymes involved in the catabolism of alternative carbon sources, such as ethanol or acetate, were synthesized to a lesser extent, while pyruvate dehydrogenase levels increased. These data confirmed the central role of PkA in carbohydrate metabolism.

The cAMP signaling pathway is mediated via G-protein coupled receptors. The regulators of G-protein (RGS) signaling negatively affect the G-protein signaling pathway, of which six are present in the *A. fumigatus* genome. One of these RGS, FlbA, was investigated by 2-D proteome analyses with a Δ*flbA* deletion mutant. The mutant showed drastically increased levels of GliT, enhanced tolerance to gliotoxin, and increased intracellular superoxide dismutase activity. FlbA-mediated signaling, therefore, downregulates these two processes, ROS and gliotoxin detoxification (Shin et al. 2013).

More recent studies on *A. fumigatus* signaling implemented LC-MS/MS-based approaches including phosphoproteomic methods to study calcium-calcineurin, MAP kinase, and PkA-mediated signaling. Titanium dioxide was mainly used for the enrichment of phosphopeptides. Alternatively, GFP-trap affinity purification was employed for the characterization of phosphosites of GFP-tagged proteins of interest. Another approach is the use of antibodies that are directed against the target motif of specific protein kinases, e.g., the PkA. Increasingly, other PTMs in fungi, such as acetylation, are also attracting the attention of researchers (Li et al. 2019).

GFP-trap affinity purification led to the discovery of phosphorylation sites on the regulatory subunit of PkA, which had not been known from the mammalian or yeast orthologues. Two phosphorylation clusters were demonstrated to play a role in hyphal growth and the cell wall stress response via site-directed mutagenesis (Shwab et al. 2017). In a follow-up study, the PkA interactome was investigated, which resulted in the discovery of new targets: the autophagy-related protein Atg24, the HapB subunit of the CCAAT-binding complex and the E3 ubiquitin ligase Not4 (Shwab et al. 2020). All three proteins, Atg24, HapB and Not4, showed to be important for hyphal growth, have a role in stress responses and contribute to virulence in immunosuppressed mice. Noteworthy, the importance of PkA in regulating autophagy-associated proteins was confirmed in a follow-up study by using different proteases for protein digest (LycC and GluC) and an anti-PKA phosphopeptide antibody enrichment strategy (Shwab et al. 2022). All these findings underline the essential role of PkA in regulating critical cellular processes in *A. fumigatus*.

The calcium (Ca^{2+})-calcineurin cascade is another central signaling pathway in fungi that orchestrates many cellular stress responses (Park et al. 2019). This pathway contributes to the paradoxical effect observed for the echinocandin antifungal drug caspofungin in *A. fumigatus*. This effect describes the phenomenon that higher concentrations of the drug reverse the growth inhibition of *A. fumigatus*. Phosphoproteomic analysis by means of GFP-trap affinity purification revealed that the protein phosphatase calcineurin is activated through phosphorylation at its serine-proline-rich region only at paradoxical growth concentrations of caspofungin (Juvvadi et al. 2015b). The same group characterized also the interaction between calcineurin and its endogenous inhibitor calcipressin CbpA and identified the domains required for this interaction (Juvvadi et al. 2015a). Such information may be useful in the design of calcineurin signaling inhibitors that can be exploited for antifungal therapy, since this pathway is important for drug resistance in diverse fungi including *A. fumigatus*. To find interactors of calcineurin during treatments with azoles and echinocandins, Juvvadi et al. (2018) employed co-immunoprecipitation with calcineurin during exposure of *A. fumigatus* to the antifungal drugs. The calcineurin binding proteins were identified by a QTOF mass spectrometer with a data-independent acquisition approach. Treatment with voriconazole revealed an interaction of calcineurin with ergosterol biosynthesis enzymes, while caspofungin or micafungin led to increased immunoprecipitation of proteins involved in cell wall biosynthesis and septation (Juvvadi et al. 2018).

The major transcription factor of calcium signaling in *A. fumigatus* is CrzA, which is activated by the phosphatase activity of calcineurin. The phosphorylation sites of CrzA were unknown. Shwab et al. (2019) set out to identify these sites by GFP-trap affinity purification using a CrzA-GFP fusion protein combined with phosphopeptide enrichment. They identified 20 phosphorylation sites, most of which were specific to filamentous fungi and were predominantly located in a serine-rich region in proximity to the conserved DNA-binding region.

The mitogen-activated protein kinase (MAPK) signaling pathways play an important role in regulating fungal physiology and adapting to the human host (May et al. 2005). When *A. fumigatus* is exposed to cell wall stress, MAPKs phosphorylate a set of target proteins that contribute to cell wall remodeling. MpkA is the central regulator of the cell wall integrity pathway, but also SakA and MpkC, both involved in osmoadaptation, contribute to the cell wall damage response. This could be demonstrated by global phosphoproteome analyses of *A. fumigatus* mutants deleted in the *sakA* or *mpkC* gene as well as the characterization of mutants lacking putative membrane receptors (Mattos et al. 2020b; Silva et al. 2020). In another study, it was shown that the SchA^{SCH9} kinase, a target of the TOR kinase, has a regulatory impact on the SakA/HOG pathway, which illustrates the complex interconnection of signaling pathways through hub molecules (Alves de Castro et al. 2016).

11.3.4 Drug-Induced Changes

As discussed above, cellular signaling has an impact on drug resistance. For a better understanding of the global stress response to antifungal drugs and the mechanisms of drug resistance numerous studies characterized the proteomic profile of *A. fumigatus* during exposure to antimycotics.

Pioneering studies were based on 2D-gel electrophoresis. Gautam et al. (2008) analyzed the proteomic changes of *A. fumigatus* treated with sublethal concentrations of amphotericin B. Eighty-five different proteins changed their abundance, predominantly with increasing quantities. These included proteins involved in ergosterol biosynthesis, cell stress responses, cell wall, and transport proteins. It is still unclear up to now if and to which extent the ergosterol biosynthesis pathway of *A. fumigatus* is involved in conferring tolerance to critical AmB concentrations and further research is needed (Alcazar-Fuoli and Mellado 2012).

From the clinical point of view, azoles are the most important drug class for the primary treatment of *A. fumigatus* infections. A 2D-gel electrophoresis study revealed increased levels of proteins involved in the oxidative and general stress response, while proteins of the translational machinery were downregulated. This reflects a stress response, which is characterized by the downregulation of energy-demanding processes such as protein translation (Amarsaikhan et al. 2017). A similar alteration of the proteome profile was also reported when *A. fumigatus* was exposed to a sublethal concentration of itraconazole: differentially expressed proteins associated with carbohydrate metabolism, cell stress, translation, and amino acid metabolism were reported (Gautam et al. 2016).

The formation of biofilms is another factor of microorganisms to drive resistance and *A. fumigatus* is known to form such structures (Morelli et al. 2021). Wei et al. (2022) performed a comparative proteomic study on the extracellular matrix (ECM) of biofilms of a sensitive and azole-resistant strain of *A. fumigatus* exposed to itraconazole. The authors extracted ECM proteins by sonication and differential centrifugation and labeled the tryptic peptides with TMT before HPLC fractionation and LC-MS/MS analysis. In total, 2102 proteins were quantified. The two *A.*

fumigatus strains showed significant differences in their ECM proteome when exposed to itraconazole. Upregulated pathways in the resistant strain comprised several amino acid metabolic processes, while downregulated pathways were related to energy metabolism. In comparison, the sensitive strain showed increased levels of ribosomal proteins and lower levels of proteins involved in cell cycle pathways. Common changes in biofilm ECM proteins were related to ergosterol biosynthesis, oxidative stress, efflux pumps, DNA repair, replication, and transcription.

The transcription factors SrbA and AtrR of *A. fumigatus* are central players in azole resistance. To find interaction partners of AtrR, Paul et al. (2022) used a TAP-tagged AtrR protein and tandem affinity purification to find AtrR-associated proteins. They successfully discovered a new coactivator protein of AtrR, which they designated as NcaA (for nuclear activator of AtrR). Deletion of the corresponding gene indeed led to a reduction in voriconazole resistance.

Echinocandins are the newest class of antifungal drugs in clinical use and the proteomic response of *A. fumigatus* to one of its representatives, caspofungin, has been characterized in several studies. iTRAQ labeling followed by LC-MS/MS analysis revealed 122 proteins with at least a twofold change in relative abundance following exposure to a sublethal caspofungin concentration. Several proteins were found which could serve as biomarkers, including a mitochondrial hypoxia-responsive domain protein (RcfB), which was significantly downregulated in the secreted fraction of a susceptible *A. fumigatus* strain after caspofungin exposure (Cagas et al. 2011a). A combined transcriptome, proteome, and secretome analysis with subsequent protein–protein interaction network analysis provided further insights into the proteomic response of *A. fumigatus* to caspofungin (Conrad et al. 2018). The authors made use of a module-detecting approach to increase the overlap between transcriptome and proteome data (Vlaic et al. 2018). Several regulatory clusters were detected including enzymes involved in the biosynthesis of the secondary metabolite fumagillin, whose production was also confirmed at the metabolite level. A phosphoproteomic study on the same topic found many kinases and transcription factors that are modulated in the fungal response to caspofungin (Mattos et al. 2020a).

Further proteomic studies investigated antifungal compounds not in clinical use to identify their molecular targets. These include the antimalarial drug artemisinin (Gautam et al. 2011), a synthetic coumarin (Singh et al. 2012) and atorvastatin, a member of the statin drug family (Ajdidi et al. 2020). It seems that the proteomic stress response to antifungal agents is rather variable and depends mainly on the type of drug, but also on drug concentrations and exposure. However, there are also some common principles, such as the downregulation of energy-consuming processes.

11.3.5 Biofilm Formation

Biofilms are defined as structured microbial communities surrounded by an extracellular matrix (ECM). Also, *A. fumigatus* makes biofilms with hyphae covered by an ECM composed of extracellular DNA, polyols, proteins, lipids, and exopolysaccharides including α-glucans, galactomannan, and galactosaminogalactan (Morelli et al. 2021; Liu et al. 2022; Beauvais and Latge 2015). As discussed above, biofilm formation is accompanied by an increased resistance to antifungal drugs. Due to the relevance of biofilm formation of *A. fumigatus* for drug resistance and virulence, the biofilm proteome was investigated from in vitro cultures grown for 24 and 48 h as a biofilm. The 2D-gel electrophoresis-based proteome data were complemented by microarray analyses (Bruns et al. 2010). Comparison with the proteome of planktonic-grown *A. fumigatus* revealed a reduction of metabolic activity in mature biofilms (24 h versus 48 h), but an increased production of the secondary metabolite gliotoxin, which was also confirmed by metabolite analyses of the culture supernatants. Due to the cytotoxic activity of gliotoxin in the host (Ye et al. 2021), enhanced production of this secondary metabolite by biofilms may also play a role under in vivo conditions and could help the fungus to ward off the attack of immune cells. In the similar manner, Muszkieta et al. (2013) investigated the global changes on the transcript and proteome levels during biofilm growth. They grew *A. fumigatus* biofilms on agar plates covered with porous cellophane for 16 hours instead of 24 or 48 h before harvesting. The proteome data confirmed changes in transcriptional and translational activities, and showed a downregulation of glycolysis, but an upregulation of the reductive arm of the TCA cycle in *A. fumigatus* biofilms. The proteome data on *A. fumigatus* biofilms suggests that the mycelium undergoes significant changes in primary and secondary metabolism during biofilm formation, which partially explains the higher resistance of biofilms against numerous external insults.

11.3.6 Proteomics of the Cell Wall, the Cell Surface, and the Cellular Membrane

The cell wall of fungi is a complex multilayered structure composed of polysaccharides, proteins, and lipids. Since the fungal cell wall is the outermost cellular structure, it is the first cellular component in contact with the host. For this reason, interest has grown for proteomic approaches to study the cell envelope of *A. fumigatus*.

Several strategies exist to extract proteins from the surface of *A. fumigatus* conidia or hyphae. Covalently linked cell wall proteins, in particular glycosylphosphatidylinositol-anchored proteins, can be chemically released with hydrofluoric acid(HF)-pyridine extraction, while alkali-sensitive proteins are liberated via NaOH treatment (de Groot et al. 2004). As an alternative to HF treatment, formic acids can be employed and were used for the characterization of the surface layer of *A. fumigatus* mutants with mutations in the *rodA* gene (Valsecchi et al. 2019). The addition of cell wall degrading enzyme to resuspended conidia or mycelium, e.g., 1,3-beta-glucanase, is an alternative approach (Asif et al. 2006). Another strategy is the tryptic digestion of live fungal cells for a short period of time followed by identification of the formed tryptic peptides by LC-MS/MS analysis (Gil-Bona et al. 2015; Hernaez et al. 2010). Other studies analyzed surface proteins by labeling them with a membrane-impermeable biotin derivative that allowed a subsequent affinity purification with avidin conjugates after cell breakage (Urban et al. 2003).

Using both HF-pyridine extraction and trypsin shaving, Voltersen et al. (2018) identified in total 148 surface proteins from dormant conidia, 43 of which contained a signal peptide for secretion. Not surprisingly, the conidial surface hydrophobin RodA was the most abundant protein. Apart from RodA, a protein of unknown function, designated as CcpA, belonged to the most highly abundant conidial proteins and was investigated further. CcpA is a small amphiphilic protein found mainly in *Aspergillus* species. The deletion of the corresponding gene led to significant changes in the surface proteome, enhanced release of proinflammatory cytokines by monocyte-derived dendritic cells, and reduced virulence in a non-neutropenic mouse model of pulmonary invasive aspergillosis. Owing to these findings, the authors suggested that CcpA serves as a conidial stealth protein, which limits detection of conidia by the innate immune system. Trypsin shaving was also applied for the comparison of the surface proteome of conidia carrying point mutations in the rodA gene or lacking the whole RodA protein (Valsecchi et al. 2020).

Another comparative study investigated the dynamic changes of the *A. fumigatus* proteome during development from dormant conidia over swollen conidia and germlings up to hyphae formed after 14 h. The proteome showed dynamic changes across germination, but at least a certain fraction of proteins was found in all four growth stages including the conidial hydrophobin RodA. Furthermore, the study demonstrated that environmental conditions during conidiation, e.g., temperature and growth time, have a great impact on the composition of the surface proteome. A comparison with other allergy-inducing molds, such as *Alternaria alternata*, *Penicillium rubens* and *Cladosporium herbarum* revealed that a certain overlap of the surface proteome exists between *A. fumigatus* and other allergic fungi (Blango et al. 2020). Another study based on surface biotinylation confirmed these significant alterations during development and the presence of constitutively, and stage-specifically exposed proteins, some of which could be targets for the design of diagnostic or immunotherapeutic tools (Jia et al. 2020).

The development of a species-specific diagnostic tool was also the motivation for the comparison of conidial surface-associated and secreted proteins of an *A. fumigatus* and *A. flavus* strain. The extraction of conidial surface proteins was performed using a highly concentrated NaCl solution (0.5 M). The authors identified a common core proteome enriched with enzymes that act on polysaccharides, but also species-specific proteins were found (either enriched or unique), which can be used to differentiate these two pathogenic fungi in infected patients (Venugopalan et al. 2023).

The growing data on fungal proteins will definitely help to find proteins of diagnostic or immunotherapeutic value. However, the high dynamic changes of the surface proteome in dependence on the environmental conditions will hamper the further development of such tools.

11.3.7 Secretome

A. fumigatus has a saprophytic lifestyle and secretes a diverse set of catabolic enzymes that allows the degradation of macromolecular biopolymers to acquire nutrients from decaying plant or animal matter. The fungus secretes also hydrolytic enzymes during the infection of the human host, which contributes to tissue damage and thus to the progression of the disease (Abad et al. 2010). It has also been shown that in particular secreted proteins trigger a Th2 response in allergic diseases caused by Aspergilli (Schwarz et al. 2023) and that secreted proteases have an impact on the development of severe asthma with fungal sensitization (Druey et al. 2020).

Proteomics can make a significant contribution to study the role of secreted proteins in virulence. One of the first proteomic characterization of the *A. fumigatus* secretome employed 2D-gel electrophoresis combined with MALDI-TOF/TOF analyses and identified 64 different proteins secreted during growth in a defined minimal medium with glucose as sole carbon and energy source (Wartenberg et al. 2011). The study was complemented with secretome profiles during growth on different protein sources as the main carbon and nitrogen sources. The alkaline serine protease Alp1 turned out to be the most abundant protease. Alp1 may have an impact on the pathogenicity of *A. fumigatus*, since it is involved in human complement evasion due to its high proteolytic activity (Behnsen et al. 2010). Besides proteases many carbohydrate polymer-degrading enzymes were found in the secretome. A remarkably high number of extracellular proteins lacked a signal peptide for protein secretion, which suggested that they are released via non-classical protein secretory pathways (Wartenberg et al. 2011). It is known that fungi secrete proteins also via extracellular vesicles (EVs). Actually, in a recent publication Rizzo et al. (2020) characterized the proteomic content of extracellular vesicles (EVs) released from *A. fumigatus* protoplasts. Freshly prepared protoplasts released EVs in total only 142 proteins, while during regeneration of protoplast the number of proteins increased to more than 2000. Many of the detected proteins were associated with lipid/sugar metabolism and cell wall biosynthesis. The number of EV-associated proteins released by growing hyphae was 60 and thus substantially lower in a study of Souza et al. (2019), but they showed immunomodulatory activity.

Proteomic methods were also used to explore targets of the two *A. fumigatus* transcription factors PrtT and XprG, which had been shown to regulate extracellular proteolysis (Hagag et al. 2012; Shemesh et al. 2017). Both studies demonstrated that PrtT and XprG regulate not only many proteases, but also other enzymes, in particular proteins involved in cell wall modification.

The secretome of *A. fumigatus* is also interesting from a biotechnological point of view, since this fungus has the potential to degrade cellulose and lignocellulosic biomass. For this reason, Adav et al. (2015b) profiled the secretome of *A. fumigatus* during growth on cellulose, xylan, and starch by iTraq-based proteomic analyses. The comparative study revealed that cellulose and xylan stimulate the secretion of specific cellulases and hemicellulases and that deamidation is an often found PTM of cellulose hydrolyzing enzymes (Adav et al. 2015a, b).

A. fumigatus is also one of the most prevalent fungal pathogens colonizing the lungs of cystic fibrosis (CF) patients and the bacterium *Pseudomonas aeruginosa* infects 70–80% of adult patients with CF. *A. fumigatus* and *P. aeruginosa* interact in a complex manner in the lungs of CF patients. In vitro studies by Margalit et al. (2020a) revealed that culture filtrates of *A. fumigatus* promote growth of *P. aeruginosa*. To get more insights into the interaction the authors investigated the secretome of *A. fumigatus* and compared changes to the proteome of *P. aeruginosa* exposed to culture filtrates of *A. fumigatus* alone, an *A. fumigatus*-*P. aeruginosa* co-culture and *P. aeruginosa* alone by label-free quantitative proteomics. *A. fumigatus* was shown to secrete a high number of proteases and peptidases. Concomitantly, *P. aeruginosa* downregulated amino ABC transporters when exposed to *A. fumigatus* culture filtrate. This suggests that *A. fumigatus* provides nutrients to *P. aeruginosa* during growth in Czapek-Dox medium and most probably also during co-infection of the CF lung.

11.3.8 Stress Responses

The ability of *A. fumigatus* to adapt to the stress conditions during the infection of the human host is a prerequisite for the pathogenicity of this fungus (Brown and Goldman 2016). One of the characteristics of *A. fumigatus* is its high thermotolerance. It can survive temperatures of up to 65 °C (Kwon-Chung and Sugui 2013). In general, stress tolerance appears to vary greatly between different strains of *A. fumigatus*. This was demonstrated by comparative proteomic analyses of two clinical *A. fumigatus* strains and two strains isolated from the International Space Station (Blachowicz et al. 2019).

A difference gel electrophoresis (DIGE)-based study on the proteomic changes of *A. fumigatus* upon heat stress revealed an increased abundance of heat shock proteins (HSPs), enzymes of the oxidative and nitrosative stress response, glycolysis, and cytoskeleton organization. Based on putative heat shock factor (Hsf) binding sequences, the authors postulated putative Hsf1 target genes, which included proteins of the oxidative stress response (Asp F3) and mannitol biosynthesis besides HSPs (Albrecht et al. 2010).

Besides elevated temperature, *A. fumigatus* is also exposed to oxygen-limiting conditions during the infection (Grahl et al. 2012). Studies using an oxygen-controlled batch-fermenter or chemostat revealed increased levels of proteins involved in glycolysis, aspartate/glutamate metabolism, and the oxidative stress response. Noteworthy, also proteins in cellular respiration increased in abundance suggests that *A. fumigatus* compensates a significant drop in oxygen levels by increasing the mitochondrial respiratory capacity (Vodisch et al. 2009; Barker et al. 2012). Hypoxia leads also to increased ROS levels and as a result to oxidative modifications of redox-sensitive cysteine residues of certain proteins, some of which are of mitochondrial origin. The authors used cysteine-reactive iodoTMT for the quantification of the *A. fumigatus* redox proteome during hypoxia in their study (Shekhova et al. 2019).

The acquisition of iron is essential for *A. fumigatus* for the establishment of an infection (Misslinger et al. 2021). Moloney et al. (2016) compared the microsomal proteome of *A. fumigatus* grown under iron-depleting versus iron-depleting conditions. More than 200 microsomal proteins altered their abundance upon iron starvation. These included proteins with siderophore-related functions, e.g., known and putative siderophore transporters. The authors suggested that targeting siderophore-mediated iron acquisition, e.g., membrane transporters, may be a promising therapeutic strategy to combat *A. fumigatus* infections.

During an infection, *A. fumigatus* has to cope not only with one but many different stress conditions. From this background, Kurucz et al. (2018) studied the proteomic response of *A. fumigatus* during single stress (iron depletion and oxidative stress) and combined stress conditions (iron depletion and oxidative stress). Under this combined stress condition, antioxidative enzymes without any iron cofactors particularly increased in their abundance to eliminate reactive oxygen species possibly due to the shortage of iron. These "iron-free" enzymes included the Cu/Zn superoxide dismutase Sod1, thioredoxin reductases, and the thioredoxin Asp F29. A link between the response to nutrient stress and iron depletion was found by another proteomic study on the TOR (target of rapamycin) signaling cascade, which regulates nutrient sensing. It revealed a repressor function of TOR in regulating ornithine biosynthesis, an important precursor of siderophore biosynthesis (Baldin et al. 2015).

The exposure of *Aspergillus fumigatus* to human serum triggers also a stress response and leads to an alteration of its metabolism and enzyme reactivity. This was shown by a study that employed activity-based profiling with chemical probes coupled to quantitative mass spectrometry (Wiedner et al. 2012). One synthesized electrophilic probe targeted cysteine and other nucleophilic amino acids, while another, more specific probe bound to serine hydrolases. The authors found a difference in stress response proteins between cultures with and without human serum addition. Furthermore, a shift from glucose catabolism to the usage of alternative carbon sources was observed in cultures of *A. fumigatus* with human serum. Interestingly, the human serum-induced changes of fungal metabolism on the protein level varied between *A. fumigatus* and the related species *Neosartorya fischeri* and *Aspergillus clavatus* (Wiedner et al. 2013).

11.3.9 Regulation and Virulence

As described above, proteomics allows for comparison of a gene knockout strain with its corresponding wild-type strain to reveal regulatory functions of the deleted gene. However, caution should be taken with the interpretation because gene deletions can lead to a rewiring of the transcriptional regulatory program that leads to many indirect effects. The advantage of a proteomic approach over transcriptomics is that also post-translational modifications with regulatory functions can be recognized.

Zhang et al. (2009) characterized an α-glucosidase (AfCwh41) of A. *fumigatus*, which plays an important role in the *N*-glycan processing and the quality control pathway of glycoproteins in the endoplasmic reticulum. The proteomic data supported the hypothesis that AfCwh41 is required for glucose trimming of cell wall synthesis and cell wall stress sensing proteins. Consequently, the AfCwh41 deletion mutant showed clear cell wall defects. Another study provided more insights into the regulon of the three developmental Velvet regulators VeA, VelB, and VosA in *A. fumigatus*. Comparative proteomic analyses revealed that VeA controls the protein level of the RNA polymerase II degradation factor DefA, which mediates the response to DNA damage (Shin et al. 2016).

Annexins are well studied in higher eukaryotes, where they are characterized by their ability to bind to negatively charged phospholipids and to contribute to organization of the cell membrane. Less is known about the function of annexins in fungi. Khalaj et al. (2011) disrupted the annexin C4 gene in *A. fumigatus* and compared the proteomes of the *anxC4*-disrupted strain with the corresponding wild type. Due to increased levels of stress response proteins in the mutant, the authors suggested that AnxC4 exhibits an anti-stress function in *A. fumigatus*.

11.3.10 Immunoproteomics

It is still a matter of debate whether the human serological response contributes to the eradication of an *A. fumigatus* infection. However, recent research provided evidence that some antibodies have an impact on the course of a fungal infection and mediate protection (Casadevall and Pirofski 2012). Apart from that, the characterization of the antibody response to *A. fumigatus* is needed for the establishment of diagnostic tools and the development of vaccination strategies. Besides protein/peptide microarrays, serological proteome analysis (SERPA) methods are applied to identify seroreactive fungal antigens. SERPA uses 2D-gel electrophoresis for the separation of protein antigens, which are subsequently transferred via blotting to either a PVDF or nitrocellulose membrane. Membranes are probed with patient sera and binding immunoglobulins are detected via labeled secondary antibodies. Proteins of interest are excised from the 2D gel, trypsin digested and identified by mass spectrometry (Macheleidt and Kniemeyer 2021).

Several groups characterized the antibody response to *A. fumigatus* in humans and animals by using the SERPA approach described above. Specific anti-*Aspergillus* antibodies are detectable in the serum of most individuals at a low level, but significantly higher titers are measured in patients with allergic (ABPA, asthma) or chronic (CPA) *A. fumigatus* diseases (Delliere and Aimanianda 2023). In particular, patients suffering from ABPA are characterized by high serum anti-*A. fumigatus* IgG/IgE antibody titers and a hyperinflammatory response after exposure to *A. fumigatus* antigens (Magee et al. 2021). Due to these facts, several studies used SERPA for the identification of immunoreactive protein antigens of *A. fumigatus*. In most studies, the immunosecretome was investigated due to the possible role of extracellular proteins in virulence and sensitization. Between 15 and 35 different secreted protein antigens were detected in several studies (Gautam et al. 2007; Kumar et al. 2011; Singh et al. 2010a). The overlap of detected antigens was not large between studies and only the catalase CatB occurred in all three. This demonstrates the high variability of the antibody response among individuals. Another study screened cytosolic proteins with ABPA patient sera for immunoreactivity (Singh et al. 2010b) and detected 3 IgG and 63 IgE-reactive proteins. These included known (Asp f12, Asp f22, Hsp70, malate dehydrogenase) and some predicted antigens (e.g., alcohol dehydrogenase AlcC).

There is a limited role for antibody testing in immunocompromised patients at risk for invasive aspergillosis (IA) due to the delayed or absent antibody response. However, the antibody response in these patients may reveal anti-*Aspergillus* antibodies with protective properties or predictive value. Seventeen different extracellular proteins separated by 2D-gel electrophoresis were detected by sera from critically ill patients with proven IA. Most identified proteins had a metabolic function (Shi et al. 2012a). The gliotoxin oxidoreductase GliT, mentioned in a previous section, was recognized by all sera, and was proved to be a potential biomarker (Shi et al. 2012b). Two other studies used protein extracts from *A. fumigatus* germlings or mycelium and sera from patients with probable or proven IA for their SERPA study. They found 40 and 49 different protein antigens, respectively. Fourteen protein antigens were detected by both studies (Teutschbein et al. 2016; Virginio et al. 2014). The work of Teutschbein et al. (2016) implemented decision trees for outcome prediction and identified several protein antigens whose recognition by patient sera were associated with survival or a fatal outcome, respectively.

Also, animal models were used in the search for *A. fumigatus* protein antigens. Multiply infected rabbits exhibited a robust immune response against proteins of *A. fumigatus* germlings in a study by Asif et al. (2010). Altogether, 59 different proteins were recognized by IgG, IgA, or IgM antibodies, many of which are involved in ROS detoxification, glycolysis, and other metabolic pathways and which were also found in studies with human patient sera.

11.3.11 Host–Pathogen Interaction

Proteome analysis of the interaction of human pathogenic fungi with the host is able to provide unprecedented insights into the molecular mechanisms of fungal pathogenicity mechanisms and the host's defense response. However, it is a demanding task, since either the fungal pathogen and the host cells have to be separated from each other before analysis or the complex proteome of the fungus and the host cell have to be analyzed simultaneously in a mixture. One of the major challenges is the fact that in most cases the protein ratio is favorable to the host proteins, which makes it difficult to dissect the fungal proteome (Krüger et al. 2015).

During a typical infection *A. fumigatus* conidia come first into contact with lung epithelial cells. Several studies characterized the proteomic profile of lung epithelial cells upon *A. fumigatus* infection, mostly using epithelial cell lines such as A549 or BEAS-2B. A study on secreted proteins revealed increased levels of lysosomal proteins (cathepsin B, cathepsin D, and HEXB) and proteins of the thioredoxin system in bronchial epithelial cells (Fekkar et al. 2012a). Others reported about a drastic change in cytoskeletal proteins in A549 cells during *A. fumigatus* interaction (Kanjanapruthipong et al. 2022), while an iTraq-based proteomic analysis reported an increase of the N-myc downstream-regulated gene 1 (NDRG1). NDRG1 is a putative stress protein and silencing of the corresponding gene results in less efficient internalization of conidia, which suggests that NDRG1 is involved in conidial uptake (Zhang et al. 2020). Margalit et al. (2020b) exposed A549 cells to both *A. fumigatus* and the bacterium *P. aeruginosa* to determine how the pulmonary epithelium responds to such co-confections, which occur in the lungs of CF patients. Distinct changes of the A549 cell proteome were observed in response to one or both microbial pathogens. Based on these data the authors concluded that *A. fumigatus* renders A549 cells in a way that the internalization of *P. aeruginosa* declines.

Primary cells get closer to in vivo predictions and several proteomic studies used them for studying the host response to *A. fumigatus*. Proteomic analysis with primary human bronchial epithelial cells demonstrated increased levels of the proteins calreticulin, a chaperon localized in the ER, and the SET nuclear proto-oncogene (Toor et al. 2018). Another proteomic study employed alveolar type II cells from mice and detected alteration of respiratory proteins and increased levels of the L-amino acid oxidase IL4I1 (42.9-fold) upon *A. fumigatus* infection. This IL4I1 regulation was confirmed with human lung tissue from aspergilloma patients (Seddigh et al. 2017). IL4I1 is likely involved in the control of the inflammatory response and signaling to T- and B lymphocytes (Cousin et al. 2015), but antimicrobial properties have also been reported (Puiffe et al. 2013). However, further studies are required to show whether IL41I1 or the other above-mentioned proteins are generally induced upon lung inflammation or whether they are specifically involved in an anti-*Aspergillus* response and could be used as potential biomarkers associated with aspergillosis.

Angioinvasion is a key feature of invasive aspergillosis and consequently *A. fumigatus* interacts with endothelial cells. Curty et al. (2014) challenged human umbilical vein endothelial cells (HUVEC) with germlings of *A. fumigatus* for a period of 16 h. Proteins involved in the immune response, cell signaling, hemostasis, coagulation, and angiogenesis (e.g., ICAM-1), increased in abundance. It is therefore plausible that *A. fumigatus* induces angiogenesis, but, in contrast, its secondary metabolites inhibit this process (Ben-Ami et al. 2009). By using a label-free quantitative proteomics approach, Neves et al. (2016, 2017) investigated the impact of the cell wall component galactosaminogalactan (GAG) on the infection response of HUVECs. They either used purified GAG or a GAG-enriched *A. fumigatus* mutant lacking the gene *umg1*. The Δ*umg1* mutant exhibited hyperadhesive phenotype and induced both a proinflammatory and cell stress response in HUVECs. Based on the proteomic data, the authors concluded that the cell wall component GAG can modulate and activate the endothelial response to *A. fumigatus*.

The innate immunity plays a key role in *A. fumigatus* eradication. In particular resident alveolar macrophages and neutrophils, which are recruited to the sites of infection, are the key drivers (Latge and Chamilos 2019). Alveolar macrophages phagocytose inhaled *A. fumigatus* conidia. However, *A. fumigatus* has developed mechanisms to resist phagolysosomal degradation, e.g., by the interaction of conidial DHN-melanin with this process. To get more insights into the underlying mechanisms, Schmidt et al. (2018) isolated conidia-containing phagolysosomes from mouse macrophages by a magnetic label-based method and analyzed the phagolysosomal protein content by LC-MS/MS-based proteomics. Two strains, the *A. fumigatus* wild type and a nonmelanized *pksP* mutant were compared with each other. In total 637 host proteins and 22 *A. fumigatus* proteins were identified. A higher number of proteins were over-represented in *pksP*-containing phagolysosomes in comparison to the wild type. The proteomic data showed that *A. fumigatus* affects intracellular processes, which include vesicle trafficking, Rab5 and Vamp8-dependent endocytosis, NADPH oxidase activity, vATPase-driven phagolysosomal acidification as well as MAPK and mTOR signaling. A higher abundance of Lamp1 in *pksP*-containing phagosomes confirmed previous data, that the melanin-free mutant *pksP* ends up to a greater extent in late phagosomes and phagolysosomes.

Neutrophils engulf *A. fumigatus* conidia and germlings, but are also able to attack *A. fumigatus* externally by degranulation and the formation of neutrophil extracellular traps (NETs). However, neutrophils release also extracellular vesicles (EVs) upon contact with *A. fumigatus.* These show some antifungal activity in comparison to spontaneously released EVs. Both types of EVs have also a different protein load as determined by LC-MS/MS-based proteomics. *A. fumigatus*-induced EVs contained more proteins with antimicrobial activity such as histones, neutrophil elastase, cathepsin G, azurocidin, and defensin 1 (Shopova et al. 2020). One major disadvantage of studying the biology of neutrophils is that no suitable cell line system for this type of immune cell exists and neutrophils have to be isolated freshly from blood. Rafiq et al. (2022) tested differentiated PLB-985 cells, a human cell line derived from a patient with acute myeloid leukemia, as alternative. Experiments with PLB-985 cells confirmed that they can be selectively employed as neutrophil-like cell lines. EVs produced by PLB-985 cells are comparable to primary neutrophils and they contain antimicrobial proteins like cathepsin G and azurocidin. Another approach is to investigate the proteomic response of A. *fumigatus* to single microbicidal compounds produced by neutrophils. This has been done for the antimicrobial peptide LL-37 and the oxidant *N*-chlorotaurine. The latter induced, in particular, an oxidative stress response (Sheehan et al. 2019), while LL-37 does not seem to have any antifungal effects, but rather stimulates growth of *A. fumigatus* (Sheehan et al. 2018a).

Single cell lines and primary cells cannot reflect the complexity of an infection process in the host. Only a few proteomic studies were also performed with animal models using mice or larvae of the greater wax moth (*Galleria mellonella*). The cell-free hemolymph of *G. mellonella* larvae was analyzed by label-free quantitative proteomics after 6 and 24 h of infection with *A. fumigatus* conidia (Sheehan et al. 2018b). After 6 hours, a rise in the abundance of antimicrobial peptides and proteins (gloverin, moricin, lysozyme, cecropin) was observed. In addition, the levels of proteins associated with the prophenoloxidase cascade increased. These proteins have analogous functions to the mammalian complement cascade. At later time points, tissue invasion was reflected by an increase of proteins associated with tissue destruction, but also proteins associated with recognition and opsonization of fungal cells were induced. Due to parallels between *G. mellonella* and the innate immune response of mammals toward *A. fumigatus* the authors concluded that *G. mellonella* represents a convenient model for studying host–pathogen interaction.

In another study, the BAL and serum of *A. fumigatus*-infected mice were studied by DIGE. The serum of infected mice contained the *A. fumigatus* allergen Aspf 2 one day after the infection, which was, however, no longer detectable after 3 days of infection. The BAL samples provided evidence for a complement evasion strategy by the cleavage of C3b (Fekkar et al. 2012b). A C3b deposition on the surface of *A. fumigatus* conidia was confirmed later by a mass spectrometric approach (Wong et al. 2020). Krel et al. (2014) used a neutropenic rabbit model instead of mice to search for host biomarkers of IPA. Haptoglobin, C-reactive protein (CRP), and annexin A1 were prominently found in the BALF of rabbits during the infection. These three putative biomarkers responded also to antifungal therapy.

The analysis of clinical samples allows to shed light on translational aspects of *Aspergillus fumigatus*-associated diseases, but the general challenge is the high variability from interpersonal diversity. Nevertheless, several studies have met the challenges. Cai et al. (2021) searched for differentially abundant proteins in serum of ABPA and asthma patients using iTraq reagents for quantification. Putative biomarkers were validated by ELISA assays. Their study found serum carnosine dipeptidase 1 (CNDPI) to be less abundant in ABPA patients in comparison to asthmatics and healthy controls. Machata et al. (2020) searched in BAL samples of mice and humans for IPA biomarkers and found more than 30 host proteins with different abundances, which also included the C-reactive protein found in the BALF of infected rabbits as described above (Krel et al. 2014). Noteworthy, also fungal proteins were detected in BAL samples including several known antigens (Aspf 2, Aspf 4, and Aspf 9). A subsequent study on host proteins in BAL fluids of patients with *A. fumigatus* infection or colonization confirmed 16 of the proteins found by Machata et al. (2020). In particular, humoral components were affected by an *A. fumigatus* infection like complement proteins, ficolins, pentraxin-3, serum protease MASP2, and surfactant protein-D. The authors proposed that humoral components completely absent or present in *Aspergillus*-positive BALF could be potentially used as immunodiagnostic markers (Delliere et al. 2021).

As shown above, proteomics can help to define biomarkers for *Aspergillus*-associated infections. However, it remains a problem that it is not easy to verify biomarkers due to the difficult diagnosis of these diseases and the limited number of well-defined cases.

11.4 Conclusion

Initially, many proteomic studies on fungal organisms have used the yeast *S. cerevisiae* due to its robustness, status as a model organism, and well-annotated genome. However, the application of established methods of proteomics to human pathogenic fungi such as *A. fumigatus* and others has increased considerably. The emergence of large-scale studies has come along with a change in biology from a classical, reductionistic approach, where individual genes or proteins are studied, to an integrative, holistic approach. This development is accompanied by efforts to study the proteome of a single cell, which is for microbial cells due to their smaller size even more challenging than for mammalian cells. However, first attempts toward single bacterium proteomics have already been reported (Vegvari et al. 2023).

Nowadays, systems biology is now an emerging field in life science. It describes and analyzes quantitatively the interaction among all the individual components of a cell. The final aim of this approach is to develop computational models that can describe and predict biological systems (Ideker et al. 2001; Albrecht et al. 2011; Schleicher et al. 2017). To reach this ambitious target, an interdisciplinary collaboration of biologists, technologists, and computational scientists is needed. Without a doubt, the application of machine learning and artificial intelligence will play an increasingly important role in the interpretation of omics data and for the development of diagnostic tools (Tran et al. 2023).

References

Abad A, Fernandez-Molina JV, Bikandi J, Ramirez A, Margareto J, Sendino J, Hernando FL, Ponton J, Garaizar J, Rementeria A (2010) What makes *Aspergillus fumigatus* a successful pathogen? Genes and molecules involved in invasive aspergillosis. Rev Iberoam Micol 27(4):155–182. https://doi.org/10.1016/j.riam.2010.10.003

Adav SS, Ravindran A, Sze SK (2015a) Data for iTRAQ secretomic analysis of *Aspergillus fumigatus* in response to different carbon sources. Data Brief 3:175–179. https://doi.org/10.1016/j.dib.2015.03.001

Adav SS, Ravindran A, Sze SK (2015b) Quantitative proteomic study of *Aspergillus fumigatus* secretome revealed deamidation of secretory enzymes. J Proteome 119:154–168. https://doi.org/10.1016/j.jprot.2015.02.007

Agarwal R, Muthu V, Sehgal IS (2023) Relationship between Aspergillus and asthma. Allergol Int 72(4):507–520. https://doi.org/10.1016/j.alit.2023.08.004

Ajdidi A, Sheehan G, Kavanagh K (2020) Exposure of *Aspergillus fumigatus* to atorvastatin leads to altered membrane permeability and induction of an oxidative stress response. J Fungi (Basel) 6(2). https://doi.org/10.3390/jof6020042

Albrecht D, Guthke R, Brakhage AA, Kniemeyer O (2010) Integrative analysis of the heat shock response in *Aspergillus fumigatus*. BMC Genomics 11:32. https://doi.org/10.1186/1471-2164-11-32

Albrecht D, Kniemeyer O, Mech F, Gunzer M, Brakhage A, Guthke R (2011) On the way toward systems biology of *Aspergillus fumigatus* infection. Int J Med Microbiol 301(5):453–459. https://doi.org/10.1016/j.ijmm.2011.04.014

Alcazar-Fuoli L, Mellado E (2012) Ergosterol biosynthesis in *Aspergillus fumigatus*: its relevance as an antifungal target and role in antifungal drug resistance. Front Microbiol 3:439. https://doi.org/10.3389/fmicb.2012.00439

Alves de Castro P, Dos Reis TF, Dolan SK, Oliveira Manfiolli A, Brown NA, Jones GW, Doyle S, Riano-Pachon DM, Squina FM, Caldana C, Singh A, Del Poeta M, Hagiwara D, Silva-Rocha R, Goldman GH (2016) The *Aspergillus fumigatus* SchA(SCH9) kinase modulates SakA(HOG1) MAP kinase activity and it is essential for virulence. Mol Microbiol 102(4):642–671. https://doi.org/10.1111/mmi.13484

Amarsaikhan N, Albrecht-Eckardt D, Sasse C, Braus GH, Ogel ZB, Kniemeyer O (2017) Proteomic profiling of the antifungal drug response of *Aspergillus fumigatus* to voriconazole. Int J Med Microbiol 307(7):398–408. https://doi.org/10.1016/j.ijmm.2017.07.011

Anagnostopoulos AK, Stravopodis DJ, Tsangaris GT (2017) Yield of 6,000 proteins by 1D nLC-MS/MS without pre-fractionation. J Chromatogr B Anal Technol Biomed Life Sci 1047:92–96. https://doi.org/10.1016/j.jchromb.2016.08.031

Antimicrobial Resistance Division CoNTD, Global Coordination and Partnership (2022) WHO fungal priority pathogens list to guide research, development and public health action

Asif AR, Oellerich M, Amstrong VW, Riemenschneider B, Monod M, Reichard U (2006) Proteome of conidial surface associated proteins of *Aspergillus fumigatus* reflecting potential vaccine candidates and allergens. J Proteome Res 5(4):954–962. https://doi.org/10.1021/pr0504586

Asif AR, Oellerich M, Amstrong VW, Gross U, Reichard U (2010) Analysis of the cellular *Aspergillus fumigatus* proteome that reacts with sera from rabbits developing an acquired immunity after experimental aspergillosis. Electrophoresis 31(12):1947–1958. https://doi.org/10.1002/elps.201000015

Bache N, Geyer PE, Bekker-Jensen DB, Hoerning O, Falkenby L, Treit PV, Doll S, Paron I, Müller JB, Meier F, Olsen JV, Vorm O, Mann M (2018) A novel LC system embeds analytes in pre-formed gradients for rapid, ultra-robust proteomics. Mol Cell Proteomics MCP 17(11):2284–2296. https://doi.org/10.1074/mcp.TIR118.000853

Baldin C, Valiante V, Kruger T, Schafferer L, Haas H, Kniemeyer O, Brakhage AA (2015) Comparative proteomics of a tor inducible *Aspergillus fumigatus* mutant reveals involvement of the Tor kinase in iron regulation. Proteomics 15(13):2230–2243. https://doi.org/10.1002/pmic.201400584

Ball B, Langille M, Geddes-McAlister J (2020) Fun(gi)omics: advanced and diverse technologies to explore emerging fungal pathogens and define mechanisms of antifungal resistance. MBio 11(5). https://doi.org/10.1128/mBio.01020-20

Barber AE, Sae-Ong T, Kang K, Seelbinder B, Li J, Walther G, Panagiotou G, Kurzai O (2021) *Aspergillus fumigatus* pan-genome analysis identifies genetic variants associated with human infection. Nat Microbiol 6(12):1526–1536. https://doi.org/10.1038/s41564-021-00993-x

Barker BM, Kroll K, Vodisch M, Mazurie A, Kniemeyer O, Cramer RA (2012) Transcriptomic and proteomic analyses of the *Aspergillus fumigatus* hypoxia response using an oxygen-controlled fermenter. BMC Genomics 13:62. https://doi.org/10.1186/1471-2164-13-62

Bauer B, Schwienbacher M, Broniszewska M, Israel L, Heesemann J, Ebel F (2010) Characterisation of the CipC-like protein AFUA_5G09330 of the opportunistic human pathogenic mould Aspergillus fumigatus. Mycoses. 53(4):296–304. https://doi.org/10.1111/j.1439-0507.2009.01718.x

Beauvais A, Latge JP (2015) Aspergillus biofilm in vitro and in vivo. Microbiol Spectr 3(4). https://doi.org/10.1128/microbiolspec.MB-0017-2015

Behnsen J, Lessing F, Schindler S, Wartenberg D, Jacobsen ID, Thoen M, Zipfel PF, Brakhage AA (2010) Secreted *Aspergillus fumigatus* protease Alp1 degrades human complement proteins C3, C4, and C5. Infect Immun 78(8):3585–3594. https://doi.org/10.1128/IAI.01353-09

Bekker-Jensen DB, Kelstrup CD, Batth TS, Larsen SC, Haldrup C, Bramsen JB, Sørensen KD, Høyer S, Ørntoft TF, Andersen CL, Nielsen ML, Olsen JV (2017) An optimized shotgun strategy for the rapid generation of comprehensive human proteomes. Cell Syst 4(6):587–599.e584. https://doi.org/10.1016/j.cels.2017.05.009

Bekker-Jensen DB, Martínez-Val A, Steigerwald S, Rüther P, Fort KL, Arrey TN, Harder A, Makarov A, Olsen JV (2020) A compact quadrupole-orbitrap mass spectrometer with FAIMS interface improves proteome coverage in short LC gradients. Mol Cell Proteomics MCP 19(4):716–729. https://doi.org/10.1074/mcp.TIR119.001906

Ben-Ami R, Lewis RE, Leventakos K, Kontoyiannis DP (2009) *Aspergillus fumigatus* inhibits angiogenesis through the production of gliotoxin and other secondary metabolites. Blood 114(26):5393–5399. https://doi.org/10.1182/blood-2009-07-231209

Bern M, Kil YJ, Becker C (2012) Byonic: advanced peptide and protein identification software. Curr Protoc Bioinformatics Chapter 13:Unit13.20. https://doi.org/10.1002/0471250953.bi1320s40

Bian Y, Zheng R, Bayer FP, Wong C, Chang YC, Meng C, Zolg DP, Reinecke M, Zecha J, Wiechmann

S, Heinzlmeir S, Scherr J, Hemmer B, Baynham M, Gingras AC, Boychenko O, Kuster B (2020) Robust, reproducible and quantitative analysis of thousands of proteomes by micro-flow LC-MS/MS. Nat Commun 11(1):157. https://doi.org/10.1038/s41467-019-13973-x

Bian Y, Bayer FP, Chang YC, Meng C, Hoefer S, Deng N, Zheng R, Boychenko O, Kuster B (2021a) Robust microflow LC-MS/MS for proteome analysis: 38 000 runs and counting. Anal Chem 93(8):3686–3690. https://doi.org/10.1021/acs.analchem.1c00257

Bian Y, The M, Giansanti P, Mergner J, Zheng R, Wilhelm M, Boychenko A, Kuster B (2021b) Identification of 7 000-9 000 Proteins from Cell Lines and Tissues by Single-Shot Microflow LC-MS/MS. Anal Chem 93(25):8687–8692. https://doi.org/10.1021/acs.analchem.1c00738

Bignell E, Cairns TC, Throckmorton K, Nierman WC, Keller NP (2016) Secondary metabolite arsenal of an opportunistic pathogenic fungus. Philos Trans R Soc Lond Ser B Biol Sci 371(1709). https://doi.org/10.1098/rstb.2016.0023

Biron DG, Brun C, Lefevre T, Lebarbenchon C, Loxdale HD, Chevenet F, Brizard JP, Thomas F (2006) The pitfalls of proteomics experiments without the correct use of bioinformatics tools. Proteomics 6(20):5577–5596. https://doi.org/10.1002/pmic.200600223

Blachowicz A, Chiang AJ, Romsdahl J, Kalkum M, Wang CCC, Venkateswaran K (2019) Proteomic characterization of *Aspergillus fumigatus* isolated from air and surfaces of the International Space Station. Fungal Genet Biol 124:39–46. https://doi.org/10.1016/j.fgb.2019.01.001

Blango MG, Pschibul A, Rivieccio F, Kruger T, Rafiq M, Jia LJ, Zheng T, Goldmann M, Voltersen V, Li J, Panagiotou G, Kniemeyer O, Brakhage AA (2020) Dynamic surface proteomes of allergenic fungal conidia. J Proteome Res 19(5):2092–2104. https://doi.org/10.1021/acs.jproteome.0c00013

Bongomin F, Gago S, Oladele RO, Denning DW (2017) Global and multi-national prevalence of fungal diseases-estimate precision. J Fungi (Basel) 3(4). https://doi.org/10.3390/jof3040057

Bonneil E, Pfammatter S, Thibault P (2015) Enhancement of mass spectrometry performance for proteomic analyses using high-field asymmetric waveform ion mobility spectrometry (FAIMS). J Mass Spectrom JMS 50(11):1181–1195. https://doi.org/10.1002/jms.3646

Brazma A, Hingamp P, Quackenbush J, Sherlock G, Spellman P, Stoeckert C, Aach J, Ansorge W, Ball CA, Causton HC, Gaasterland T, Glenisson P, Holstege FC, Kim IF, Markowitz V, Matese JC, Parkinson H, Robinson A, Sarkans U, Schulze-Kremer S, Stewart J, Taylor R, Vilo J, Vingron M (2001) Minimum information about a microarray experiment (MIAME)-toward standards for microarray data. Nat Genet 29(4):365–371. https://doi.org/10.1038/ng1201-365

Brown NA, Goldman GH (2016) The contribution of *Aspergillus fumigatus* stress responses to virulence and antifungal resistance. J Microbiol 54(3):243–253. https://doi.org/10.1007/s12275-016-5510-4

Brown GD, Denning DW, Gow NA, Levitz SM, Netea MG, White TC (2012) Hidden killers: human fungal infections. Sci Transl Med 4(165):165rv113. https://doi.org/10.1126/scitranslmed.3004404

Brown KA, Tucholski T, Eken C, Knott S, Zhu Y, Jin S, Ge Y (2020) High-throughput proteomics enabled by a photocleavable surfactant. Angew Chem Int Ed Eng 59(22):8406–8410. https://doi.org/10.1002/anie.201915374

Bruns S, Seidler M, Albrecht D, Salvenmoser S, Remme N, Hertweck C, Brakhage AA, Kniemeyer O, Muller FM (2010) Functional genomic profiling of *Aspergillus fumigatus* biofilm reveals enhanced production of the mycotoxin gliotoxin. Proteomics 10(17):3097–3107. https://doi.org/10.1002/pmic.201000129

Budnik B, Levy E, Harmange G, Slavov N (2018) SCoPE-MS: mass spectrometry of single mammalian cells quantifies proteome heterogeneity during cell differentiation. Genome Biol 19(1):161. https://doi.org/10.1186/s13059-018-1547-5

Cagas SE, Jain MR, Li H, Perlin DS (2011a) Profiling the *Aspergillus fumigatus* proteome in response to caspofungin. Antimicrob Agents Chemother 55(1):146–154. https://doi.org/10.1128/AAC.00884-10

Cagas SE, Jain MR, Li H, Perlin DS (2011b) The proteomic signature of *Aspergillus fumigatus* during early development. Mol Cell Proteomics MCP 10(11). https://doi.org/10.1074/mcp.M111.010108

Cai H, Shuai D, Xue X, Mo Y, Song X, Ye L, Li S, Wang D, Wang Y, Jin M (2021) Proteomic analysis of serum differentially expressed proteins between allergic bronchopulmonary aspergillosis and asthma. Mycopathologia 186(1):1–13. https://doi.org/10.1007/s11046-020-00506-0

Cao Z, Evans AR, Robinson RA (2015) MS(3)-based quantitative proteomics using pulsed-Q dissociation. Rapid Commun Mass Spectrom RCM 29(11):1025–1030. https://doi.org/10.1002/rcm.7192

Casadevall A, Pirofski LA (2012) Immunoglobulins in defense, pathogenesis, and therapy of fungal diseases. Cell Host Microbe 11(5):447–456. https://doi.org/10.1016/j.chom.2012.04.004

Chen T, Ma J, Liu Y, Chen Z, Xiao N, Lu Y, Fu Y, Yang C, Li M, Wu S, Wang X, Li D, He F, Hermjakob H, Zhu Y (2022) iProX in 2021: connecting proteomics data sharing with big data. Nucleic Acids Res 50(D1):D1522–D1527. https://doi.org/10.1093/nar/gkab1081

Chi H, Chen H, He K, Wu L, Yang B, Sun RX, Liu J, Zeng WF, Song CQ, He SM, Dong MQ (2013) pNovo+: de novo peptide sequencing using complementary HCD and ETD tandem mass spectra. J Proteome Res 12(2):615–625. https://doi.org/10.1021/pr3006843

Chua XY, Mensah T, Aballo T, Mackintosh SG, Edmondson RD, Salomon AR (2020) Tandem mass tag approach utilizing pervanadate BOOST channels delivers deeper quantitative characterization of the tyrosine phosphoproteome. Mole Cell Proteomics

MCP 19(4):730–743. https://doi.org/10.1074/mcp.TIR119.001865

Clauser KR, Baker P, Burlingame AL (1999) Role of accurate mass measurement (+/- 10 ppm) in protein identification strategies employing MS or MS/MS and database searching. Anal Chem 71(14):2871–2882. https://doi.org/10.1021/ac9810516

Cong Y, Liang Y, Motamedchaboki K, Huguet R, Truong T, Zhao R, Shen Y, Lopez-Ferrer D, Zhu Y, Kelly RT (2020) Improved single-cell proteome coverage using narrow-bore packed NanoLC columns and ultrasensitive mass spectrometry. Anal Chem 92(3):2665–2671. https://doi.org/10.1021/acs.analchem.9b04631

Conrad T, Kniemeyer O, Henkel SG, Kruger T, Mattern DJ, Valiante V, Guthke R, Jacobsen ID, Brakhage AA, Vlaic S, Linde J (2018) Module-detection approaches for the integration of multilevel omics data highlight the comprehensive response of *Aspergillus fumigatus* to caspofungin. BMC Syst Biol 12(1):88. https://doi.org/10.1186/s12918-018-0620-8

Cousin C, Aubatin A, Le Gouvello S, Apetoh L, Castellano F, Molinier-Frenkel V (2015) The immunosuppressive enzyme IL4I1 promotes FoxP3(+) regulatory T lymphocyte differentiation. Eur J Immunol 45(6):1772–1782. https://doi.org/10.1002/eji.201445000

Couvillion SP, Zhu Y, Nagy G, Adkins JN, Ansong C, Renslow RS, Piehowski PD, Ibrahim YM, Kelly RT, Metz TO (2019) New mass spectrometry technologies contributing towards comprehensive and high throughput omics analyses of single cells. Analyst 144(3):794–807. https://doi.org/10.1039/c8an01574k

Cox J, Neuhauser N, Michalski A, Scheltema RA, Olsen JV, Mann M (2011) Andromeda: a peptide search engine integrated into the MaxQuant environment. J Proteome Res 10(4):1794–1805. https://doi.org/10.1021/pr101065j

Craig R, Beavis RC (2004) TANDEM: matching proteins with tandem mass spectra. Bioinformatics (Oxford, England) 20(9):1466–1467. https://doi.org/10.1093/bioinformatics/bth092

Culibrk L, Croft CA, Tebbutt SJ (2016) Systems biology approaches for host-fungal interactions: an expanding multi-omics frontier. OMICS 20(3):127–138. https://doi.org/10.1089/omi.2015.0185

Curtis A, Ryan M, Kavanagh K (2023) Exposure of *Aspergillus fumigatus* to Klebsiella pneumoniae culture filtrate inhibits growth and stimulates gliotoxin production. J Fungi (Basel) 9(2). https://doi.org/10.3390/jof9020222

Curty N, Kubitschek-Barreira PH, Neves GW, Gomes D, Pizzatti L, Abdelhay E, Souza GH, Lopes-Bezerra LM (2014) Discovering the infectome of human endothelial cells challenged with *Aspergillus fumigatus* applying a mass spectrometry label-free approach. J Proteome 97:126–140. https://doi.org/10.1016/j.jprot.2013.07.003

Damerval C, De Vienne D, Zivy M, Thiellement H (1986) Technical improvements in two-dimensional electrophoresis increase the level of genetic variation detected in wheat-seedling proteins. Electrophoresis 7(1):52–54. https://doi.org/10.1002/elps.1150070108

Dancík V, Addona TA, Clauser KR, Vath JE, Pevzner PA (1999) De novo peptide sequencing via tandem mass spectrometry. J Comput Biol 6(3-4):327–342. https://doi.org/10.1089/106652799318300

de Godoy LM, Olsen JV, Cox J, Nielsen ML, Hubner NC, Fröhlich F, Walther TC, Mann M (2008) Comprehensive mass-spectrometry-based proteome quantification of haploid versus diploid yeast. Nature 455(7217):1251–1254. https://doi.org/10.1038/nature07341

de Groot PW, de Boer AD, Cunningham J, Dekker HL, de Jong L, Hellingwerf KJ, de Koster C, Klis FM (2004) Proteomic analysis of *Candida albicans* cell walls reveals covalently bound carbohydrate-active enzymes and adhesins. Eukaryot Cell 3(4):955–965. https://doi.org/10.1128/EC.3.4.955-965.2004

Delliere S, Aimanianda V (2023) Humoral immunity against *Aspergillus fumigatus*. Mycopathologia 188(5):603–621. https://doi.org/10.1007/s11046-023-00742-0

Delliere S, Duchateau M, Wong SSW, Giai Gianetto Q, Guegan H, Matondo M, Gangneux JP, Aimanianda V (2021) Proteomic analysis of humoral immune components in bronchoalveolar lavage of patients infected or colonized by *Aspergillus fumigatus*. Front Immunol 12:677798. https://doi.org/10.3389/fimmu.2021.677798

Desiere F, Deutsch EW, King NL, Nesvizhskii AI, Mallick P, Eng J, Chen S, Eddes J, Loevenich SN, Aebersold R (2006) The PeptideAtlas project. Nucleic Acids Res 34(Database issue):D655–D658. https://doi.org/10.1093/nar/gkj040

Deutsch EW, Albar JP, Binz PA, Eisenacher M, Jones AR, Mayer G, Omenn GS, Orchard S, Vizcaíno JA, Hermjakob H (2015) Development of data representation standards by the human proteome organization proteomics standards initiative. J Am Med Inform Assoc JAMIA 22(3):495–506. https://doi.org/10.1093/jamia/ocv001

Deutsch EW, Orchard S, Binz PA, Bittremieux W, Eisenacher M, Hermjakob H, Kawano S, Lam H, Mayer G, Menschaert G, Perez-Riverol Y, Salek RM, Tabb DL, Tenzer S, Vizcaíno JA, Walzer M, Jones AR (2017) Proteomics standards initiative: fifteen years of progress and future work. J Proteome Res 16(12):4288–4298. https://doi.org/10.1021/acs.jproteome.7b00370

Distler U, Łącki MK, Schumann S, Wanninger M, Tenzer S (2019) Enhancing sensitivity of microflow-based bottom-up proteomics through postcolumn solvent addition. Anal Chem 91(12):7510–7515. https://doi.org/10.1021/acs.analchem.9b00118

Doellinger J, Schneider A, Hoeller M, Lasch P (2020) Sample preparation by easy extraction and digestion (SPEED)—a universal, rapid, and detergent-free protocol for proteomics based on acid extraction. Mol

Cell Proteomics MCP 19(1):209–222. https://doi.org/10.1074/mcp.TIR119.001616

Donnelly JP, Chen SC, Kauffman CA, Steinbach WJ, Baddley JW, Verweij PE, Clancy CJ, Wingard JR, Lockhart SR, Groll AH, Sorrell TC, Bassetti M, Akan H, Alexander BD, Andes D, Azoulay E, Bialek R, Bradsher RW, Bretagne S, Calandra T, Caliendo AM, Castagnola E, Cruciani M, Cuenca-Estrella M, Decker CF, Desai SR, Fisher B, Harrison T, Heussel CP, Jensen HE, Kibbler CC, Kontoyiannis DP, Kullberg BJ, Lagrou K, Lamoth F, Lehrnbecher T, Loeffler J, Lortholary O, Maertens J, Marchetti O, Marr KA, Masur H, Meis JF, Morrisey CO, Nucci M, Ostrosky-Zeichner L, Pagano L, Patterson TF, Perfect JR, Racil Z, Roilides E, Ruhnke M, Prokop CS, Shoham S, Slavin MA, Stevens DA, Thompson GR, Vazquez JA, Viscoli C, Walsh TJ, Warris A, Wheat LJ, White PL, Zaoutis TE, Pappas PG (2020) Revision and update of the consensus definitions of invasive fungal disease from the European Organization for Research and Treatment of Cancer and the Mycoses Study Group Education and Research Consortium. Clin Infect Dis 71(6):1367–1376. https://doi.org/10.1093/cid/ciz1008

Dorfer V, Pichler P, Stranzl T, Stadlmann J, Taus T, Winkler S, Mechtler K (2014) MS Amanda, a universal identification algorithm optimized for high accuracy tandem mass spectra. J Proteome Res 13(8):3679–3684. https://doi.org/10.1021/pr500202e

Dorfer V, Maltsev S, Winkler S, Mechtler K (2018) CharmeRT: boosting peptide identifications by chimeric spectra identification and retention time prediction. J Proteome Res 17(8):2581–2589. https://doi.org/10.1021/acs.jproteome.7b00836

Doyle S, Jones GW, Dolan SK (2018) Dysregulated gliotoxin biosynthesis attenuates the production of unrelated biosynthetic gene cluster-encoded metabolites in *Aspergillus fumigatus*. Fungal Biol 122(4):214–221. https://doi.org/10.1016/j.funbio.2017.12.007

Druey KM, McCullough M, Krishnan R (2020) *Aspergillus fumigatus* protease alkaline protease 1 (Alp1): a new therapeutic target for fungal asthma. J Fungi (Basel) 6(2). https://doi.org/10.3390/jof6020088

Edman P (1950) Method for determination of the amino acid sequence in peptides. Acta Chem Scand 4:283–293. https://doi.org/10.3891/acta.chem.scand.04-0283

Eng JK, McCormack AL, Yates JR (1994) An approach to correlate tandem mass spectral data of peptides with amino acid sequences in a protein database. J Am Soc Mass Spectrom 5(11):976–989. https://doi.org/10.1016/1044-0305(94)80016-2

Eng JK, Jahan TA, Hoopmann MR (2013) Comet: an open-source MS/MS sequence database search tool. Proteomics 13(1):22–24. https://doi.org/10.1002/pmic.201200439

Espina V, Wulfkuhle JD, Calvert VS, VanMeter A, Zhou W, Coukos G, Geho DH, Petricoin EF 3rd, Liotta LA (2006) Laser-capture microdissection. Nat Protoc 1(2):586–603. https://doi.org/10.1038/nprot.2006.85

Fekkar A, Balloy V, Pionneau C, Marinach-Patrice C, Chignard M, Mazier D (2012a) Secretome of human bronchial epithelial cells in response to the fungal pathogen *Aspergillus fumigatus* analyzed by differential in-gel electrophoresis. J Infect Dis 205(7):1163–1172. https://doi.org/10.1093/infdis/jis031

Fekkar A, Pionneau C, Brossas JY, Marinach-Patrice C, Snounou G, Brock M, Ibrahim-Granet O, Mazier D (2012b) DIGE enables the detection of a putative serum biomarker of fungal origin in a mouse model of invasive aspergillosis. J Proteome 75(9):2536–2549. https://doi.org/10.1016/j.jprot.2012.01.040

Fisher MC, Alastruey-Izquierdo A, Berman J, Bicanic T, Bignell EM, Bowyer P, Bromley M, Bruggemann R, Garber G, Cornely OA, Gurr SJ, Harrison TS, Kuijper E, Rhodes J, Sheppard DC, Warris A, White PL, Xu J, Zwaan B, Verweij PE (2022) Tackling the emerging threat of antifungal resistance to human health. Nat Rev Microbiol 20(9):557–571. https://doi.org/10.1038/s41579-022-00720-1

Frank A, Pevzner P (2005) PepNovo: de novo peptide sequencing via probabilistic network modeling. Anal Chem 77(4):964–973. https://doi.org/10.1021/ac048788h

Furtwängler B, Üresin N, Motamedchaboki K, Huguet R, Lopez-Ferrer D, Zabrouskov V, Porse BT, Schoof EM (2022) Real-time search-assisted acquisition on a tribrid mass spectrometer improves coverage in multiplexed single-cell proteomics. Mol Cell Proteomics MCP 21(4):100219. https://doi.org/10.1016/j.mcpro.2022.100219

Futcher B, Latter GI, Monardo P, McLaughlin CS, Garrels JI (1999) A sampling of the yeast proteome. Mol Cell Biol 19(11):7357–7368. https://doi.org/10.1128/mcb.19.11.7357

Gabelica V, Shvartsburg AA, Afonso C, Barran P, Benesch JLP, Bleiholder C, Bowers MT, Bilbao A, Bush MF, Campbell JL, Campuzano IDG, Causon T, Clowers BH, Creaser CS, De Pauw E, Far J, Fernandez-Lima F, Fjeldsted JC, Giles K, Groessl M, Hogan CJ Jr, Hann S, Kim HI, Kurulugama RT, May JC, McLean JA, Pagel K, Richardson K, Ridgeway ME, Rosu F, Sobott F, Thalassinos K, Valentine SJ, Wyttenbach T (2019) Recommendations for reporting ion mobility mass spectrometry measurements. Mass Spectrom Rev 38(3):291–320. https://doi.org/10.1002/mas.21585

Gao Y, Ping L, Duong D, Zhang C, Dammer EB, Li Y, Chen P, Chang L, Gao H, Wu J, Xu P (2021) Mass-spectrometry-based near-complete draft of the Saccharomyces cerevisiae proteome. J Proteome Res 20(2):1328–1340. https://doi.org/10.1021/acs.jproteome.0c00721

Gautam P, Sundaram CS, Madan T, Gade WN, Shah A, Sirdeshmukh R, Sarma PU (2007) Identification of novel allergens of *Aspergillus fumigatus* using immunoproteomics approach. Clin Exp Allergy 37(8):1239–1249. https://doi.org/10.1111/j.1365-2222.2007.02765.x

Gautam P, Shankar J, Madan T, Sirdeshmukh R, Sundaram CS, Gade WN, Basir SF, Sarma PU (2008) Proteomic and transcriptomic analysis of *Aspergillus fumigatus* on exposure to amphotericin B. Antimicrob Agents Chemother 52(12):4220–4227. https://doi.org/10.1128/AAC.01431-07

Gautam P, Upadhyay SK, Hassan W, Madan T, Sirdeshmukh R, Sundaram CS, Gade WN, Basir SF, Singh Y, Sarma PU (2011) Transcriptomic and proteomic profile of *Aspergillus fumigatus* on exposure to artemisinin. Mycopathologia 172(5):331–346. https://doi.org/10.1007/s11046-011-9445-3

Gautam P, Mushahary D, Hassan W, Upadhyay SK, Madan T, Sirdeshmukh R, Sundaram CS, Sarma PU (2016) In-depth 2-DE reference map of *Aspergillus fumigatus* and its proteomic profiling on exposure to itraconazole. Med Mycol 54(5):524–536. https://doi.org/10.1093/mmy/myv122

Geer LY, Markey SP, Kowalak JA, Wagner L, Xu M, Maynard DM, Yang X, Shi W, Bryant SH (2004) Open mass spectrometry search algorithm. J Proteome Res 3(5):958–964. https://doi.org/10.1021/pr0499491

Getz EB, Xiao M, Chakrabarty T, Cooke R, Selvin PR (1999) A comparison between the sulfhydryl reductants tris(2-carboxyethyl)phosphine and dithiothreitol for use in protein biochemistry. Anal Biochem 273(1):73–80. https://doi.org/10.1006/abio.1999.4203

Giansanti P, Tsiatsiani L, Low TY, Heck AJ (2016) Six alternative proteases for mass spectrometry-based proteomics beyond trypsin. Nat Protoc 11(5):993–1006. https://doi.org/10.1038/nprot.2016.057

Gil-Bona A, Parra-Giraldo CM, Hernaez ML, Reales-Calderon JA, Solis NV, Filler SG, Monteoliva L, Gil C (2015) *Candida albicans* cell shaving uncovers new proteins involved in cell wall integrity, yeast to hypha transition, stress response and host-pathogen interaction. J Proteome 127(Pt B):340–351. https://doi.org/10.1016/j.jprot.2015.06.006

Gillet LC, Navarro P, Tate S, Röst H, Selevsek N, Reiter L, Bonner R, Aebersold R (2012) Targeted data extraction of the MS/MS spectra generated by data-independent acquisition: a new concept for consistent and accurate proteome analysis. Mol Cell Proteomics MCP 11(6). https://doi.org/10.1074/mcp.O111.016717

Gouw JW, Krijgsveld J, Heck AJ (2010) Quantitative proteomics by metabolic labeling of model organisms. Mol Cell Proteomics MCP 9(1):11–24. https://doi.org/10.1074/mcp.R900001-MCP200

Grahl N, Shepardson KM, Chung D, Cramer RA (2012) Hypoxia and fungal pathogenesis: to air or not to air? Eukaryot Cell 11(5):560–570. https://doi.org/10.1128/EC.00031-12

Grosse C, Heinekamp T, Kniemeyer O, Gehrke A, Brakhage AA (2008) Protein kinase A regulates growth, sporulation, and pigment formation in *Aspergillus fumigatus*. Appl Environ Microbiol 74(15):4923–4933. https://doi.org/10.1128/AEM.00470-08

Gutierrez DB, Gant-Branum RL, Romer CE, Farrow MA, Allen JL, Dahal N, Nei YW, Codreanu SG, Jordan AT, Palmer LD, Sherrod SD, McLean JA, Skaar EP, Norris JL, Caprioli RM (2018) An integrated, high-throughput strategy for multiomic systems level analysis. J Proteome Res 17(10):3396–3408. https://doi.org/10.1021/acs.jproteome.8b00302

Gygi SP, Rist B, Gerber SA, Turecek F, Gelb MH, Aebersold R (1999) Quantitative analysis of complex protein mixtures using isotope-coded affinity tags. Nat Biotechnol 17(10):994–999. https://doi.org/10.1038/13690

Hagag S, Kubitschek-Barreira P, Neves GW, Amar D, Nierman W, Shalit I, Shamir R, Lopes-Bezerra L, Osherov N (2012) Transcriptional and proteomic analysis of the *Aspergillus fumigatus ΔprtT* protease-deficient mutant. PLoS One 7(4):e33604. https://doi.org/10.1371/journal.pone.0033604

HaileMariam M, Eguez RV, Singh H, Bekele S, Ameni G, Pieper R, Yu Y (2018) S-trap, an ultrafast sample-preparation approach for shotgun proteomics. J Proteome Res 17(9):2917–2924. https://doi.org/10.1021/acs.jproteome.8b00505

Hebert AS, Prasad S, Belford MW, Bailey DJ, McAlister GC, Abbatiello SE, Huguet R, Wouters ER, Dunyach JJ, Brademan DR, Westphall MS, Coon JJ (2018) Comprehensive single-shot proteomics with FAIMS on a hybrid orbitrap mass spectrometer. Anal Chem 90(15):9529–9537. https://doi.org/10.1021/acs.analchem.8b02233

Heil LR, Damoc E, Arrey TN, Pashkova A, Denisov E, Petzoldt J, Peterson AC, Hsu C, Searle BC, Shulman N, Riffle M, Connolly B, MacLean BX, Remes PM, Senko MW, Stewart HI, Hock C, Makarov AA, Hermanson D, Zabrouskov V, Wu CC, MacCoss MJ (2023) Evaluating the performance of the astral mass analyzer for quantitative proteomics using data-independent acquisition. J Proteome Res 22(10):3290–3300. https://doi.org/10.1021/acs.jproteome.3c00357

Hermjakob H (2006) The HUPO proteomics standards initiative—overcoming the fragmentation of proteomics data. Proteomics 6(Suppl 2):34–38. https://doi.org/10.1002/pmic.200600537

Hernaez ML, Ximenez-Embun P, Martinez-Gomariz M, Gutierrez-Blazquez MD, Nombela C, Gil C (2010) Identification of *Candida albicans* exposed surface proteins in vivo by a rapid proteomic approach. J Proteome 73(7):1404–1409. https://doi.org/10.1016/j.jprot.2010.02.008

Horta MAC, Steenwyk JL, Mead ME, Dos Santos LHB, Zhao S, Gibbons JG, Marcet-Houben M, Gabaldon T, Rokas A, Goldman GH (2022) Examination of genome-wide ortholog variation in clinical and environmental isolates of the fungal pathogen *Aspergillus fumigatus*. mBio 13(4):e0151922. https://doi.org/10.1128/mbio.01519-22

Hsu JL, Huang SY, Chow NH, Chen SH (2003) Stable-isotope dimethyl labeling for quantitative proteomics. Anal Chem 75(24):6843–6852. https://doi.org/10.1021/ac0348625

Hu Q, Noll RJ, Li H, Makarov A, Hardman M, Graham Cooks R (2005) The Orbitrap: a new mass spectrometer. J Mass Spectrom JMS 40(4):430–443. https://doi.org/10.1002/jms.856

Huesgen PF, Lange PF, Rogers LD, Solis N, Eckhard U, Kleifeld O, Goulas T, Gomis-Rüth FX, Overall CM (2015) LysargiNase mirrors trypsin for protein C-terminal and methylation-site identification. Nat Methods 12(1):55–58. https://doi.org/10.1038/nmeth.3177

Hughes CS, Moggridge S, Müller T, Sorensen PH, Morin GB, Krijgsveld J (2019) Single-pot, solid-phase-enhanced sample preparation for proteomics experiments. Nat Protoc 14(1):68–85. https://doi.org/10.1038/s41596-018-0082-x

Iacob RE, Murphy JP 3rd, Engen JR (2008) Ion mobility adds an additional dimension to mass spectrometric analysis of solution-phase hydrogen/deuterium exchange. Rapid Commun Mass Spectrom RCM 22(18):2898–2904. https://doi.org/10.1002/rcm.3688

Ibarrola N, Molina H, Iwahori A, Pandey A (2004) A novel proteomic approach for specific identification of tyrosine kinase substrates using [13C]tyrosine. J Biol Chem 279(16):15805–15813. https://doi.org/10.1074/jbc.M311714200

Ideker T, Galitski T, Hood L (2001) A new approach to decoding life: systems biology. Annu Rev Genomics Hum Genet 2:343–372. https://doi.org/10.1146/annurev.genom.2.1.343

Izumikawa K, Tashiro T, Tashiro M, Takazono T, Kosai K, Morinaga Y, Kurihara S, Nakamura S, Imamura Y, Miyazaki T, Tsukamoto M, Kakeya H, Hayashi T, Yanagihara K, Nagayasu T, Kohno S (2014) Pathogenesis and clinical features of chronic pulmonary aspergillosis—is it possible to distinguish CNPA and CCPA clinically? J Infect Chemother 20(3):208–212. https://doi.org/10.1016/j.jiac.2013.10.016

Ji Y, Liu M, Bachschmid MM, Costello CE, Lin C (2015) Surfactant-induced artifacts during proteomic sample preparation. Anal Chem 87(11):5500–5504. https://doi.org/10.1021/acs.analchem.5b00249

Jia LJ, Kruger T, Blango MG, von Eggeling F, Kniemeyer O, Brakhage AA (2020) Biotinylated surfome profiling identifies potential biomarkers for diagnosis and therapy of *Aspergillus fumigatus* infection. mSphere 5(4). https://doi.org/10.1128/mSphere.00535-20

Juvvadi PR, Ma Y, Richards AD, Soderblom EJ, Moseley MA, Lamoth F, Steinbach WJ (2015a) Identification and mutational analyses of phosphorylation sites of the calcineurin-binding protein CbpA and the identification of domains required for calcineurin binding in *Aspergillus fumigatus*. Front Microbiol 6:175. https://doi.org/10.3389/fmicb.2015.00175

Juvvadi PR, Munoz A, Lamoth F, Soderblom EJ, Moseley MA, Read ND, Steinbach WJ (2015b) Calcium-mediated induction of paradoxical growth following caspofungin treatment is associated with calcineurin activation and phosphorylation in *Aspergillus fumigatus*. Antimicrob Agents Chemother 59(8):4946–4955. https://doi.org/10.1128/AAC.00263-15

Juvvadi PR, Moseley MA, Hughes CJ, Soderblom EJ, Lennon S, Perkins SR, Thompson JW, Geromanos SJ, Wildgoose J, Richardson K, Langridge JI, Vissers JPC, Steinbach WJ (2018) Scanning quadrupole data-independent acquisition, part B: application to the analysis of the calcineurin-interacting proteins during treatment of *Aspergillus fumigatus* with azole and echinocandin antifungal drugs. J Proteome Res 17(2):780–793. https://doi.org/10.1021/acs.jproteome.7b00499

Kang UB, Yeom J, Kim H, Lee C (2010) Quantitative analysis of mTRAQ-labeled proteome using full MS scans. J Proteome Res 9(7):3750–3758. https://doi.org/10.1021/pr9011014

Kanjanapruthipong T, Sukphopetch P, Reamtong O, Isarangkul D, Muangkaew W, Thiangtrongjit T, Sansurin N, Fongsodsri K, Ampawong S (2022) Cytoskeletal alteration is an early cellular response in pulmonary epithelium infected with *Aspergillus fumigatus* rather than *Scedosporium apiospermum*. Microb Ecol 83(1):216–235. https://doi.org/10.1007/s00248-021-01750-7

Kawashima Y, Nagai H, Konno R, Ishikawa M, Nakajima D, Sato H, Nakamura R, Furuyashiki T, Ohara O (2022) Single-shot 10K proteome approach: over 10,000 protein identifications by data-independent acquisition-based single-shot proteomics with ion mobility spectrometry. J Proteome Res 21(6):1418–1427. https://doi.org/10.1021/acs.jproteome.2c00023

Kelani AA, Bruch A, Rivieccio F, Visser C, Kruger T, Weaver D, Pan X, Schauble S, Panagiotou G, Kniemeyer O, Bromley MJ, Bowyer P, Barber AE, Brakhage AA, Blango MG (2023) Disruption of the *Aspergillus fumigatus* RNA interference machinery alters the conidial transcriptome. RNA 29(7):1033–1050. https://doi.org/10.1261/rna.079350.122

Kellermann J, Lottspeich F (2012) Isotope-coded protein label. Methods Mol Biol (Clifton, NJ) 893:143–153. https://doi.org/10.1007/978-1-61779-885-6_11

Keri VC, Chandrasekar PH (2023) Polymerase chain reaction in the diagnosis of invasive aspergillosis: approaches for appropriate use. Leuk Lymphoma 64(7):1330–1334. https://doi.org/10.1080/10428194.2023.2209683

Khalaj V, Azarian B, Enayati S, Vaziri B (2011) Annexin C4 in *A. fumigatus*: a proteomics approach to understand the function. J Proteome 74(10):1950–1958. https://doi.org/10.1016/j.jprot.2011.05.018

Kim S, Mischerikow N, Bandeira N, Navarro JD, Wich L, Mohammed S, Heck AJ, Pevzner PA (2010) The generating function of CID, ETD, and CID/ETD pairs of tandem mass spectra: applications to database search. Mole Cell Proteomics MCP 9(12):2840–2852. https://doi.org/10.1074/mcp.M110.003731

Kniemeyer O, Schmidt AD, Vodisch M, Wartenberg D, Brakhage AA (2011) Identification of virulence determinants of the human pathogenic fungi *Aspergillus fumigatus* and *Candida albicans* by proteomics. Int J Med Microbiol 301(5):368–377. https://doi.org/10.1016/j.ijmm.2011.04.001

Köcher T, Pichler P, Swart R, Mechtler K (2012) Analysis of protein mixtures from whole-cell extracts by

single-run nanoLC-MS/MS using ultralong gradients. Nat Protoc 7(5):882–890. https://doi.org/10.1038/nprot.2012.036
Kong AT, Leprevost FV, Avtonomov DM, Mellacheruvu D, Nesvizhskii AI (2017) MSFragger: ultrafast and comprehensive peptide identification in mass spectrometry-based proteomics. Nat Methods 14(5):513–520. https://doi.org/10.1038/nmeth.4256
Kovanich D, Cappadona S, Raijmakers R, Mohammed S, Scholten A, Heck AJ (2012) Applications of stable isotope dimethyl labeling in quantitative proteomics. Anal Bioanal Chem 404(4):991–1009. https://doi.org/10.1007/s00216-012-6070-z
Krel M, Petraitis V, Petraitiene R, Jain MR, Zhao Y, Li H, Walsh TJ, Perlin DS (2014) Host biomarkers of invasive pulmonary aspergillosis to monitor therapeutic response. Antimicrob Agents Chemother 58(6):3373–3378. https://doi.org/10.1128/AAC.02482-14
Krüger T, Lehmann T, Rhode H (2013) Effect of quality characteristics of single sample preparation steps in the precision and coverage of proteomic studies—a review. Anal Chim Acta 776:1–10. https://doi.org/10.1016/j.aca.2013.01.020
Krüger T, Luo T, Schmidt H, Shopova I, Kniemeyer O (2015) Challenges and strategies for proteome analysis of the interaction of human pathogenic fungi with host immune cells. Proteomes 3(4):467–495. https://doi.org/10.3390/proteomes3040467
Kulak NA, Pichler G, Paron I, Nagaraj N, Mann M (2014) Minimal, encapsulated proteomic-sample processing applied to copy-number estimation in eukaryotic cells. Nat Methods 11(3):319–324. https://doi.org/10.1038/nmeth.2834
Kumar A, Ahmed R, Singh PK, Shukla PK (2011) Identification of virulence factors and diagnostic markers using immunosecretome of *Aspergillus fumigatus*. J Proteome 74(7):1104–1112. https://doi.org/10.1016/j.jprot.2011.04.004
Kurucz V, Kruger T, Antal K, Dietl AM, Haas H, Pocsi I, Kniemeyer O, Emri T (2018) Additional oxidative stress reroutes the global response of *Aspergillus fumigatus* to iron depletion. BMC Genomics 19(1):357. https://doi.org/10.1186/s12864-018-4730-x
Kwon-Chung KJ, Sugui JA (2013) *Aspergillus fumigatus*—what makes the species a ubiquitous human fungal pathogen? PLoS Pathog 9(12):e1003743. https://doi.org/10.1371/journal.ppat.1003743
Lass-Florl C (2019) How to make a fast diagnosis in invasive aspergillosis. Med Mycol 57(suppl_2):S155–S160. https://doi.org/10.1093/mmy/myy103
Latge JP, Chamilos G (2019) *Aspergillus fumigatus* and Aspergillosis in 2019. Clin Microbiol Rev 33(1). https://doi.org/10.1128/CMR.00140-18
Lau E, Lam MP, Siu SO, Kong RP, Chan WL, Zhou Z, Huang J, Lo C, Chu IK (2011) Combinatorial use of offline SCX and online RP-RP liquid chromatography for iTRAQ-based quantitative proteomics applications. Mol BioSyst 7(5):1399–1408. https://doi.org/10.1039/c1mb05010a
Li H, Hwang KB, Mun DG, Kim H, Lee H, Lee SW, Paek E (2014) Estimating influence of cofragmentation on peptide quantification and identification in iTRAQ experiments by simulating multiplexed spectra. J Proteome Res 13(7):3488–3497. https://doi.org/10.1021/pr500060d
Li Y, Li H, Sui M, Li M, Wang J, Meng Y, Sun T, Liang Q, Suo C, Gao X, Li C, Li Z, Du W, Zhang B, Sai S, Zhang Z, Ye J, Wang H, Yue S, Li J, Zhong M, Chen C, Qi S, Lu L, Li D, Ding C (2019) Fungal acetylome comparative analysis identifies an essential role of acetylation in human fungal pathogen virulence. Commun Biol 2:154. https://doi.org/10.1038/s42003-019-0419-1
Lindemann C, Leichert LI (2012) Quantitative redox proteomics: the NOxICAT method. Methods Mol Biol (Clifton, NJ) 893:387–403. https://doi.org/10.1007/978-1-61779-885-6_24
Link AJ, LaBaer J (2011) Trichloroacetic acid (TCA) precipitation of proteins. Cold Spring Harb Protoc 2011(8):993–994. https://doi.org/10.1101/pdb.prot5651
Link AJ, Eng J, Schieltz DM, Carmack E, Mize GJ, Morris DR, Garvik BM, Yates JR 3rd (1999) Direct analysis of protein complexes using mass spectrometry. Nat Biotechnol 17(7):676–682. https://doi.org/10.1038/10890
Liu S, Le Mauff F, Sheppard DC, Zhang S (2022) Filamentous fungal biofilms: Conserved and unique aspects of extracellular matrix composition, mechanisms of drug resistance and regulatory networks in *Aspergillus fumigatus*. NPJ Biofilms Microbiomes 8(1):83. https://doi.org/10.1038/s41522-022-00347-3
Lo A, Tang Y, Chen L, Li L (2013) Automation of dimethylation after guanidination labeling chemistry and its compatibility with common buffers and surfactants for mass spectrometry-based shotgun quantitative proteome analysis. Anal Chim Acta 788:81–88. https://doi.org/10.1016/j.aca.2013.05.060
Lou R, Liu W, Li R, Li S, He X, Shui W (2021) DeepPhospho accelerates DIA phosphoproteome profiling through in silico library generation. Nat Commun 12(1):6685. https://doi.org/10.1038/s41467-021-26979-1
Ma B, Zhang K, Hendrie C, Liang C, Li M, Doherty-Kirby A, Lajoie G (2003) PEAKS: powerful software for peptide de novo sequencing by tandem mass spectrometry. Rapid Commun Mass Spectrom RCM 17(20):2337–2342. https://doi.org/10.1002/rcm.1196
Ma X, Liu J, Zhang Z, Bo T, Bai Y, Liu H (2017) Drift tube ion mobility and four-dimensional molecular feature extraction enable data-independent tandem mass spectrometric 'omics' analysis without quadrupole selection. Rapid Commun Mass Spectrom RCM 31(1):33–38. https://doi.org/10.1002/rcm.7767
Machata S, Muller MM, Lehmann R, Sieber P, Panagiotou G, Carvalho A, Cunha C, Lagrou K, Maertens J, Slevogt H, Jacobsen ID (2020) Proteome analysis of bronchoalveolar lavage fluids reveals host and fungal proteins highly expressed during invasive pul-

monary aspergillosis in mice and humans. Virulence 11(1):1337–1351. https://doi.org/10.1080/21505594.2020.1824960

Macheleidt J, Kniemeyer O (2021) Serological Proteome Analysis for the Characterization of Secreted Fungal Protein Antigens. Methods Mol Biol (Clifton, NJ) 2260:15–26. https://doi.org/10.1007/978-1-0716-1182-1_2

Magee LC, Louis M, Khan V, Micalo L, Chaudary N (2021) Managing fungal infections in cystic fibrosis patients: challenges in clinical practice. Infect Drug Resist 14:1141–1153. https://doi.org/10.2147/IDR.S267219

Makarov A (2019) Orbitrap journey: taming the ion rings. Nat Commun 10(1):3743. https://doi.org/10.1038/s41467-019-11748-y

Mao Y, Tipton JD, Blakney GT, Hendrickson CL, Marshall AG (2011) Valence parity to distinguish c′ and z• ions from electron capture dissociation/electron transfer dissociation of peptides: effects of isomers, isobars, and proteolysis specificity. Anal Chem 83(20):8024–8028. https://doi.org/10.1021/ac201619t

Margalit A, Carolan JC, Sheehan D, Kavanagh K (2020a) The Aspergillus fumigatus secretome alters the proteome of *Pseudomonas aeruginosa* to stimulate bacterial growth: implications for co-infection. Mol Cell Proteomics MCP 19(8):1346–1359. https://doi.org/10.1074/mcp.RA120.002059

Margalit A, Kavanagh K, Carolan JC (2020b) Characterization of the proteomic response of A549 cells following sequential exposure to *Aspergillus fumigatus* and *Pseudomonas aeruginosa*. J Proteome Res 19(1):279–291. https://doi.org/10.1021/acs.jproteome.9b00520

Margalit A, Sheehan D, Carolan JC, Kavanagh K (2022) Exposure to the *Pseudomonas aeruginosa* secretome alters the proteome and secondary metabolite production of *Aspergillus fumigatus*. Microbiology (Reading) 168(3). https://doi.org/10.1099/mic.0.001164

Martínez-Bartolomé S, Deutsch EW, Binz PA, Jones AR, Eisenacher M, Mayer G, Campos A, Canals F, Bech-Serra JJ, Carrascal M, Gay M, Paradela A, Navajas R, Marcilla M, Hernáez ML, Gutiérrez-Blázquez MD, Velarde LF, Aloria K, Beaskoetxea J, Medina-Aunon JA, Albar JP (2013) Guidelines for reporting quantitative mass spectrometry based experiments in proteomics. J Proteome 95:84–88. https://doi.org/10.1016/j.jprot.2013.02.026

Mattos EC, Palmisano G, Goldman GH (2020a) Phosphoproteomics of *Aspergillus fumigatus* exposed to the antifungal drug caspofungin. mSphere 5(3). https://doi.org/10.1128/mSphere.00365-20

Mattos EC, Silva LP, Valero C, de Castro PA, Dos Reis TF, Ribeiro LFC, Marten MR, Silva-Rocha R, Westmann C, da Silva C, Taft CA, Al-Furaiji N, Bromley M, Mortensen UH, Benz JP, Brown NA, Goldman GH (2020b) The *Aspergillus fumigatus* phosphoproteome reveals roles of high-osmolarity glycerol mitogen-activated protein kinases in promoting cell wall damage and caspofungin tolerance. mBio 11(1). https://doi.org/10.1128/mBio.02962-19

May GS, Xue T, Kontoyiannis DP, Gustin MC (2005) Mitogen activated protein kinases of *Aspergillus fumigatus*. Med Mycol 43(Suppl 1):S83–S86. https://doi.org/10.1080/13693780400024784

Mayer G, Montecchi-Palazzi L, Ovelleiro D, Jones AR, Binz PA, Deutsch EW, Chambers M, Kallhardt M, Levander F, Shofstahl J, Orchard S, Vizcaíno JA, Hermjakob H, Stephan C, Meyer HE, Eisenacher M (2013) The HUPO proteomics standards initiative-mass spectrometry controlled vocabulary. Database (Oxford) 2013:bat009. https://doi.org/10.1093/database/bat009

McAlister GC, Nusinow DP, Jedrychowski MP, Wühr M, Huttlin EL, Erickson BK, Rad R, Haas W, Gygi SP (2014) MultiNotch MS3 enables accurate, sensitive, and multiplexed detection of differential expression across cancer cell line proteomes. Anal Chem 86(14):7150–7158. https://doi.org/10.1021/ac502040v

Meier F, Beck S, Grassl N, Lubeck M, Park MA, Raether O, Mann M (2015) Parallel accumulation-serial fragmentation (PASEF): multiplying sequencing speed and sensitivity by synchronized scans in a trapped ion mobility device. J Proteome Res 14(12):5378–5387. https://doi.org/10.1021/acs.jproteome.5b00932

Meier F, Geyer PE, Virreira Winter S, Cox J, Mann M (2018) BoxCar acquisition method enables single-shot proteomics at a depth of 10,000 proteins in 100 minutes. Nat Methods 15(6):440–448. https://doi.org/10.1038/s41592-018-0003-5

Meier F, Park MA, Mann M (2021) Trapped ion mobility spectrometry and parallel accumulation-serial fragmentation in proteomics. Mol Cell Proteomics MCP 20:100138. https://doi.org/10.1016/j.mcpro.2021.100138

Merrill AE, Hebert AS, MacGilvray ME, Rose CM, Bailey DJ, Bradley JC, Wood WW, El Masri M, Westphall MS, Gasch AP, Coon JJ (2014) NeuCode labels for relative protein quantification. Mol Cell Proteomics MCP 13(9):2503–2512. https://doi.org/10.1074/mcp.M114.040287

Misslinger M, Hortschansky P, Brakhage AA, Haas H (2021) Fungal iron homeostasis with a focus on *Aspergillus fumigatus*. Biochim Biophys Acta Mol Cell Res 1868(1):118885. https://doi.org/10.1016/j.bbamcr.2020.118885

Moloney NM, Owens RA, Meleady P, Henry M, Dolan SK, Mulvihill E, Clynes M, Doyle S (2016) The iron-responsive microsomal proteome of *Aspergillus fumigatus*. J Proteome 136:99–111. https://doi.org/10.1016/j.jprot.2015.12.025

Moon H, Han KH, Yu JH (2022) Upstream regulation of development and secondary metabolism in Aspergillus species. Cells 12(1). https://doi.org/10.3390/cells12010002

Morelli KA, Kerkaert JD, Cramer RA (2021) *Aspergillus fumigatus* biofilms: Toward understanding how growth as a multicellular network increases antifungal resistance and disease progression. PLoS Pathog 17(8):e1009794. https://doi.org/10.1371/journal.ppat.1009794

Müller T, Winter D (2017) Systematic evaluation of protein reduction and alkylation reveals massive unspecific side effects by iodine-containing reagents. Mol Cell Proteomics MCP 16(7):1173–1187. https://doi.org/10.1074/mcp.M116.064048

Müller JB, Geyer PE, Colaço AR, Treit PV, Strauss MT, Oroshi M, Doll S, Virreira Winter S, Bader JM, Köhler N, Theis F, Santos A, Mann M (2020) The proteome landscape of the kingdoms of life. Nature 582(7813):592–596. https://doi.org/10.1038/s41586-020-2402-x

Muszkieta L, Beauvais A, Pahtz V, Gibbons JG, Anton Leberre V, Beau R, Shibuya K, Rokas A, Francois JM, Kniemeyer O, Brakhage AA, Latge JP (2013) Investigation of *Aspergillus fumigatus* biofilm formation by various "omics" approaches. Front Microbiol 4:13. https://doi.org/10.3389/fmicb.2013.00013

Neilson KA, Ali NA, Muralidharan S, Mirzaei M, Mariani M, Assadourian G, Lee A, van Sluyter SC, Haynes PA (2011) Less label, more free: approaches in label-free quantitative mass spectrometry. Proteomics 11(4):535–553. https://doi.org/10.1002/pmic.201000553

Neves GW, Curty N, Kubitschek-Barreira PH, Fontaine T, Souza GH, Cunha ML, Goldman GH, Beauvais A, Latge JP, Lopes-Bezerra LM (2016) Dataset of differentially regulated proteins in HUVECs challenged with wild type and UGM1 mutant *Aspergillus fumigatus* strains. Data Brief 9:24–31. https://doi.org/10.1016/j.dib.2016.07.062

Neves GW, Curty NA, Kubitschek-Barreira PH, Fontaine T, Souza GH, Cunha ML, Goldman GH, Beauvais A, Latge JP, Lopes-Bezerra LM (2017) Modifications to the composition of the hyphal outer layer of *Aspergillus fumigatus* modulates HUVEC proteins related to inflammatory and stress responses. J Proteome 151:83–96. https://doi.org/10.1016/j.jprot.2016.06.015

Nickerson JL, Doucette AA (2020) Rapid and quantitative protein precipitation for proteome analysis by mass spectrometry. J Proteome Res 19(5):2035–2042. https://doi.org/10.1021/acs.jproteome.9b00867

Nielsen ML, Vermeulen M, Bonaldi T, Cox J, Moroder L, Mann M (2008) Iodoacetamide-induced artifact mimics ubiquitination in mass spectrometry. Nat Methods 5(6):459–460. https://doi.org/10.1038/nmeth0608-459

Okuda S, Watanabe Y, Moriya Y, Kawano S, Yamamoto T, Matsumoto M, Takami T, Kobayashi D, Araki N, Yoshizawa AC, Tabata T, Sugiyama N, Goto S, Ishihama Y (2017) jPOSTrepo: an international standard data repository for proteomes. Nucleic Acids Res 45(D1):D1107–D1111. https://doi.org/10.1093/nar/gkw1080

Olsen JV, Ong SE, Mann M (2004) Trypsin cleaves exclusively C-terminal to arginine and lysine residues. Mol Cell Proteomics MCP 3(6):608–614. https://doi.org/10.1074/mcp.T400003-MCP200

Olsen JV, Macek B, Lange O, Makarov A, Horning S, Mann M (2007) Higher-energy C-trap dissociation for peptide modification analysis. Nat Methods 4(9):709–712. https://doi.org/10.1038/nmeth1060

Ong SE, Mann M (2007) Stable isotope labeling by amino acids in cell culture for quantitative proteomics. Methods Mol Biol (Clifton, NJ) 359:37–52. https://doi.org/10.1007/978-1-59745-255-7_3

Ong SE, Blagoev B, Kratchmarova I, Kristensen DB, Steen H, Pandey A, Mann M (2002) Stable isotope labeling by amino acids in cell culture, SILAC, as a simple and accurate approach to expression proteomics. Mol Cell Proteomics MCP 1(5):376–386. https://doi.org/10.1074/mcp.m200025-mcp200

Owens RA, Hammel S, Sheridan KJ, Jones GW, Doyle S (2014) A proteomic approach to investigating gene cluster expression and secondary metabolite functionality in *Aspergillus fumigatus*. PLoS One 9(9):e106942. https://doi.org/10.1371/journal.pone.0106942

Park HS, Lee SC, Cardenas ME, Heitman J (2019) Calcium-calmodulin-calcineurin signaling: a globally conserved virulence cascade in eukaryotic microbial pathogens. Cell Host Microbe 26(4):453–462. https://doi.org/10.1016/j.chom.2019.08.004

Patterson SD, Aebersold RH (2003) Proteomics: the first decade and beyond. Nat Genet 33(Suppl):311–323. https://doi.org/10.1038/ng1106

Paul S, Ror S, McDonald WH, Moye-Rowley WS (2022) Biochemical identification of a nuclear coactivator protein required for AtrR-dependent gene regulation in *Aspergillus fumigatus*. mSphere 7(6):e0047622. https://doi.org/10.1128/msphere.00476-22

Paulo JA, Gygi SP (2019) mTMT: an alternative, nonisobaric, tandem mass tag allowing for precursor-based quantification. Anal Chem 91(19):12167–12172. https://doi.org/10.1021/acs.analchem.9b03162

Perez-Riverol Y, Bai J, Bandla C, García-Seisdedos D, Hewapathirana S, Kamatchinathan S, Kundu DJ, Prakash A, Frericks-Zipper A, Eisenacher M, Walzer M, Wang S, Brazma A, Vizcaíno JA (2022) The PRIDE database resources in 2022: a hub for mass spectrometry-based proteomics evidences. Nucleic Acids Res 50(D1):D543–D552. https://doi.org/10.1093/nar/gkab1038

Perkins DN, Pappin DJ, Creasy DM, Cottrell JS (1999) Probability-based protein identification by searching sequence databases using mass spectrometry data. Electrophoresis 20(18):3551–3567. https://doi.org/10.1002/(sici)1522-2683(19991201)20:18<3551::Aid--elps3551>3.0.Co;2-2

Puiffe ML, Lachaise I, Molinier-Frenkel V, Castellano F (2013) Antibacterial properties of the mammalian

L-amino acid oxidase IL4I1. PLoS One 8(1):e54589. https://doi.org/10.1371/journal.pone.0054589

Rafiq M, Rivieccio F, Zimmermann AK, Visser C, Bruch A, Kruger T, Gonzalez Rojas K, Kniemeyer O, Blango MG, Brakhage AA (2022) PLB-985 neutrophil-like cells as a model to study *Aspergillus fumigatus* pathogenesis. mSphere 7(1):e0094021. https://doi.org/10.1128/msphere.00940-21

Rappsilber J, Mann M, Ishihama Y (2007) Protocol for micro-purification, enrichment, pre-fractionation and storage of peptides for proteomics using StageTips. Nat Protoc 2(8):1896–1906. https://doi.org/10.1038/nprot.2007.261

Rhodes J, Abdolrasouli A, Dunne K, Sewell TR, Zhang Y, Ballard E, Brackin AP, van Rhijn N, Chown H, Tsitsopoulou A, Posso RB, Chotirmall SH, McElvaney NG, Murphy PG, Talento AF, Renwick J, Dyer PS, Szekely A, Bowyer P, Bromley MJ, Johnson EM, Lewis White P, Warris A, Barton RC, Schelenz S, Rogers TR, Armstrong-James D, Fisher MC (2022) Population genomics confirms acquisition of drug-resistant *Aspergillus fumigatus* infection by humans from the environment. Nat Microbiol 7(5):663–674. https://doi.org/10.1038/s41564-022-01091-2

Rizzo J, Chaze T, Miranda K, Roberson RW, Gorgette O, Nimrichter L, Matondo M, Latge JP, Beauvais A, Rodrigues ML (2020) Characterization of extracellular vesicles produced by *Aspergillus fumigatus* protoplasts. mSphere 5(4). https://doi.org/10.1128/mSphere.00476-20

Roboubi A, Audousset C, Frealle E, Brun AL, Laurent F, Vitte J, Mortuaire G, Lefevre G, Cadranel J, Chenivesse C (2023) Allergic bronchopulmonary aspergillosis: A multidisciplinary review. J Mycol Med 33(3):101392. https://doi.org/10.1016/j.mycmed.2023.101392

Ross PL, Huang YN, Marchese JN, Williamson B, Parker K, Hattan S, Khainovski N, Pillai S, Dey S, Daniels S, Purkayastha S, Juhasz P, Martin S, Bartlet-Jones M, He F, Jacobson A, Pappin DJ (2004) Multiplexed protein quantitation in Saccharomyces cerevisiae using amine-reactive isobaric tagging reagents. Mol Cell Proteomics MCP 3(12):1154–1169. https://doi.org/10.1074/mcp.M400129-MCP200

Russell WK, Park ZY, Russell DH (2001) Proteolysis in mixed organic-aqueous solvent systems: applications for peptide mass mapping using mass spectrometry. Anal Chem 73(11):2682–2685. https://doi.org/10.1021/ac001332p

Saleh AA, Jones GW, Tinley FC, Delaney SF, Alabbadi SH, Fenlon K, Doyle S, Owens RA (2018) Systems impact of zinc chelation by the epipolythiodioxopiperazine dithiol gliotoxin in *Aspergillus fumigatus*: a new direction in natural product functionality. Metallomics 10(6):854–866. https://doi.org/10.1039/c8mt00052b

Samodova D, Hosfield CM, Cramer CN, Giuli MV, Cappellini E, Franciosa G, Rosenblatt MM, Kelstrup CD, Olsen JV (2020) ProAlanase is an effective alternative to trypsin for proteomics applications and disulfide bond mapping. Mol Cell Proteomics MCP 19(12):2139–2157. https://doi.org/10.1074/mcp.TIR120.002129

Sandberg A, Branca RM, Lehtiö J, Forshed J (2014) Quantitative accuracy in mass spectrometry based proteomics of complex samples: the impact of labeling and precursor interference. J Proteome 96:133–144. https://doi.org/10.1016/j.jprot.2013.10.035

Scharf DH, Brakhage AA, Mukherjee PK (2016) Gliotoxin—bane or boon? Environ Microbiol 18(4):1096–1109. https://doi.org/10.1111/1462-2920.13080

Scharmann U, Verhasselt HL, Kirchhoff L, Furnica DT, Steinmann J, Rath PM (2023) Microbiological non-culture-based methods for diagnosing invasive pulmonary aspergillosis in ICU patients. Diagnostics (Basel) 13(16). https://doi.org/10.3390/diagnostics13162718

Schleicher J, Conrad T, Gustafsson M, Cedersund G, Guthke R, Linde J (2017) Facing the challenges of multiscale modelling of bacterial and fungal pathogen-host interactions. Brief Funct Genomic 16(2):57–69. https://doi.org/10.1093/bfgp/elv064

Schmidt H, Vlaic S, Kruger T, Schmidt F, Balkenhol J, Dandekar T, Guthke R, Kniemeyer O, Heinekamp T, Brakhage AA (2018) Proteomics of *Aspergillus fumigatus* conidia-containing phagolysosomes identifies processes governing immune evasion. Mol Cell Proteomics MCP 17(6):1084–1096. https://doi.org/10.1074/mcp.RA117.000069

Schoch CL, Ciufo S, Domrachev M, Hotton CL, Kannan S, Khovanskaya R, Leipe D, McVeigh R, O'Neill K, Robbertse B, Sharma S, Soussov V, Sullivan JP, Sun L, Turner S, Karsch-Mizrachi I (2020) NCBI Taxonomy: a comprehensive update on curation, resources and tools. Database (Oxford). https://doi.org/10.1093/database/baaa062

Schoof EM, Furtwängler B, Üresin N, Rapin N, Savickas S, Gentil C, Lechman E, Keller UAD, Dick JE, Porse BT (2021) Quantitative single-cell proteomics as a tool to characterize cellular hierarchies. Nat Commun 12(1):3341. https://doi.org/10.1038/s41467-021-23667-y

Schwanhäusser B, Gossen M, Dittmar G, Selbach M (2009) Global analysis of cellular protein translation by pulsed SILAC. Proteomics 9(1):205–209. https://doi.org/10.1002/pmic.200800275

Schwarz C, Eschenhagen P, Schmidt H, Hohnstein T, Iwert C, Grehn C, Roehmel J, Steinke E, Stahl M, Lozza L, Tikhonova E, Rosati E, Stervbo U, Babel N, Mainz JG, Wisplinghoff H, Ebel F, Jia LJ, Blango MG, Hortschansky P, Brunke S, Hube B, Brakhage AA, Kniemeyer O, Scheffold A, Bacher P (2023) Antigen specificity and cross-reactivity drive functionally diverse anti-*Aspergillus fumigatus* T cell responses in cystic fibrosis. J Clin Invest 133(5). https://doi.org/10.1172/JCI161593

Searle BC, Swearingen KE, Barnes CA, Schmidt T, Gessulat S, Küster B, Wilhelm M (2020) Generating high quality libraries for DIA MS with empirically corrected peptide predictions. Nat Commun 11(1):1548. https://doi.org/10.1038/s41467-020-15346-1

Seddigh P, Bracht T, Molinier-Frenkel V, Castellano F, Kniemeyer O, Schuster M, Weski J, Hasenberg A, Kraus A, Poschet G, Hager T, Theegarten D, Opitz CA, Brakhage AA, Sitek B, Hasenberg M, Gunzer M (2017) Quantitative analysis of proteome modulations in alveolar epithelial type II cells in response to pulmonary *Aspergillus fumigatus* infection. Mol Cell Proteomics MCP 16(12):2184–2198. https://doi.org/10.1074/mcp.RA117.000072

Sharma V, Eckels J, Schilling B, Ludwig C, Jaffe JD, MacCoss MJ, MacLean B (2018) Panorama public: a public repository for quantitative data sets processed in skyline. Mol Cell Proteomics MCP 17(6):1239–1244. https://doi.org/10.1074/mcp.RA117.000543

Sheehan G, Bergsson G, McElvaney NG, Reeves EP, Kavanagh K (2018a) The human cathelicidin antimicrobial peptide LL-37 promotes the growth of the pulmonary pathogen *Aspergillus fumigatus*. Infect Immun 86(7). https://doi.org/10.1128/IAI.00097-18

Sheehan G, Clarke G, Kavanagh K (2018b) Characterisation of the cellular and proteomic response of *Galleria mellonella* larvae to the development of invasive aspergillosis. BMC Microbiol 18(1):63. https://doi.org/10.1186/s12866-018-1208-6

Sheehan G, Nagl M, Kavanagh K (2019) Exposure to N-chlorotaurine induces oxidative stress responses in *Aspergillus fumigatus*. J Med Microbiol 68(2):279–288. https://doi.org/10.1099/jmm.0.000900

Shekhova E, Ivanova L, Kruger T, Stroe MC, Macheleidt J, Kniemeyer O, Brakhage AA (2019) Redox proteomic analysis reveals oxidative modifications of proteins by increased levels of intracellular reactive oxygen species during hypoxia adaptation of *Aspergillus fumigatus*. Proteomics 19(5):e1800339. https://doi.org/10.1002/pmic.201800339

Shemesh E, Hanf B, Hagag S, Attias S, Shadkchan Y, Fichtman B, Harel A, Kruger T, Brakhage AA, Kniemeyer O, Osherov N (2017) Phenotypic and proteomic analysis of the *Aspergillus fumigatus ΔPrtT*, *ΔXprG* and *ΔXprG/ΔPrtT* protease-deficient mutants. Front Microbiol 8:2490. https://doi.org/10.3389/fmicb.2017.02490

Sheridan KJ, Lechner BE, Keeffe GO, Keller MA, Werner ER, Lindner H, Jones GW, Haas H, Doyle S (2016) Ergothioneine biosynthesis and functionality in the opportunistic fungal pathogen, *Aspergillus fumigatus*. Sci Rep 6:35306. https://doi.org/10.1038/srep35306

Shevchenko A, Tomas H, Havlis J, Olsen JV, Mann M (2006) In-gel digestion for mass spectrometric characterization of proteins and proteomes. Nat Protoc 1(6):2856–2860. https://doi.org/10.1038/nprot.2006.468

Shi LN, Li FQ, Huang M, Lu JF, Kong XX, Wang SQ, Shao HF (2012a) Immunoproteomics based identification of thioredoxin reductase GliT and novel *Aspergillus fumigatus* antigens for serologic diagnosis of invasive aspergillosis. BMC Microbiol 12:11. https://doi.org/10.1186/1471-2180-12-11

Shi LN, Li FQ, Lu JF, Kong XX, Wang SQ, Huang M, Shao HF, Shao SH (2012b) Antibody specific to thioredoxin reductase as a new biomarker for serodiagnosis of invasive aspergillosis in non-neutropenic patients. Clin Chim Acta 413(9-10):938–943. https://doi.org/10.1016/j.cca.2012.02.011

Shibasaki S, Karasaki M, Aoki W, Ueda M (2018) Molecular and physiological study of *Candida albicans* by quantitative proteome analysis. Proteomes 6(3). https://doi.org/10.3390/proteomes6030034

Shin KS, Park HS, Kim YH, Yu JH (2013) Comparative proteomic analyses reveal that FlbA down-regulates *gliT* expression and SOD activity in *Aspergillus fumigatus*. J Proteome 87:40–52. https://doi.org/10.1016/j.jprot.2013.05.009

Shin KS, Kim YH, Yu JH (2015) Proteomic analyses reveal the key roles of BrlA and AbaA in biogenesis of gliotoxin in *Aspergillus fumigatus*. Biochem Biophys Res Commun 463(3):428–433. https://doi.org/10.1016/j.bbrc.2015.05.090

Shin KS, Park HS, Kim Y, Heo IB, Kim YH, Yu JH (2016) *Aspergillus fumigatus* spore proteomics and genetics reveal that VeA represses DefA-mediated DNA damage response. J Proteome 148:26–35. https://doi.org/10.1016/j.jprot.2016.07.022

Shishkova E, Hebert AS, Coon JJ (2016) Now, more than ever, proteomics needs better chromatography. Cell Syst 3(4):321–324. https://doi.org/10.1016/j.cels.2016.10.007

Shopova IA, Belyaev I, Dasari P, Jahreis S, Stroe MC, Cseresnyes Z, Zimmermann AK, Medyukhina A, Svensson CM, Kruger T, Szeifert V, Nietzsche S, Conrad T, Blango MG, Kniemeyer O, von Lilienfeld-Toal M, Zipfel PF, Ligeti E, Figge MT, Brakhage AA (2020) Human neutrophils produce antifungal extracellular vesicles against *Aspergillus fumigatus*. mBio 11(2). https://doi.org/10.1128/mBio.00596-20

Shvartsburg AA, Smith RD (2008) Fundamentals of traveling wave ion mobility spectrometry. Anal Chem 80(24):9689–9699. https://doi.org/10.1021/ac8016295

Shwab EK, Juvvadi PR, Waitt G, Soderblom EJ, Moseley MA, Nicely NI, Asfaw YG, Steinbach WJ (2017) A novel phosphoregulatory switch controls the activity and function of the major catalytic subunit of protein kinase A in *Aspergillus fumigatus*. mBio 8(1). https://doi.org/10.1128/mBio.02319-16

Shwab EK, Juvvadi PR, Waitt G, Soderblom EJ, Barrington BC, Asfaw YG, Moseley MA, Steinbach WJ (2019) Calcineurin-dependent dephosphorylation of the transcription factor CrzA at specific sites controls conidiation, stress tolerance, and virulence of *Aspergillus fumigatus*. Mol Microbiol 112(1):62–80. https://doi.org/10.1111/mmi.14254

Shwab EK, Juvvadi PR, Waitt G, Shaheen S, Jt A, Soderblom EJ, Bobay BG, Asfaw YG, Moseley MA, Steinbach WJ (2020) The protein kinase A-dependent phosphoproteome of the human pathogen *Aspergillus fumigatus* reveals diverse virulence-associated kinase targets. mBio 11(6). https://doi.org/10.1128/mBio.02880-20

Shwab EK, Juvvadi PR, Shaheen SK, Allen J, Waitt G, Soderblom EJ, Asfaw YG, Moseley MA, Steinbach WJ (2022) Protein kinase A regulates autophagy-associated proteins impacting growth and virulence of *Aspergillus fumigatus*. J Fungi (Basel) 8(4). https://doi.org/10.3390/jof8040354

Silva LP, Frawley D, Assis LJ, Tierney C, Fleming AB, Bayram O, Goldman GH (2020) Putative membrane receptors contribute to activation and efficient signaling of mitogen-activated protein kinase cascades during adaptation of *Aspergillus fumigatus* to different stressors and carbon sources. mSphere 5(5). https://doi.org/10.1128/mSphere.00818-20

Singh B, Oellerich M, Kumar R, Kumar M, Bhadoria DP, Reichard U, Gupta VK, Sharma GL, Asif AR (2010a) Immuno-reactive molecules identified from the secreted proteome of *Aspergillus fumigatus*. J Proteome Res 9(11):5517–5529. https://doi.org/10.1021/pr100604x

Singh B, Sharma GL, Oellerich M, Kumar R, Singh S, Bhadoria DP, Katyal A, Reichard U, Asif AR (2010b) Novel cytosolic allergens of *Aspergillus fumigatus* identified from germinating conidia. J Proteome Res 9(11):5530–5541. https://doi.org/10.1021/pr100605c

Singh S, Gupta S, Singh B, Sharma SK, Gupta VK, Sharma GL (2012) Proteomic characterization of *Aspergillus fumigatus* treated with an antifungal coumarin for identification of novel target molecules of key pathways. J Proteome Res 11(6):3259–3268. https://doi.org/10.1021/pr300006j

Sinitcyn P, Hamzeiy H, Salinas Soto F, Itzhak D, McCarthy F, Wichmann C, Steger M, Ohmayer U, Distler U, Kaspar-Schoenefeld S, Prianichnikov N, Yılmaz Ş, Rudolph JD, Tenzer S, Perez-Riverol Y, Nagaraj N, Humphrey SJ, Cox J (2021) MaxDIA enables library-based and library-free data-independent acquisition proteomics. Nat Biotechnol 39(12):1563–1573. https://doi.org/10.1038/s41587-021-00968-7

Souza JAM, Baltazar LM, Carregal VM, Gouveia-Eufrasio L, de Oliveira AG, Dias WG, Campos Rocha M, Rocha de Miranda K, Malavazi I, Santos DA, Frezard FJG, de Souza DDG, Teixeira MM, Soriani FM (2019) Characterization of *Aspergillus fumigatus* extracellular vesicles and their effects on macrophages and neutrophils functions. Front Microbiol 10:2008. https://doi.org/10.3389/fmicb.2019.02008

Specht H, Emmott E, Petelski AA, Huffman RG, Perlman DH, Serra M, Kharchenko P, Koller A, Slavov N (2021) Single-cell proteomic and transcriptomic analysis of macrophage heterogeneity using SCoPE2. Genome Biol 22(1):50. https://doi.org/10.1186/s13059-021-02267-5

Spitzer MH, Nolan GP (2016) Mass cytometry: single cells, many features. Cell 165(4):780–791. https://doi.org/10.1016/j.cell.2016.04.019

Stadlmann J, Hudecz O, Krššáková G, Ctortecka C, Van Raemdonck G, Op De Beeck J, Desmet G, Penninger JM, Jacobs P, Mechtler K (2019) Improved sensitivity in low-input proteomics using micropillar array-based chromatography. Anal Chem 91(22):14203–14207. https://doi.org/10.1021/acs.analchem.9b02899

Steenwyk JL, Mead ME, Knowles SL, Raja HA, Roberts CD, Bader O, Houbraken J, Goldman GH, Oberlies NH, Rokas A (2020) Variation among biosynthetic gene clusters, secondary metabolite profiles, and cards of virulence across Aspergillus species. Genetics 216(2):481–497. https://doi.org/10.1534/genetics.120.303549

Stejskal K, Op de Beeck J, Dürnberger G, Jacobs P, Mechtler K (2021) Ultrasensitive NanoLC-MS of subnanogram protein samples using second generation micropillar array LC technology with Orbitrap Exploris 480 and FAIMS PRO. Anal Chem 93(25):8704–8710. https://doi.org/10.1021/acs.analchem.1c00990

Suh MJ, Fedorova ND, Cagas SE, Hastings S, Fleischmann RD, Peterson SN, Perlin DS, Nierman WC, Pieper R, Momany M (2012) Development stage-specific proteomic profiling uncovers small, lineage specific proteins most abundant in the *Aspergillus fumigatus* conidial proteome. Proteome Sci 10(1):30. https://doi.org/10.1186/1477-5956-10-30

Sui X, Wu Q, Cui X, Wang X, Zhang L, Deng N, Bian Y, Xu R, Tian R (2022) Robust capillary- and micro-flow liquid chromatography-tandem mass spectrometry methods for high-throughput proteome profiling. J Proteome Res. https://doi.org/10.1021/acs.jproteome.2c00405

Suttapitugsakul S, Xiao H, Smeekens J, Wu R (2017) Evaluation and optimization of reduction and alkylation methods to maximize peptide identification with MS-based proteomics. Mol BioSyst 13(12):2574–2582. https://doi.org/10.1039/c7mb00393e

Taylor JA, Johnson RS (2001) Implementation and uses of automated de novo peptide sequencing by tandem mass spectrometry. Anal Chem 73(11):2594–2604. https://doi.org/10.1021/ac001196o

Taylor CF, Hermjakob H, Julian RK Jr, Garavelli JS, Aebersold R, Apweiler R (2006) The work of the Human Proteome Organisation's Proteomics Standards Initiative (HUPO PSI). OMICS 10(2):145–151. https://doi.org/10.1089/omi.2006.10.145

Tekaia F, Latge JP (2005) *Aspergillus fumigatus*: saprophyte or pathogen? Curr Opin Microbiol 8(4):385–392. https://doi.org/10.1016/j.mib.2005.06.017

Teutschbein J, Albrecht D, Potsch M, Guthke R, Aimanianda V, Clavaud C, Latge JP, Brakhage AA, Kniemeyer O (2010) Proteome profiling and functional classification of intracellular proteins from conidia of the human-pathogenic mold *Aspergillus fumigatus*. J Proteome Res 9(7):3427–3442. https://doi.org/10.1021/pr9010684

Teutschbein J, Simon S, Lother J, Springer J, Hortschansky P, Morton CO, Loffler J, Einsele H, Conneally E, Rogers TR, Guthke R, Brakhage AA, Kniemeyer O (2016) Proteomic profiling of serological responses to *Aspergillus fumigatus* antigens in patients with invasive aspergillosis. J Proteome Res

15(5):1580–1591. https://doi.org/10.1021/acs.jproteome.6b00042

Thompson A, Schäfer J, Kuhn K, Kienle S, Schwarz J, Schmidt G, Neumann T, Johnstone R, Mohammed AK, Hamon C (2003) Tandem mass tags: a novel quantification strategy for comparative analysis of complex protein mixtures by MS/MS. Anal Chem 75(8):1895–1904. https://doi.org/10.1021/ac0262560

Tiwary S, Levy R, Gutenbrunner P, Salinas Soto F, Palaniappan KK, Deming L, Berndl M, Brant A, Cimermancic P, Cox J (2019) High-quality MS/MS spectrum prediction for data-dependent and data-independent acquisition data analysis. Nat Methods 16(6):519–525. https://doi.org/10.1038/s41592-019-0427-6

Toby TK, Fornelli L, Srzentić K, DeHart CJ, Levitsky J, Friedewald J, Kelleher NL (2019) A comprehensive pipeline for translational top-down proteomics from a single blood draw. Nat Protoc 14(1):119–152. https://doi.org/10.1038/s41596-018-0085-7

Toor A, Culibrk L, Singhera GK, Moon KM, Prudova A, Foster LJ, Moore MM, Dorscheid DR, Tebbutt SJ (2018) Transcriptomic and proteomic host response to *Aspergillus fumigatus* conidia in an air-liquid interface model of human bronchial epithelium. PLoS One 13(12):e0209652. https://doi.org/10.1371/journal.pone.0209652

Tran NK, Kretsch C, LaValley C, Rashidi HH (2023) Machine learning and artificial intelligence for the diagnosis of infectious diseases in immunocompromised patients. Curr Opin Infect Dis 36(4):235–242. https://doi.org/10.1097/QCO.0000000000000935

Tsou CC, Avtonomov D, Larsen B, Tucholska M, Choi H, Gingras AC, Nesvizhskii AI (2015) DIA-Umpire: comprehensive computational framework for data-independent acquisition proteomics. Nat Methods 12(3):258–264, 257 p following 264. https://doi.org/10.1038/nmeth.3255

Tyanova S, Temu T, Cox J (2016) The MaxQuant computational platform for mass spectrometry-based shotgun proteomics. Nat Protoc 11(12):2301–2319. https://doi.org/10.1038/nprot.2016.136

Ullmann AJ, Aguado JM, Arikan-Akdagli S, Denning DW, Groll AH, Lagrou K, Lass-Florl C, Lewis RE, Munoz P, Verweij PE, Warris A, Ader F, Akova M, Arendrup MC, Barnes RA, Beigelman-Aubry C, Blot S, Bouza E, Bruggemann RJM, Buchheidt D, Cadranel J, Castagnola E, Chakrabarti A, Cuenca-Estrella M, Dimopoulos G, Fortun J, Gangneux JP, Garbino J, Heinz WJ, Herbrecht R, Heussel CP, Kibbler CC, Klimko N, Kullberg BJ, Lange C, Lehrnbecher T, Loffler J, Lortholary O, Maertens J, Marchetti O, Meis JF, Pagano L, Ribaud P, Richardson M, Roilides E, Ruhnke M, Sanguinetti M, Sheppard DC, Sinko J, Skiada A, Vehreschild M, Viscoli C, Cornely OA (2018) Diagnosis and management of Aspergillus diseases: executive summary of the 2017 ESCMID-ECMM-ERS guideline. Clin Microbiol Infect 24(Suppl 1):e1–e38. https://doi.org/10.1016/j.cmi.2018.01.002

Urban C, Sohn K, Lottspeich F, Brunner H, Rupp S (2003) Identification of cell surface determinants in *Candida albicans* reveals Tsa1p, a protein differentially localized in the cell. FEBS Lett 544(1-3):228–235. https://doi.org/10.1016/s0014-5793(03)00455-1

Urfer W, Grzegorczyk M, Jung K (2006) Statistics for proteomics: a review of tools for analyzing experimental data. Proteomics 6(Suppl 2):48–55. https://doi.org/10.1002/pmic.200600554

Valsecchi I, Lai JI, Stephen-Victor E, Pille A, Beaussart A, Lo V, Pham CLL, Aimanianda V, Kwan AH, Duchateau M, Gianetto QG, Matondo M, Lehoux M, Sheppard DC, Dufrene YF, Bayry J, Guijarro JI, Sunde M, Latge JP (2019) Assembly and disassembly of *Aspergillus fumigatus* conidial rodlets. Cell Surf 5:100023. https://doi.org/10.1016/j.tcsw.2019.100023

Valsecchi I, Stephen-Victor E, Wong SSW, Karnam A, Sunde M, Guijarro JI, Rodriguez de Francisco B, Kruger T, Kniemeyer O, Brown GD, Willment JA, Latge JP, Brakhage AA, Bayry J, Aimanianda V (2020) The role of RodA-conserved cysteine residues in the *Aspergillus fumigatus* conidial surface organization. J Fungi (Basel) 6(3). https://doi.org/10.3390/jof6030151

van Oss CJ (1989) On the mechanism of the cold ethanol precipitation method of plasma protein fractionation. J Protein Chem 8(5):661–668. https://doi.org/10.1007/bf01025606

Vegvari A, Zhang X, Zubarev RA (2023) Toward single bacterium proteomics. J Am Soc Mass Spectrom 34(10):2098–2106. https://doi.org/10.1021/jasms.3c00242

Venable JD, Dong MQ, Wohlschlegel J, Dillin A, Yates JR (2004) Automated approach for quantitative analysis of complex peptide mixtures from tandem mass spectra. Nat Methods 1(1):39–45. https://doi.org/10.1038/nmeth705

Venugopalan LP, Aimanianda V, Namperumalsamy VP, Prajna L, Kuppamuthu D, Jayapal JM (2023) Comparative proteome analysis identifies species-specific signature proteins in *Aspergillus* pathogens. Appl Microbiol Biotechnol 107(12):4025–4040. https://doi.org/10.1007/s00253-023-12559-4

Virginio ED, Kubitschek-Barreira PH, Batista MV, Schirmer MR, Abdelhay E, Shikanai-Yasuda MA, Lopes-Bezerra LM (2014) Immunoproteome of *Aspergillus fumigatus* using sera of patients with invasive aspergillosis. Int J Mol Sci 15(8):14505–14530. https://doi.org/10.3390/ijms150814505

Vizcaíno JA, Deutsch EW, Wang R, Csordas A, Reisinger F, Ríos D, Dianes JA, Sun Z, Farrah T, Bandeira N, Binz PA, Xenarios I, Eisenacher M, Mayer G, Gatto L, Campos A, Chalkley RJ, Kraus HJ, Albar JP, Martinez-Bartolomé S, Apweiler R, Omenn GS, Martens L, Jones AR, Hermjakob H (2014) ProteomeXchange provides globally coordinated proteomics data submission and dissemination. Nat Biotechnol 32(3):223–226. https://doi.org/10.1038/nbt.2839

Vlaic S, Conrad T, Tokarski-Schnelle C, Gustafsson M, Dahmen U, Guthke R, Schuster S (2018)

ModuleDiscoverer: Identification of regulatory modules in protein-protein interaction networks. Sci Rep. 8(1):433. https://doi.org/10.1038/s41598-017-18370-2

Vodisch M, Albrecht D, Lessing F, Schmidt AD, Winkler R, Guthke R, Brakhage AA, Kniemeyer O (2009) Two-dimensional proteome reference maps for the human pathogenic filamentous fungus *Aspergillus fumigatus*. Proteomics 9(5):1407–1415. https://doi.org/10.1002/pmic.200800394

Voltersen V, Blango MG, Herrmann S, Schmidt F, Heinekamp T, Strassburger M, Kruger T, Bacher P, Lother J, Weiss E, Hunniger K, Liu H, Hortschansky P, Scheffold A, Loffler J, Krappmann S, Nietzsche S, Kurzai O, Einsele H, Kniemeyer O, Filler SG, Reichard U, Brakhage AA (2018) Proteome analysis reveals the conidial surface protein CcpA essential for virulence of the pathogenic fungus *Aspergillus fumigatus*. mBio 9(5). https://doi.org/10.1128/mBio.01557-18

Walker SWC, Anwar A, Psutka JM, Crouse J, Liu C, Le Blanc JCY, Montgomery J, Goetz GH, Janiszewski JS, Campbell JL, Hopkins WS (2018) Determining molecular properties with differential mobility spectrometry and machine learning. Nat Commun 9(1):5096. https://doi.org/10.1038/s41467-018-07616-w

Wang M, Wang J, Carver J, Pullman BS, Cha SW, Bandeira N (2018) Assembling the community-scale discoverable human proteome. Cell Syst 7(4):412–421.e415. https://doi.org/10.1016/j.cels.2018.08.004

Wartenberg D, Lapp K, Jacobsen ID, Dahse HM, Kniemeyer O, Heinekamp T, Brakhage AA (2011) Secretome analysis of *Aspergillus fumigatus* reveals Asp-hemolysin as a major secreted protein. Int J Med Microbiol 301(7):602–611. https://doi.org/10.1016/j.ijmm.2011.04.016

Wei T, Zheng N, Zheng H, Chen Y, Hong P, Liu W, Liu M (2022) Proteomic perspective of azole resistance in *Aspergillus fumigatus* biofilm extracellular matrix in response to itraconazole. Med Mycol. https://doi.org/10.1093/mmy/myac084

Wekesa JS, Kimwele M (2023) A review of multi-omics data integration through deep learning approaches for disease diagnosis, prognosis, and treatment. Front Genet 14:1199087. https://doi.org/10.3389/fgene.2023.1199087

Wessel D, Flügge UI (1984) A method for the quantitative recovery of protein in dilute solution in the presence of detergents and lipids. Anal Biochem 138(1):141–143. https://doi.org/10.1016/0003-2697(84)90782-6

Wiedner SD, Burnum KE, Pederson LM, Anderson LN, Fortuin S, Chauvigne-Hines LM, Shukla AK, Ansong C, Panisko EA, Smith RD, Wright AT (2012) Multiplexed activity-based protein profiling of the human pathogen *Aspergillus fumigatus* reveals large functional changes upon exposure to human serum. J Biol Chem 287(40):33447–33459. https://doi.org/10.1074/jbc.M112.394106

Wiedner SD, Ansong C, Webb-Robertson BJ, Pederson LM, Fortuin S, Hofstad BA, Shukla AK, Panisko EA, Smith RD, Wright AT (2013) Disparate proteome responses of pathogenic and nonpathogenic *aspergilli* to human serum measured by activity-based protein profiling (ABPP). Mol Cell Proteomics MCP 12(7):1791–1805. https://doi.org/10.1074/mcp.M112.026534

Wijayawardene NN, Boonyuen N, Ranaweera CB, de Zoysa HKS, Padmathilake RE, Nifla F, Dai DQ, Liu Y, Suwannarach N, Kumla J, Bamunuarachchige TC, Chen HH (2023) OMICS and other advanced technologies in mycological applications. J Fungi (Basel) 9(6). https://doi.org/10.3390/jof9060688

Wilkins MR, Sanchez JC, Gooley AA, Appel RD, Humphery-Smith I, Hochstrasser DF, Williams KL (1996) Progress with proteome projects: why all proteins expressed by a genome should be identified and how to do it. Biotechnol Genet Eng Rev 13:19–50. https://doi.org/10.1080/02648725.1996.10647923

Wilkins MR, Appel RD, Van Eyk JE, Chung MC, Görg A, Hecker M, Huber LA, Langen H, Link AJ, Paik YK, Patterson SD, Pennington SR, Rabilloud T, Simpson RJ, Weiss W, Dunn MJ (2006) Guidelines for the next 10 years of proteomics. Proteomics 6(1):4–8. https://doi.org/10.1002/pmic.200500856

Wiśniewski JR, Zougman A, Nagaraj N, Mann M (2009) Universal sample preparation method for proteome analysis. Nat Methods 6(5):359–362. https://doi.org/10.1038/nmeth.1322

Wong SSW, Daniel I, Gangneux JP, Jayapal JM, Guegan H, Delliere S, Lalitha P, Shende R, Madan T, Bayry J, Guijarro JI, Kuppamuthu D, Aimanianda V (2020) Differential interactions of serum and bronchoalveolar lavage fluid complement proteins with conidia of airborne fungal pathogen *Aspergillus fumigatus*. Infect Immun 88(9). https://doi.org/10.1128/IAI.00212-20

Wu F, Sun D, Wang N, Gong Y, Li L (2011) Comparison of surfactant-assisted shotgun methods using acid-labile surfactants and sodium dodecyl sulfate for membrane proteome analysis. Anal Chim Acta 698(1-2):36–43. https://doi.org/10.1016/j.aca.2011.04.039

Yang Y, Liu X, Shen C, Lin Y, Yang P, Qiao L (2020) In silico spectral libraries by deep learning facilitate data-independent acquisition proteomics. Nat Commun 11(1):146. https://doi.org/10.1038/s41467-019-13866-z

Ye W, Liu T, Zhang W, Zhang W (2021) The toxic mechanism of gliotoxins and biosynthetic strategies for toxicity prevention. Int J Mol Sci. 22(24):13510. https://doi.org/10.3390/ijms222413510

Yi L, Tsai CF, Dirice E, Swensen AC, Chen J, Shi T, Gritsenko MA, Chu RK, Piehowski PD, Smith RD, Rodland KD, Atkinson MA, Mathews CE, Kulkarni RN, Liu T, Qian WJ (2019) Boosting to amplify signal with isobaric labeling (BASIL) strategy for comprehensive quantitative phosphoproteomic characterization of small populations of cells. Anal Chem 91(9):5794–5801. https://doi.org/10.1021/acs.analchem.9b00024

Zhang X (2015) Less is more: membrane protein digestion beyond urea-trypsin solution for next-level proteomics. Mol Cell Proteomics MCP 14(9):2441–2453. https://doi.org/10.1074/mcp.R114.042572

Zhang R, Sioma CS, Wang S, Regnier FE (2001) Fractionation of isotopically labeled peptides in quantitative proteomics. Anal Chem 73(21):5142–5149. https://doi.org/10.1021/ac010583a

Zhang L, Feng D, Fang W, Ouyang H, Luo Y, Du T, Jin C (2009) Comparative proteomic analysis of an *Aspergillus fumigatus* mutant deficient in glucosidase I (AfCwh41). Microbiology (Reading) 155(Pt 7):2157–2167. https://doi.org/10.1099/mic.0.027490-0

Zhang X, He D, Gao S, Wei Y, Wang L (2020) iTRAQ-based proteomic analysis of the interaction of A549 human lung epithelial cells with *Aspergillus fumigatus* conidia. Mol Med Rep 22(6):4601–4610. https://doi.org/10.3892/mmr.2020.11582

Zhang S, Larsen B, Colwill K, Wong CJ, Youn JY, Gingras AC (2022) Mapping protein-protein interactions using data-dependent acquisition without dynamic exclusion. Anal Chem 94(30):10579–10583. https://doi.org/10.1021/acs.analchem.2c00755

Zhao W, Panepinto JC, Fortwendel JR, Fox L, Oliver BG, Askew DS, Rhodes JC (2006) Deletion of the regulatory subunit of protein kinase A in *Aspergillus fumigatus* alters morphology, sensitivity to oxidative damage, and virulence. Infect Immun 74(8):4865–4874. https://doi.org/10.1128/IAI.00565-06

Zhu Y, Clair G, Chrisler WB, Shen Y, Zhao R, Shukla AK, Moore RJ, Misra RS, Pryhuber GS, Smith RD, Ansong C, Kelly RT (2018a) Proteomic analysis of single mammalian cells enabled by microfluidic nanodroplet sample preparation and ultrasensitive NanoLC-MS. Angew Chem Int Ed Eng 57(38):12370–12374. https://doi.org/10.1002/anie.201802843

Zhu Y, Piehowski PD, Zhao R, Chen J, Shen Y, Moore RJ, Shukla AK, Petyuk VA, Campbell-Thompson M, Mathews CE, Smith RD, Qian WJ, Kelly RT (2018b) Nanodroplet processing platform for deep and quantitative proteome profiling of 10-100 mammalian cells. Nat Commun 9(1):882. https://doi.org/10.1038/s41467-018-03367-w

Zhu Y, Zhao R, Piehowski PD, Moore RJ, Lim S, Orphan VJ, Paša-Tolić L, Qian WJ, Smith RD, Kelly RT (2018c) Subnanogram proteomics: impact of LC column selection, MS instrumentation and data analysis strategy on proteome coverage for trace samples. Int J Mass Spectrom 427:4–10. https://doi.org/10.1016/j.ijms.2017.08.016

Zhu H, Tamura T, Hamachi I (2019) Chemical proteomics for subcellular proteome analysis. Curr Opin Chem Biol 48:1–7. https://doi.org/10.1016/j.cbpa.2018.08.001

Zolg DP, Gessulat S, Paschke C, Graber M, Rathke-Kuhnert M, Seefried F, Fitzemeier K, Berg F, Lopez-Ferrer D, Horn D, Henrich C, Huhmer A, Delanghe B, Frejno M (2021) INFERYS rescoring: Boosting peptide identifications and scoring confidence of database search results. Rapid Commun Mass Spectrom. https://doi.org/10.1002/rcm.9128

Index

A. A. Brakhage et al. (eds.), *Human and Animal Relationships*, The Mycota 6,
https://doi.org/10.1007/978-3-031-64853-3

GPSR Compliance

The European Union's (EU) General Product Safety Regulation (GPSR) is a set of rules that requires consumer products to be safe and our obligations to ensure this.

If you have any concerns about our products, you can contact us on ProductSafety@springernature.com

In case Publisher is established outside the EU, the EU authorized representative is:

Springer Nature Customer Service Center GmbH
Europaplatz 3
69115 Heidelberg, Germany

Batch number: 09565659

Printed by Printforce, the Netherlands